World and Logic

World and Logic

Jens Lemanski

© Individual author and College Publications 2021. All rights reserved.

Published with the financial support of the FernUniversität in Hagen and the Schopenhauer Gesellschaft e.V.

ISBN 978-1-84890-384-5

College Publications
Scientific Director: Dov Gabbay
Managing Director: Jane Spurr

http://www.collegepublications.co.uk

Cover produced by Laraine Welch

All rights reserved. No part of this publication may be reproduced, stored in a retrieval system or transmitted in any form, or by any means, electronic, mechanical, photocopying, recording or otherwise without prior permission, in writing, from the publisher.

Contents

Preface 1
1 The World and its Representationalist Interpretation 5
   1.1 Interpretive Approaches 7
      1.1.1 Unity: One Single Thought, One Single World 7
      1.1.2 Multiplicity: The Organic System 12
      1.1.3 Interpretations: The Descriptive Approach 15
      1.1.4 Aporias: Alleged Contradictions 21
   1.2 The System of WWR 27
      1.2.1 Research Literature Regarding the System 29
      1.2.2 The Preface 33
      1.2.3 Book I: Theory of Cognition (Representation) 35
      1.2.4 Book II: Metaphysics (Will) 50
      1.2.5 Book III: Aesthetics (Representation) 57
      1.2.6 Book IV: Ethics (Will) 62
      1.2.7 Evaluation 71
   1.3 The Status of Logic in WWR and WWR2 77
      1.3.1 The Logica Minor of WWR 80
      1.3.2 The System of WWR2 93
      1.3.3 The Logica Major of WWR2 101
2 Logic and its Geometrical Interpretation 111
   2.1 Semantics – Context Principle, Use Theory and Representationalism 113
      2.1.1 The Kant/ Frege-Thesis 117
      2.1.2 The Context Principle in 19th-Century Neo-Aristotelism 124
      2.1.3 The Context Principle in Early Analytic Philosophy 131
      2.1.4 The Schopenhauer/ Wittgenstein Theses 140
      2.1.5 Schopenhauer's Use Theory of Meaning and the Context Principle 149
      2.1.6 Representationalism and Contextualism 157

2.2 Analyticity – Analytic judgements, Containment Metaphors and Logic Diagrams 165
    2.2.1 The State of Research on the History of Analytical Diagrams............. 175
    2.2.2 Logic Diagrams from Antiquity to Early Modern Times..................... 182
    2.2.3 Analytical Diagrams of Geometric Logic 196
    2.2.4 Kant's "Notion of Containment"... 217
    2.2.5 Schopenhauer's Geometrical Doctrine of Judgement 226
    2.2.6 The Principle of Extensionality and the Context Principle 239
2.3 Proof – Elementary Geometry, Syllogistic and Intuitive Proof Theory 247
    2.3.1 Geometria more syllogismorum? The Controversy of Leibnizians and Kantians 256
    2.3.2 Conjuring Tricks, Mousetraps and Stilted Proofs 272
    2.3.3 Reception and Evaluation of Schopenhauer's Philosophy of Geometry 285
    2.3.4 Assessments of Geometric Logic from Reimers to Maaß 303
    2.3.5 Logica More Geometrico versus Geometria More Syllogismorum 318
    2.3.6 A Quite Exact Analogy to the Circumference of Concepts.................. 331

3 Logic and World 337
  3.1 Inferentialisms 340
    3.1.1 Inferentialism of Grounding 341
    3.1.2 Inferentialism of Matter 350
    3.1.3 Inferentialism of Form 358
  3.2 Rational Representationalism 372
    3.2.1 Abstraction and Being 375
    3.2.2 Intuition and Concept 397
    3.2.3 Translation and Transfer 418

Appendix 427
  Bibliography 427
  Abbreviated Sources 464
  List of Abbreviations 466

Preface

World and Logic is an expression of philosophical revisionism. In this book, I will attempt to demonstrate how it is possible to argue for the thesis that the space of reasons must be broader than the space of concepts without falling into a causal and naïve representationalism. In other words, this thesis is the expression of a non-naïve or rational representationalism. It could be understood as meaning that our logic has always had basic transfers of worldly forms, and is, therefore, a complement to the rationalist picture, according to which the world would always have been logically constructed if both pictures were not over-simplifications of our far too complex concepts of the world and logic.

In this context, the two-part picture seems to me to be in need of revision, which is systematically represented by contemporary rationalism and projected onto the history of philosophy and science: On the one hand, it naively characterises those research programmes that express that the world can be represented with the help of logic and language without claiming that the world must already be completely opened up by logic; and on the other hand, contemporary rationalism claims that this characterisation applies to all representationalist approaches before the beginning of the paradigm of language-philosophy-logicism and still shows itself in different approaches today.

The rationalism addressed here falls into two areas: On the one hand, into inferentialism, which claims that everything that has meaning in our world has received this meaning through the practical role in our always already inferentially structured language; on the other hand, into neologicism, which assumes that the objects, assertions and structures with which we understand our world can be traced back to logic. The world of inferentialism is the world created by our everyday language, the world of neologicism is the quantitatively captured world.

The battlefield of current rationalism against naïve representationalism are the testaments of the so-called 'mighty dead' such as Aristotle, Leibniz, Kant, Frege or Wittgenstein. These 'heroes' are used by various inferentialist and neologicist programmes as forerunners and sources of ideas. Since I believe that I would be ill-advised to carry out my revisionist thesis interpretatively on this unmanageable battlefield, I have decided to consult, above all, the writings of many forgotten anti-heroes, which show a recognizable proximity to the rational representationalism represented here. In addition to anti-heroes such as Bacon, Reimers, Weigel, Grosser, Euler or McCulloch, the focus is especially on Arthur Schopenhauer's hitherto forgotten lectures on logic, which, on the one hand, represent the foundation of his representationalist system and, on the other hand, critically anticipate with a geometric logic many semantic foundations of today's neologicist-inferentialist philosophies. In this respect, these lectures offer the historical starting point for a programme that is both representationalist and rational and on which a modern philosophy with the same claims and characteristics can be built.

As the introductory Chapter 1 will show, in contrast to the prevailing research opinion (1.1), Schopenhauer's logic lectures (1.3) are an essential part of representationalism (1.2). In Chapter 2 the view is justified that this representationalism does not have to be regarded as naïve, since it cuts through the two-part picture of modern rationalism: Schopenhauer represents as the starting point of his system precisely those semantic principles (2.1.4–2.1.6) on which the modern inferentialist-neologicist paradigm is built, in particular the context principle and the use theory of meaning (2.1.1–2.1.3). This semantics is also the starting point for an explanation of the notions of containment and circumference problematised in modern logic and philosophy of language: I argue that the established metaphors of containment, which play a central role in distinguishing analytical and synthetic judgements, in transcendental philosophy emerge from the semantics of geometric logic (2.2.4–2.2.6) and that geometric logic results from a history of development whose beginnings reach back to medieval philosophy (2.2.1–2.2.3). Finally, it shall be shown into which problems proof theories in geometry (2.3.1–2.3.3) and logic (2.3.4–2.3.6) get into when they follow a rigorous logicist programme. As a helpful way out of the problem of reasoning of logicism and neologicism, there is a proof theory representing intuition, which is discussed based on elementary geometry and of a fragment of first-order logic. This helpful way of geometric logic is based on the central insight of Chapter 2, that logic only has to resort to forms of intuition if it comes under pressure to justify itself.

Chapters 1 and 2 may give the impression as if I wanted to make Schopenhauer, usually perceived as an anti-hero, the hero of this book. But this is not the case. As much as Chapters 1 and 2 argue that Schopenhauer's lectures on logic should be used to revise the two-part picture of modern rationalist historiography and the systematics associated with it, I firmly believe, for the reasons given in Chapter 1, that many parts of his system can no longer meet the systematic requirements of the present age. (That we, moreover, occasionally encounter opinions in the testaments of the ancients that do not correspond to our ethical, moral or political principles is something I take for granted.) Rather, the arguments elaborated in Chapter 2 have convinced me that we need a modernised version of semantics and the foundation of logic from the spirit of rational representationalism.

But based on the knowledge of what a rational or non-naïve representationalism has looked like historically, which gives a promising picture of the relationship between the world and logic, Chapter 3 tries to explain why the concept 'space of concepts' cannot be congruent and completely overlapping with the sphere of the concept 'space of reasons'. Following those anti-heroes of this book who can be described as geometric logicians, my answer can probably be described as Aristotelian: Namely, because even in the space of concepts there are inauthentic transfers from the intuitive given world, which play an essential role in the game of giving and demanding of reasons. Non-naïve or rational representationalism can thus be called a representationalism that need not go beyond the sphere comprising the space of concepts in order to argue, within its borders, that the sphere of the space of reasons must be the broader of the two. In other words, this representationalism is rational because, without leaving the space of concepts, it can explain why in it representations are necessarily expressed that are not founded in itself.

This central thesis is supported by several arguments which are, on the one hand, considered in the historical material of Chapter 2, but which are not yet emphasised, elaborated and, of course, updated in the form in which they are finally to be brought to bear in Chapter 3. Nevertheless, Chapter 3 is not the complete programme of a representationalism itself, but it only presents the semantic conditions of the possibility of representing a representationalism that does not conflict with the requirements of modern rationalism.

Already at the end of Chapter 2.3, it is argued that logicism or neologicism is subject to serious philosophical problems of reasoning. Based on a critique of modern inferentialism in Chapter 3.1, Chapter 3.2.1 will take up a core element of neologicism, namely the theory of abstraction, and present it independently. Here a new perspective on the abstraction theory of meaning is argued for, which should make the strict distinction between singular and general terms logical and, above all, ontologically dispensable. Such a theory of meaning explains the different roles of conceptual contents in judgment solely by the degree of abstraction, which in turn can be represented by geometric logic. Chapter 3.2.2 will show this by an exemplary geometric logic. Here the diagrammatic logic shall mediate between the intuitive and the conceptual world. Finally, Chapter 3.2.3 will present the central thesis of this book, which offers an explanation why, on the one hand, so many of the forms of geometric logic presented in Chapter 2 exist in the history of philosophy and mathematics and why, on the other hand, it may be useful for rationalism to recognise the representationalist thesis that the space of reasons is larger than the space of concepts.

I have four groups of readers in mind who could benefit from reading this book: The group that expects, above all, a systematic answer to the question of why the space of reasons must be larger than the space of concepts will begin to read Chapter 3 and will perceive Chapter 2 and finally Chapter 1 as explanations and remarks. The group that anticipates that the basis for the answer to this question is already inherent in the history of its problem will begin in Chapter 2 and perceive Chapter 1 as a justification of different historical views. The third group is the one that starts with Chapter 1 because on the one hand, they assume that by receiving a voluminous book they have also gained the preservation of the time to read it and because on the other hand, they are not afraid of an anti-hero like Schopenhauer. The fourth group is the one that is not interested in a historically or systematically structured defence of the main thesis, but only in individual topics, such as the context principle, the use theory of meaning

(Chapters 2.1.4ff.) and its history (2.1.1ff.), geometric logic (2.2.1ff., 2.3.4), for analytical and synthetic judgments (2.2.4ff.), for proof theories and grounding in elementary geometry (2.3.1ff.) and logic (2.3.4ff.), for a non-individual abstraction theory of meaning and a critique of singular terms (3.2.1), or even for the attempt to use geometric logic to represent the steps of abstraction from the intuitive world to the most abstract concepts (3.2.2).

Many excerpts in this book have been accompanied by lectures or published studies. For the present version, the topics and theses have in part been extensively revised, in part written from scratch, and in part reproduced in a shortened form. Where papers transpose similar content in a different form, this is pointed out.

The fact that I have been allowed to present my topics, theses and arguments on geometric logic, on philosophy of language and metaphysics in the last 10 years at many personal talks, seminars, workshops, conferences and congresses on four continents has contributed significantly to the book I am able to present here. I would like to thank all the participants of these events, all colleagues and friends who have encouraged me to present many of the theses that can be found in this paper, and who have also saved me from having to defend some problematic theses with more conviction than I once thought I should. My thanks go to Dieter Birnbacher, Hubertus Busche and Eberhard Knobloch, who read and commented on a first draft of the German version of this book. I am particularly indebted to Judith Werntgen-Schmidt and Theo Berwe, who proofread the German version, and to Sean Murphy, who proofread the English version of the manuscript. All remaining errors are of course my own.

1 The World and its Representationalist Interpretation

How can the world be represented linguistically? What does a linguistically adequate description of all components of the world look like? Numerous philosophers, schools, and fields of research have sought to answer these questions up to the present day employing various systematic representations. If one looks only at modern times, one can recognise such representationalist approaches in very different programs, e.g. in Nicolaus Reimer's *Metamorphosis Logicae* in the 16th century, in Francis Bacon's *Advancement of Learning* in the 17th century, in the book on Weltweisheit (world wisdom) and encyclopedias of the 18th century, in Rudolf Carnap's *Logical Structure of the World* in the 20th century or in the branch of research on knowledge representation in the present. Thus, one could show in numerous writings or research programs that the concept of representationalism is too narrowly defined when it is reduced to very specific theories of intentionality or consciousness.[1] In this Chapter 1, I will use the writings of a classical 19th-century philosopher as an example to show what a systematic approach to a representationalist programme looks like, namely Arthur Schopenhauer's main work *The World as Will and Representation* (= WWR).

Although Schopenhauer is usually not included in the canon of system philosophers of classical German philosophy or German idealism, in his major work he, like many of his contemporaries, speaks explicitly of having a system of philosophy or of his philosophy being a system.[2] The system of WWR has the claim to provide a complete representation of the world in a few abstract concepts. Even if the location of Schopenhauer's system in all his writings will prove problematic in the course of the following chapters, both Schopenhauer's explicit statements and the history of reception demand that the system is given in his main work, WWR. It is only in Chapters 1.3 and 2 that it is argued that there are good reasons to consult Schopenhauer's *Berlin Lectures* as a more coherent system instead of WWR.

But even an interpretation that restricts the Schopenhauer system to his main work proves difficult: Ernst Bloch had already pointed out that the interpretation of WWR had become a "prime example of a 'terrible simplification'"[3] because the structure and content of this work are far more complex than the philosophical historiography, influenced by the interpretation of the educated middle-class milieu, still convey today. I am of the opinion that the dominance of the biased historiography of philosophy also undermines the revisionist approaches of many interpreters since it influences the

---

[1] What connection the writings or research programmes mentioned here have to representationalism and what exactly the term representationalism used here in a broad sense encompasses will become apparent in the course of Chapters 1 and 2.
[2] Vide infra, Chapter 1.2.2.
[3] Ernst Bloch: Leipziger Vorlesungen zur Geschichte der Philosophie (1950–1956). Vol. 4. Frankfurt/Main 1985, p. 368.

reader to look for only well-known motives such as 'subjective idealism', 'metaphysical irrationalism', 'ideological pessimism' or 'nihilistic mysticism' in Schopenhauer's work.[4]

For those readers who are not familiar with Schopenhauer's system, I recommend that they begin in Chapter 1.2.2 if they are primarily interested in representationalism, or in Chapter 1.3.1 if their interest lies primarily in the field of logic. Both readers should consider the previous chapters as additional information after reading Chapter 1.3. However, for readers who cannot imagine that 200-year-old texts by an author like Schopenhauer can make any meaningful contribution to today's debates, I recommend a direct jump to Chapter 2 and assume that they will find their way back to Chapter 1 at some point. All other readers who do not find themselves in any of the groups mentioned so far should be well prepared for successive readings of the following chapters.

In Chapter 1, an attempt will be made to show how differently Schopenhauer's complete works are interpreted in research, starting from the system of WWR. First of all, the different approaches to interpreting Schopenhauer's system are presented in relevant passages (Chapter 1.1). Then, the WWR is used to give an overview of how Schopenhauer's system is structured and which topics and components it comprises (Chapter 1.2). Finally, the role of the so-called 'logica minor' (short logic) in the system of WWR is shown and the differences to the 'logica major' (great logic) of the *Berlin Lectures* are worked out (Chapter 1.3), which is then examined in more detail in Chapter 2. In all three subchapters mentioned, the thesis is advanced that Schopenhauer's system is the expression of a representationalism that aims to describe the world with the help of logic.

---

[4] Cf. Otto Friedrich Gruppe: Gegenwart und Zukunft der Philosophie in Deutschland. Berlin 1855, p. 151: "Schopenhauer [...], whose philosophical doctrine is subjective idealism"; Otto Jenson: Die Ursache der Widersprüche im Schopenhauerschen System. Rostock 1906, p. 34: "a negation-philosophy, like the one of Schopenhauer".

1.1 Interpretive Approaches

Schopenhauer's system of WWR begins with two traditional emotive words, `thought` and `system`, which are still intensively discussed in philosophical research today. What exactly Schopenhauer means is controversial when he claims at the beginning of WWR that his work contains only (one) single thought and that his philosophy is not architectural but organic. Since, in my opinion, both questions can only be answered from the context of the system, which will only be developed in the following chapters, I will first only present the research opinions on the 'single thought' (Chapter 1.1.1) and the 'organic system' (Chapter 1.1.2); thereby I dogmatically anticipate my opinion that is justified only in Chapter 1.2.

1.1.1 Unity: One Single Thought, One Single World

The preface to the 1st edition of WWR begins with the following two sentences:

> What I propose to do here is to specify how this book is to be read so as to be understood. – It aims to convey a single thought.[1]

(1) The interpretation that focuses on the second sentence of the quote is widely popular. Representatives of this reading claim that this second sentence explains what the author's intention is and thus the content of the whole book (WWR). Since, according to the prevailing opinion, a single thought must be expressed – if not in inferences or theories, then at least – in the form of judgements, there are representatives of this interpretation who *heuristically* examine individual propositions of the WWR to see whether they could be an abstract summary of the entire system. (2) Representatives of a similar reading, who also focus on the second sentence of the given quotation, but who are not of the opinion that such a thought, as described by Schopenhauer, can at least be expressed in the form of a judgment, see this thought in more than just a single proposition. This *hermeneutic* approach attempts to go beyond the propositional content of a judgement. (3) I believe, however, that not the second, but the first sentence is the central theme of the preface. In my opinion, Schopenhauer's intention and motivation for writing WWR is only presented in § 15 of WWR. It is only from the content of § 15 that the *holistic* interpretation of the single thought results.

(1) *Heuristic Interpretation*: Schopenhauer himself does not explicitly mention at any point in his oeuvre what the single thought is.[2] This has motivated scholars to

---
[1] WWR I, p. 5.
[2] Cf. John Atwell: Schopenhauer on the Character of the World. The Metaphysics of Will. Berkeley 1995, p. 18; Christopher Janaway: Introduction, in: The Cambridge Companion to Schopenhauer. Ed. by Christopher Janaway. Cambridge 1999, pp. 1–18, here: p. 4.

1 The World and its Representationalist Interpretation

offer various speculations as to the judgment within the WWR or the complete oeuvre in which the single thought is to be found. Rudolf Malter, for example, points out that a distinction is to be made between propositions and thoughts, and that the one thought "although no proposition itself, is present only in propositions consisting of abstract representations";[3] nevertheless Malter sees the one thought finally in the following proposition: "The world is the self-cognition of the will."[4] This judgment goes beyond the WWR since it is found in a manuscript by Schopenhauer from the year 1817, in which it says: "The whole of my philosophy can be condensed into one expression, namely: the world is the will's knowledge of itself."[5] Jochem Hennigfeld, however, names another judgment as a candidate for the single thought, since this has an axiomatic character:[6] "As a thing in itself, the will constitutes the true, inner and indestructible essence of the human being".[7]

This heuristic of the single thought within the judgments of Schopenhauer's complete works, however, also finds critics within this line of interpretation: Schopenhauer's main work is divided into four books, which are usually summarised by the keywords (B1) 'idealistic epistemology', (B2) 'voluntaristic metaphysics', (B3) 'contemplative aesthetics' and (B4) 'will-negating ethics'.[8] John Atwell now explains that although the above judgments are a summary of the first two books of WWR, they do not take into account the decisive findings of the third and fourth books.[9] After discussing some candidates for the single thought and its consequences, he finally names the following judgement as an improved candidate from the ranks of the heuristic reading:

> The double-sided world is the striving of the will to become fully conscious of itself so that, recoiling in horror its inner, self-divisive nature, it may annul itself and thereby its self-affirmation, and then reach salvation.[10]

This judgment is intended to provide a summary of all four books, as can already be seen from the catchphrases: (B1) "fully conscious of itself" refers to the idealistic epistemology; (B2) "striving of the will" refers to voluntaristic metaphysics; and (B3)

---

[3] Rudolf Malter: Arthur Schopenhauer. Transzendentalphilosophie und Metaphysik des Willens. Stuttgart-Bad Cannstatt 1991, p. 47.
[4] WWR I, p. 437; cf. Rudolf Malter: Der eine Gedanke. Hinführung zur Philosophie Arthur Schopenhauers. Darmstadt 2010, p. 32; Peter Welsen: Schopenhauers Theorie des Subjekts: ihre transzendentalphilosophischen, anthropologischen und naturmetaphysischen Grundlagen. Würzburg 1995, p. 156.
[5] MR I, p. 512 (No. 662).
[6] Jochem Hennigfeld: Metaphysik und Anthropologie des Willens. Methodische Anmerkungen zur Freiheitsschrift und zur Welt als Wille und Vorstellung, in: Die Ethik Arthur Schopenhauers im Ausgang vom Deutschen Idealismus (Fichte/Schelling). Ed. by Lore Hühn. Würzburg 2006, pp. 459–473, here: p. 465.
[7] WWR II (1844), p. 212 (= Chapter 19).
[8] Vide infra, Chapter 1.1.3.
[9] Cf. John Atwell: Schopenhauer on the Character of the World, p. 30; Christopher Janaway: Introduction, p. 5.
[10] John Atwell: Schopenhauer on the Character of the World, p. 31.

1.1 Interpretive Approaches

and (B4) "recoiling in horror its inner ..." are associated with the contemplative and will-denying features of aesthetics and ethics.

In my opinion, Atwell's approach is instructive in several respects, because, on the one hand, he shows the weaknesses of the above-mentioned heuristic attempts and, on the other hand, he involuntarily demonstrates the basic problem of heuristic interpretations with a self-made example: Atwell's criticism of the above-mentioned heuristic interpretation is justified because e.g. Malter or Hennigfeld cannot explain with their respective judgements why Schopenhauer's WWR includes more than just two books. Atwell himself, however, tries to square the circle: he tries to combine four books with many different topics in one judgement, but he cannot justify why, on the one hand, he has chosen to include some aspects listed separately in WWR (e.g. B3 and B4) into a single consequence ("so that"), but explicitly separates other aspects (B1 and B2) as two sides of an antecedent and why, on the other hand, he does not name central aspects of the work at all (e.g. the difference between understanding and reason, the hierarchy of will, the hierarchy of art).

(2) *Hermeneutic Interpretation*: Overall, it also remains questionable how the heuristic reading can integrate passages in Schopenhauer's work which emphasise that a distinction must be made between the communicated thoughts as parts of the single thought and the one thought itself.[11] For some scholars, these passages suggest that the one thought should not be understood as an abstraction of the individual parts of the system, but that the work has a performative trait of its own, which rather authorises the one thought only in the sense of autonomous and lively thinking and reflection. The question would therefore be whether the assumptions of the above-mentioned authors of the heuristic interpretation are correct, namely that firstly the one thought is abstract and directly communicable and that secondly, it is the summary of the individual parts of the system.[12] A frequently used quotation along these lines comes from Matthias Koßler and states that the one thought "is to be sought at the centre of the intersecting but not converging directions".[13] One can probably understand this quotation of Koßler to mean that he wants to criticise a unilateral reading that always emphasises only individual aspects of Schopenhauer's philosophy, but ignores others. What is performative, however, is to think and follow the transitions between the individual system components (or "directions").[14]

Daniel Schubbe also speaks of an explicitly performative interpretation of the single thought. Accordingly, the single thought does not guarantee content, but the unity of

---

[11] Cf. MR I, p. 428.
[12] Cf. Daniel Schubbe: Philosophie des Zwischen. Hermeneutik und Aporetik bei Schopenhauer. Würzburg 2010, p. 51f.
[13] Matthias Koßler: Schopenhauer als Philosoph des Übergangs, in: Nietzsche und Schopenhauer. Rezeptionsphänomene der Wendezeiten. Ed. by Marta Kopij, Wojciech Kunicki. Leipzig 2006, p. 375.
[14] Cf. also David G. Carus: Die Gründung des Willensbegriffs. Die Klärung des Willens als rationales Strebevermögen in einer Kritik an Schopenhauer und die Ergründung des Willens in einer Auseinandersetzung mit Aristoteles. Wiesbaden 2016, p. 61.

1 The World and its Representationalist Interpretation

the work itself. In his opinion, the WWR presents different perspectives on the relationship between man and world, the unity of which is formulated by the specification of the single thought: the unity demanded by the single thought should, according to Schubbe, "be understood as the commonality of the different perspectives or areas of reality".[15]

In my opinion, these hermeneutical interpretations are heading in the right direction, as they are distinct from heuristic interpretations, which show too much association of individual aspects and themes from the overall work. In my opinion, however, the metaphors and transfers of the individual hermeneutical interpretations are problematic, as they do not evoke helpful clarity, nor do they reveal a specific conceptual or metaphorical tradition, and in the end, they are rarely oriented on statements by Schopenhauer: What are the characteristics of so-called 'directions' that cross but do not converge, and what is the difference between crossing and converging? What exactly is the common ground between the different perspectives and areas of reality, and what is the advantage in terms of understanding or application of the much-discussed performance that scholars of the hermeneutic interpretation emphasise?

(3) *Holistic Interpretation*: The interpretation to be further substantiated in the following chapters, which I favour, on the other hand, does not locate the objective of WWR from Schopenhauer's own statements in the preface, but only at the end of § 15. With the '(one) single thought', the preface represents only an argument of tradition, which (3.1) must be interpreted in the historical context and (3.2) is only instrumental in answering the question of how the book is to be read and what formal content it is able to communicate.

(3.1) The lexeme of a `single thought` in connection with `organism` is not an unprecedented oddity within Schopenhauer's system, as Rudolf Malter has claimed.[16] Fichte had already used such phrases in a prominent place, in the first paragraph of his *Characteristics of the Present Age*, in the same sense as Schopenhauer did at the beginning of WWR I:

> We now enter upon a series of meditations which, nevertheless, at bottom contains *only a single thought*, constituting of itself *one organic whole*. If I could at once communicate to you this *single thought* in the same clearness with which it must necessarily be present to my own mind before I begin my undertaking, and with which it must guide me in every word which I have now to *address* to you, then from the first step of our progress, perfect light would overspread the whole path which we have to pursue together. But I am compelled gradually, and in your own sight, *to build up this single thought out of its several parts*, disengaging it at the same time from

---
[15] Daniel Schubbe: Philosophie des Zwischen, p. 195.
[16] Cf. Rudolf Malter: Arthur Schopenhauer, pp. 44f.

> various modifying elements: this is the *necessary condition of every communication of thought*, and only by this its fundamental law *does that which in itself is but one single thought become expanded and broken up into a series of thoughts and meditations.*[17]

The paragraph or sentence co-occurrences that I emphasised in this quotation prove the high probability with which Schopenhauer took over the wording from Fichte: In the first three paragraphs of the preface to the first edition of the WWR, Schopenhauer also uses the lexemes "one single thought" ("ein einziger Gedanke"), "to communicate" ("mitzutheilen"), "divided up in order to be communicated" ("zum Behuf seiner Mittheilung, sich in Theile zerlegen"), "the various parts must still be organically coherent" ("Zusammenhang dieser Theile ein organischer").[18]

Even the synonymous use of the single thought with the phrase "a single intuition" opens up a history of metaphor and ideas that goes back deep into the early modern era:[19] the author creates a unified philosophical work *sub specie unitatis*, which she can only communicate to the recipient *sub specie diversitatis*. In the context of Romantic philosophy, the expression of a single thought fulfils the central function of pointing to an author-creator analogy: Just as the world before creation was uniform in God, so was the *liber mundi* before it was written uniform in the mind of the author.[20] Although the speech acts should express the same content as the thought act, both are different in form.

(3.2) The metaphor of the single thought also has the function of indicating which form of communication is conditioned by the written form and how the book should therefore be read: it must first be received *sub specie diversitatis* so that the recipient can then understand it *sub specie unitatis* as a single thought. The written multiplicity should communicate the unity of a thought. The dialectic of unity and multiplicity becomes clear once again in the relationship of the author and the reader: the reader receives the book *sub specie diversitatis* so that she can then understand the thought of the author *sub specie unitatis*. The metaphor of a single thought thus has the primary function of announcing the unified, but also holistic character of the work and

---

[17] Johann Gottlieb Fichte: Characteristics of the Present Age (1806). In: The popular works of Johann Gottlieb Fichte, Vol. II, translated by William Smith, 4th ed. London: Trübner & Co., 1889. (My emphasis – J.L.)

[18] Vide infra, Chapter 1.1.2.

[19] Cf. Jens Lemanski: Christentum im Atheismus. Spuren der mystischen Imitatio Christi-Lehre in der Ethik Schopenhauers. Vol. 2. London 2011, p. 316; Matthias Koßler: Die eine Anschauung – der eine Gedanke. Zur Systemfrage bei Fichte und Schopenhauer, in: Die Ethik Arthur Schopenhauers im Ausgang vom Deutschen Idealismus (Fichte/Schelling). Ed. by Lore Hühn. Würzburg 2006, pp. 349–364; Friedrich Schleiermacher: Kurze Darstellung des theologischen Studiums zum Behuf einleitender Vorlesungen. Berlin 1811, p. 45 (= II.2): "as one single intuition" ("als Eine einzige Anschauung").

[20] Cf. Hans Blumenberg: Die Lesbarkeit der Welt. Frankfurt/Main 1986.

at the same time, through the religious author-creator analogy, to increase the expectations of the audience and to demand patience from a presumably overburdened reader.[21]

1.1.2 Multiplicity: The Organic System

Also in the first preface to WWR, a few sentences after the formulation of the 'single thought', Schopenhauer addresses the relationship between the unity and multiplicity of the thought or thoughts developed in a book. The interpretation of the metaphors of multiplicity, namely 'architectural', 'systematic' and 'organic', has also given rise to a debate in research. The controversial passage in the text reads:

> A *system of thoughts* must always have an architectonic coherence, i.e. a coherence in which one part always supports another without the second supporting the first, so the foundation stone will ultimately support all the parts without itself being supported by any of them, and the summit will be supported without itself supporting anything. A *single thought*, on the other hand, however comprehensive it might be, must preserve the most perfect unity. If it is divided up in order to be communicated, the various parts must still be organically coherent, i.e. each part containing the whole just as much as it is contained by the whole [...].[22]

What is disputed is (1) whether Schopenhauer uses 'system' synonymously with 'architectural' and contrary to 'organic' or (2) whether he treats 'system' as a generic term for the two contradictory terms 'architectural' or 'organic'.

The fact that Schopenhauer uses the term 'system' only once in this quotation, speaks in favour of (1), namely at the beginning of the first sentence ("A system of thoughts"). The second sentence shows a significant demarcation from the content of the first sentence ("A single thought, *on the other hand*..."); furthermore, the term 'system' is not explicitly repeated in it. In the first sentence, Schopenhauer speaks of an 'architectural coherence', but in the separate context he says that "the various parts must still be organically coherent".

If one does not become irritated by the word "still" in the last sentence, and instead emphasises "on the other hand" (in German it is the adversative conjunction "Hingegen"), then according to the prevailing opinion one will be able to interpret "architecturally" and "organically" as contradictions: If something is not architectural,

---

[21] This overburdening becomes particularly clear when comparing the different approaches to logic, vide infra, Chapter 1.3.
[22] WWR I, p. 5.

1.1 Interpretive Approaches

then it must be organic, vice versa. It remains disputed, however, whether the concept 'system' is reserved for the term 'architectural' alone. Daniel Schubbe, for example, interprets the quote cited in such a way that Schopenhauer distinguishes "an organism sharply from the idea of a system".[23] For Schubbe, 'system' and 'architectural' are synonyms, whereas 'system and 'organic' are antonyms.

(2) According to Christian Strub, however, Schopenhauer contrasts "the 'organic' system concept with an architectural one".[24] There are, in my opinion, several arguments in favour of Strub's conceptual scheme, according to which 'system' is the generic term and 'architectural' and 'organic' are the two subordinate terms: On the one hand, Schopenhauer uses expressions such as 'my system' in the WWR or in other works,[25] and on the other hand, etymology and the history of the concept suggest that 'system' and 'coherence' ('Zusammenhang') should be understood as synonyms in this quotation.[26] A substitution test also shows that 'system' and 'context' can be replaced *salva significatione et veritate*: If [the single thought] is divided up in order to be communicated, the various parts [of the system] must still be organically coherent.' In general terms, then, it can be said that Schopenhauer explicitly separates the architectural and the organic, and there is much to be said for taking the 'architectural' and the 'organic' as sub-concepts of the concept 'system'.

Although the meaning of the metaphor of the above-given quote ('to support/being supported', 'containing/is contained', 'foundation stone', 'summit', etc.) can never be determined due to the lack of possibilities for contextualisation, attempts at interpretation suggest that the difference between the two might initially lie in the attribution and the relationship of justification: The architectural system is assigned to the "system of thoughts" (plural!), the organic system to the "one single thought" (singular!). The architectural system consists of at least one element that only 'supports' (the foundation stone) and at least one element that is exclusively 'being supported' (the summit). The organic system, on the other hand, stands for the mutual implications of its parts, in that each part contains or receives all other parts (the whole) and the other parts (the whole) contain or receive each individual part.

The organic system, with its mutual implications ('containing/ being contained'), seems to have the argumentative and inferential justification advantage of considering individual parts and propositions as dispensable or not strictly truth-conservative since there are no other parts of the system that depend solely on only one part or one proposition. The mutual implications of the organic system, however, have the explanatory

---

[23] Daniel Schubbe: Philosophie des Zwischen, p. 50.
[24] Christian Strub: Weltzusammenhänge. Kettenkonzepte in der europäischen Philosophie. Würzburg 2011, p. 106; Cf. also Ernst Bloch: Leipziger Vorlesungen zur Geschichte der Philosophie. Vol. 4, p. 369.
[25] E.g. WWR I (Pref. 2nd ed.), p. 15f.: "When I had the strength originally to grasp the basic idea of *my system*, to follow it immediately through its four ramifications, a to return from these to the unity of the trunk from which these four branches emerged, and then to give a clear presentation of the whole […]"; PP I (1851), p. 121: "One could call my system an *immanent dogmatism* […]".
[26] Cf. Otto Ritschl: System und systematische Methode in der Geschichte des wissenschaftlichen Sprachgebrauchs und der philosophischen Methodologie. Bonn 1906.

disadvantage of not being successively or 'linearly' ascertainable. The architectural system, on the other hand, has the explanatory advantage of being able to be understood by the recipient in a stringent, linear and sequential way. However, it has the argumentative and inferential explanatory disadvantage that each individual part and each sentence is indispensable since each sentence or each part is directly justified only by another. With regard to mediated parts of the system, however, this disadvantage of reasoning disappears bottom-up: While the 'foundation stone' still justifies everything directly or indirectly and is not justified by anyone, the 'summit' is fully justified but does not justify anything itself.

The architectural system shows an analogy to the reciprocity law of traditional conceptual logic with regard to this function of justification:[27] just as in the architectural system, the elements in ascending order provide less and less justification but are justified more and more, so in traditional conceptual logic a concept contains the less in itself, the more it contains under itself. A further allusion to the so-called 'containment' metaphors[28] of the traditional logical doctrines of concept and judgement can also be found in the introductory sentence to the organic system: "A single thought, on the other hand, however *comprehensive* it might be...". Schopenhauer is thus alluding to his quantitative concept of the world, which is explained in more detail below.[29] The contradictoriness of this multiplicity embraced in the unity of thought dissolves with reference to the analogous conceptual logic: just as a single thought can contain or comprehend a multiplicity, so, for example, a single abstract generic concept contains or comprehends many concrete concepts of species within itself. Thus, for example, a single generic term such as 'system' can contain or comprehend the multiplicity of specific concepts such as 'architectural', 'organic system', etc.

The prevailing opinion on this quote is that Schopenhauer rejects the architectural system and advocates the organic system. The argumentation for this is usually similar to the following: If the WWR communicates only one single thought (cf. Chapter 1.1.1), and if the context of the one single thought (in its multiplicity only broken down for the purpose of communication) is an organic one, then the WWR is also organically composed. This line of argument also implies that if the concept of 'organic' excludes the concept of 'architectural' (as the classifications 'organic' = 'thought, singular', 'architectural' = 'thoughts, plural' as well as the phrase 'on the other hand' prove), and if the WWR is organic, then the WWR cannot be architectural.

As shown in Chapter 1.1.1, however, the first premise of the first argument is open to attack: Although the aim of WWR may be to communicate a single thought, however comprehensive it may be, this can only be achieved by means of the multiplicity of thoughts. This plurality is also announced by the antecedent of the anankastic conditional in the last sentence of the quote: "If it is divided up in order to be

---

[27] Vide infra, Chapter 1.3.1.
[28] Vide infra, Chapter 2.2.
[29] Vide infra, Chapter 1.2.3.

communicated, the various parts must …".³⁰ The antecedent shows, on the one hand, that the theme of the preface is still (as in the first paragraph of the Preface) communication or readability and, on the other hand, that this communication can only succeed through a multiplicity of parts into which the comprehensive unity (of the single thought) is 'divided'. The consistency of the conditional is, however, problematic in several respects: "…the various parts must still be organically coherent". It remains incomprehensible, however, why this coherence is *still* organic and why it *has to* be organic at all. In my opinion, neither the quotation nor the context can explain the repetition ("still") or the necessity ("must").

An explanation for the repetition and the necessity could perhaps be provided by a picture theory, but no evidence for this can be found in the context or in the quote: If the unity of thought that comprehends the multiplicity can only be communicated through multiplicity (by dividing the unity into parts), then multiplicity must be, as far as possible, the repetition or representation of the unity in multiplicity. In other words: the multiplicity contained in the unity must again be represented by a unity in the multiplicity communicated. The organic system, in which everything is directly and nothing indirectly in a relationship of justification, can possibly better represent this unity in the multiplicity than the architectural system, which for the most part consists only of indirect relationships. – However, this whole argumentation is pure speculation and lacks any textual basis.

What is certain, however, is that the overarching theme of the quotes from the first preface so far is communication or readability. As ornate and metaphorically overloaded as the preface is, all the text sections discussed ultimately lead to a recommendation to the recipient to read the WWR twice. According to Schopenhauer, the division of the work into four parts is thus not a matter of substance, but of communication or readability. Finally, a similar basic statement can be found in the quote of Fichte given in Chapter 1.1.1. Since a book must have a first and last line (just as communication has a beginning and an end) and therefore resembles an architectural system, there is no other way than to read the book successively and linearly. But this should not be confused with the object, with the single thought itself, which the book intends to communicate in form through its multiplicity. From the context of the two quotes discussed, and especially from the metaphors 'architectural' and 'organic', the basic ways of interpreting and dealing with the contradictions or aporias prominent in the work are developed.

1.1.3 Interpretations: The Descriptive Approach

Two opposing interpretative approaches to Schopenhauer's philosophy can be found in current scholarship: (1) the still dominant approach of the so-called 'normative interpretation' is close to the mediating linearity of the architectural system, whereas

---

[30] On anankastic conditionals vide infra, Chapter 1.2.6.

1 The World and its Representationalist Interpretation

(2) the reading that has emerged in recent decades emphasises Schopenhauer's 'descriptive approach' and is close to the immediate plurality of the organic system.

(1) In the history of Schopenhauer's reception, a direction of interpretation can be found early on which approaches an architectural metaphor in that it determines the position of certain themes within the work: some interpreters claim that the beginning of WWR with epistemology is not simply arbitrary, because "each transcendental dogmatism should be avoided",[31] or because it "is the part of the representation of the processual event, through which this event is opened up".[32] Representatives of this reading occasionally refer to Schopenhauer's statement that "every philosophy must commence with an examination of the cognitive faculty, its form and laws, as well as their validity and limitations."[33]

For many performers, the end of the WWR seems to be equally predetermined. Particularly relevant to this position was Franz Rosenzweig's talk of Schopenhauer's innovation of a "system-generated saint of the final paragraphs", who "closed the system arch, really closed it as a keystone, not as an ethical ornament or appendage".[34] Eduard von Hartmann also speaks of an emphasis on nothingness, which Schopenhauer "repeatedly and emphatically described as the summit not only of his ethics but also of his entire philosophical system".[35] According to Hans Zint, the philosophy of religion and especially the sacred and the nothingness, which Schopenhauer discusses at the end of the fourth book of the WWR, thus become the "shining end of Schopenhauer's entire philosophy".[36] As Rudolf Neidert says, Schopenhauer's ethical principles, the affirmation and negation of the will to life, correspond to the Christian doctrine of sin and redemption, and thus Schopenhauer's "anti-deed redemption doctrine" is the "quietistic vanishing point towards which all lines of his ethics ultimately converge"[37]. Similarly, Klamp also points out that the third book can only be an 'introduction' for the "impressive final parts" of the fourth book.[38]

Already in the early 19th century, the architectural arrangement of themes, which started from the idealistic-subjective epistemology and led to mystical nihilism, encouraged most scholars to adopt a linear-normative interpretation: When Schopenhauer, at the end of his main work, describes the ascetic and her escape into nothingness, it was obvious to many interpreters that the author "demands his reader

---

[31] Volker Spierling: Arthur Schopenhauer. Philosophie als Kunst und Erkenntnis. Frankfurt/Main 1994, p. 49.
[32] Rudolf Malter: Arthur Schopenhauer. Transzendentalphilosophie und Metaphysik des Willens. Stuttgart-Bad Cannstatt 1991, p. 53.
[33] PP II, p. 21 (= § 21).
[34] Franz Rosenzweig: Stern der Erlösung. Frankfurt/Main 1921, p. 8f.
[35] Eduard von Hartmann: Phänomenologie des sittlichen Bewusstseins: Prolegomena zu jeder künftigen Ethik. Berlin 1879, p. 41.
[36] Hans Zint: Das Religiöse bei Schopenhauer, in: 17. Schopenhauer-Jahrbuch (1930), p. 63.
[37] Rudolf Neidert: Die Rechtsphilosophie Schopenhauers und ihr Schweigen zum Widerstandsrecht. Tübingen 1966, p. 184.
[38] Gerhard Klamp: Die Architektonik im Gesamtwerk Schopenhauers, in: 41. Schopenhauer-Jahrbuch (1960), p. 83.

to deny the will to live [...]".³⁹ This reading was particularly prevalent in early Hegelianism and the schools that emerged from it. Johann Carl Friedrich Rosenkranz, for example, declared that Schopenhauer would lull his readers into "death orgies of Indian passivity" and spread a "longing for non-existence". His conclusion was therefore: "Instead of this philosophy of death, let us stick to Kant's philosophy of life [...]."⁴⁰ For Karl Kautsky, Schopenhauer's "new doctrine of salvation" leads to an "ossified Chinoiserie" or – according to the wording of the Munich Philistines – to a philosophy of " Leave Me Be!" ("I will mei Ruh hab'n!")⁴¹

In the late 19th century, this interpretation was advocated in particular by critics of the New Kantians in the pessimism controversy⁴² and became the prevailing opinion both among the general public and the early Schopenhauer scholarship. Although Schopenhauer scholars of the early 20th century were already aware of the one-sidedness of such arguments, they adopted this interpretation, sometimes without questioning it. This paradoxical reading, according to which Schopenhauer claimed something explicitly, but probably could not mean it, becomes particularly clear in a quote from Paul Deussen, the founder of the Schopenhauer Society:

> Schopenhauer fights the imperative form of Kantian ethics without seeing that his, like all ethics, has an imperative form. For him, it lies in the fact that he consistently contrasts the negation of the will to life with the affirmation as the higher, better, as he even calls it in his first manuscripts with a comparative expression 'the better consciousness'.⁴³

Jan Garewicz reinforces Deussen's opinion by accusing Schopenhauer of an unintentional is-ought problem. Although Schopenhauer only wants to represent what is, Hume's law slips away from him in such a way that he can only focus on what should be. The supposed representationalism thus becomes an unconscious doctrine of redemption or soteriology, which is already inherent in the epistemology of the first book of WWR:

> Here Schopenhauer violates his own rule of always making statements only about what is and never about what should be. In my [sc. Garewicz'] opinion, this is no coincidence: the whole system is

---

[39] Georg Weigelt: Zur Geschichte der neueren Philosophie: Populäre Vorträge. Hamburg 1855, p. 156.
[40] Johann Carl Friedrich Rosenkranz: Zur Charakteristik Schopenhauer's, in: Deutsche Wochenschrift von Karl Goedeke 22 (1854), p. 684.
[41] Karl Kautsky: Arthur Schopenhauer (Schluß), in: Die neue Zeit. Revue des geistigen und öffentlichen Lebens 6:3 (1888), pp. 97–109.
[42] Cf. Frederick C. Beiser: Weltschmerz: Pessimism in German Philosophy, 1860–1900. Oxford 2016.
[43] Paul Deussen: Allgemeine Geschichte der Philosophie mit besonderer Berücksichtigung der Religionen. Vol. II/3: Die neuere Philosophie von Descartes bis Schopenhauer. Leipzig 1917, p. 555.

based from the very beginning on the foundation of the ideal of holiness.[44]

Similarly, the soteriological reading followed by Rudolf Malter explains the main book as a process of liberation guided by the author: "The formulaic naming of the one thought indicates a process: the process in which the liberation of the subject from its negative state takes place."[45] According to Malter, the process progresses through various crises to redemption:

> Schopenhauer's philosophy can only articulate itself as soteriology [...] because the liberating, redeeming moment is already originally inherent in the subject. To trace how it is activated and how the will – despite its own substantiality – no longer determines the subject is the goal towards which Schopenhauer's system moves thanks to the transcendentalism that guides it.[46]

This quote shows that the architectural connection between 'transcendentalism' (initial idea) and 'doctrine of redemption' (goal) leads to a linear interpretation of the work: The end of the WWR ("the liberating, redeeming moment") is already predicted in the first book ("is already originally inherent in the subject"). The linear-successive movement from the first to the last book is teleologically determined ("goal towards which Schopenhauer's system moves") but is constantly regulated by the initial architecture ("thanks to the transcendentalism that guides it").

Similarly, architectural coherence and linear method are combined when Alfred Schmidt writes: "Resignation is the elusive basic mood into which Schopenhauer's thinking flows".[47] Martin Booms also combines linearity and architectonics, for it seems "by no means coincidental that Schopenhauer's philosophy, according to the facts of the case, [...] comes down to a theme of suffering and redemption".[48] The linearity and normativity can thus be asserted and read out of stylistic analyses, by interpreting the final passages of the main work and from Schopenhauer's later self-characterizations.

(2) Both (a) Schopenhauer himself in many text passages and (b) more recently, an ever-increasing number of scholars have adhered to the reading recommendation and the adoption of the metaphor of the organism for his work. (a) Schopenhauer says: "As this one thought is considered from different sides, it reveals itself respectively

---

[44] Jan Garewicz: Erkennen und Erleben: Ein Beitrag zu Schopenhauers Erlösungslehre, in: 70. Schopenhauer-Jahrbuch (1989), pp. 75–83, here: p. 76.
[45] Rudolf Malter: Arthur Schopenhauer, p. 52.
[46] Rudolf Malter: Arthur Schopenhauer, p. 55.
[47] Alfred Schmidt: Die Wahrheit im Gewande der Lüge. Schopenhauers Religionsphilosophie. München 1986, p. 75.
[48] Martin Booms: Aporie und Subjekt. Die erkenntnistheoretische Entfaltungslogik der Philosophie Schopenhauers. Würzburg 2003, p. 312.

1.1 Interpretive Approaches

as what has been called metaphysics, what has been called ethics, and what has been called aesthetics".[49] For this reason, Robert Jan Berg believes that there are "in principle arbitrary ways of access" to the organism.[50] Also, Schopenhauer's famous metaphor of Thebes in the preface to *On the Will in Nature* from 1836 says that the entry into the system is arbitrary because one can get to the centre from anywhere. In this respect the metaphor of Thebes supports this interpretation:

> If ever the time will come when people read me, they will find that my philosophy is like Thebes with a hundred gates: one can enter from all sides and reach the centre point on a straight path through all of them.[51]

Although in the first book of the WWR Schopenhauer develops the 'world as will' factually from the 'world as representation',[52] a reader can just as well start with the second book, because Schopenhauer there, the other way round, also explains the world as representation from the world as will genetically.

This plurality of entry or access to the system might remind some readers of the famous choice of the basis of the constitutional system in Carnap's *The Logical Structure of the World* (especially §§ 59ff.). Before the outline of a constructional system, Carnap discusses what the system should actually begin with, the physical or the psychological. Carnap chooses the autopsychological as the basis but emphasises that a constructional system including the physical as the basis is also conceivable. Whereas Carnap makes the choice for his exemplary system himself, Schopenhauer, with the metaphor of Thebes, wants to leave this choice to his reader, although as the author of a book, he has to make the same decision in fact as Carnap concerning the starting point of the system. In Carnap's terminology, Schopenhauer also initially decides on the autopsychological and only later develops the physical (or even metaphysical). But in contrast to Carnap, this is not intended to be a choice of reduction: For Schopenhauer, the autopsychological (the world as representation) can be reduced to the (meta-)physical (the world as will), *and vice versa*.[53]

(b) I have described this figure of thought in another place with the expression 'mutual epiphenomenalism':[54] In the first book, the factually existing, objective world as will seems to be the product of the subjective cognition of the world as representation, while in the second book, ontogenesis (the world as representation) seems to be only

---

[49] WWR I, p. VII (= Pref.).
[50] Robert J. Berg: Objektiver Idealismus und Voluntarismus in der Metaphysik Schellings und Schopenhauers. Würzburg 2003, p. 99.
[51] FW, p. 6 (Pref.).
[52] I will discuss how to understand the reference to these apparently two worlds in Chapter 1.2.2.
[53] On Schopenhauer and Carnap vide infra, Chapter 2.3.3.
[54] Jens Lemanski: Schopenhauers hagioethischer Konsequentialismus im System der Welt als Wille und Vorstellung. In: 93. Schopenhauer-Jahrbuch (2012), pp. 485–503.

a product of the factually generating phylogeny (the world as will).[55] Each of the two worlds is only a contingent by-product from the perspective of the other world. Only intersections, such as the subjectively and objectively experienceable body ('Leib'), go beyond this impression of accidental side effects.[56]

The extent to which the expression 'mutual epiphenomenalism' is appropriate for the relationship between the world as representation and the world as will, as presented in the first two books of the WWR, in the face of such intersections may be debatable. Nevertheless, this expression gives a name to the figure of thought which, on the one hand, philosophers of the present day continue to emphasise with reference to classical German philosophy and early linguistic-analytical philosophy,[57] and which, on the other hand, is an indication of the Thebes-like plurality of access paths to Schopenhauer's system. After all, it is irrelevant whether the world as representation first establishes the world as will or vice versa, since both mutual modes of justification were separated only for the purpose of communication.

The almost arbitrary position of the individual books and the topics collected in them becomes particularly clear in a comparison between the WWR and the new version of the WWR presented in the *Berlin Lectures* of the 1820s: whereas the linear interpretation considers it relevant that Schopenhauer began his main work with the sentence "The world is my representation" and ended with the concept "nothing", the *Berlin Lectures* show the arbitrariness of special positions of this textual fragments. The new version of WWR for lecture purposes does indeed contain the two relevant phrases; however, neither of them form the beginning or end of the system but are in each case after or before metaphilosophical reflections.[58] The linearity which interpreters such as Rosenzweig, Hartmann, Zint, Klamp, Malter and many others justify by the special position of books, themes, phrases and concepts in the WWR loses its textual basis in the new version of WWR. Within the organic reading, an alternative version of WWR would thus also be conceivable, which does not begin with the world

---

[55] Whether the concepts 'ontogenesis' and 'phylogeny' are appropriate or only metaphorical for the facts described here, I do not want to discuss at this point. An intensive discussion can be found in Jens Lemanski: Die 'Evolutionstheorien' Goethes und Schopenhauers. Eine kritische Aufarbeitung des wissenschaftsgeschichtlichen Forschungsstandes. In: Schopenhauer und Goethe. Biographische und philosophische Perspektiven. Ed. by Daniel Schubbe, Søren R. Fauth. Hamburg 2016, pp. 247–295.
[56] Cf. the volume Philosophie des Leibes. Die Anfänge bei Schopenhauer und Feuerbach. Ed. by Matthias Koßler, Michael Jeske. Würzburg 2012.
[57] Vide infra, Chapter 2.1.4 and Chapter 3.1.
[58] The article by Thomas Regehly, *Die Berliner Vorlesungen*, shows how shockingly normative interpreters twist the facts (Thomas Regehly: Die Berliner Vorlesungen: Schopenhauer als Dozent. In: Schopenhauer-Handbuch. Leben – Werk – Wirkung. Ed. by Daniel Schubbe, Matthias Koßler. Weimar 2014, pp. 171–180.): Regehly shows especially on p. 171 (and further on p. 179) that he follows Malter in the architectural-normative interpretation and then explains on p. 175, the "Lecture begins like the main work with the sentence 'The world is my representation' [...]". A quick glance at Chapter 1 of the *Berlin Lectures* is enough to falsify this statement: The lecture does not begin like the WWR with the sentence 'The world is my representation'. Regehly also suppresses in his account on p. 179 that Schopenhauer conceived the end of the lectures differently than the main work. Regehly's otherwise very meritorious survey article thus shows how advocates of the normative interpretation fall for their own prejudices and expectations of a text.

as representation in the first book, but with the world as a will in the second, or ends with the affirmation and not with the negation of the will in the fourth book.

Representatives of this pluralistic descriptiveness, for which I also take a position here, refer in contrast to the initial passages of the fourth book of the WWR, in which Schopenhauer explains that his ethics also remain only theoretical and do not recommend prescribing anything.[59] For Matthias Koßler, this is the reason to speak of an "empirical ethics" and to emphasise several times that Schopenhauer also "understands ethics as 'descriptive' rather than prescriptive".[60] The so-called 'morphological interpretation' also ties in with this aspect of descriptiveness and rejects any linearity and normativity.[61] Rather, this school of interpretation emphasises those statements by Schopenhauer in which he conceives his work as a representational description of the one world. According to this view, the four books do not linearly follow one another, but stand parallel to one another and explain the world or the one thought, but do not prescribe how one should behave in or towards this world.

1.1.4 Aporias: Alleged Contradictions

Very early – i.e. already in 1819 by an anonymous reviewer – the Schopenhauer audience drew attention to aporias or contradictions in his work.[62] Schopenhauer opposed these accusations and repeatedly emphasised that his system was free of contradictions and rather uniform, or that the aporias were only based on misunderstandings of the interpreters.[63] Almost a century after the first accusation of aporias, in 1906, Otto Jenson wrote a dissertation on the subject, in which he provides a tabular overview of the fourteen fundamental contradictions he found in the works of almost 25 relevant Schopenhauer interpreters. A complete literature review even reveals a total of 52 "inconsistencies or impossibilities of thinking", and this list is by no means exhausted.[64] As the overviews and treatises on the aporias in the course of the 20th and at the beginning of the 21st century show, the topics of discussion have

---

[59] WWR I, p. 297ff.
[60] Matthias Koßler: Empirische Ethik und christliche Moral. Zur Differenz einer areligiösen und einer religiösen Grundlegung der Ethik am Beispiel der Gegenüberstellung Schopenhauers mit Augustinus, der Scholastik und Luther. Würzburg 1999, p. 434.
[61] Cf. Daniel Schubbe: Formen der (Er-)kenntnis. Ein morphologischer Blick auf Schopenhauer, in: Der Besen, mit dem die Hexe fliegt. Wissenschaft und Therapeutik des Unbewussten. Vol. 1: Psychologie als Wissenschaft der Komplementarität. Ed. by Günter Gödde, Michael B. Buchholz. Gießen 2012, pp. 359–385.
[62] Cf. Anonymous Reviewer: Arthur Schopenhauers Die Welt als Wille und Vorstellung, in: Literarisches Wochenblatt 4:30 (Weimar 1819) (also in Schopenhauer-Jahrbuch 6 (1917), pp. 81–85).
[63] A summary of these statements by Schopenhauer can be found at Otto Jenson: Die Ursache der Widersprüche im Schopenhauerschen System, p. 8.
[64] Otto Jenson: Die Ursache der Widersprüche im Schopenhauerschen System, p. 23, also p. 29.

1 The World and its Representationalist Interpretation

shifted in part, but have lost nothing of their explosiveness:[65] While (1) some interpreters deplore the "contradictions" or "paradoxes" inherent in Schopenhauer's work, (2) the opposite side tries to expose these evaluations as misunderstandings of the accusers. (1) The first direction of interpretation is either voluntarily or involuntarily closer to the linear, architectural and normative reading, (2) while the second claims an either singular or plural aspect of organic descriptiveness for itself. Both readings, (1) and (2), are in turn divided into an affirmative (aI) and a negative interpretation (nI):

(1) Within the first line of interpretation, the discussed contradictions are evaluated either, in the sense of (nI), as an expression of a failed theory or, in the sense of (aI), as a constitutive, positive component of Schopenhauer's thought.[66] In addition to Schopenhauer's critics, scholars who do not see the system as a "balanced, smooth, secure edifice of thought" can be added to the negative direction of interpretation.[67] Due to the inherent contradictions in the system, one of the earliest interpreters of the system writes that one could hardly find a "more contradictory philosopher [...]".[68] According to Vittorio Hösle, this is due to the fact that Schopenhauer "did not have the intelligence of theoretical justification" that the great philosophers, from Plato to Hegel, had.[69] Booms is much milder in his judgement and sees a fragility in the circularity and in the contradictions of the system, which, however, could only be fixed by interpretation.[70]

The Jenson study mentioned above forms the transition from (nI) to (aI) within the direction of interpretation that sees contradictions in Schopenhauer's system. Jenson believes that Schopenhauer's organic system cannot arrange the uniform overall impression it heralds. Schopenhauer's demand to read the WWR several times was even a disservice to his own system, as each time one reads it again, more and more contradictions become apparent.[71] Nevertheless, and with this comes the turn to (aI), these aporias constitute the "mystical attraction" of Schopenhauer's system.[72] To recognise the value of Schopenhauer's system despite these antinomies, it is crucial to read Schopenhauer more as an artist and less as a researcher.[73]

Remnants of an (nI) can also be seen in Volker Spierling's approach, which repeatedly assigns Schopenhauer a "self-misunderstanding", but attempts to cure this by so-

---

[65] To a list of authors who have commented on the topic, cf. Rudolf Malter: Arthur Schopenhauer, p. 48, Fn. 25; for the following systematisation cf. Martin Booms: Aporie und Subjekt, p. 25f.
[66] Cf. e.g. Volker Spierling: Arthur Schopenhauer. Philosophie als Kunst und Erkenntnis, pp. 223–240; Daniel Schubbe: Philosophie des Zwischen, Chapter 1.
[67] Gisela Sauter-Ackermann: Erlösung durch Erkenntnis? Studien zu einem Grundproblem der Philosophie Schopenhauers. Cuxhaven 1994, p. 131.
[68] Rudolf Seydel: Schopenhauers philosophisches System. Leipzig 1857, p. 7.
[69] Vittorio Hösle: Zum Verhältnis von Metaphysik des Lebendigen und allgemeiner Metaphysik. Betrachtungen in kritischem Anschluss an Schopenhauer. In: Metaphysik. Herausforderungen und Möglichkeiten. Ed. by Vittorio Hösle. Stuttgart-Bad Cannstatt 2002, pp. 59–97, here: p. 61f.
[70] For example, Martin Booms: Aporie und Subjekt, e.g. p. 153ff.
[71] Cf. Otto Jenson: Die Ursache der Widersprüche im Schopenhauerschen System, p. 12ff.
[72] Otto Jenson: Die Ursache der Widersprüche im Schopenhauerschen System, p. 33.
[73] Cf. Otto Jenson: Die Ursache der Widersprüche im Schopenhauerschen System, pp. 55ff.

called "Copernican turns".[74] According to Spierling, the paradoxes in Schopenhauer's work, which are shown by means of the "turn", are ultimately an advantage of the system, as they oppose a dogmatic and absolute standpoint, which in the view of the philosopher should be avoided. Thus, in Schopenhauer one recognises a philosopher "who reflects prudently, who methodically remembers the difference between concept and thing, who puts a stop to a priori idealistic thinking about identity".[75]

Following on from Spierling, Daniel Schubbe, in his hermeneutical-phenomenological reading, also understood aporias not as a deficiency but as "the key to Schopenhauer's work".[76] The decisive factor is not the focus on the respective antinomic poles of the aporias, paradoxes and contradictions, but rather the concentration on "the in-between" ("das Zwischen") that connects the respective poles. This concentration on the "in-between" emphasises the rare moment in which the familiar appears anew and the conceptual aporias point to the nonconceptual.[77] All the authors of this (aI) have a strategy in common, which consists in reinterpreting the disadvantage of the contradictions to an advantage by means of an interpretation that is predominantly external to the system.

(2) Within the second school of interpretation, the discussion of the contradictions in Schopenhauer's work is largely unanimously rejected, either in the sense of (aI) being of the opinion that the contradictions are based on a bad or wrong interpretation on the part of the accusers, or in the sense of (nI) believing that the contradictions are in part only due to formal-philological inaccuracies or generally not a decisive evaluation criterion in dealing with historical philosophers. The affirmative direction of interpretation is completely reactionary since it alone defends the attacks on Schopenhauer's system on the part of the interpreters subsumed under (1) and (nI). The most stubborn defence of Schopenhauer can probably be found in Wilhelm Gwinner, Paul Deussen and Arthur Hübscher: Gwinner tries to prove, for example, by also criticizing Johann Friedrich Herbart, that many critics often misunderstood Schopenhauer.[78] Paul Deussen goes even further. He speaks of a "perfection of critical philosophy by Schopenhauer" and sets himself the goal of "proving everywhere that Schopenhauer's procedure is strictly methodical and scientific and of invalidating the assertions of those who enjoy discovering all kinds of contradictions in Schopenhauer's system".[79] Hübscher, too, first shows how Schopenhauer, in his lifetime, refuted accusations of contradictions in his system in his written correspondence with critics. After Schopenhauer's death, "the search for contradictions was sometimes made surprisingly easy", and Hübscher, therefore, comes to the conclusion: "Enough! The search for

---

[74] Volker Spierling: Arthur Schopenhauer: Philosophie als Kunst und Erkenntnis.
[75] Volker Spierling: Arthur Schopenhauer, Philosophie als Kunst und Erkenntnis, p. 240.
[76] Daniel Schubbe: Philosophie des Zwischen, esp. pp. 21ff.
[77] Cf. Daniel Schubbe: Philosophie des Zwischen, p. 60, p. 142.
[78] Wilhelm Gwinner: Schopenhauer's Leben. Arthur Schopenhauer aus persönlichem Umgange dargestellt. 2nd ed. Leipzig 1878, p. 267ff.
[79] Paul Deussen: Allgemeine Geschichte der Philosophie, Vol. II/3, p. 430.

inconsistencies and contradictions that fills a large part of the literature on Schopenhauer by no means reaches the whole of his doctrine, which was still encountered in the second half of the 19th century in a strange helplessness by many interpreters."[80]

Representatives of (aI) are motivated by Schopenhauer's own attempts at refutation, which he brought up against his own critics. What all representatives of (aI) have in common is that they are convinced of Schopenhauer's uniformity, infallibility and consistency, which they defend against all attacks.

The representatives of (nI) within this line of interpretation, to which I willingly submit myself, complain that all the above-mentioned lines of interpretation place the uniformity of Schopenhauer's system postulated in terms of content above their own scholarship. Schlüter and, above all, Lovejoy have demonstrated that this begins, especially in the philological exact processing of the texts within the different creative phases. Kuno Fischer had already pointed out that Schopenhauer's philosophy had "changed its character" over the decades.[81] Robert Schlüter's examination of Schopenhauer's philosophically relevant letters leads him to the conclusion that Schopenhauer's systemic concepts have changed in general, but especially in detail, or at least have been strongly modified.[82] In doing so, he attacks the "fairy tale of the absence of any development in Schopenhauer's doctrine".[83] Schlüter sees Schopenhauer's correspondence with his friends in particular as a process in which Schopenhauer had repeatedly led to new ideas, modifications and extensions.

Whereas with Schlüter and Fischer, flanked by authors such as Jacob Mühlethaler, Oscar Janzens or Harald Høffding, the main representatives of a revised doctrine in Schopenhauer's work in the paradigm of Neo-Kantianism are named,[84] the question of system development lost importance in German post-war philosophy due to the dogmatic assertion of a unified system and a glorification of last-hand editions of Schopenhauer's writings. Although today there are various successful approaches to reconstruct the developments in Schopenhauer's work, especially on the basis of manuscripts, in my opinion, many systematic treatises are subject to the difficulty that they collect context-free statements from the entire work, the uniformity of which is partly only guaranteed by the author's name 'Schopenhauer'.

Against such a method, Arthur Lovejoy's study on Schopenhauer's philosophy of nature is particularly noteworthy, as he convincingly demonstrated that supposed contradictions in Schopenhauer's work can be resolved by first interpreting thesis and

---

[80] Arthur Hübscher: Denker gegen den Strom. Schopenhauer: Gestern – Heute – Morgen. Bonn 1973, pp. 256–259.
[81] Kuno Fischer: Schopenhauers Leben, Werke und Lehre. (Geschichte der neuern Philosophie IX) 3rd. ed. Heidelberg 1908, p. 530 (= 21.3.5), also: p. 273 (= 8.1.3).
[82] Cf. Robert Schlüter: Schopenhauers Philosophie in seinen Briefen. Leipzig 1900, pp. 37ff., p. 43, p. 72.
[83] Robert Schlüter: Schopenhauers Philosophie in seinen Briefen, p. 5.
[84] Cf. Jacob Mühlethaler: Die Mystik bei Schopenhauer. Berlin 1910, p. 147f.; Harald Høffding: Geschichte der neueren Philosophie. Eine Darstellung der Geschichte der Philosophie von dem Ende der Renaissance bis zum Schlusse des 19. Jahrhunderts. Vol. II. Leipzig 1896, esp. p. 247ff. On Oscar Jansen's thesis of Schopenhauer's revised geometrical theory vide infra, Chapter 2.3.5.

1.1 Interpretive Approaches

antithesis separately within their respective contexts and not starting from the thesis of a unified system in which no Schopenhauer sentence – no matter from which work and historical context it is torn – is allowed to contradict another.[85] Lovejoy has thus attempted to prove that many of the supposed contradictions in natural philosophy are based on the fact that they originate from different periods of work and that these were written in periods of different scientific paradigms. It can thus be said that Schopenhauer did not do his work any favours by merely supplementing the system in later years and not fundamentally revising it.

I particularly agree with Lovejoy's demand for a separate analysis of individual writings and statements, as such a method need not necessarily contradict the unifying idea of the organic system: If, for example, a separately analysed text from the 1810s coincides in content with a similarly analysed text from the 1850s, then this does not contradict the system unity postulated by Schopenhauer. However, if such texts show contradictions, it should first be clarified to what extent the required consistency and uniformity of these texts are justified and also whether the thesis and antithesis depend on factors external and internal to the text.[86]

Irrespective of the philological preparatory work, which in my opinion has a considerable influence on the way many of the apparent aporias and contradictions are dealt with, the question of the deeper philosophical meaning of the intensive discussion of aporias in Schopenhauer scholarship does arise. I can probably only approach this question from my standpoint: The reading I propose here is related to the (aI) of (1) in that it is less concerned with the validity of Schopenhauer's historical system than with one's point of view, which is gained by examining this system. In contrast to the affirmative interpretation of (1), however, the new is not gained through the formulation of an externally proposed interpretation scheme, but rather through the elaboration of individual parts of Schopenhauer's conceptual scheme that generates his system. While the (nI) of (1) and the (nI) of (2) argue about the validity of Schopenhauer's system, the other two positions seem to be concerned at most with the fact that there are purely factual contradictions. The (aI) of (1), however, is, in my opinion, more interested in the justification of its own metatheory, which is developed using the example of Schopenhauer's philosophy. With the (aI) of (1), however, I more or less share the view that they are not seriously attacking or defending Schopenhauer's system and arguing about its validity or invalidity, but rather making a historical system and its conceptual scheme readable in the paradigm of history of philosophy and systematics that is oriented towards facts and argumentation.

Within this book, I combine three different ways of dealing with Schopenhauer's philosophy: Chapter 1 is initially interested in a historically exact classification of logic (Chapter 1.3) in Schopenhauer's system (Chapter 1.2) (as far as one is possible).

---

[85] Cf. Arthur O. Lovejoy: Schopenhauer as an Evolutionist. In: The Monist 21:2 (1911), pp. 195–222.
[86] Cf. the detailed description of this method at the end of Jens Lemanski: Die 'Evolutionstheorien' Goethes und Schopenhauers.

This will show that Schopenhauer, although he hardly ever reworked the structure of his system, did, however, rework the logic within it several times and that this logic is most extensively present in the work of which I have good reason to claim that philosophers should regard it as the actual main work – namely the *Berlin Lectures*. Only Chapter 2 will argue for the topicality of individual systematic themes of this actual major work. In Chapter 3, I will take the liberty of formulating my point of view, which, while building on the results developed in Chapters 1 and 2, will place them in a modern theoretical context, which Schopenhauer, of course, could not have included due to his historical position.

1.2 The System of WWR

Chapter 1.1 presented different interpretative positions on four thematic areas of the WWR, which are represented in research. The position has already been adopted and explained that there are good reasons to place the concept of the world at the centre of Schopenhauer's representationalism and to understand its organic system as an expression of a plural descriptiveness. The system is best thought of as descriptive because even in ethics it does not make any normative claim, but only describes actions and assigns concepts to them. Although the world as an organic entity is to be conveyed through the WWR, the entrance to the book can be described as 'plural', since it seems to be completely irrelevant with which topic one finds an entrance to the WWR and the world depicted in it. Finally, Chapter 1.1 contained a plea to approach the aporias in Schopenhauer's work, which have been intensively discussed in research, with certain philological maxims and, if this strategy fails, to recognise it as a deficit of a philosophical system, but not to overestimate it (after all, there is probably no scientific system that can absolve itself of this).

The objective of Chapter 1.2 is to give an overview of Schopenhauer's system in order to be able to better name the status of logic within Schopenhauer's work later on (Chapter 1.3). In Chapter 1.2.1, it will become clear that there is little preliminary work on the structure of the system in Schopenhauer and that the most important work on this topic introduces unreflected interpretative premises which I cannot approve from the remarks already made in Chapter 1.1. But first of all, I share the assessment with the few preliminary studies that it is advantageous to work out the system structure at WWR. Only beginning with Chapter 1.3, several reasons are given for locating Schopenhauer's central system not in WWR, but in the *Berlin Lectures*.

I have decided to proceed in seven steps: First, I will present the research literature that has dealt with the concept of the system (1.2.1) in order to be able to draw on its theses in the following chapters and to examine it critically. The subsequent chapters are based on the classification of the first volume of WWR (= WWR I): In Chapter 1.2.2 the preface to WWR I is examined and in Chapters 1.2.3 to 1.2.6 one of each of the four books of WWR I is presented. Finally, I will evaluate this presentation in Chapter 1.2.7, allowing myself to make a judgement about the research theses in particular. Although Schopenhauer already presented logic in Book I (= B I), I do not want to limit myself to the examination of B I alone, since only the overall concept of WWR, i.e. Schopenhauer's location and evaluation of logic in the system, makes its status completely transparent.

My demand for a precise interpretation of the system, which was particularly motivated by my criticism of the philological fuzziness in previous research, as outlined in Chapter 1.1.4, has forced me to make some methodological *restrictions* and *compromises*. As Chapter 1.3 will show that logic played a special role in Schopenhauer's work around 1820, and that one must speak of a revised doctrine from 1844 onwards,

1 The World and its Representationalist Interpretation

I will use the first edition of the WWR from 1819 as the textual basis of Schopenhauer's system, and limit myself almost exclusively to this edition. Consequently and in contrast to many other Schopenhauer studies – and I cannot exclude some of my early treatises on Schopenhauer –, I will not use Schopenhauer's complete oeuvre in order to search for quotations that meet my sense of interpretation.

However, the restriction to Schopenhauer's early writings also gives rise to problems that have forced me to make the compromises already announced: Although the first edition of WWR has been freely and unproblematically available as a digital copy for several years, scholars do not read and use it, but almost exclusively the third edition or even the last hand edition, which was no longer published during Schopenhauer's lifetime. Scholars may object that Schopenhauer did not make many changes in the second and third editions, but since I still advocate a close reading in chronological order for the philological reasons outlined in Chapter 1.1.4, part of the compromise is to use only the first edition of WWR. (For the English version of *World and Logic* some quotations had therefore to be adapted to the English translation used.)

In the preface to the second edition, Schopenhauer had stated that he had "numbered the sections which were separated only by lines in the first edition" in order to make it easier to name them.[1] This reference represents the other part of the compromise, which is intended to benefit those recipients who are not familiar with the 1819 edition but with later editions. Since, in my view, it is irrelevant whether the 70 dividing lines are numbered consecutively or whether the later numbering of the 71 paragraphs is taken over, the decision seems obvious to focus on the first edition as the textual basis, but to quote the 71 sections marked by dividing lines as paragraphs: On the one hand, this allows me to name the sections of the WWR in the main text in a precise and uncomplicated way, and on the other hand, it allows the recipient of this publication, who only has the later editions at hand, to better understand the numbering of the section.

A further methodological limitation also means that the following analysis of the systematic structure of the WWR does not necessarily correspond to what is usually expected from a presentation of the WWR. I will speak at several points of an argumentative course or a content-thematic elaboration etc. and contrast these with reflexive, system- or structure-related text passages. While research on the WWR as well as introductions to this work usually attempt to explain content-thematic relations – for example between concepts such as 'will', 'thing in itself', 'ideas' etc. and their semantics – the following chapters are limited primarily to text passages in which Schopenhauer explains what he has done, what he is doing and what he intends to do. Text passages that discuss thematic breaks, digressions, beginnings and endings of sections or even evaluations of system contents are more important to me than the

---

[1] WWR I, p. 17.

contents and topics of these treatises and of parts of the system. This approach is motivated by the fact that I hope that a preliminary interpretation of the form of the system will lead to a more reliable analysis of the individual contents and arguments in the WWR – here especially of logic – than an interpretation that presents contents and arguments in the WWR without having thought about the form of the system, the structure of the book and the order of the argumentation.

1.2.1 Research Literature Regarding the System

Even a brief glance at the bibliographies on Schopenhauer shows that the topic of 'the system' has been given little attention so far. In the chronological order of publication, the title of Rudolf Seydel's monograph *Schopenhauers philosophisches System* (engl. *Schopenhauer's Philosophical System*) announces a contribution to this topic as early as 1857, but in terms of content, it discusses almost exclusively contradictions and inconsistencies in the overall work and to what extent these endanger the systematic character.[2] William Caldwell published a study in 1896 entitled *Schopenhauer's System in its Philosophical Significance*, but it was rather a unique interpretation of the main themes of the WWR in relation to Hegel and von Hartmann. Otto Jenson's book *Die Ursache der Widersprüche im Schopenhauerschen System* (*The Reason for the Contradictions in Schopenhauer's System*) also ties in more with the aporia discussion in Schopenhauer's work, which had already been greatly inflamed at the time, than with the systematic character of the texts.[3] Since Jenson's investigation leads to the thesis that Schopenhauer was more an artist than a scientifically interested philosopher, the question regarding the character of the system becomes completely obsolete for the author.

Until 1960 the phrase 'Schopenhauer's system' does not seem to play a significant role in book titles, articles or book chapters. The term 'Schopenhauer's System' was then only used as a synonym for 'Schopenhauer's philosophy', 'Schopenhauer's works' or similar. It was Gerhard Klamm who first drew attention to the concept of the system again in 1960 in a paper which has remained an isolated case of research to this day.[4] In view of the prevailing opinion that Schopenhauer rejects the term 'architectural' but favours 'organic' with regard to the concept of the system,[5] the title of Klamm's paper, i.e. *Die Architektonik im Gesamtwerk Schopenhauers* (*Architectonics in the Complete Works of Schopenhauer*) seems problematic. Contrary to the prevailing opinion, Klamp makes offensive use of the architectural metaphor and even reinforces it by speaking of 'arched arches', which are (1) between the four books of

---

[2] Cf. Rudolf Seydel: Schopenhauers philosophisches System, p. VIff.
[3] Cf. Otto Jenson: Die Ursache der Widersprüche im Schopenhauerschen System.
[4] The paper, Hans Margolius: System und Aphorismus. In: 41. Schopenhauer-Jahrbuch (1960), pp. 117–124, does not provide any new insights into Schopenhauer's system.
[5] Vide supra, Chapter 1.1.2.

WWR I, (2) between WWR I and WWR II, (3) between WWR I/II and Schopenhauer's four monographs remaining,[6] and (4) between the aforementioned works and PP I and PP II.[7]

Above all, Klamp's pioneering work has the value of having emphasised the systematic character of Schopenhauer's philosophy for the first time. A cursory glance at the four arched arches mentioned above shows that Klamp was particularly concerned with the "external structure", but less with the "internal design" of the system.[8] In addition, the external structure shows that the WWR is given a central position in the overall work, which Klamp describes at the beginning with the following words:

> What immediately catches the eye when looking at the whole is the peculiar symmetrical correspondence of the partial units of a building front, which is comprehensive in both width and height, and which fit together harmoniously into a uniform whole. In the final analysis, it is a single thought that has taken shape in ever-changing form, as it were, in a multi-storey giant building, and of which the "whole" is evident in every detail. Schopenhauer himself expressly pointed this out in the preface to the 1st edition of his major work. The title itself: "World as Will and Representation", expresses this basic idea in keywords in a way that is not to be misunderstood. It is spread out in four (!) "books", each in two (!) extensive volumes, [...] in clearly structured form, piece by piece in front of our inner eye [...].[9]

The quote shows that for Klamp the architectural metaphor as an instrument to describe the system does not seem to contradict the first preface of WWR I. The uniformity hinted at there by the 'one single thought' is also evident in the harmonious and symmetrical 'giant building' of the overall system. According to Klamp in another paper, the symmetry addressed in this architecture is explained by the evenness repeatedly found in Schopenhauer's way of thinking, which is particularly evident in two- or four-part constructions.[10] Klamp's proclaimed dominance of the WWR is not only visible in the arched arches listed above, but also in the quote: the title of the main work and the division of a single thought into two volumes indicates the division

---

[6] This refers to the four monographs *On Vision and Colours*, *On the Fourfold Root of the Principle of Sufficient Reason*, *On the Will in Nature* and *The Two Fundamental Problems of Ethics*.
[7] Gerhard Klamp: Die Architektonik im Gesamtwerk Schopenhauers, pp. 82–98.
[8] Ibid., p. 82.
[9] Ibid., p. 82f.
[10] Cf. Gerhard Klamp: Das Streitgespräch zwischen Becker und Schopenhauer. In: 39. Schopenhauer-Jahrbuch (1958), p. 71; Cf. also Margit Ruffing: Die 1, 2, 3/4-Konstellation bei Schopenhauer. In: Die Macht des Vierten. Über eine Ordnung der europäischen Kultur. Eb. by Reinhard Brandt. Hamburg 2014, pp. 329–349.

1.2 The System of WWR

into two, the division of the two volumes of the WWR indicates the respective four-part division.

Although Klamp's paper was rarely quoted and was not critically reviewed or continued, he has nevertheless – especially through the inclusion of Klamp's theses in Malter's work – shaped the image in research to this day that Schopenhauer's system is centred around the WWR.[11] Neither Klamp nor any of his successors have been able to present an explicit study on "the inner design".

What is particularly problematic about Klamp's research is his unquestioned affinity for normative interpretation[12] and his philological bias – which is quite common in Schopenhauer scholarship – not to develop Schopenhauer's work chronologically on the basis of the genesis of the system, but based on the last hand edition that is dogmatically considered complete, perfect and uniform. Klamp interprets the relationship between the four books (= B) of the WWR I and WWR II, for example, in such a way that B III provides the "introduction" ("Vorschule")[13] for the "impressive final paragraphs" of B IV.[14] The philological bias is not only evident in Klamp's assumption that the WWR was divided "into two (!) extensive volumes", but also in the other arch arches that exist between WWR I and WWR II and the other works that were written in Schopenhauer's later period. In short: Klamp interprets Schopenhauer's system chronologically, but from the unhistorical view of the finished last hand edition. This is particularly unsatisfactory as Klamp was intensively involved in the research on Schopenhauer's first system (entitled *Das Systemchen,* the German diminutive of 'system') and was therefore close to research on the history of origins.[15]

With regard to the WWR-internal arch arches, Klamp has worked out six pair-wise groupings between the four books ("Buch") of the work, which are numbered in Fig. 1.

According to Klamp, the six arches of the scheme can be interpreted as follows:

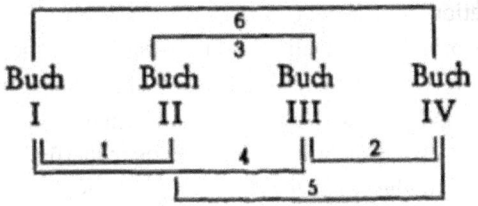

Fig. 1

Gerhard Klamp: Die Architektonik im Gesamtwerk Schopenhauers. In: Schopenhauer-Jahrbuch 41 (1960), pp. 82–98, here: p. 85. (Buch = Book)

---

[11] Cf. Rudolf Malter: Arthur Schopenhauer, esp. pp. 44ff.
[12] Vide supra, Chapter 1.1.3.
[13] Rudolf Malter: Arthur Schopenhauer, p. 83.
[14] Rudolf Malter: Arthur Schopenhauer, p. 85.
[15] Gerhard Klamp: Zur Zeit- und Wirkungsgeschichte Schopenhauers. In: 40. Schopenhauer-Jahrbuch (1959), pp. 1–23.

1 The World and its Representationalist Interpretation

(1) B I and B II are linked by the concept of the world, once as representation (I), the other time as will (II).
(2) B III and B IV are linked by the concept of the negation of the will, which is sometimes thought out aesthetically (III), and sometimes thought through to its ethical perfection (IV).
(3) B II and B III are linked by the concept of idea, which is sometimes presented in general terms (II) and sometimes in concrete terms (IV).
(4) B I and B III are linked by the concept of representation, which is sometimes subject to the principle of sufficient reason (I), and at other times is independent of it (III).
(5) B II and B IV are linked by the concept of will, which is presented once as objectivation and once as self-knowledge.
(6) B I and B IV represent the "dual unity" ("Zwei-Einheit") of the whole system, which begins with the subjective construction of the world and vanishes into nothingness when the subject's will is denied.

Although there has been no direct discussion of Klamp's scheme in research to date, several of these six arches have been indirectly criticised on several occasions: (1) has been indirectly questioned by many interpreters who have reproduced the individual titles of the four books of the WWR more precisely and, above all, in their entirety. For the concept `world` is the unifying title of all four books, B I and B III read: "The World as Representation"; books II and IV read: "The World as Will". Thus the complete reference to the concept of the world in (1) already implies the arch arches of (4) and (5). (2) and (6) are the expression of a normative reading that Klamp tacitly presupposes to be valid and which is questioned especially by representatives of the descriptive interpretation.

Indirectly, Klamp's principle of duality in (1)–(6) has been criticised in recent years, especially by Margit Ruffing, who has used Daniel Schubbe's organic and descriptive interpretation as a starting point to formulate a new form of linear reading with a new pair of opposites (1,2,3/4) – namely in such a way that "the first three books are presented as a world understanding of affirmation, the fourth as a self-understanding from negation".[16] According to Ruffing, the aporias of the first three books would be dissolved by the fourth book. As Ruffing suggests, Schopenhauer would "certainly [...] not have considered this 1,2,3/4 structure to be meaningful".[17] This makes it clear that Ruffing's approach does indeed follow Schubbe in that both ("Between", "1,2,3/4 structure") have developed their philosophical approach beyond Schopenhauer's text, but with the help of the same.[18]

---

[16] Margit Ruffing: Die 1,2,3/4-Konstellation bei Schopenhauer, p. 331.
[17] Margit Ruffing: Die 1,2,3/4-Konstellation bei Schopenhauer, p. 331.
[18] Vide supra, Chapter 1.1.3.

1.2 The System of WWR

In the following chapters, I will first follow Klamp and the prevailing opinion of Schopenhauer scholarship by considering Schopenhauer's main work as the central system of his philosophy. In contrast to Klamp, however, I do not approach the system by analysing the "external structure", which omits the "internal design", but try to open up the external structure of the system from the internal design. The central task, however, is not seen in developing a complete history of origins on the system, but rather in highlighting the philosophical motivation of the system and determining the place of logic. For this purpose and as an introduction to the genesis of the system, I present the internal design of WWR I following the first edition published in 1819, in order to be able to compare it in Chapter 1.3 with Schopenhauer's contemporary reformulation of the system in the *Berlin Lectures*.

1.2.2 The Preface

This thesis is trivial: Philosophical books are not created out of nothing; they are written by an author, and in the best case the text explains to its recipient the author's motivation and intention or the aim of the presented treatise. Philosophical books are usually subject to their own dynamics, since their questions, arguments, answers, can often only be understood in the context of their respective predecessors, the contemporary debate or, more generally, the zeitgeist. For example, more than two hundred years after the publication of the *Critique of Pure Reason* (= CpR), it is still a research dispute as to what intentions, objectives and purposes are expressed in this work itself.[19] As trivial as it may seem to the philosopher or scholar, it can be important to be aware that many philosophical texts have an objective that goes beyond the conviction of the recipient; after all, depending on the determination of the objective, the interpretation of a work can also shift.

Schopenhauer's system seems to announce such an intention of his author directly in the first sentence of the preface of WWR I: "It [sc. this book, i.e. WWR I] aims to convey a single thought. But in spite of all my efforts, I could not find a shorter way of conveying the thought than the whole of this book."[20] The content of the book is structured in the form of an 'organic system' in which "each part containing the whole just as much as it is contained by the whole, with no part first and no part last, the whole thought rendered more distinct through each part, and even the smallest part incapable of being fully understood without a prior understanding of the whole."[21] In contrast, the form of the book resembles more an 'architectural system' "in which one

---

[19] Cf. Jens Lemanski: Die Königin der Revolution. Zur Rettung und Erhaltung der Kopernikanischen Wende. In: Kant-Studien 103:4 (2012), pp. 448–471; Id.: Galilei, Torricelli, Stahl. Zur Wissenschaftsgeschichte der Physik in der B-Vorrede zu Kants *Kritik der reinen Vernunft*. In: Kant-Studien 107:3 (2016), pp. 451–484.
[20] WWR I, p. 5.
[21] WWR I, p. 5. Vide supra, Chapter 1.1.2.

part always supports another without the second supporting the first, so the foundation stone will ultimately support all the parts without itself being supported by any of them, and the summit will be supported without itself supporting anything".[22] For since a book – analogous to the metaphor of the 'foundation stone' and the 'summit' – must have a "first line and a last", it "will always be very different from an organism".[23] Due to the "organic rather than chainlike" construction of the whole, it is on the one hand, inevitable for the reader to read the book twice, and on the other hand, impossible for the author "to divide the work into chapters and paragraphs" – as will be shown, this remark is only partially comprehensible.[24]

Schopenhauer first indicates that the addressee of the book is herself a philosopher. This is implied in the phrase "philosopher, because he [or she] is one himself [or herself]", and especially on the last pages of the work it becomes clear that this is not a mere phrase of politeness, since Schopenhauer distances himself and the reader from the insight of the ascetic, saint and mystic: "But *we* who are firmly entrenched in the standpoint of philosophy [...]".[25] Schopenhauer recommends the reader to acquire further basic philosophical knowledge by reading the writings *On the Fourfold Root of the Principle of Sufficient Reason* and *On Vision and Colors* and to become familiar with Kant, Plato and the Vedas.

The references to the additions and to the desired previous knowledge of the reader are not insignificant. The fact that Schopenhauer regarded WWR I as the central system is made clear by his explicit indications as to which work is a preliminary work and which is an addition to the main work: Schopenhauer names *On the Fourfold Root of the Principle of Sufficient Reason* written in 1813 as the introduction and *On Vision and Colors* of 1816 as an addition to the system of WWR I. The prefaces to the second and third editions of WWR I also illustrate the links between WWR and the works written later. In the second preface, Schopenhauer explains WWR II as a supplement to WWR I. In the third edition, PP I and PP II are proclaimed as "additions to the systematic presentation of my philosophy".[26]

But while the reader of the first edition may already be discouraged by these reading demands, there are, as Schopenhauer expressly emphasises, other ways of using the book. By listing the possibilities of what one could do with his book instead of reading it, Schopenhauer is explicitly allowing himself a "joke",[27] which is not uninteresting, however, in that it defines the circle of addressees more precisely. According to Schopenhauer, if one does not like to read the book oneself, one can "can leave it in the dressing room or on the tea table of his educated lady friend".[28] What is still expressly

---

[22] WWR I, p. 5.
[23] WWR I, p. 6.
[24] WWR I, p. 7. Vide infra, Chapter 1.2.3.
[25] WWR I, p. 438 (§ 71), also p. 406 (§ 68).
[26] WWR I p. 22 (2nd Pref., 1844).
[27] WWR I p. 10 (1st Pred., 1819).
[28] WWR I, p. 10.

meant as a joke here, however, is confirmed as a seriously meant reference, especially in the explanations on logic in WWR I:[29] the book is not only addressed to academic philosophers, but to a wider, educated middle-class audience (the Bildungsbürgertum).

In the course of the work, it becomes clear that the inclusion of the recipient in the circle of philosophers and the reference to the broader reading public are not necessarily mutually exclusive. The encyclopaedic character of the work, which is to be demonstrated in the following chapters, aims at forming the educated reader of the work into a universal scholar and philosopher. Schopenhauer's main work acquaints the reader with the faculties of understanding and reason (B I), the levels of nature (B II), the forms of art (B III) and the modes of action (B IV). In the process, the WWR passes through all the disciplines of theoretical and practical philosophy, so that the educated reader should ultimately be able to determine the position of philosophy immanent in language and argumentation in contrast to all other transcendent positions.

1.2.3 Book I: Theory of Cognition (Representation)

B I of the WWR is entitled "The world as representation, first consideration // Representation subject to the principle of sufficient reason: the object of experience and science".[30] B I is divided into sixteen parts, which are separated by lines in the first edition and numbered as paragraphs from the second edition onwards. There are therefore good reasons to claim that the concept of order through paragraphs, which was introduced from the second edition onwards, was already indirectly present in the first edition through the lines.

The new English edition of WWR I by Richard E. Aquila offers a particularly helpful research achievement. For each paragraph, keywords have been placed in brackets that reflect the content of the respective paragraph. As far as I know, there is no other edition or research work that offers the reader this reading aid. Unfortunately, it is problematic that Aquila has tried to cite almost every topic that is treated in detail without indicating the function of the respective topic (main topic, superordinate/subordinate topic, note, excursus, etc.). Thus Aquila offers with its edition a thematic reading aid, which is particularly useful for those who are not yet familiar with the text; however, the undifferentiated arrangement of keywords is not a particularly helpful aid for the purposes of systematic orientation.

Here, as in Chapters 1.2.4 to 1.2.6, I will compile the main themes of the respective books and justify these provisions by the regulatory notes that Schopenhauer gives in the book itself. B I, divided into sixteen parts or paragraphs, includes the following main topics:

---

[29] Vide infra, Chapter 1.3.3.
[30] WWR I, p. 23.

1 The World and its Representationalist Interpretation

§§	Topics
1–2	Introduction (Representation, Subject – Object)
3	Time & Space
4	Matter = Causality
5	Reality of the External world
6	Grades of the Understanding (Plants, Animals, Intuition)
7	Materialism & Idealism
8	Reflection & Reason
9	Language (Logic & Dialectic)
11–13	Excursus: Relationship between the Understanding & Reason
10, 14–15	Theory of Science
16	Practical Reason

Before I explain this table in more detail, I would like to make a few remarks on the basic structure of B I. B I is divided into two main parts or sections: The first section ranges from § 3 to § 7 and is subsumed under the concept 'understanding', which is synonymous with 'intuitive representation/ cognition'; section II comprises §§ 8 to 16 and is marked with the term 'reason', which is also used synonymously with 'abstract representation/ cognition'.[31] At the beginning of § 3, Schopenhauer writes: "Later we will consider these abstract representations on their own; but first we will discuss only intuitive representations."[32] The term 'later' refers to the paragraphs up to the end of § 7, where Schopenhauer announces the second section of B I:

> The next Book will establish this [sc. the will] by means of a fact just as immediately certain to every living being as the fact of representation. Before we can do this, however, another class of representations, belonging only to human beings, must be considered: their material is the concept and their subjective correlative is reason, just as the representations we have considered up to this point have had as their correlatives understanding and sensibility, which we share with the animals.[33]

Although in his later works (from the second edition of WWR I onwards) Schopenhauer integrates §§ 1–7, all of which are supposed to refer to the understanding, into

---

[31] Cf. WWR I, p. 42f. (§ 6), p. 79f (§ 12).
[32] WWR I, p. 27.
[33] WWR I, p. 57 (§ 7).

the first half of B I of WWR II,[34] there are several arguments that suggest that in his more recent phase of work (at the time of the first edition of WWR I) Schopenhauer only ascribes §§ 3–7 to the section on the understanding. The first argument for this assertion is supported by the beginning of § 3, which has already been partially mentioned above, since only there, for the first time in the work, is the division between the understanding and reason made:

> The most important division among all our representations is between the intuitive and the abstract. The latter form only one group of representations, namely concepts. Of all the creatures on earth, only human beings possess concepts, and the ability to conceptualize (which has always been referred to as reason∗) distinguishes humans from all animals.[35]

Only here, at the beginning of § 3, Schopenhauer points out the "most important division" between the understanding and reason, which at the same time denotes the difference between animal and human beings. That this distinction was not yet made at the beginning of B I is proven by the first sentences of § 1 of WWR I, which are at the same time a second argument against the classification of the late Schopenhauer:

> 'The world is my representation': – this holds true for every living, cognitive being, although only a human being can bring it to abstract, reflective consciousness:[36]

As is claimed in the first sentence after the hyphen, the statement of the preceding quasi-quotation applies to both mere intuitive and rational beings, i.e. to animals and humans. The phrase following "although", on the other hand, excludes mere intuitive beings (animals) and only refers to rational beings that possess a reflective, abstract consciousness (humans). In the further course of the first paragraph of WWR I, Schopenhauer also speaks of "abstract or intuitive" in the sense of an inclusive disjunction and makes assertions that are valid for both animals and humans.[37] In my opinion, these quotes serve as arguments for the assertion that §§ 1 and 2 represent an introduction to B I and that the division into the two main sections of B I only takes place from § 3 onwards.

A further argument can even be found at the beginning of § 7, in which Schopenhauer again reflects on the structure of B I:

---

[34] Cf. WWR I, p. 57, Fn. (§ 7): "The first four chapters of the first book of the supplementary volume [sc. WWR II] belong with the first seven paragraphs of this volume [sc. WWR I]."; WWR II, p. 3: "On the First Book, First Half, The Doctrine of Intuitive Representation (Concerning §§ 1–7 of the First Volume)".
[35] WWR I, p. 27.
[36] WWR I, p. 23.
[37] WWR I, p. 23.

1 The World and its Representationalist Interpretation

> We must still make the following remark concerning our entire discussion so far: we did not start with either the subject or the object, but rather from the representation, and this already includes and presupposes the other two, because the subject/ object dichotomy is the primary, most universal and essential form of representation. So we began by considering this form as such; only then (while referring the reader to the introductory essay for the main point) did we consider its other, subordinate, forms, forms that concern only the object: time, space and causality.[38]

The expression "introductory essay" ("einleitende Abhandlung") which appears in the parenthesis of the quote is ambiguous; it may refer either to the monograph *On the Fourfold Root of the Principle of Sufficient Reason* or to §§ 1, 2 of WWR I or even to both the monograph and the paragraphs. However, even if there were arguments for the fact that the expression can only mean the independent monograph, at least the phrase "we began [at first, zuerst]" in the third sentence refers to the beginning of B I, i.e. to §§ 1 and 2 of WWR I. In these paragraphs, Schopenhauer took the representation as his starting point. Only in a further step is the division of subject and object, from which time, space (§ 3) and causality (§ 4) are then derived. Since time, space and causality are only precisely addressed and subsumed under the concept 'understanding' from § 3 onwards, the first main section of B I, which deals with the understanding, also begins only from § 3 onwards.

As important as the last three paragraphs of the section on the understanding (§§ 5–7) may be for understanding Schopenhauer's philosophy, they have no systemic relevance. In terms of content alone, they deal with topics such as the reality of the external world, dream and actuality (§ 5), soul abilities (§ 6) and one-sided approaches to philosophy such as materialism and idealism (§ 7). They are additions to the section on the understanding, as they do not play a constitutive role in the system for the justification of other parts of the system.

According to the above quote from § 7, the pure concept of representation presented in § 1 is again divided into 'object' and 'subject'. Already in § 2, Schopenhauer announces that time, space and causality, i.e. the faculties of the understanding, are the "most essential – and therefore most general – forms of all objects".[39] With this, he anticipates the content of § 3 on the one hand, but on the other hand, he only draws attention to the fact that the faculties of the understanding are object-constitutive; more interesting is the concept of form in § 2, which at first seems difficult to interpret, but which reveals a more precise system structure through its correlative. Schopenhauer explains the expression of form more precisely at the beginning of § 4 and at

---
[38] WWR I, p. 47. Cf. also p. 57.
[39] WWR I, p. 26.

1.2 The System of WWR

the end of § 7: The concepts of the understanding (Verstandesbegriffe), i.e. 'time', 'space' and 'causality' are as forms in correlation with the object attributions 'succession', 'position' and 'matter'.[40] This makes the general structure of the first section of B I in WWR I more explicit: § 1 deals with the undivided representation, § 2 with the division into object and subject, which is explained in §§ 3–7 as object recognition employing the subjective faculty of the understanding.

The second section of B I begins with § 8. This section has only an introductory function and reports that so far only intuitive representation has been discussed. It is noticeable in a passage from § 8 summarising the contents of §§ 2–7 that Schopenhauer does not always clearly differentiate the object- and subject-related concepts of the section on the understanding:

> So far we have discussed only representations whose composition allows them to be traced back to time and space and matter (when we consider the object) or sensibility and understanding, i.e. cognition of causes (when we consider the subject); but in human beings alone, out of all the inhabitants of the earth, another cognitive power has appeared and a completely novel consciousness has arisen. This is very fittingly and correctly known as *reflection*.[41]

According to the conceptual scheme worked out so far, it would probably have been more explicit if Schopenhauer had chosen the following expression in the quote cited: "succession, position and matter (when we consider the object) or time, space and causality (when we consider the subject)". The concept of 'sensibility' (Sinnlichkeit), which was introduced in the last sentence of § 7 as a synonym for 'time and space' and which was added to 'the understanding' (Verstand, synonym for 'causality'), is particularly problematic in this quote. Since the concept of 'sensibility' is problematic in that it weakens the otherwise strict division of understanding (section I) and reason (section II) within B I of the WWR. If one takes the concept of sensibility seriously, B I, apart from the introduction (§§ 1, 2), would consist of a three-part division: sensibility (§ 3), the understanding (§ 4), reason (§§ 8ff.).

However, Schopenhauer had already pointed out in § 4 that the word 'sensibility' was a foreign element adopted from Kant. The word 'sensibility' was retained, since "Kant broke new ground here, [...] though it is not quite appropriate, since sensibility already presupposes matter."[42] If matter or causality is now the pure function of the understanding, and if time and space as sensibility basically presupposes matter or causality, then time and space as sensuality can actually be subsumed under the concept of the understanding. The division into three, which seems confusing, especially

---

[40] WWR I, p. 29, p. 57.
[41] WWR I, p. 59. (In the 1819-edition the sentence ends here.)
[42] WWR I, p. 32.

from § 7 on, is thus ultimately only due to Kant, whereas Schopenhauer's main focus is on the division of the understanding and reason. Despite the confusing assignment of sensibility to the understanding, the above-given quote from § 8 nevertheless indicates a dichotomy: This dichotomy is maintained by the distinction between human beings, who alone possess the faculty of reason or reflection, and the other "inhabitants of the earth". This division has already been discussed several times above.

The quote from § 8 also announces that the rest of B I is about the abstract or reflective reason. "[R]eason has only one function: the formation of concepts".[43] Because of this function, the second section of B I deals first with language, and this treatise (§ 9) is again divided into two parts: firstly, logic, and secondly, dialectics.[44] It is only at the beginning of § 10 that Schopenhauer indirectly points out to his reader that the section on reason of B I is divided into three parts:

> All this brings us closer and closer to the question how we can achieve *certainty*; how we can ground *judgements*; and what knowledge and science consist in. For these are acclaimed as the third greatest advantage conferred on us by reason, after language and circumspection in our actions.[45]

The fact that this quote presents the ordering scheme of the second section of B I of the WWR is only made clear by three parallel passages at the beginning of § 9, § 14 and § 16 respectively.[46] Only the close reading of these four passages indicates to the recipient that a three-part division of reason is also implied: 1. language (§ 9), 2. science (§§ 10, 14, 15), 3. practical reason (§ 16). These are the three "advantages" of human beings over animals.

That §§ 11–13 are excursuses that deal with the relationship between the understanding and reason is shown, on the one hand, by the fact that they do not fit into any of the three sections on reason; on the other hand, this is also indicated by the first parallel text passage, located at the beginning of § 10, that deals with the division into three sections. And similarly, Schopenhauer emphasises this at the beginning of § 14:

> I hope all these various enquiries have completely clarified the distinction and relation between, on the one hand, reason's mode of cognition (knowledge and concepts) and, on the other hand, immediate cognition in pure, sensible, mathematical intuition and apprehension in the understanding; moreover, our account of the peculiar relation between these two kinds of cognition almost

---

[43] WWR I, p. 62 (§ 8).
[44] WWR I, pp. 62–70 (Logic); pp. 70–74 (Dialectics).
[45] WWR I, p. 75.
[46] WWR I, p. 62, pp. 86f., p. 110.

1.2 The System of WWR

> unavoidably led us to parenthetical discussions about feeling and laughter. I now return from all this to a further discussion of science, the third most important advantage conferred upon us by reason, after language and judicious actions. We must now give a general account of science, which will concern partly its form, partly the ground of its judgements, and finally its content.[47]

This quotation from the beginning of § 14 is in many respects illuminating for the system construction of the section on reason in WWR I: Schopenhauer says *he returns* from the "various enquiries" and further from the "parenthetical discussions". The various enquiries refer to the "relation between, on the one hand, reason's mode of cognition [der Erkenntnißweise der Vernunft] [...] and, on the other hand, immediate cognition [...] in the understanding [unmittelbaren Erkenntniß durch den Verstand]". This suggests that these various enquiries, which are found in §§ 11–12, are not an essential component of the section on reason of B I, but rather excursuses. The "parenthetical discussions about feeling and laughter" in § 13 also only deal with the "peculiar relation between these two kinds of cognition" – in Kant's sense, they are therefore only psychological, metaphysical and anthropological supplements to logic, which have crept into this discipline through "some moderns".[48]

Already in the very short § 10, which first explained the structure of the section on reason, Schopenhauer had hinted at a transition from logic to a theory of science.[49] Schopenhauer must have assumed that this transition was made within § 10 and that the end of this short paragraph could already be attributed to the theory of science. This conviction is expressed in the above quote from § 14, in which Schopenhauer explicitly states that he will "now return from all this to a further discussion of science". Since this discussion of science is described as "the third most important advantage conferred upon us by reason, after language and judicious actions", the trichotomic division of the section on reason has been considered once again in § 14. Thus, a diairetic – for the most part dichotomic, but also partially polyadic – structure results for B I according to the following intuitive schema:[50]

---

[47] WWR I, p. 86f.
[48] CpR, p. 105 (B VIII).
[49] The later Schopenhauer explicitly uses the term 'Wissenschaftslehre' (Doctrine of Science) to describe this part of the system, cf. WWR II (1844), pp. 128ff. (Chapter 12).
[50] On the interpretation of these tree diagrams vide infra, Chapter 2.2.2.

1 The World and its Representationalist Interpretation

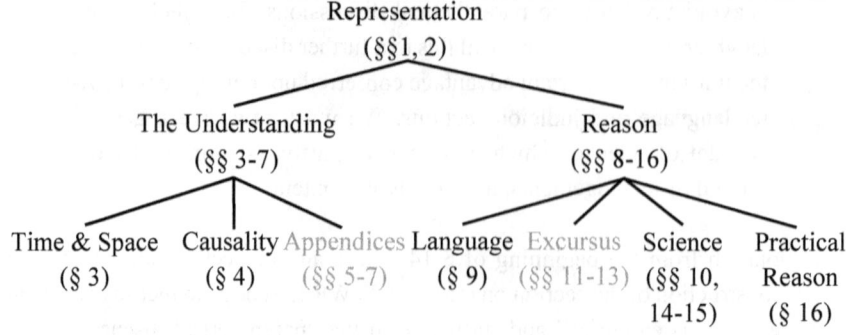

In this tree diagram, the appendices (§§ 5–7) and the excursus (§§ 11–13) are shaded in grey, as neither of them is of systemic relevance: The appendices deal with contents which, although they are argumentatively related to the previously developed paragraphs, are not constitutive components of the faculty of the understanding. The excursus is excluded, since, as already explained, it cannot be assigned to the three-part part structure of the faculty of reason.

The branches of the tree diagram – the mathematician would probably rather speak of nodes or vertices here – are not completely indicated in the scheme. For example, as already mentioned, § 9 could be divided into at least two further components: Logic and dialectics. As will become clearer, logic and dialectic are also much more precisely divided.[51] Another good example for the possibility of a more exact subdivision is provided by the treatise on science in § 14: According to the dichotomic method Schopenhauer first wins the principle pair 'subordination and coordination', of which the subordination is further subdivided into 'induction and deduction'. From this formal methodology, he derives a material theory of science, in which almost all branches or nodes of natural science (zoology, botany, physics, chemistry, astronomy etc.) and 'sciences of mind' (history, mathematics, philosophy) are classified and evaluated.

In § 15, arithmetic, geometry and philosophy are then examined in more detail, as they achieved an extraordinary status within science – similar to history as a science only using the method of coordination. Schopenhauer explains here that although mathematics and philosophy can be related when they are presented in a subordinated way (as in the case of Euclid or Spinoza, for example),[52] they do not necessarily have to be subordinating sciences that proceed deductively or inductively or that establish axioms and derive theorems from them.

Since Schopenhauer demonstrates this assertion at the end of § 15, presenting the method of his philosophy, this is the most important section of the text for understanding WWR. The reflection on the aim and purpose of the entire WWR begins with the

---

[51] Vide infra, Chapter 1.3.
[52] Cf. also Chapters 2.2.2 and 2.2.3.

words: "The present philosophy at least is...".[53] From this introductory phrase to the end of § 15, the recipient experiences for the first time the meaning of the entire system of WWR or the purpose of Schopenhauer's book. Only here one can find an answer to the trivial question of what the aim of WWR is. Although the significance of the entire section has hardly been sufficiently perceived in the history of the reception of the work to date, I would like to take out only six quotes in the following, which, however, reveal one single explicit and uniform objective:

(1) "The present philosophy at least is [...] remotely concerned [...] only with what it [sc. the world] is."
(2) "But such cognition is intuitive, concrete cognition: philosophy's task is to reproduce this in the abstract, to elevate [it] into permanent knowledge."
(3) "Accordingly, philosophy must be an abstract statement of the essence of the entire world"
(4) "philosophy must make use of abstraction [...], condensing all the variety in the world as a whole into a few abstract concepts in accordance with its nature, and handing them over to knowledge."
(5) "philosophy will be a complete recapitulation, a reflection, as it were, of the world, in abstract concepts."
(6) "Bacon of Verulam already set this as the task for philosophy."[54]

All six quotes are found in the text – albeit with several insertions – in the order given and build on each other in terms of content. (1) names the basic task of philosophy; (2) explains how this task is to be accomplished: according to Schopenhauer, every human being knows "what the world is" intuitively and in concreto; but only philosophy presents this abstractly in concepts. This achievement of abstraction is taken up not only in (2) but also in (3)–(5): Schopenhauer names abstraction not only as the result of philosophy or philosophising ("abstract statement", "abstract concepts") but also as its necessary method ("must be an abstract statement"; "must make use of abstraction"). By abstraction, the world can be handed "over to knowledge", "to permanent knowledge". With this method, (4) offers an answer to the question that arises in (2), how philosophy can raise the intuitive, concrete cognition to permanent knowledge. The abstract concept, Schopenhauer adds in the context of (4), defines the essence of the world. The impermanence and transience of intuitive cognition are preserved in a constant and permanent form. The intuitive and concrete world is elevated to an abstract and conceptual form by means of abstraction, and this is precisely the necessary condition to memorize and communicate cognition. Representationalism, which depicts the concrete in the abstract, is explicitly expressed in the metaphors of

---

[53] WWR I, p. 108.
[54] All quotes (without emphasises) taken from WWR I, pp. 108–109.

(2), (4) and especially (5): 'reproduction', 'handing over', 'recapitulation', 'reflection'.

(6) occupies an extraordinary position. The reference to Francis Bacon has several functions: On the one hand, it is an essential component of a *lingua franca* based on the history of philosophy in order to simplify communication between philosophers (here: author – recipient),[55] and on the other hand, it can be interpreted as an indication of school affiliation.[56] Although Schopenhauer had recommended the reader in the preface to become acquainted with Kant, the attentive recipient will recognise after reading § 15 that this recommendation was probably based primarily on familiarity with the conceptual scheme, but less on the manner in which Kant's philosophical aims were pursued. Whereas Kant, Plato and others may have been the source of ideas and concepts for Schopenhauer, it is clear from (1)-(6) of § 15 that the aim and method of the WWR is an empirical and representationalist one. Bacon stands in (6) as a cypher for empiricism and representationalism.

But the name 'Francis Bacon' has a further function: not only does it indicate school membership, which is also related to the keywords 'empiricism' and 'representationalism', but it also recalls three metaphors closely related to Bacon: 1) the mirror metaphor, 2) the readability metaphor and 3) the encyclopaedia metaphor, which were worked out with reference to Bacon by Richard Rorty, Meyer Howard Abrams, Hans Blumenberg and Ulrich Gottfried Leinsle respectively.

1) The mirror metaphor was already mentioned above in Schopenhauer's quote (5). It refers to the function between the concrete world in everyday cognition and the abstract world in philosophical language. The book about the world is an image of the intuitive given world itself and language is the medium of reflection. Even today, many scholars emphasise that the prominent mirror metaphor of Enlightenment has become a symbol of impressionism and representationalism and stands in opposition to the Romantic lamp and tool metaphor, which functions as a symbol of expressivism: Whereas the lamp produces light, the mirror reflects the light. In terms of the philosophy of language, the mirror metaphor expresses the selfless depiction of the given reality, whereas the lamp and tool metaphor is an expression of one's own creativity and expressiveness.

2) The representational function of the mirror metaphor is ultimately hypostasised in the metaphor of readability, which is alluded to in (3), (4) and (5) of the Schopenhauer quote by means of the juxtaposition of world and statement or concept. The great book of nature, as it is later stated in §§ 44 of the WWR I, demands "deciphering

---

[55] Cf. Wilfrid Sellars: Sensibility and Understanding. In: Ib.: Science and Metaphysics. Variation on Kantian Themes. London 1968, pp. 1–31.
[56] Regarding this function cf. Jens Lemanski, Konstantin Alogas: The Function of Decadence and Ascendance in Analytic Philosophy. In: Decadence in Literature and Intellectual Debate since 1945. Ed. by Diemo Landgraf. New York 2014, pp. 49–65.

1.2 The System of WWR

the true Signatura rerum [signature of all things]".[57] The deciphering takes place employing the philosophical translation of the intuition into the conceptual. This transmission aims at completeness: just as judgements express the facts of the world, so books express the totality of the world. The world or its representation fixed in a single book becomes readable. An emphatic emphasis on the first two words in the title of Schopenhauer's main work also makes this explicit.

3) This manifoldness, unified in the concept of the world, is most recently reflected in the encyclopaedia metaphor, which stands for the presentation of the entire educational circle of the present time in an ordered panorama. This ordered panorama of all the stocks of knowledge is the WWR itself. That between two book covers the whole world should be compressed into a single text is expressed particularly in the quantitative aspects of (3), (4) and (5): The concept 'world' can be understood qualitatively in the sense of 'what of the world' or 'essence of the world', but also quantitatively: for Schopenhauer, after all, speaks explicitly of 'the entire world', of 'all the variety in the world', of 'a whole', of a 'condensation', of something 'complete'.

All three metaphors can be seen in the Bacon quote reproduced by Schopenhauer, which immediately follows (6):

ea demum vera est philosophia, quae mundi ipsius voces fidelissime reddit, et veluti dictante mundo conscripta est, et nihil aliud est, quam ejusdem *simulacrum et reflectio*, neque addit quidquam de proprio, sed tantum iterat et resonat. (de augm. scient. L. 2, cap. 13)[58]	for that alone is true Philosophy, which doth faithfully render the very words of the world; and which is written, no otherwise, than the world doth dictate; and is nothing else than *the image and reflexion* thereof; and addeth nothing of its owne, but only iterates, and resounds. (Adv. of L. 2, cap. XIII)

The readability of the world is here even initiated by the world itself: it makes statements, words (voces); the task of true philosophy is to write down this dictate of the world as precisely as possible, without adding anything of its own. The Bacon quote, which comes from *De augmentis scientiarum*, Bacon's encyclopaedia, defines true philosophy (vera philosophia) as an image, reflection, iteration and echo. Hans Blumenberg announced this quotation as follows: "It is the basic idea of empiricism that nature would tell its own story if only it were allowed to [...]."[59]

Book II, Chapter 13 of Bacon's *De augmentis scientiarum*, in which the quote is found, opens up yet another horizon of interpretation. Here Bacon deals with parabolic poetry and takes the parable of Pan as an example: Pan, as the Greek name says, stands

---

[57] WWR I, p. 245.
[58] WWR I, p. 109. English translation taken from Francis Bacon: Of the advancement and proficience of learning; or, The partitions of sciences, IX bookes. Oxford 1640, p. 119.
[59] Hans Blumenberg: Die Lesbarkeit der Welt, p. 86.

for the universe or for the totality of things ("Pan (ut & Nomen ipsum etiam sonat) Vniversum sive Vniversitatem Rerum repræsentat & proponit."[60]). His appearance is a symbol of nature itself; his dichotomous nature reflects the dominance of the human over the animal, the vegetable and the mineral; his flute stands for the harmony and unity of nature; the panic frights or horror that nature has instilled in man protects man from excess, on the one hand, and limits the endeavours, on the other; the shape of the upward-pointing horns symbolise the logical order of the world ("Cornua autem mundo attribuuntur") – divided into individuals, species and genera; Pan's relationship to Echo finally reflects the representationalist relationship between world and logic and introduces the above-given quote. As will be shown in the following chapters, all these aspects of Pan's work are also dealt with in different parts of Schopenhauer's main work.

The empiricist aspect in Schopenhauer's work was first intensively emphasised by Matthias Koßler in the book *Empirische Ethik und christliche Moral* (*Empirical Ethics and Christian Morality*), which serves as the paradigmatic work for a descriptive interpretation of Schopenhauer.[61] This descriptive interpretation emphasises Schopenhauer's empiricist approach and emphasises that the evaluations and contradictions in Schopenhauer's work are not an addendum by the author, but that the world expresses and conceptually repeats itself in form of contradictions and judgments that are in the world and about the world.[62]

The fact that I prefer the concept of representationalism to terms such as 'empiricism' or even 'realism' in the context presented here is mainly because it has less problematic connotations than the other two: For the modern concept of empiricism connotes, among other things, an experimentalism that is not to be found in (1)–(6) of Schopenhauer's quotes; and the expression of the real that is central to realism

---

[60] Francis Bacon: De dignitate et augmentis scientiarum. Argentoratum [Strassburg] 1654, p. 116 (= II 13). The Parable of Pan was first published by Bacon in *De Sapientia Veterum VI* in 1609.

[61] Cf. Matthias Koßler: Empirische Ethik und christliche Moral. This change in research can be seen, for example, in the fact that Rudolf Malter, in a paper 'The Essence of the World', had asserted a decade earlier: "He [sc. Schopenhauer] is not an empiricist (in the sense of the realistic ontologist) in so far as his orientation towards the empirical is geared towards world interpretation [...]". (Schopenhauer und die Biologie. Metaphysik der Lebenskraft auf empirischer Grundlage. In: Berichte Zur Wissenschaftsgeschichte 6 (1983), pp. 41–58, here p. 43). Since, according to my usage, 'world interpretation' aims at a qualitative definition of essence and 'world description' takes into account a – in this case quantitative – complete sum of manifold data, WWR must be more than just an interpretation of the world, otherwise it could be reduced to the beginning of Book II. (vide infra, Chapter 1.2.4.) This is also evident in Schopenhauer's – although not always uniform – use of language: 'World', as in the title 'WWR', has the quantitative connotation of completeness and unity of the manifold, while 'Essence of the World' can only be set synonymously with 'Will', 'Thing in itself' etc. in almost all text passages of the first edition, or even becomes explicit. (vide infra, Chapter 1.3.2) Schopenhauer's orientation towards empiricism is also an interpretation of the world, but this is only one side of his philosophy.

[62] Paul Deussen: Allgemeine Geschichte der Philosophie mit besonderer Berücksichtigung der Religionen, vol. II/3, p. 430f. tried to explain the contradictions by means of the empiricist theory of reflection. In detail, however, there are many differences to the new research approach, cf. Jens Lemanski: The Denial of the Will-to-Live in Schopenhauer's World and his Association of Buddhist and Christian Saints. In: Understanding Schopenhauer through the Prism of Indian Culture. Philosophy, Religion and Sanskrit Literature. Ed. by Arati Barua, Michael Gerhard, Matthias Koßler. Berlin 2013, pp. 149–187.

1.2 The System of WWR

is indefinite, since it can be understood both internally and externally, and Schopenhauer alludes at most in (2) to a certain position of the intuitive real, but often takes a hard line with the real that is absolute or the real in itself.[63]

Representationalism also explains the equality of the system parts that can be found at each level of the polyadic system. From a conceptual and argumentative point of view, the structure of the WWR in its present form may make sense; but for reasons inherent in the system, there is no precedence of the section on the understanding over the section on reason in B I or precedence of B I over B II etc., since there can be no reasons for a preferential or exceptional position of any part of the system in the world itself. The order and classification is one which the author has brought to the world and serves only to represent the world by using concepts. Only the empiricist or representationalist method of abstraction and selection, which 'elevates' from the concrete to the abstract (2), gives an indication as to why it must be more understandable for rational beings to start bottom-up at the level of the understanding (or sensuality) than top-down by deducing all propositions of a science from a small set of axioms (as Euclid or Spinoza tried). This argument for representationalism announces already indirectly Schopenhauer's criticism of the unnaturalness of logicism.[64]

The unilateral history of the reception of Schopenhauer's WWR consists in a normative interpretation that selectively emphasises parts of the system on the basis of its architectural structure: Thus, so-called 'transcendental idealism' and a kind of 'nihilism' were emphasised because B I begins with the quasi-quotation "The world is my representation" and because B IV ends with the word "nothing". Until today, the word 'pessimism', taken out of Schopenhauer's purely descriptive conceptual scheme, is still applied by many scholars to Schopenhauer's philosophy itself. The reason often given for this is that the author emphasises the allegedly cruel forces of will in B II and the seemingly ascetic negation of the will in Book IV. But the fact that Schopenhauer also emphasises the beauty of the arts and points out at the end of § 15 that the unity of the "single thought" in the system of WWR originates from "harmony and unity of the intuitive world itself" has often been ignored in interpretations and research.[65]

Representatives of the normative reading often stress the 'single thought' from the preface to explain what the intention and motivation of the WWR is. However, it is often overlooked that the purely formal reference to the 'single thought' is not explained in terms of content until the end of § 15, subsequent to the Bacon quote. There Schopenhauer also alludes to almost all aspects of the preface: the organic system, in which one "be derived from others and this must in fact always be reciprocal"; the

---

[63] Cf. Valentin Pluder: "Skitze einer Geschichte der Lehre vom Idealen und Realen". In: Schopenhauer-Handbuch. Leben – Werk – Wirkung. Ed. by Daniel Schubbe, Matthias Koßler. Weimar 2014, pp. 124–129.
[64] Vide infra, Chapter 2.3.
[65] WWR I, p. 110.

incomprehensibility of the first reading of the WWR, since the task of the system "can only become fully clear when it has been completed"; and finally "the unity of a *single* thought,", which results from the fact that "all the parts and aspects of the world agree with each other" and therefore "be rediscovered in philosophy's abstract copy of the world".[66] The thought that communicates itself in the WWR is a single one because the intuitive world is also one and forms a unity.[67] The content of the one single thought is thus the world, and 'world' is the conceptus summus of Schopenhauer's systematic conceptual scheme, which, depending on the perspective, shows itself "as will" or "as representation".

Before the next chapter deals with the world from the perspective of the will, § 16 and its outlook on B IV will be discussed in the following. In his arched arches theory, Klamp had established, among others, a connection between B I and B IV.[68] This consists in the fact that B I describes the subject's world view, which in B IV disappears into nothingness with the negation of the will. For him, this connection between B I and B IV results in a 'dual unity' of the whole system. This interpretation, however, is, above all, due to Klamp's implicitly normative interpretation, which is problematic in that it does not understand concepts such as the subject as general terms of the descriptive conceptual scheme of Schopenhauer, but always reifies concepts such as 'subject' or 'nothingness'. According to Klamp, Schopenhauer tells a story that begins with becoming a subject and ends with the destruction of this subject. Normative interpreters following Klamp believe that a 'soteriological determination' can be derived from the connection between § 16 and B IV, according to which B I already announces the desire for redemption, which is then fulfiled in B IV.

I agree with Klamp that there is a link between B I and B IV. In my opinion, however, this shows itself less in linear-teleological ontogenesis (from becoming a subject to overcoming the subject), but rather systematically in the division of the system, and the stoic sage in § 16 plays a decisive role in this. Generally, however, one can first of all say that § 16 deals with the third "advantage" of the faculty of reason, namely practical reason. Schopenhauer explains at the beginning of the paragraph that practical reason was discussed particularly in contrast to the Kantian view in the treatise that is attached to the WWR and therefore, he has not "much more to say here about the actual influence of reason, in the true sense of the word, on action."[69] Furthermore, he had already said "at the outset of our discussion of reason" something about the practical action of man in comparison to the animal. Through this, attention is again drawn to the dichotomy of B I, since Schopenhauer points to the beginning of § 8 in the passage quoted.[70]

---

[66] WWR I, p. 110.
[67] This is also demonstrated by Schopenhauer's criticism of unique existential quantification; vide infra, Chapters 2.2.5, 2.2.6, also Chapter 1.3.1.
[68] Vide supra, Chapter 1.2.1.
[69] WWR I, p. 110.
[70] WWR I, p. 59.

1.2 The System of WWR

§ 16 is difficult to classify systematically or argumentatively due to the many retrospectives and outlooks. First of all, Schopenhauer takes up again the theme of the purely deductive mathematical and philosophical systems, which were also dealt with in § 15; in addition, he contrasts the difference between the intuitive given world *in concreto*, which "possessed us completely and moved us deeply", whereas in the conceptual world *in abstracto* "we are simply onlookers and spectators".[71] This contrast, in which "the distinction between humans and animals shows itself most clearly", is represented in an ideal form by the Stoic sage.[72]

Many representatives of the normative interpretation have interpreted the expression of the ideal as if it had been used by Schopenhauer in this paragraph as an example of appropriate or correct action. However, the context given shows that the ideal presented in § 16 was not indicated as a recommendation for action, but rather as an explanation of the difference between the pure form of the understanding and the faculty of reason, between animal and human beings.[73] This can also be verified employing a substitution test: The term 'ideal' can be replaced salva congruitate by 'extraordinary' or 'extreme', but not with 'recommendation for action' or similar.

Schopenhauer explains this point in more detail after his description of the Stoic sage: He had to present Stoic philosophy in order to show in an exemplary way "what reason is and what it can achieve".[74] According to Schopenhauer's announcement, many of the aspects of practical reason presented in § 16 will be substantiated and coherently presented in B IV. In order to motivate the reader to continue reading until B IV, he explains at the end of § 16 – which also concludes B I at the same time – that the philosophy of the Stoics and "their ideal, the Stoic sage", is only "stiff and wooden, a mannequin that no one can engage with" since he never reached the "inner poetic truth" of the Indian and Christian "penitents who overcome the world".[75]

This outlook and the issues addressed in § 16 illustrate the link between B I and B IV already mentioned by Klamp. Klamp's linear interpretation derives an ontogenetic connection from the content of the representation alone, from the construction of the world to overcome the world, from transcendental philosophy to soteriology. However, according to the descriptive-systematic interpretation presented here, the connection between B I and B IV lies in the representation of practical reason. B IV, as will be shown in Chapter 1.2.6, is a detailed and differentiated representation of § 16, i.e. a clarification of practical reason, but from the perspective of the world as will. In both parts of the system, in § 16 and in B IV, it is, above all, a matter of showing

---

[71] WWR I, p. 112.
[72] WWR I, p. 113.
[73] WWR I, p. 113.
[74] WWR I, p 117.
[75] WWR I, p. 118.

1 The World and its Representationalist Interpretation

that reason is a means that can be instrumentalised for quite different purposes, in the service "of noble intentions as it does in the service of bad intentions".[76]

Due to the connection between § 16 and B IV, explicitly mentioned by Schopenhauer and implicitly imposed by the themes discussed, it can be argued, on the one hand, against Klamp's normative-architectural interpretation that a 'dual unity' in WWR I, spanned from becoming a subject to the destruction of the subject, is only the product of the interpreter and contradicts the organic-systematic character of the work. On the other hand, it is precisely this organic-systematic character that is evident in the connection between § 16 and B IV: the content of B IV does not necessarily have to conclude the book entitled WWR, but could have been integrated with good reason, for example, into B I or appended to §16 as B II. That the young Schopenhauer himself was flexible in the arrangement of the individual parts of the system will be shown in Chapter 1.3 by comparing WWR with the *Berlin Lectures*.

1.2.4 Book II: Metaphysics (Will)

B II is entitled "The World as Will, First Consideration: // The Objectivation of Will" and is divided into twelve paragraphs. A first look at the main themes of B II does not initially indicate an order that could correspond to the diairetic structure of B I:

§§	Topics
17	Meaning of Representation
18	Will
19	Double Cognition
20	Characterology
21, 22	Will as Thing in Itself
23	Principium Individuationis
24	Philosophy and Aetiology
25–27	Gradations of the Will's Objectivation
28	Teleology
29	Summary

§ 17 begins with an introduction, the significance and interpretation of which is highly controversial:

---

[76] WWR I, p. 112f. This quote shows why representatives of the critical theory and discourse ethics see their central thesis anticipated in Schopenhauer's work. However, since many of them also tend towards a normative reading (vide supra, Chapter 1.1.3), their interpretation often reveals both a closeness and a strange distance to Schopenhauer.

1.2 The System of WWR

> In the First Book [sc. B I] we considered representation only as such, which is to say only with respect to its general form. Of course, when it comes to *abstract representations* (concepts), we are familiar with their content as well, since they acquire this content and *meaning* only through their *connection to intuitive representation* and would be worthless and empty without it. This is why we will have to focus exclusively on intuitive representation in order to learn anything about its content, its more precise determinations, or the configurations it presents to us. We will be particularly *interested in discovering the true meaning of intuitive representation*; we have only ever felt this meaning before, but this has ensured that the images do not pass by us strange and meaningless as they would otherwise necessarily have done; rather, they speak and are immediately understood and have an interest that engages our entire being.[77]

The quotation begins with a reflection on the content of B I. Schopenhauer claims in the first sentence that he has so far only dealt with the representation in terms of form, but he immediately makes a restriction in the second sentence: the abstract representation (reason) has already received "content and meaning" through its relationship to intuitive representation (the understanding). This is reminiscent of the well-known Kantian formula that meaning only emerges from the interplay of intuition and concept.[78] Whereas the second sentence was based on reason, Schopenhauer refers from the third sentence onwards only to the understanding ("focus exclusively on intuitive representation"). Here, as in the following sentence, Schopenhauer seems to determine the task of B II, namely "discovering the true meaning of intuitive representation".

At the end of the quote, two possibilities are distinguished: 1) If we did not discover the meaning of the forms, determinations, or images, they would pass us by completely strange and meaningless. 2) But since we will receive information about these forms, determinations, or images, the case occurs that they "speak and are immediately understood and have an interest that engages our entire being". Daniel Schubbe understood the expressions 'understanding', 'speaking', 'having interest', 'engages our entire being' as signs of a so-called 'hermeneutics of existence' (Daseinshermeneutik) and related them to authors like Karl Jaspers or Hans-Georg Gadamer. If I understand Schubbe correctly, in his opinion all three expressions – similar to the

---
[77] WWR I, p. 119.
[78] Cf. CpR, p. 193f. (A 51/B 75).

previous "discovering" ('Aufschluss erhalten') – are interchangeable words, thus synonyms and to be understood literally.[79]

Schubbe's interpretation is quite plausible, and the reference to modern hermeneutics of existence certainly shed light on this passage of Schopenhauer's text, which can certainly not be understood as unambiguous. However, I would like to point out in this respect that the expressions under discussion can be understood not only literally but also metaphorically, and that they tie in with the expressions found at the end of § 15 and the beginning of § 16 of B I. If one interprets the expressions 'understanding', 'speaking', 'having interest' ('Ansprechen', 'in Anspruch nehmen', 'Interesse erhalten') metaphorically, there is a parallel to the Baconian quote given § 15, in which philosophy *understand* the very word of the world (mundus ipsius voces), *speaks* what the world does dictate (dictante mundo conscripta est) and thus is made interesting without giving it itself (neque addit quidquam de proprio).[80]

At the beginning of § 16, Schopenhauer explains that "concepts have such a sweeping and significant influence on our whole existence" that this influence alone justifies the difference between animal and human.[81] Both interpretations, the representationalist and the hermeneutical one, agree that these expressions indicate a reaction of the philosopher to the world: But while the literal interpretation of the three expressions means 'active participation in…', the metaphorical interpretation, on the other hand, indicates a 'passive reception of…'.

Schubbe's approach has the merit of having pointed out and draw attention to the concept of meaning in B II. This concept is also profitable for the analysis of the system structure: § 17 reclassifies the natural sciences, which were already presented in §§ 14 and 15. Schopenhauer no longer uses the dichotomies 'subordination/ coordination' and 'induction/ deduction'[82] for this classification, but the term pair 'morphology/ aetiology'. Morphology is the "description of forms", aetiology deals with the "explanation of alterations".[83] Geology, mineralogy, botany and zoology are now classified under morphology, whereas mechanics, chemistry, physics and physiology are classified under aetiology.

Both methods of natural science fail, however, in the attempt to 'acquire the content' or to 'discover the true meaning' (Aufschluss erhalten) of the determinations, forms and images: If one looks at the results of natural science, one soon realises that the information one is mainly looking for "does not belong to aetiology any more than it belongs to morphology".[84] However, both scientific methods do "not shed any light

---

[79] Daniel Schubbe: "…welches unser ganzes Wesen in Anspruch nimmt" – Zur Neubesinnung philosophischen Denkens bei Jaspers und Schopenhauer. In: 89. Schopenhauer-Jahrbuch (2008), pp. 19–40.
[80] Vide supra, Chapter 1.2.3.
[81] WWR I, p. 110.
[82] Vide supra, Chapter 1.2.3.
[83] WWR I, p. 120.
[84] WWR I, p. 121.

1.2 The System of WWR

at all [nicht den mindesten Aufschluß] on the inner essence of any of these appearances" by their research on forms or alterations.[85] They "give us the sort of elucidation" beyond representation (hinausführenden Aufschluß) and therefore seem to be captured in an infinite regress of their strategies of justification.[86]

After this explanation of why scientific procedures fail to explain meanings, Schopenhauer returns to the central question of B II towards the end of § 17: "We want to know the meaning of those representations: we ask if this world is nothing more than representation;"[87] With this question, Schopenhauer again connects the two metaphors of 'insight' (Aufschluss) and 'meaning' (Bedeutung) into one context and continues this in § 18. The philosopher can only gain information (Aufschluss erhalten) about the meaning of the phenomena, as stated in § 18, by means of the word 'will':

> This and this alone [sc. the word 'Will'] gives him [sc. the subject] the key to his own appearance, reveals to him the meaning and shows him the inner workings of his essence, his deeds, his movements.[88]

Whereas the term 'Aufschluss' ('insight', 'disclosure') was previously used in the sense of 'explaining' a question or 'solving' a riddle (such as *solutio aenigmatis*), Schopenhauer now uses the combination of 'Aufschluss' and 'key' to indicate the second meaning of the term in the sense of 'unlocking' or 'opening' (such as *ianua aperienda*).

§ 19 explains this will by the "double cognition we have of our own body".[89] Our body (Leib) can be seen as the "key to the essence of every appearance in nature"[90] since it shows itself once as a direct object (representation) and once as an indirect object (will). The fact that 'meaning' also has a metaphorical connotation for Schopenhauer only becomes clear in § 24 based on a hypothetical-anankastic conditional (Sollten... , so müßten...):

> Now if the objects that appear in these forms are not just empty phantoms, that is, if they are to be significant, then they need to signify or express something that is not just another object or representation (as they are), something whose existence is not just relative to a subject;[91]

---

[85] WWR I, p. 121.
[86] WWR I, p. 122.
[87] WWR I, p. 123.
[88] WWR I, p. 124.
[89] WWR I, p. 128.
[90] WWR I, p. 129.
[91] WWR I, p. 144.

This quote can be seen as a further confirmation of the prevailing research opinion that Schopenhauer is a consistent representative of naïve-representationalist semantics.[92] For it is probably in the spirit of both Schopenhauer's quotation and the theories of language mentioned above that the meaning of 'meaning' is not ascribed a function beyond indexicality and representation of the concept, but rather that meaning is always connected with deictical terms ("need to signify", "auf etwas deuten"). However, Schopenhauer's concept of meaning goes a step further here: in the anankastic consequent of the conditional cited in the quote, Schopenhauer speaks of a condition that must be fulfiled in order to arrive at the meaningful deictic expression mentioned in the hypothetical antecedent. The condition for meaning is that there is something that goes beyond the infinite regress of representation, namely in such a way that the deictical term is at the same time an expression ("express something", "Ausdruck von etwas seyn"): meaning thus not only indicates something but is at the same time the expression of something.

The word 'meaning' becomes a metaphor in that it is more than a linguistically nebulous property that is attributed to concepts or judgments which, by virtue of that property, fulfil a referencing function. Rather, meaning fuels the expectation of an involvement of a mental, a psychological or generally an expressing instance, which Schopenhauer defines here, however, only in relation to objects. 'Having meaning' means that things and living beings are more than a requisite (whether one thinks of Goldman barns, Cartesian stoves, p-zombies etc. here), and this is in turn assured by the sign or expression of 'the will'.

The words 'Bedeutung', 'Aufschlüsselung', 'Ansprechen', 'Wille' ('meaning', 'disclosure', 'speaking', 'will') are metaphors that indicate the continuation of the representationalist approach explained in B I, but also go beyond it on a conceptual level. The metaphor 'will' is at the centre of this conceptual scheme: Schopenhauer transfers the term 'will' from the sphere of human experience to all beings and world phenomena, thus giving the term 'will' "a broader scope than it has had before".[93] Because of this broad scope of the concept, he takes up the same position in B II, which in B I had the concept 'representation', which in turn had 'subject' and 'object', 'understanding' and 'reason' as its content.

If one searches in B II for a similar structure of the system as in B I, i.e. for the content of the concept of will, one will initially be discouraged. The end of § 19 makes for the first time an argumentative structure of B II explicit. Schopenhauer says that he wants to "take what has been presented so far in a broad and provisional way and establish and justify it more clearly and in greater detail, developing it to its fullest extent."[94] Based on this quotation, it can be said that § 17 has raised the question of

---

[92] Vide infra, Chapter 2.1.4.
[93] WWR I, p. 135.
[94] WWR I, p. 156.

1.2 The System of WWR

meaning, which is broken down in §§ 18 and 19 and deepened in the following paragraphs (§§ 20–29). As informative and interesting as §§ 20–24 may be, from a system-related perspective they actually only represent a deepening of what has been discussed so far.

Only §§ 25 to 27 are systematically relevant since they describe the four levels of the objectivation/objectification of will, which constitute the content of the concept of will: 1. the human (highest level), 2. the animal, 3. the vegetative, 4. the mineral kingdom (lowest level).[95] Although the will is only one, it manifolds itself in appearance. In other words: every appearance is an expression of a will or every appearance gives independent signs of being meaningful or subject to the will at the respective level of objectivation. Each of these manifold forms can be classified into one of the four levels of objectivation, even if there are typical transitional forms, such as the crystal, which has characteristics of both the 'realm of inorganic nature' and the 'plant kingdom'.[96]

In the two passages above, where the four levels are listed, Schopenhauer uses a new criterion in each case to justify the assignment of man to the highest level and of mineral to the lowest level: In § 26, Schopenhauer explains the order of levels top-down based on the criterion of individuality already mentioned in § 23 (and also § 16). The following rule applies: "The further we descend the more any trace of individual character is lost in the generality of the species." [97] Human beings possess the highest proportion of individuality with their character, whereas in the inorganic realm all individuality has disappeared. In § 28, however, the criterion is participation in the Platonic idea, which, like the expression 'thing in itself', is usually used and described synonymously with 'will': Human beings have the highest participation in the idea, which extends "by the stepwise descent through all [...] forms", and only all forms taken together represent the complete objectivation of the will.[98]

Since §§ 26ff. are broken through many digressions due to the argumentation and justification of the system, a strict assignment of these four levels to the individual parts of the text is difficult. Nevertheless, a rough structure is suggested at the beginning of § 26 with the words "The most universal forces of nature present themselves as the lowest levels of objectivation of the will."[99] If one follows this hint as a bottom-up explanation of the four-level sequence, one can recognise a structure in the text of § 26ff. despite all difficulties of classification: § 26 begins with the 'inorganic realm' and especially represents forces and laws of nature. § 27 takes up the criticism of the purely aetiological and morphological philosophy of nature of § 17 at the beginning.

---

[95] WWR I, p 156f., also p. 178. Cf. Christian R. Steppi: Der Mensch im Denken Arthur Schopenhauers. Eine Anatomie der fundamentalen Aspekte philosophischer Anthropologie in des Denkers Konzeption als kritische und systematische Würdigung. Frankfurt/ Main et al. 1987, pp. 343–365.
[96] WWR I, p. 157ff., also p. 286.
[97] WWR I, p. 156.
[98] WWR I, p. 178.
[99] WWR I, p. 155.

The paragraph "Sometimes several of the appearances..."[100] begins with an examination of general phenomena of emergence and saltationism in phylogeny, which then leads to an examination of a specific dynamism in the vegetable and animal kingdom,[101] which can already be observed on an ontogenetic level. The penultimate section of § 27 in the first edition of WWR I ("The will is at work in the plant kingdom, [...]") then leads to the human kingdom.[102]

This penultimate section of § 27 draws attention to the fact that the essential characteristic of the human being in this sequence of levels is the individual character, which is then examined also at the end of § 28 and was already dealt with intensively in § 20. This allows a rough structure of B II to be shown in an overview: The final sections of §§ 27 and 28 focus on the highest level of objectivity, together with § 20; § 27 and the aetiological explanations, especially on geology, botany and zoology in § 17, form the second and third highest level of objectivity, while § 26 deals with the lowest level.

This fourfold division shows the superordinate system structure of B II, which can be further differentiated in each case: In the inorganic realm, for example, natural forces and material phenomena such as "rigidity, fluidity, elasticity, electricity, magnetism, chemical properties" etc. can be classified according to the principle pair 'repulsion/ attraction'.[103] Since this detailed spelling out of the system is not my concern here, but my interest is in the representation of the general structure of the system, I would like to point out a passage from § 26, which first gives a reflection on the contents of B I and then outlines a general overview of both books:

> All this is only a passing reminder of what was covered in the First Book [sc. B I of WWR I]. The two Books can be rendered completely intelligible only by paying close attention to their inner agreement: this is because the will and representation, which are inseparably united as the two sides of the real world, are torn apart from each other in these two Books in order to examine each more clearly in isolation.[104]

From the uniformity of the world, of B I and B II, as well as the quadripartition of B II described above, it is possible to anticipate an overarching structure that takes into account the overall system of the WWR:

---

[100] WWR I, p. 169.
[101] Cf. Jens Lemanski: Die 'Evolutionstheorien' Goethes und Schopenhauers.
[102] WWR I, p. 174. A further paragraph was added in subsequent editions.
[103] WWR I, p. 155, also p. 147, p. 174.
[104] WWR I, p. 160, also p. 144.

1.2 The System of WWR

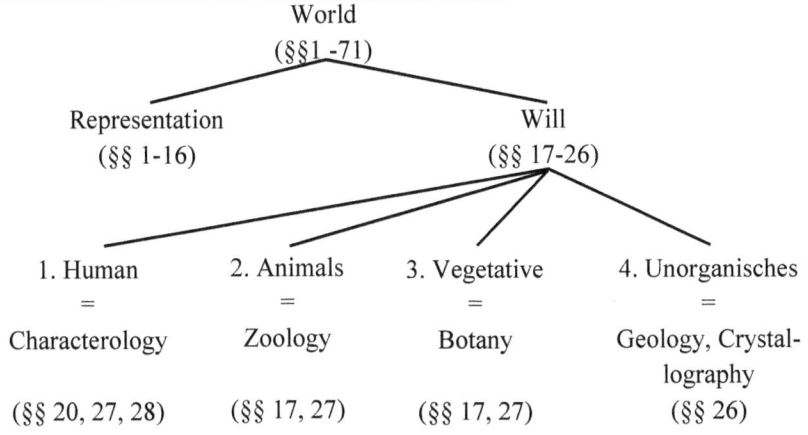

At this point, I would like to point out that this schematic presentation will be further modified and partly revised in the following. What becomes obvious here, however, is the conceptual structure,[105] according to which the 'world' as the highest concept (*conceptus summus*) includes all other structural concepts of B I and B II as lower concepts (*conceptus inferiores*). At least one science was added to each of the four levels of objectivity of the will, which Schopenhauer cites in B II as an example, although – in contrast to his metaphysical approach – these were treated morphologically or aetiologically, especially in the scientific studies of his time. As an example: The essential characteristic of human beings is their character, and the science which deals with this character is characterology. However, Schopenhauer already points out here that characterology will mainly be treated in B IV. Before I sketch this parallel between B II and B IV in more detail, I will first examine the structure of B III.

1.2.5 Book III: Aesthetics (Representation)

B III is entitled "The world as representation, second consideration // Representation independent of the principle of sufficient reason: the Platonic Idea: the object of art"[106] and is divided into 22 Paragraphs, that treat the following topics:

§§	Topics
30–32	Introduction
33–38	Contemplation and Genius
39, 40	The Sublime and the Stimulating

---

[105] Vide supra, Chapter 1.2.3.
[106] WWR I (1819), p. 241.

	1 The World and its Representationalist Interpretation
41, 42	The Idea and the Beautiful
43	Architecture and Artistic Fountainry
44	Garden Art, Animal Sculpture
45–47	Sculpture
48	Historical Painting
49–51	Poetry
52	Music

Klamp had already pointed out three links between B III and the other books of WWR:[107] B I and B III are connected by the concept of representation; B II and B III are connected by the concept of idea; B III and B IV are connected by the concept of negation of the will. The connection between B I and B III, as stated by Klamp, is already apparent from the title of the two books and is not further questioned here. Whether and to what extent B III and B IV are connected by the negation of the will is to be discussed in the next chapter (1.2.6). I would first like to focus on Klamp's assertion that B II and B III are connected by the concept of the idea. Klamp himself explains this connection in the following words:

> Book II speaks more generally, i.e. basically, of the will and its self-representation in phenomena, specifying then the 'ideas' as the 'fixed, certain levels of the objectivation of the will'. Directly following on from these remarks, Book III then develops a completely philosophical-metaphysical theory of art, of the beautiful at all, including the beautiful in nature, whereby both books constitute a further subunit.[108]

In my view, it is not clear from the quote how Klamp exactly justifies the link between B II and III which is claimed here. The first sentence of the quote says that B II speaks more in general terms; one might think that B III deals with the same subject, only in more concrete or detailed terms. However, this is not explicitly evident from the second sentence of the quote, and particularly emphasised expressions such as 'completely', 'at all' rather indicate that Klamp also understands B III as a general consideration. The 'direct connection' claimed at beginning of the second sentence also remains unexplained: Due to external reasons, it should be a matter of course that a third book follows on from a second one. Platonists may recognise the only relationship in terms of content that could be read out of the two sentences of Klamp's work in the relationship between the expressions 'idea' and 'the beautiful' – but whether this was really Klamp's intention when writing the text remains questionable.

---

[107] Siehe auch oben, Kap. 1.2.1.
[108] Gerhard Klamp: Die Architektonik im Gesamtwerk Schopenhauers, p. 84.

1.2 The System of WWR

In another paper, however, Klamp points out that it is the fourfoldness that Schopenhauer associates with Plato: the cardinal virtues or the order of the polis in Plato's work correspond to the preference for fourness that can also be found in Schopenhauer's work.[109] Indeed, in the previous Chapter 1.2.5, I have worked out and established a preliminary scheme that would correspond to this fourfoldness: 1. the human, 2. the animal, 3. the vegetable and 4. the mineral kingdom. If the connoisseur of Plato now thinks of the idea of beauty, she will probably remember a similar number of ideas and order in the middle Platonic dialogues, which is not directly connected with the cardinal virtues or the estates, but with the so-called 'theory of ideas'.

I believe that Klamp is right in his two basic assertions, but that the justification and the details of his judgments are insufficient. Like Klamp, I also recognise a very close connection between B II and B III, based on the theory of ideas. Indeed, I assume that Schopenhauer in both books follows a structure based on the Platonic dialogues, which I would not describe as a four-structure, but rather – following Margit Ruffing – as a '1,2,3,4/5-' or '4+1-scheme'.[110]

Even a brief glance at the table above for B III, in particular §§ 43 ff., reveals a discrepancy between B III and the fourfold constellation elaborated in the previous Chapter 1.2.4. Schopenhauer describes in B III several art forms which outline the ideas of the stages of objectivity presented in B II: 1. architecture shows the idea of inorganic matter, 2. garden and landscape art show the idea of the plant kingdom, 3. animal sculptures and animal painting shows the idea of the animal kingdom, and 4. human sculptures, history painting and poetry show the idea of rational begins.[111] There are also transitional forms, such as artistic fountainry (Wasserleitungskunst), which contain inorganic matter, but which also symbolise the reproduction and movement that plants and animals are only capable of. However, a problem of classification into this scheme of four, which has been shown so far, arises when music is taken into account, as Schopenhauer notes in § 52:

> Now that we have considered all the fine arts with the universality proper to our point of view, beginning with fine architecture (whose

---

[109] Gerhard Klamp: Das Streitgespräch zwischen Becker und Schopenhauer, p. 72: "Incidentally, Schopenhauer and Plato are largely related in spirit, also in formal terms. A popular way of thinking in Plato's work, just like in the former, is the two or fourfold division, e.g. the dichotomous division of concepts and number ideas according to the dialectical method in the late dialogues, or: the fourfold of the Platonic cardinal virtues (wisdom, courage, temperance, justice), which were later elevated to the canon of ancient ethics. Plato's distinction between the ideal state (the 'Politeia') and the second-best state (the 'Laws'), i.e. his theory of the state in two versions, according to which he calls for a fourfold class division of the polis: a) he distinguishes the great mass of the population or citizens b) from the class or caste of warriors and from both c) the officials, but, above all, three is d) the elite of the dominating class. (For details cf. Hans Leisegang: Denkformen)."

[110] Vide supra, Chapter 1.2.1.

[111] For a detailed analysis of the art forms and the hierarchical ladder, cf. Sandra Shapshay: Schopenhauer's Aesthetics. In: The Stanford Encyclopedia of Philosophy (Summer 2018 Edition). Ed. by Edward N. Zalta, URL = https://plato.stanford.edu/archives/sum2018/entries/schopenhauer-aesthetics/.

1 The World and its Representationalist Interpretation

> goal as such is to make the objectivation of the will clear at the lowest level of its visibility, where it shows itself as the dull striving of mass, conforming to law but with no cognition, but nonetheless still revealing self-dichotomy and struggle, namely between gravity and rigidity) – and concluding our investigation with tragedy at the highest level of the objectivation of the will, and which puts that very schism before our eyes in fearful grandeur and clarity; – we find that one fine art still remained, and must remain excluded from our consideration since there was absolutely no suitable place for it in the systematic context of our presentation: and this is music. It stands completely apart from all the others. What we recognise in it is not an imitation or repetition of some Idea of the essence of the world:[112]

The quotation, first of all, illustrates once again the analogies drawn so far between the four levels shown in B II and the fine arts shown from § 43 onwards. At the same time, however, it also emphasises the extraordinary position of music, since this art form is not a reflection of any of the four stages in the world. As Schopenhauer explains in the following, "music is an unmediated objectivation and copy of the entire will, just as the world itself is".[113] This results in the '1,2,3,4/5-' or '4+1-scheme' already mentioned: All arts correspond to one of the four inner-worldly levels of objectivation, with the exception of music, which corresponds to the entire will itself:

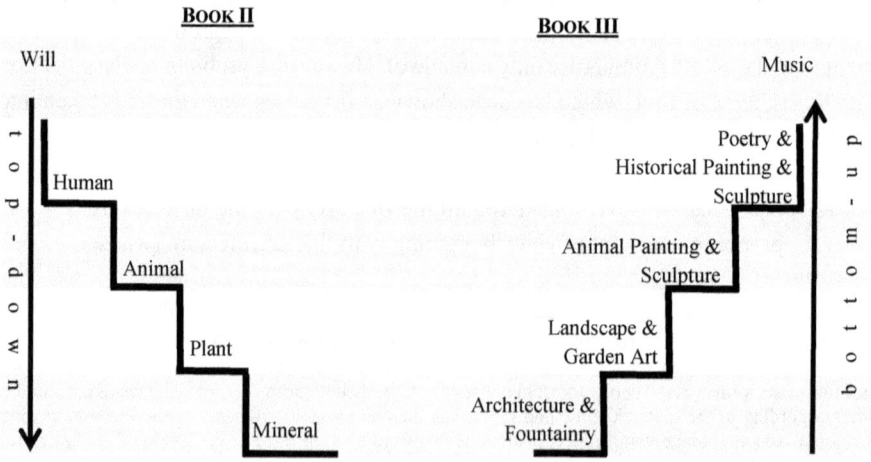

In my opinion, this is one of the most interesting system structures that can be found in the WWR, as it shows several relationships between the individual books: (1) First,

---
[112] WWR I, p. 282f.
[113] WWR I, p. 285.

1.2 The System of WWR

there is an obvious connection between several individual paragraphs of B II and B III: the lowest level (§ 26 = § 43), the two middle levels (§§ 17, 27 = § 44), and the highest level (§§ 45–51 = §§ 20, 27, 28). (2) In addition, there is a fifth level, 'the entire will' or 'music', which was not obviously recognisable in B II, but can be clearly assigned retrospectively (§ 52 = § 18). (3) The top-down structure already mentioned in the previous chapter is reinforced by the correspondence between § 18 and § 52: the top-down structure of B II corresponds to a bottom-up structure in B III (from the lowest level in § 43 to the highest § 51). Those who do not shy away from any interpretation effort can even establish the Neoplatonic scheme 'unity (§ 18) – procession (§§ 25–29) – return (§§ 43–52)' in this process.[114] (4) A similar '1,2,3,4/5-' or '4+1-scheme' can also be found in central passages of the middle period of Plato's work, in which four inner-worldly levels and a fifth transcendent stage, the idea or the unhypothesized principle, are also sketched.[115]

Schopenhauer's depiction of the fine arts is the most important passage in B III, primarily because of the first three connections that have been worked out. Nevertheless, Schopenhauer gives the hint in at least three passages in §§ 38, 39 and 41 that B III is divided into two sections and that each of these two sections is in turn divided into two parts.[116] In the following, I will only quote the first of these three passages:

> But before we turn to a closer examination of this objective side and its contribution to art, it is more to the purpose to remain with the subjective side of aesthetic pleasure somewhat longer, in order to conclude our examination of it by discussing the impression of the sublime, which depends on it alone and arises through a modification of it. After this, an examination of the objective side will complete our investigation of aesthetic pleasure.[117]

If one compares all three passages in the text which give indications of the structure of B III, one becomes aware of a threefold division, which was most explicitly mentioned in the above quote from § 38. §§ 33–38 deal with contemplation and genius and belong to the 'subjective side of aesthetic pleasure'. §§ 39 and 40 deal with the sublime and the charming, which, according to the above quotation and the beginning of paragraph 41, are "only a special modification of this subjective side".[118] In this respect, the first section of B III is again divided into two parts: the main part and a subsequent modification. §§ 41 and 42, on the other hand, deal with the objective side of aesthetics and address the idea and the beautiful in general.

---

[114] Cf. Jens Lemanski: Summa und System. Historie und Systematik vollendeter bottom-up- und top-down-Theorien. Münster 2013, pp. 85–163.
[115] Cf. Jens Lemanski: Summa und System, pp. 57–77.
[116] WWR I, p. 223, p. 224, p. 233. Cf. also Sandra Shapshay: Schopenhauer's Aesthetics, sect. 3.
[117] WWR I, p. 223.
[118] WWR I, p. 233.

At the end of § 42 Schopenhauer writes: "We will now review the arts one by one, which will lend completion and clarity to the theory of beauty we have presented."[119] This quote can be understood as an indication that the second section of B III is also divided into two parts, since B III first deals with the idea in general terms and then concretizes it in the part that I have discussed intensively in this chapter in connection with B II.

1.2.6 Book IV: Ethics (Will)

B IV is entitled "The world as will, second consideration With the achievement of self-knowledge, affirmation and negation of the will to life"[120] and is divided into 18 paragraphs. The main topics and numerous excursuses can be summarised as follows:

§§	Topics
53	Introduction
54	Affirmation & Negation of the Will to Life
55	Excursus I: The Necessity of the Will
56–59	Excursus II: Life
60	Affirmation of the Will to Life
61	Egoism
62	Temporal Justice
63, 64	Eternal Justice
65, 66	Good & Evil
67	Excursus: Compassion
68	Negation of the Will to Life
69	Excursus: Suicide
70	Excursus: Freedom of the Will
71	Excursus: Ontology

Before I discuss the given table and the structure of B IV in more detail, I would first like to talk about Klamp's thesis, which was omitted in the previous chapter.[121] Klamp had asserted that B III was only "a kind of introduction [Vorschule] for the serious doctrine of the 'negation of the will to life', which is explained in detail in the fourth book".[122] In contrast to Klamp, the descriptive reading I have adopted here defends the thesis that there is no precedence of the doctrine of the negation of the will to live over any other main theme in B IV or in the entire work.

---

[119] WWR I, p. 238.
[120] WWR I, p. 297.
[121] Vide supra, Chapters 1.2.5 and 1.2.1.
[122] Gerhard Klamp: Die Architektonik im Gesamtwerk Schopenhauers, p. 83.

1.2 The System of WWR

Klamp is right that Schopenhauer announces the "most serious" consideration of the entire WWR directly in the first sentence of B IV. However, Schopenhauer does not say that this seriousness only refers to the negation of the will, as Klamp explains. On the contrary, the relevant sentence reads:

> The final part of our discussion [sc. B IV] declares that it will be the most serious, since it deals with human actions, which are of direct concern to everyone; no one is unfamiliar with or indifferent to such a topic. In fact, it is so natural for people to relate everything to action that they will always consider that part of any systematic discussion which concerns deeds to be the culmination of the whole work, at least to the extent that it is of interest to them, and will accordingly pay serious attention to this part, if to no other.[123]

The quote shows that the seriousness relates to the whole content of B IV and not only to a part of it.[124] Schopenhauer's explanation of this seriousness in the second part of the quote (especially "direct concern to everyone", "no one is unfamiliar with or indifferent to such a topic") relates to those passages in §§ 16 and 17 which I have already discussed in Chapter 1.2.4 above.

Following on from his assertion made above, Klamp explained that the denial or negation of the will in B IV corresponds to the "creation of art and genuine art experience" in B III – apart from the fact, however, that these forms of experience are "only occasional and limited in time" in comparison to the negation of will.[125] Since the creation of art and the experience of art were described in the 'subjective' or first section of B III, one must therefore search for the parallel between the first section of B III and the doctrine of negation in B IV. In fact, in the first section of B III, one finds several passages that point to a parallel to the negation of the will since Schopenhauer speaks there several times of the liberation of individuality and a revised mode of cognition. The argumentation of these passages is as follows:

(1) The idea is not subject to the principle of reason.
(2) The subject (i.e. individual cognition) is always bound to the principle of reason.
(3) "Thus, if the Idea is supposed to be the object of cognition, then cognition will be possible only when individuality is suppressed in the cognitive subject."[126]

---

[123] WWR I, p. 297.
[124] Not even the parallel passage at the end of B III, § 52 (WWR I, p. 294f.) shows that Schopenhauer limits seriousness to the doctrine of negation.
[125] Gerhard Klamp: Die Architektonik im Gesamtwerk Schopenhauers, p. 83.
[126] WWR I, p. 244. In later editions: "…the Ideas are…".

1 The World and its Representationalist Interpretation

We can assume that Klamp's thesis is based on judgements such as (3), which can be found several times in B III. (3) has the form of a so-called 'anankastic conditional',[127] which was widely used in the paradigm of Kantian philosophy as a logical means of expressing transcendental arguments.[128] The anankastic form of the conditional is made clearer by the typical form 'If it should be..., then must...':

> (3') "Thus, if the idea should be the object of cognition, then individuality must be suppressed in the cognitive subject."

(3) and (3') are substitutable in that the modality in the antecedent and in the consequent remains the same: The antecedent is hypothetical (possible), the consequent anankastic or categorical (necessary). According to a strict interpretation, conditionals can only be classified as anankastic "if and only if they are understood as conveying that the complement of the modal in the consequent is a necessary precondition for the complement of the desire predicate in the antecedent to be realized".[129] Such a precondition is mentioned in (3) and (3'): The liberation or suppression of individuality is the necessary precondition for the idea to become an object of knowledge (the desire predicate).

Whereas Klamp already sees a normativity and factuality in such statements by Schopenhauer, he overlooks, in my opinion, that anankastic or rather hypothetical-anankastic conditionals (desire predicate + necessary precondition) express only a condition of possibility but no normativity. (3) and (3') say nothing about whether the idea should become an object of cognition at all. In the language of transcendental philosophy, (3) and (3') only say: the suppression of individuality is the condition of the possibility that the idea becomes the object of cognition; but this does not (yet) express the demand that the idea should actually become the object of cognition. Since the realisation of the possibility is not the subject of this statement, however, interpretations such as Klamp's are based on a deontic fallacy that confuses the hypothetical 'Sollen' ('should') with a compulsory action verb or misunderstands the hypothetical-anankastic conditional as a conditional imperative. This can be seen from the fact that conditional imperatives or judgements with an obligatory 'Sollen' ('shall', 'must') such as the following (3") are not substitutable with (3):

> (3") "The idea shall be the object of cognition, thus individuality must be suppressed in the cognitive subject."

---

[127] Cf. Georg Henrik von Wright: Norm and Action. A Logical Enquiry. London 1963; Kjell Johan Sæbø: Notwendige Bedingungen im Deutschen. Zur Semantik modalisierter Sätze. Arbeitspapiere des Sonderforschungsbereiches 99, No. 108. Konstanz 1985.

[128] Cf. Jens Lemanski: Summa und System, p. 211, p. 225, p. 235f.

[129] Cleo Condoravdi und Sven Lauer: Anankastic conditionals are just conditionals. In: Semantics & Pragmatics 9:8 (2016), pp. 1–60, here: p. 3.

The parallel passages show that (3) was not formulated arbitrarily by Schopenhauer in a hypothetical-anankastic form. At the beginning of § 33 one finds the same transcendental argument in nested clauses "[…] *it is certain that, if it is possible for us to raise ourselves from cognition of particular things to cognition of the Ideas, this can only take place by means* of an alteration in the subject […]"[130] And also at the beginning of § 34, Schopenhauer once again points out the hypothetical character of the antecedent of (3): "As we have said, it is *possible* – although only in exceptional cases – to go from the ordinary cognition of particular things to cognition of the Idea."[131]

The beginning of B IV also makes it explicit that Schopenhauer is not concerned with a normative but rather with an approach that continues to pursue the descriptive-representationalist objective of Bacon's philosophy of § 15 and further § 16. Since I do not wish to reproduce the content of the introductory § 53 in its entirety, only those quotations should be mentioned that name the method and the goal of B IV:

(4) "But in my opinion, philosophy is always theoretical, since what is essential to it is that it treats and investigates its subject-matter (whatever that may be) in a purely contemplative manner, describing without prescribing."

(5) "Philosophy can never do more than to interpret and explain what is present, to bring the essence of the world – that essence which speaks intelligibly to everyone in a concrete fashion, which is to say as a feeling – to the clear and abstract cognition of reason, […]."

(6) "I will remain strictly faithful to the method we have been using so far, […] thus doing everything I can to communicate this idea as fully as possible."

(7) "The perspective we have adopted and the method we have specified should discourage any expectation that this ethical Book will contain precepts or a doctrine of duty;"

(8) "In so doing, our philosophy will continue to assert the same immanence that it has maintained all along:"

(9) "Now, since the real, cognizable world will continue to provide as rich a source of material and reality for our ethical investigations as it did for our previous investigations, it will be entirely unnecessary for us to take refuge in insubstantial negative concepts, […]."[132]

All six objectives are linked to the representationalist position of § 15: The analysis is purely observational (4), furthermore also interpretative (5); its starting point and basis are concrete phenomena and not the semantically broadest possible concept or formula in a set of logical axioms (9); it is a repetition and reflection of the concrete

---

[130] WWR I, p. 198f. (My emphasis – J.L.)
[131] WWR I, p. 200. (My emphasis – J.L.)
[132] All quotes are (without emphasis): WWR I, p. 297–299.

representation by using abstract concepts (5); the repetition of the concrete in the abstract aims at completeness, which will be concluded with the analysis of human action (6);[133] ethics does not go beyond the observation, description and interpretation of empirical phenomena (8); it, therefore, does not prescribe, is not normative (4), (7).

I would now like to turn to the structure of B IV. As I have just mentioned, § 53 is an introduction that follows methodically and thematically from §§ 15 and 16 (Representationalism and practical reason/ ethics). § 54 then introduces the content of the topic of B IV. At the beginning of § 54, Schopenhauer announces: "But we want to look at life philosophically, i.e. according to its Ideas, [...]."[134] In the course of this consideration of the pair of opposites 'life and death', two central principles emerge for Schopenhauer: "the perspective of the complete *affirmation of the will to life*. [...] The opposite [Gegentheil] of this, the *negation of the will to life* [...]".[135] In principle, § 54 could already be concluded with B IV, if on the one hand, these two principles were not difficult to understand concepts and if, on the other hand, no further topics could be centred around both concepts, which would further guarantee the completeness of the system.

The rest of § 54 gives further information on the concept of B IV and the WWR in general. Schopenhauer reflects once again on his representationalist method, then gives some indications of the structure of B IV and also takes up the content of the preface again. The methodological digression specifies in a few sentences the representationalist approach of § 53. Since both principles, affirmation and negation, are "expressed only through deeds and behaviour", it is the purpose of the analysis to "to present both and bring them to the clear [deutlichen] cognition of reason".[136] Deeds and behaviour of human beings are thus described and assigned to the two general principles. The aim is not to focus on "prescribing or recommending one or the other [sc. affirmation or negation]".[137]

Schopenhauer then indicates that he would insert two more excursuses, namely 'general' and 'helpful' treatises on freedom and necessity (§ 55) and on life (§§ 56–59) before he reaches the actually announced content of B IV (affirmation and negation). The last section of § 54 recapitulates once again the content of the preface (a single thought, organic system, division into four books for the purpose of communication, reading instructions) and points out, above all, that the presentation "does not by any means allow for a linear progression" due to the mutual preconditions of the theses.[138]

---

[133] Against Klamp, it should be pointed out that Schopenhauer also emphasised this objective in B III, namely that "philosophy is nothing other than a complete and correct repetition and expression of the essence of the world in very general concepts". (WWR I, p. 292)
[134] WWR I, p. 301.
[135] WWR I, p. 311.
[136] WWR I, p. 311.
[137] WWR I, p. 311.
[138] WWR I, p. 312.

1.2 The System of WWR

The fact that §§ 55–59 are indeed excursuses can only be ascertained by the fact that Schopenhauer speaks in § 56 and at the beginning of § 60 of a discussion "intervened" or "to intervene" ("dazwischen getretenen Betrachtungen", "Dazwischentreten") regarding these treatises between §§ 54 and § 60.[139] It can be argued that these digressions appear necessary with regard to the argumentation, the conceptual scheme or the completeness of the system; nevertheless, they only take up again – partly explicitly marked –[140] topics especially of B II, e.g. characterology (§§ 55, 58 = §§ 20, 28), teleology (§ 56 = § 29) or dynamism (§ 56 = § 27).

Only with § 60 begins the announced first section, which deals with the main theme of the affirmation of the will to live. This main principle is examined between § 60 and § 67, whereby § 60 is a general presentation of the affirmation of the will and the following paragraphs examine more detailed phenomena which must be assigned to the same main theme. The second section, which deals with the second main theme of B IV, namely the negation of the will to life, begins with § 68, which is also rather general. Similar to B II, Schopenhauer asks at the beginning of each of the two sections (§§ 60 and 68) about the "meaning" of the affirmation and negation of the will to life.[141] Again, 'meaning' has a representationalist connotation, as it is subject to a bottom-up theory of action, which attributes to the two internal principles from the observations of deeds and behaviour:[142]

> We have completed the two discussions that needed to intervene, the first concerning the freedom of the will in itself along with the necessity of its appearance, the second concerning its fate in the world that mirrors its essence, given that it has to affirm or negate itself based on cognition of this world; now that this has been accomplished, we can further clarify the nature of the affirmation and negation themselves, having mentioned and explained them above [sc. § 54] in only very general terms; we will do this by looking at ways of acting (since this is the only way in which affirmation and negation are expressed) and by regarding this action with respect to its inner meaning.[143]

---

[139] WWR I, p. 334, p. 352.
[140] Cf. e.g. WWR I (1819), p. 335: "To begin with, I would like to recall the discussion from the end of the Second Book […]."
[141] WWR I, p. 352, p. 405.
[142] Cf. on such theories of action, e.g. Steven A. Sloman, Philip M. Fernbach, Scott Ewing: A Causal Model of Intentionality Judgment. In: Mind & Language 27:2 (2012), pp. 154–180. A detailed presentation of Schopenhauer's theory of action can also be found in Matthias Koßler: Empirische Ethik und christliche Moral, pp. 422–460.
[143] WWR I, p. 352.

1 The World and its Representationalist Interpretation

Schopenhauer reflects on his theory of action in other passages (e.g. §§ 55, 62)[144] and illustrates it with some exemplary views of tort law.[145] Although Schopenhauer's theory of action has not yet been given much consideration, there are many studies on Schopenhauer's so-called 'ethics of compassion'. This is astonishing in so far as § 67 represents explicitly an excursus in the system of WWR I. Schopenhauer makes this explicit in the first sentence of § 68: "After this *digression* on how pure love is identical with compassion [...]".[146]

The last section, which deals with the negation of the will to life, consists in the narrower sense only of § 68 (and possibly also § 69), which is also followed by several excursuses. Schopenhauer also tries to remind the reader at the beginning of this section of his representationalist approach. He says at the beginning that he

> *will take up the thread of our earlier discussion of the ethical meaning of action*; I will now show how from the *same source* that gives rise to all goodness, love, virtue and nobility there *ultimately* emerges also what I call the negation of the will to life.[147]

On the one hand, the reference to the "thread" explicitly refers to the division between the two sections of B IV, since this thread was considered as an objective ("interpretation...") in §§ 53 and 54, was continued in § 60 and is now taken up again in § 68; on the other hand, the word "ultimately" also refers to the conclusion of the system. Schopenhauer had already announced this second section at the beginning of § 65 as "our final discussion [...] as part of our central line of thought".[148] And also at the end of § 66 Schopenhauer had addressed "the final part of my presentation" and pointed out that the excursus on love and compassion in § 67 guaranteed the completeness of the system.[149] As the above-given quote from § 68 further indicates, both parts, affirmation and negation, depend on "the same source", namely reason. Such system-related statements, in turn, suggest the connection between § 16 and Book IV, which has already been discussed several times, but without a normative-soteriological interpretation being confirmed.[150]

I have already discussed the three-part structure of § 68 in detail elsewhere.[151] If I go into more detail in the following about the structure of the second section of B IV than about the first part, it is not because the latter is more important than the former,

---

[144] WWR I, p. 328, p. 370.
[145] WWR I, p. 370; cf. Rudolf Neidert: Die Rechtsphilosophie Schopenhauers und ihr Schweigen zum Widerstandsrecht.
[146] WWR I, p. 405. (My emphasis – J.L.)
[147] WWR I, p. 352. Hervorhebung von mir – J.L.
[148] WWR I, p. 386.
[149] WWR I, p. 401.
[150] Vide supra, Chapter 1.2.3.
[151] Cf. Jens Lemanski: Christentum und Mystik. In: Schopenhauer-Handbuch. Leben – Werk – Wirkung. Ed. by Daniel Schubbe, Matthias Koßler. Weimar 2014, pp. 201–208, here: p. 206.

1.2 The System of WWR

but because I believe that the second part has been received much more strongly in research and has so far been received without a system-related context, which has often led to a normative interpretation. The first part of § 68 deals with the "deeds and conduct" of ascetics, saints etc., so that "the inner nature of holiness, self-denial, asceticism, and the mortification of one's own will has been expressed abstractly, cleansed of all mythology" and subordinated under the concept 'negation of the will to life'.[152] In this section, too, Schopenhauer gives a reference to his representationalist approach, which explicitly recalls Bacon's empiricism from § 15:

> To use concepts that abstractly, universally and clearly reflect the whole essence of the world, and to transcribe a reflected image of the world into permanent concepts that are always available to reason: this and nothing else is philosophy. I recall the passage from Bacon of Verulam quoted in the First Book.[153]

Following this quote, Schopenhauer identifies two problems of his representationalist approach: on the one hand, his depiction of the negation of the will is "only abstract, universal, and therefore cold"[154] and, on the other hand, the object of his empirical observation, namely the saints and ascetics, cannot be found "in everyday experience".[155] For this reason, Schopenhauer implores his readers to read hetero- and autobiographies or books in general instead of direct experience of the world:[156] "Just read the (often poorly written) biographies of the people who are sometimes termed holy souls, sometimes pietists, quietists, pious enthusiasts, etc."[157]

The writings by and about saints, mystics and ascetics mentioned here thus have a system-relevant function: they serve as concrete evidence of the abstract theory of the negation of the will. The second part of § 68 then begins with the following passage:

> We will go a long way towards a fuller and more detailed understanding of what we are calling (in the abstraction and universality of our mode of presentation) the negation of the will to life, if we

---

[152] WWR I, p. 410.
[153] WWR I, p. 410. Cf. also p. 409f.: "When it comes to cognition of the essence of the world, there is a wide gulf between the two kinds of cognition [sc. the intuitive and abstract one] that only philosophy can traverse. In fact, everyone is conscious of all philosophical truths on an intuitive level or in concrete fashion: but to bring these truths to abstract knowledge, to reflection, is the business of philosophers, who should do, and can do, nothing else."
[154] WWR I, p. 410.
[155] WWR I, p. 411.
[156] Cf. Georg Misch: Geschichte der Autobiographie. Vol. 4/2: Von der Renaissance bis zu den autobiographischen Hauptwerken des 18. und 19. Jahrhunderts. Frankfurt/ Main 1969, p. 752. In more detail Heinz G. Ingenkamp: Plutarch und das Leben der Heiligen. In: Valori letterari delle opere di Plutarco. Ed. by Aurelio Pérez Jiménez, Frances Bonner Titchener. Málaga 2005, pp. 225–242.
[157] WWR I, p. 553.

also consider the ethical injunctions issued in this regard by people filled with its spirit.[158]

Following this objective of the second part of § 68, Schopenhauer examines the maxims and dogmas of the saints and ascetics of the Occident and Orient, which, for example, lead Christian mystics from the charity (lowest level) to the imitation of Christ (highest level).[159] As he explains after the quote given, although the concept of 'negation of will' is new, its content is old familiar through the actions and the behaviour of the saints and ascetics. The new concept has only been established for the purpose of subsumption, in order to be able to denote many concrete descriptions of actions with an abstract and broad concept.

The third part of § 68 begins with the words "I have now provided the sources [...]" and provides mixed comments on the "general description" of the state and the nature of the deniers: the conversion to the denial or negation of the will, the duration of the conversion and the literary sources for it.[160]

It is difficult to determine whether the last section of B IV ends with the beginning of § 69 or § 70. At the beginning of § 69, Schopenhauer speaks of that this should now "suffice [...] for a description of the negation of the will to life".[161] This suggests an exhaustive description of the second section of B IV so that the remaining paragraphs would be addenda or additamenta to what has been said so far. But at the beginning of § 70 Schopenhauer writes also: "We have now finished presenting what I have been calling the negation of the will".[162] Although this confirms the end of the general analysis, the "nunmehr" (now), which can be found in both quotes of the German edition given at the beginning of § 69 and at the beginning of § 70, is confusing. But since § 69 deals with suicide, which is not part of the negation of the will, § 69 can, in my opinion, be classified as an excursus.[163] In any case, it is certain that §§ 70 and 71 are only supplements to the system: § 70 deals, says Schopenhauer at the beginning and end of the paragraph, with a possible incompatibility and "apparent contradiction" of the second part with the excursus of § 55.[164] § 71 also deals with a possible "objection", which consists in the fact that the second part possibly appears as a "transition into an empty nothing".[165]

The fact that normative interpreters such as Klamp evoke emphatic sounds and linear approaches here, which are heading for § 71, thus seems extremely surprising on closer reading of the system structure. In my opinion, Schopenhauer constructs in §

---

[158] WWR I, p. 413.
[159] Cf. Jens Lemanski: Christentum im Atheismus. Spuren der mystischen Imitatio Christi-Lehre in der Ethik Schopenhauers. Vol. 2. London 2011.
[160] WWR I, p. 416.
[161] WWR I, p. 425.
[162] WWR I, p. 429.
[163] Cf. Jean-Yves Béziau: O suicídio segundo Arthur Schopenhauer. In: Discurso 28 (1997), pp. 127–143.
[164] WWR I, p. 430, p. 435.
[165] WWR I, p. 436.

70 and § 71 an apparent contradiction and a possible objection, on the one hand, to enrich his system, which has been shown to be complete, with dogmatics, church history and ontology – and on the other hand, a book that begins with "The world is my representation" and ends with "nothing" seems to appeal to a broader audience than academics, which are used to have sober chapter headings such as "On Transcendental Philosophy" or "On Ontology". Finally, it should also be remembered that there is a famous model for the WWR which adds an appendix on the concept 'nothing' for the sake of "the completeness of the system": The Transcendental Analytic of the *Critique of Pure Reason*.[166]

1.2.7 Evaluation

The previous chapters may have shown that Klamp's general assessment regarding a link between all four books of the WWR is justified, but that his arguments are incomplete in many places or based on interpretative premises that not all interpreters are obliged to share. If one looks at the six "arched arches" described by Klamp in Chapter 1.2.1 above, it can be said that (1), (4) and (5) are trivial, since they can already be read out of the title of Books I–IV. (2) and (6) are due to the prejudice of the normative reading that Klamp tacitly presupposes as a premise for interpretation. (3) is, in my opinion, correct, but the argument is not precise enough, since in B III almost only the second part corresponds to the last paragraphs of B II.

On a general level, the explicit division of B I, B III and B IV is striking: In B I, the first section deals with the understanding (§§ 3–7), the second section with reason (§§ 8–16); In B III, Schopenhauer distinguishes the objective side (§§ 33–40) from a subjective side (§§ 41–52); B IV deals first with affirmation (§§ 60–67) and finally with negation (§ 68). Only B II breaks with this symmetry since only at the end of § 19 is there any indication that Schopenhauer considers the preceding paragraphs of B II to be given "in a broad and provisional way"; the paragraphs that follow are, however, only intended to "establish and justify it more clearly and in greater detail, developing it to its fullest extent".[167]

In my opinion, the quotation from § 19 does not justify speaking of a consistent division of all four books of the WWR. Even if such a continuous division of the four books had satisfied the recipient's need for aesthetic symmetry and thus been a well-structured approach, this division could not have claimed the same systemic relevance for all books. For although the division of B I into understanding and reason and of B IV into affirmation and negation, for example, reflect the most general concepts of the system – in modern terms, they represent the top-level domain – the distinction between the objective and subjective sides of aesthetics in B III has

---

[166] CpR, p. 382f. (A290, B346)
[167] WWR I, p. 130.

no such function: Understanding and reason, affirmation and negation form faculties and principles within the system, whereas objective and subjective side are thematic classifications that combine faculties and principles taken up in the system into a text section.

But also the equation of the principles of B I and B IV seems problematic if one considers the connection between § 16 and B IV: § 16 had discussed practical reason as the last of the three faculties of reason dealt with in B I. In § 16, Schopenhauer had given an outlook on B IV, which can be interpreted in such a way that the negation and affirmation of the will to life as principles are subordinate to the capacity of reason. If this subsumption of the negation and affirmation of the will to life is recognised under practical reason, then the division of B I (understanding/ reason) and B IV (affirmation/ negation) cannot represent an equivalent dichotomy. Rather, B IV would then be a more detailed description of a rational faculty already systematised in § 16.

A determination of the sub-level-domain – or classically speaking the *conceptus inferiores* – proves to be difficult insofar as there are no explicit indications given by Schopenhauer as to how this should be taken from the text of the WWR. The mere outward emphasis of these concrete concepts and the way they are embedded in their context seems to me to be a sign of their purpose. This becomes particularly clear in the excursuses, as they do not always include concrete concepts in their argumentation. If one takes § 13, which deals with humour, as a good example, one finds several concepts given e.g. in italics, namely 'laughter', 'wit', 'foolishness', 'pedantry', which take the role of the lower or lowest terms in the concept scheme. It is noticeable that Schopenhauer, at the points where these concepts are present, embeds them in a context that includes a definition, e.g. 'laughter', 'wit', 'foolishness', 'pedantry': "[...] this type of the ridiculous is called wit"; "[...] this type of the ridiculous is called foolishness" or similar.[168]

A consistent expression of representationalism can also be seen in such passages: the task of WWR I is not to offer the recipient a way out of the world by using a soteriological approach, but rather a logically constructed concept scheme with which one can orientate conceptually in the world. The representationalism of WWR I is thus not only an empirical representation theory but also a semantic project or a conceptual scheme justified with arguments. What is representationalistic (and thus not rationalistic) about this project is that it does not start from the concept of 'world' and analyse it top-down, but rather composes and assembles it from the bottom up from the individual parts with the help of subordinate concepts in the same way as it is already available to intuition as an unreflected unity.

On the basis of the text passages examined in Chapters 1.2.3 to 1.2.6, I have tried to design a tree diagram, as was still found in encyclopaedias in Schopenhauer's time,

---

[168] WWR I, p. 84f.

1.2 The System of WWR

and which is still used today, especially in the field of Knowledge Representation.[169] The development of such a tree diagram is difficult, and although I can claim to present a diagram for which one can argue plausibly, it is certainly not satisfactory. Whether this shortcoming stems from Schopenhauer's system or from my interpretation is something that ultimately only the reader can decide. In my opinion, it is caused by the fact that Schopenhauer's project was too 'extensive' for a single person in terms of its representationalist and encyclopaedic aspirations, but its argumentative stringency of justification would have made it difficult to distribute the task among several collaborators. In addition, I believe that Schopenhauer – in keeping with the zeitgeist – after the 1830s emphasised more the late idealistic and pessimistic tendencies in his work and only in the last years of his life did he begin to return to his original project.[170]

There are several reasons why I am presenting such a tree diagram here, despite the problems already announced: Firstly, it concretises the expression of a representationalist programme which, apparently due to the argumentative process, is not as obvious in the text of WWR I as one might expect. It thus clarifies what the descriptive interpretation emphasises and how it differs from the normative reading. Furthermore, the diagram shows not only a possible structure of the representationalist conceptual scheme, but also its problems and weaknesses. Finally, it also shows why it is possible to say in a representationalist system that logic is a part of the world and that 'world' is a concept that can only be put into relation to other concepts by using the ordering instrument of logic.

---

[169] Cf. e.g. John F. Sowa: Knowledge Representation. Logical, Philosophical, and Computational Foundations. Pacific Grove, Calif. 1999. I avoid, where possible, the concept 'ontology' in the philosophical-classifying sense of language and prefer to speak of 'conceptual scheme' etc., in order to avoid a possible confusion with classical ontology (as given e.g. in WWR I, § 71) and the modern term.
[170] Cf. Jens Lemanski: The Denial of the Will-to-Live in Schopenhauer's World.

1 The World and its Representationalist Interpretation

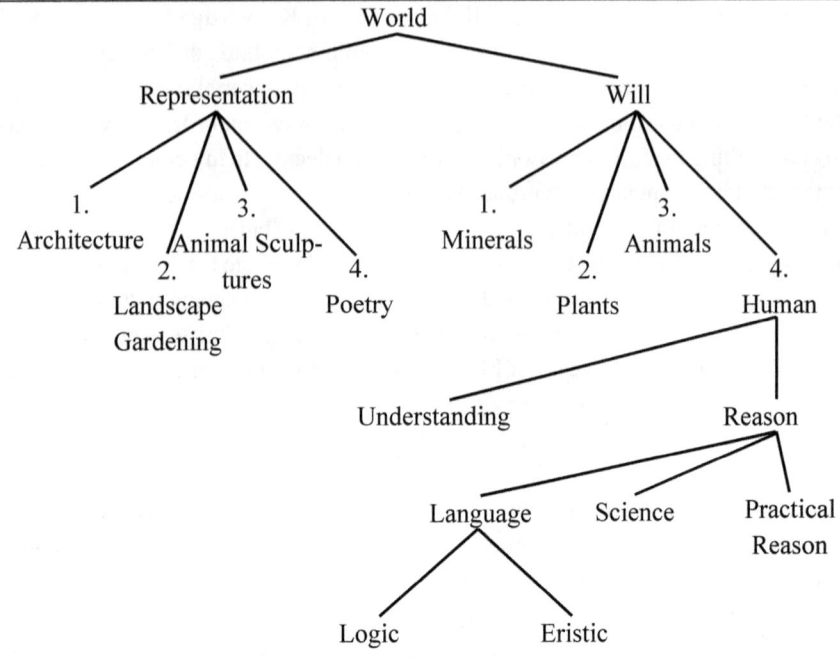

I think I need to go into this last point in more detail: The diagram shows that logic, on the one hand, has a firm place within the second section (reason) of the first book of Schopenhauer's system. This is remarkable in that Schopenhauer defines the goal of the system as the reflection of the world in abstract concepts. Logic thus becomes, on the one hand, a semantic subset of the concept of the world. On the other hand, however, it has also been shown that the project of reflecting the world in abstract concepts as visualised here is based, for example, on logical principles of conceptual order or inferential-argumentative structures of justification. Logic is thus presupposed by the system. Only through logic can the intuitive given world be reflected in abstract concepts at all. Whereas naïve representationalism wrongly emphasises only the first role of logic, I also see programs in need of revision that emphasise only the second role of logic.[171]

But I now return to the problem of such a diagram, which in my opinion lies in the analysis and evaluation of the status of individual concepts: In Chapter 1.2.3, I had claimed that world is the *conceptus summus*, which, depending on the perspective, shows itself as will or as representation. It may seem reasonable to use the concepts will and representation as the second-highest level, but this immediately raises the question of what function these two concepts have as a system: Are they components of the system or are they perspectives on the components of the system? For both functions I will mention only one argument (of many possible ones):

---

[171] Vide infra, Chapter 3.2.

1.2 The System of WWR

(1) `Will` and `representation` are only perspectives since they deal with the four levels of the system in B II and B III either as object levels of nature (will) or as manifestations of art (representation). (2) `Will` and `representation` are system components since the `whole will` of B II is represented as the 'fifth level' in the music of B III; likewise, `representation` in B I is the whole world, which only in B II a 'second world' as a counterpart.[172]

Arguments for both modes of operation were given in particular in Chapter 1.2.5. The fact that the two concepts can be interpreted so differently does not necessarily lead to aporias or contradictions, but it does lead to problems with a diagrammatic representation of the system and thus with the semantic assessment of its use and function. In the diagram given, I have chosen to regard `will` and `representation` as system components that convey the highest concept `world` with the four stages of B II and B III. However, such a decision does not make it possible to visualise the correspondence of the four stages in a single and two-dimensional tree diagram in a meaningful way, as I have tried to do with the level diagram given in Chapter 1.2.5.

Moreover, the following question may also arise: Do the concepts `understanding` and `reason` from B I not also depend on the concept of `representation`? How can this be connected with the four objectivations of art that are also derived from representation? My solution to this problematic question shows a certain ambiguity of the top-level domain: Since `understanding` and `reason` belong to the human being, who is the fourth level of the objectivity of the will, the two concepts mentioned can also be brought under the concept `human being` and not necessarily under the concept `representation`. I have opted for the first possibility in the diagram given above; but I can well imagine that this could be interpreted as a tendency towards objective idealism, and that other interpreters will find good reasons to derive the four levels of objectivity only from the concept of representation, or even to apply a diagram only to each book (as it was largely done in Chapters 1.2.3 to 1.2.6).

In the following, it will become clear that many of these problems and decision-making questions depend on Schopenhauer's view on logic, even if Schopenhauer's logic cannot completely solve these problems and decision-making questions. Moreover, it is not the aim of the present book to solve these problems; nor do I believe that we would do well to try to solve these problems by interpreting Schopenhauer's text. Instead of correcting Schopenhauer's system interpretatively until we have a consistent basis for how to reflect the concrete world in the first half of the 19th century without contradiction, we should rather reflect on how and why this philosophical

---

[172] I am obviously using the word 'world' metaphorically at this point and, unlike authors such as Atwell, I believe that Schopenhauer, too, did not want this expression to be understood *sensu literalis* in WWR I.

project should be implemented in the present day. This does not mean that Schopenhauer's representationalism has been completely written off; for, as will be shown in Chapter 1.3 and especially in Chapter 2, there are semantic, analytical and proof-theoretical elements in his doctrine, which can still make a valuable contribution to problems in modern philosophy and research today. A critique of historical representationalism is thus not an attempt to rescue it, but rather a decision on what can and cannot still claim to be the truth for justified reasons.

1.3 The Status of Logic in WWR and WWR2

In the previous Chapter 1.2, I have argued that it is justified to regard WWR I first of all as the main work and thus as the centre of Schopenhauer's system. In fact, this impression has been confirmed by the fact that in the prefaces, but also in the four books of WWR I, Schopenhauer repeatedly refers to other parts of his oeuvre, which can be seen as a supplement, modification or explanation of his short, sometimes very cryptic presentation of individual system components. In addition, in Chapter 1.1.4, I have argued together with Arthur Lovejoy that Schopenhauer did not do his work any favours by always commenting on, supplementing and modifying the original system of WWR I of 1819 in later years without fundamentally revising it.

The reason for declaring this procedure to be unobliging lies in Schopenhauer's representationalist method, which was particularly emphasised in Chapter 1.2.3: WWR I is the attempt to reflect the whole of the intuitive world in abstract concepts. As was shown in Chapter 1.2.4, the book *The World* (*as Will and Representation*) thus becomes a mirror of the world itself, allegedly even without the author's own intervention. In doing so, Schopenhauer, as shown in Chapter 1.2.6, also includes written opinions about the world as part of the intuitive world itself. The system is not only a reflection of the world, but also a reflection of the opinions held in the world, and this includes opinions that are formed about the world itself.

If Schopenhauer attempts to describe the characteristics of his present age with the help of this representationalist theory in various decades, then contradictions and semantic inconsistencies arise, which are particularly due to the fact that the world and the opinions about the world prevailing in it have changed. This can be seen, as for example, Lovejoy has shown, especially in many excursions on natural philosophy. To put it simply: the world of the year 1819 is different from the world of 1859; and for this reason, Schopenhauer did not do his system and his recipient any favours by presenting his observations about the world and the opinions about the world from the different decades to his reader as one single abstract and uniform system.

However, Schopenhauer had good reason to supplement and change his system. His more famous contemporaries accused each other of the incompleteness of their systems.[1] They mainly focused on enumerative induction in classical logic, according to

---

[1] Cf. e.g. Friedrich H. Jacobi: Über die Lehre des Spinoza in Briefen an den Herrn Moses Mendelssohn. Hamburg 2000, pp. 191f.; Wilhelm Traugott Krug: Briefe über den neuesten Idealismus. Eine Fortsetzung der Briefe über die Wissenschaftslehre. Leipzig 1801, p. 74f.; Friedrich Schleiermacher: Schriften aus der Berliner Zeit, 1800–1802. In: Kritische Gesamtausgabe. Ed. by Hans-Joachim Birkner et al. Berlin et al. 1988, Vol. 1/3, p. 320 [No. 149]; id.: Review. In: Jenaische Allgemeine Literatur-Zeitung 1 (1804), Vol. 2, No. 96–97, pp. 137–151, here: p. 138, p. 147; Jakob Friedrich Fries: Letter to Jacobi, December 10, 1807. In: Hegel in Berichten seiner Zeitgenossen. Ed. by Günther Nicolin. Hamburg 1970, p. 87f.; ibid.: Die Geschichte der Philosophie dargestellt nach den Fortschritten ihrer wissenschaftlichen Entwicklung. Vol. 1. Halle 1837, pp. 672ff.; Friedrich Wilhelm Joseph von Schelling: Zur Geschichte der neueren Philosophie.

which completeness claims are already problematic if, within the scope of a universal quantifier, only one proposition is found, which can be interpreted as unknown, not mentioned, contradictory etc.² In the following years, Schopenhauer thus supplemented his system of WWR I of 1819, which claimed to be complete, with propositions that could have been interpreted by contemporaries as a subset or a partition of a set called 'world' but could not be found in the original system.³

As will be shown in Chapter 1.3.1, especially § 9 of WWR I, which deals with the logic and also the dialectic in all three editions of the work on less than 20 pages each, is extremely cryptic, incomplete and therefore vulnerable to attack. This shows a significant deficit of the system. But I will also show that the additions to logic in Chapters 9 and 10 of WWR II, which Schopenhauer first published in 1844, cannot eliminate this shortcoming. On the contrary, the chapters from WWR II announced as additions indicate that Schopenhauer abandoned his original project of a logic based mainly on so-called 'analytical diagrams' or 'Eulerian diagrams' in the years after 1830 and instead favoured a different form of logic in later years, which he probably never elaborated. The chapters of WWR II announced as additions thus themselves become evidence of a deficit that can be found in the logic of WWR I in all published editions. Schopenhauer has only marginally revised the logic of WWR I in the second and third editions of the work in terms of rhetoric, but not in terms of content.

In addition, the shortcomings of Schopenhauer's logic and philosophy of language have also been reflected in the history of reception and in the research literature: according to the title of his 1979 paper, the Tübingen linguist Eugenio Coseriu made a relevant judgement after a five-page paper in which he listed "essentially everything" that "Schopenhauer had to say about language in general": *The Schopenhauer Case – A Dark Chapter in German Philosophy of Language.*⁴ Wolfgang Weimer confirmed this impression in his comparison of Schopenhauer's and Wittgenstein's philosophy of language,⁵ and even the rather benevolent dissertation *Schopenhauer als Logiker* of Adolf Kewe does not necessarily paint a better picture of Schopenhauer as a whole.⁶

(Aus dem handschriftlichen Nachlaß). In: Id.: SämmtlicheWerke, Vol. 1/10. Ed. by K. W. A. Schelling. Stuttgart 1861, pp. 1–201, here: p. 137–140.
² Cf. Jens Lemanski: Summa und System; ibid.: Vom Alles zum Nichts oder die Überwindung des dogmatischen Spinozismus in der Ethik Schopenhauers. In: Schopenhauer-Jahrbuch 90 (2009), pp. 19–44.
³ The extent to which this discussion of completeness has been conducted beyond Schopenhauer's lifetime can be seen, for example, very well in Paul Deussen's 1917 thesis of the "Completion of critical philosophy by Schopenhauer" (Paul Deussen: Allgemeine Geschichte der Philosophie mit besonderer Berücksichtigung der Religionen. Vol. II/3, pp. 376–443). Even in the 1960s Rudolf Neidert still tried to complete the ius resistendi missing in Schopenhauer's doctrine on natural law by bequest manuscript remains on the tyrannicidium (cf. Rudolf Neidert: Die Rechtsphilosophie Schopenhauers).
⁴ Eugenio Coseriu: Der Fall Schopenhauer. Ein dunkles Kapitel in der deutschen Sprachphilosophie. In: Integrale Linguistik. Festschrift für Helmut Gipper. Ed. by Edeltraut Bülow, Peter Schmitter. Amsterdam 1979, pp. 13–19.
⁵ Vide infra, Chapter 2.1.4.
⁶ Vide infra, Chapter 1.3.1.

1.3 The Status of Logic in WWR and WWR2

As Chapter 2 will show, only a few of his contemporaries have received Schopenhauer's logic, and even today, opinions about Schopenhauer's logic are ambivalent at best: Whereas many modern philosophers of language repeat classical prejudices about Schopenhauer or ignore him completely, only logicians of the last decades have re-emphasised Schopenhauer's value for individual areas of contemporary geometry and logic. What all of the disputes with Schopenhauer have in common, however, is their limited knowledge of Schopenhauer's theory of language, logic and geometry; almost exclusively they refer to paragraphs and sections of texts which Schopenhauer later considered in need of revision, although he never carried out and implemented his plans, corrections and modifications in publications.

Even today, Schopenhauer is still ridiculed by logicians, philosophers of language and analytical philosophers: if Kant or Hegel with their logical and semantic contributions offer a good basis for modernisation or their inclusion in current debates, Schopenhauer is only considered a bizarre marginal figure. Probably the only logician who had ever looked through Schopenhauer's complete oeuvre on logic before the mid-2010s was Albert Menne.[7] Although Menne wrote only a few sentences on Schopenhauer's logic, one of them, however, gives rise to strong hopes of revising Schopenhauer's image as a logician and philosopher of language: "Schopenhauer has an excellent command of the rules of formal logic (much better than Kant, for example)."[8]

But how does Menne reach this opinion? As will be shown in Chapter 1.3.2, Schopenhauer was already aware of a certain deficiency of individual paragraphs of WWR I at the time of publication of the first edition. For his *Berlin Lectures*, Schopenhauer also took WWR I as the textual basis, but due to the incompleteness of individual passages and the darkness of the overall approach, he revised it strongly in terms of argumentation, content and system. In contrast to the second and third editions of WWR I, which offer few additions compared to the first edition, these *Berlin Lectures* are therefore the only heavily revised and expanded version of his main work. In order to mark the textual basis as well as its extensions and modifications with an expression, I have decided to call the system of the *Berlin Lectures* 'WWR2'.

The extensions and modifications concern two main points: Firstly, the lectures relativise the impression of a linear and normative system, as Schopenhauer further intensifies the empirical aspect of WWR and relativises the relevance of parts of the system that previously seemed to be emphasised. Due to the modification of the system, on the one hand, and the partial relativisation of individual system parts, on the other hand, I will only name the essential differences between WWR and the system of WWR2 in Chapter 1.3.2. Since Chapters 1.3ff. are particularly concerned with

---

[7] On the reception of Schopenhauer's logic, vide infra, Chapter 2.2.5.
[8] Alfred Menne: Arthur Schopenhauer. In: Klassiker des philosophischen Denkens. Vol. 2. Ed. by N. Hoerster. 7th ed. München 2003, pp. 194–230, here: p. 201. I am grateful to Andrea Reichenberger for pointing out the Schopenhauer manuscript by Heinrich Scholz and the dissseration by Edith Matzun. This reception of Schopenhauer should be examined more closely elsewhere.

Schopenhauer's logic, I will largely restrict myself to the doctrine of reason and its context in this comparison.

As Chapter 1.3.3 will show, the logic of the original system (WWR I) is modified and expanded in WWR2. This is also the reason for Menne's above-mentioned opinion. Schopenhauer elaborates § 9 of WWR I, which comprises less than twenty printed pages, into a logic of almost 200 pages. For this reason, the term 'logica minor' is used in the following for the logic in the WWR (§ 9 of WWR I, Chapters 9 and 10 of WWR II) and the term 'logica major' for the approximately 200-page logic of WWR2. Since the logica minor in WWR I is almost exclusively a conceptual logic, Schopenhauer in WWR2 supplements this conceptual logic with a doctrine of judgments and inferences, which is introduced and concluded by approaches that can be attributed to the philosophy of language or the philosophy of logic. As Chapter 2 will show, the text passages on the philosophy of language and logic given in the logica major still contain today worthy of discussion, usable and profitable approaches to semantics, to the doctrine of judgement and also to proof theory; and the entire logica major is permeated by an independent preoccupation with geometric logic, which Schopenhauer links, above all, to the writings of Lambert, Ploucquet and Euler.[9]

1.3.1 The Logica Minor of WWR

If the terms 'logica major/ minor', which I introduced into Schopenhauer research, may at first associated with Hegel or Nietzsche,[10] this way of speaking does not go back to the 19th century, but to the scholastic era, in which bachelors had to complete a minor course on basics of logic (logica minor, parva logicalia) and magistrands a major course on logic (logica major, logica magna) aiming for completeness.[11] This was soon reflected in the names of the textbooks, which – in addition to the names mentioned – had titles such as 'Logica major' or 'Summa logicae' and thus differed from 'Logica minor/ brevis/ elementaris', 'Summulae' etc.[12] That both terms retained their qualitative connotation in the 18th and 19th centuries cannot only be seen from the wording of the early Hegelians but already from Gottsched's synonymisation of 'great logic' and 'extensive logic':

---

[9] Vide infra, Chapter 2.2.5.
[10] Cf. e.g. Carl F. Bachmann: Ueber Hegel's System und die Notwendigkeit einer nochmaligen Umgestaltung der Philosophie. Leipzig 1833, p. 103; Friedrich Nietzsche: The Case Wagner. In: The Complete Works, Vol. VIII. Ed. by Oscar Levy. Edinburgh, London 1911, pp. 1–53, here: p. 32 (= Chapter 10).
[11] Arno Seifert: Logik zwischen Scholastik und Humanismus. Das Kommentarwerk Johann Ecks. München 1978, esp. p. 14ff., also pp. 49ff.
[12] Cf. Leonhard Rabus: Logik und Metaphysik. Vol. 1: Erkenntnisslehre, Geschichte der Logik, System der Logik, nebst einer chronologisch gehaltenen Uebersicht über die logische Literatur und einem alphabetischen Sachregister. Erlangen 1868, pp. 196ff.

1.3 The Status of Logic in WWR and WWR2

> In my opinion, a great logic is as much of a nuisance to a beginner as a great grammar. For just as a person who knows the whole grammar by heart, and therefore does not yet have the language under control, so too is a person who has had an extensive logic explained to her or him for a whole year, and therefore is not yet a master of common sense.[13]

In my opinion, also the minor logic or *logica minor* of Schopenhauer cannot transform a beginner into a master of common sense: Much too short and much too cryptic, Schopenhauer presented an outline of individual themes of logic in the first part of § 9 in less than 20 printed pages in the first edition of WWR I. Even the additions in Chapters 9 and 10 of WWR II do not help to go beyond this shortcoming. The end of the 19th century has tried to turn this around in a positive way: Nietzsche's dictum "Schopenhauer the Simplifier" ("Schopenhauer der Vereinfacher")[14] was applied to logic by Kuno Fischer and Adolf Kewe. Attempts were made to reinterpret brevity and compactness as an advantage. In the following, I will illustrate the place and evaluation of logic in WWR I and then in WWR II, which together make up the 'simplified', minor logic. In the course of this analysis, I will not interpret against Kewe and Fischer the shortcomings of the logica minor as an original simplification but will argue that they are due to Schopenhauer's expectations of the addressees of WWR I and, furthermore, of WWR II itself.

The logica minor consists essentially of § 9 of the first edition of the WWR I and Chapters 9 and 10 of the first edition of the WWR II, both of which have been adopted largely unchanged in their respective later editions. § 9 is in B I of WWR I. As described in Chapter 1.2.3, this book is divided into two sections: It first deals with the understanding (§§ 3–7), including its cognitive faculties of space, time and causality, and then with the reason (§§ 8–16), including its faculties or "advantages" of language, science and practical reason. Logic (analytics) together with eristics (dialectics) form the subjects of § 9 of the WWR I, and both parts of the paragraph fall within the rational area of language. In the first German edition of WWR I, the first part of § 9, which deals with logic, is eight pages long;[15] the second part of § 9, which deals with dialectics, is nine pages long.[16]

---

[13] Johann Christoph Gottsched: Erste Gründe der gesammten Weltweisheit: darinn alle philosophische Wissenschaften in ihrer natürlichen Verknüpfung abgehandelt werden, zum Gebrauche academischer Lectionen. 2nd ed. Leipzig 1735, p. **3.
[14] Cf. the compilation of quotes in Adolf Kewe: Schopenhauer als Logiker. Bonn 1907, p. 92.
[15] In the english edition of WWR I, pp. 62–65.
[16] In the english edition of WWR I, pp. 62–72.

1 The World and its Representationalist Interpretation

The part on logic of § 9 consists of (1) a linguistic-philosophical introduction on two pages,[17] (2) excursory remarks on the concept of reflection on three pages[18] and treatises (3) on *abstracta* and *concreta*,[19] (4) on concept extension/ intension, (5) on possibilities of composing two concepts[20] or (6) three concepts,[21] and (7) on logical rules.[22]

In general, the first part of § 9 is a mere conceptual logic (3, 4, 5, 6) with several additions (1, 2, 7). Whoever interprets favourably reads a compositionalist logic on judgments in (5) and a similar doctrine of inferences in (6). My justification for describing the logica minor of WWR I as too short and much too cryptic, is at first based on the quantity: (5) consists (in the German edition) of just under three pages, (6) consists of only two sentences, (7) of only one sentence. As I have said, anyone who considers this to be a complete logic written at the beginning of the 19th century provides a very charitable interpretation. The following brief summaries of (1)–(7) show that quantity is reflected in quality:

(1) Philosophy of language: Schopenhauer begins in § 9 with reflections on the functioning of the concept and suggests that the concept is expressed in the three cognitive advantages or faculties of the human being (language, knowledge, practical reason) and can thus be experienced.[23] In addition, Schopenhauer takes up the issue of the relationship between conceptual and non-conceptual content and argues that a simultaneous translation of speech into purely intuitive or non-conceptual content is unusual. The speech is therefore similar to a perfect telegraph, which "communicates arbitrary signs with the greatest speed and the finest nuance", which in turn are interpreted just as directly by the recipient.[24]

(2) Reflection: Reflection is defined as a necessary "copy or repetition of the original intuitive world".[25] However, this relationship of reflection to the intuitive world does not happen directly but is mediated through (conceptual) intermediate levels. For this reason, reflexive concepts by Schopenhauer are also called "representations of representations".

(3) Abstracta/ Concreta: Schopenhauer uses this digression on reflection to lead to the first topic, which one would probably classify by a majority as a typical topic of logic of concept.[26] He classifies concepts according to their relation to intuition: concepts that refer directly to intuition are called *concreta* (e.g. 'person, 'stone', 'horse'); concepts that refer to intuition only through one or more concrete concepts are called

---

[17] In the english edition of WWR I, pp. 62–63.
[18] In the english edition of WWR I, pp. 63–64.
[19] In the english edition of WWR I, p. 64f.
[20] In the english edition of WWR I, pp. 65–67.
[21] In the english edition of WWR I, p. 67f.
[22] In the english edition of WWR I, p. 70.
[23] Cf. WWR I, pp. 62
[24] WWR I, p. 62.
[25] WWR I, p. 63.
[26] Cf. e.g. William Hamilton: Lectures on Metaphysics and Logic. 4 Vols. Ed. by H. L. Mansel, J. Veitch. London 1860, Vol. IV, p. 239.

1.3 The Status of Logic in WWR and WWR2

abstracta (e.g. 'relation', 'virtue', 'investigation'). Basically, all concepts are abstractions from intuition and are general. Even if only one single real object can be thought through them, concreta also retain the same generality as abstracta in the way they are used – a thesis on the basis of which Kuno Fischer described Schopenhauer as the 'most pronounced nominalist' in the succession of Bacon, Locke and others.[27] The distinction between *abstracta* and *concreta* serves, above all, to describe the topological position of concepts (c. superior, inferior, infimus, supremus etc.) within a primarily vertically organised scheme of concepts. Schopenhauer himself uses the picture of a conceptual edifice in which the abstracta occupy the upper stories, the concreta the ground floor.[28]

(4) Subordination: Schopenhauer discusses two metaphors whose use already seemed problematic in Kant's time and which have explicitly become a research topic in today's metaphysics and logic:[29] Subordination and comprehension or 'to fall/ subsumed under something' and 'to be contained in a scope/ be comprehended by a sphere'. The reference to the vertically arranged conceptual scheme in (3) serves as a starting point for Schopenhauer to discuss the first metaphor in (4). Following on from this, he explains the metaphors of comprehension and containment in (5). Concerning subordination, Schopenhauer makes it explicit that every concept is a "representation of representations", from which the condition of possibility (not as a constant factuality) is derived that several or many things fall under one concept. If we look at the Schopenhauerian system building of WWR I,[30] we can say that everything in this system falls under the concept 'world', but only the topics to be discussed here, i.e. (1) – (7), fall under the concept of 'logic'.

(5) Possibilities of composing two concepts: Schopenhauer sees as a consequence of the subordination metaphor, which implies a verticality of the conceptual scheme, the metaphors of comprehension or containment, which imply a certain semantic interpretation of a space:[31] Each concept has an extension or a sphere which, according to the speaker's intention, should contain at least one object, even if, regardless of the speaker's intention, several objects are always designated by the corresponding concept. The comparison of two conceptual spheres expresses the relationship between subject and predicate: "To recognize this relation is *to judge.*"[32] Schopenhauer names

---

[27] Cf. Kuno Fischer: Schopenhauers Leben, Werke und Lehre, p. 215 (= 5.2.1). This nominalism is discussed in more detail below in Chapter 2.2.6 and I will discuss it further in Chapter 3.2.1.

[28] Cf. WWR I, p. 64. Schopenhauer avoids the traditional term 'individua' for the lowest level of the concept, as he only attributes to terms the function of designating a quantitative particularity or generality (see below, chapter 2.2.5).

[29] Regarding logic vide infra, Chapter 2.2. Regarding metaphysics cf. e.g. Peter van Inwagen & Meghan Sullivan: Metaphysics (Art.). In: *The Stanford Encyclopedia of Philosophy* (Spring 2016 Edition). Ed. by Edward N. Zalta, Chapter 2.2.

[30] Cf. the tree diagram in Chapter 1.2.7.

[31] For the metaphors of comprehension or containment vide infra, Chapter 2.2.

[32] WWR I, p. 66f.

1 The World and its Representationalist Interpretation

five possibilities of composing two concepts and illustrates four of them with circle diagrams, so-called 'analytical diagrams' or 'Eulerian diagrams':[33]

1) "The spheres of two concepts are exactly equal."
2) "The sphere of one concept completely encloses the sphere of another." E.g. `Pferd` = Horse; `Thier` = Animal.

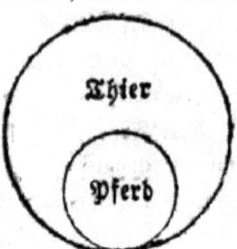

3) "A sphere includes two or more further spheres, which are mutually exclusive and at the same time exhaust the first sphere." E.g. `rechter Winkel` = right angle; `spitzer Winkel` = acute angle; `stumpfer Winkel` = obtuse angle.

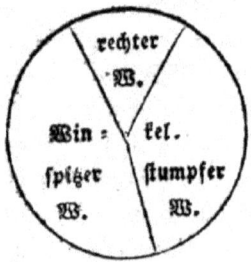

4) Two spheres each include a part of the other. E.g. `Blume` = Flower; `roth` = red.

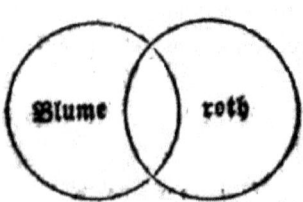

---

[33] For the historical context of these diagram forms vide infra, Chapters 2.2.2–2.2.4. Cf. also Amirouche Moktefi: Schopenhauer's Eulerian Diagrams. In: Language, Logic, and Mathematics in Schopenhauer. Ed. by Jens Lemanski. Cham 2020, pp. 111–129; Lorenz Demey: From Euler Diagrams in Schopenhauer to Aristotelian Diagrams in Logical Geometry. In: Language, Logic, and Mathematics in Schopenhauer. Ed. by Jens Lemanski. Cham 2020, pp. 181–205.

1.3 The Status of Logic in WWR and WWR2

5) Two spheres lie inside a third, but do not exhaust it. E.g. `Materie` = Matter; `Wasser` = Water; `Erde` = Earth.

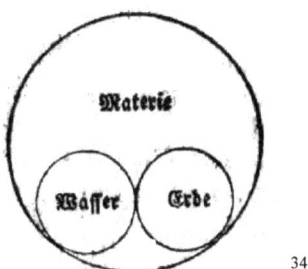

[34]

Following this quote, Schopenhauer explains: "All combinations of concepts may be reduced to these cases, and the entire doctrine of judgement [...] can be derived from them." [35] Thus, for Schopenhauer, the doctrine of judgement has essentially been reduced to a geometric logic of concepts.

(6) Possibilities of composing several concepts: These relational figures shown in (5) "can themselves be combined with each other in various ways", so that "[l]ong chains of syllogisms arise" (sorites); this shows that the geometric relational figures or diagrams are suitable "to ground the doctrine of judgement as well as the whole of syllogistic logic".[36]

(7) Rules of logic and laws of thought: Schopenhauer adds that it is not necessary to learn rules of inference, since they can be observed, deduced and explained ("einsehen, ableiten und erklären") from the "schematism of concepts" given in (5). In this respect there is only one sentence – namely in the second part of § 9 – which lists laws of thought and rules of logic within a historical digression:

> they [sc. people] gradually discovered more or less complete formulations for the fundamental principles of logic, such as the laws of non-contradiction, sufficient reason, and the excluded middle, the maxim of all and none [*dictum de omni et nullo*], as well as for the special rules of syllogisms, e.g. 'nothing follows from merely particular or negative premises' [*ex meris particularibus aut negativis nihil sequitur*], 'inference from the consequent to the ground is not valid' [*a rationato ad rationem non valet consequentia*], etc.[37]

---

[34] WWR I, p. 63f. (All Diagrams taken from the 1813 orginal German edition of WWR.)
[35] WWR I, p. 65.
[36] WWR I, p. 68.
[37] WWR I, p. 68, p. 71.

These briefly presented points (1)–(7) contain the central assertions in WWR I on the logic of concepts, judgements and inferences. I believe that my – albeit abridged – presentation of the first part of § 9 in seven points should justify my thesis that Schopenhauer's logica minor is far too short and cryptic and so in need of explanation that it cannot succeed in transforming a beginner into a master of common sense. How is it possible for a beginner in logic to make sense of what the rules and laws in (7) mean or how they could be derived from (5)?

Nevertheless, it is astonishing that, especially in recent times, there have been research approaches to individual points that have produced interesting and insightful results: For example, Sascha Dümig has concluded that, on the one hand, the analogy between speech and a telegraph in (1) is in line with modern cognitive processing models and representationalism in the sense of Jerry Fodor, and that, on the other hand, Schopenhauer represents an organon model that differs significantly from that of Karl Bühler.[38] Michał Dobrzański has seen in (3) anticipation of Tadeusz Kotarbińskis' Reism,[39] and Lorenz Demey has discovered in (5) anticipation of current logical geometry.[40]

Both interpretations, that of Dümig and that of Dobrzański meet my assertion that the structure of Schopenhauer's logica minor points to the principle of representationalist compositionality: Representationalism stands for the approach, particularly discussed in Chapter 1.2.3, of depicting the empirical-concrete world by means of logic using abstract concepts. The intuition of the empirical world forms the scientific basis of all rational knowledge. Compositionality is shown, above all, by the logical structure of the world grounded in atomic parts. The smallest units are the objects or phenomena of intuition, which are represented by *concreta* and *abstracta* and represented geometrically in analytical or Eulerian diagrams: Several concepts form five relational figures, from which the whole logic of judgements and inferences is then to be constructed. The affinity to representationalism and compositionality that can be seen here with Schopenhauer can be read out of Kotarbińskis Reism and found explicitly in Fodor.[41]

The hypothesis of an affinity to representationalism and compositionality is also confirmed by the interpretation given at the beginning of this chapter by Fischer and Kewe, who transferred Nietzsche's word of "Schopenhauer the simplifier" to logic

---

[38] Cf. Sascha Dümig: Lebendiges Wort? Schopenhauers und Goethes Anschauungen von Sprache im Vergleich. In: Schopenhauer und Goethe. Biographische und philosophische Perspektiven. Ed. by Daniel Schubbe, Søren R. Fauth. Hamburg 2016, pp. 150–183; Sascha Dümig: The World as Will and I-Language. Schopenhauer's Philosophy as Precursor of Cognitive Sciences. In: Language, Logic, and Mathematics in Schopenhauer. Ed. by Jens Lemanski. Basel 2020, pp. 85–95.

[39] Cf. Michał Dobrzański: Begriff und Methode bei Arthur Schopenhauer. Würzburg 2017, pp. 292–295; Jens Lemanski & Michał Dobrzański: Reism, Concretism and Schopenhauer Diagrams. In: Studia Humana 9:3/4 (2020), pp. 104–119 (also in: Judgments and Truth: Essays in Honour of Jan Woleński. (Tributes, Vol. 43). Ed. by Andrew Schumann. London 2020, pp. 105–131).

[40] Cf. Lorenz Demey: From Euler Diagrams in Schopenhauer to Aristotelian Diagrams in Logical Geometry.

[41] Cf. e.g. Jerry A. Fodor, Ernest LePore: The Compositionality Papers. Oxford 2002.

1.3 The Status of Logic in WWR and WWR2

and thus interpreted Schopenhauer's logic as a unique and original form of reductionism. Compositionality and reductionism express the same thing, but look at the structure of logic from different perspectives: From an atomistic point of view, logic is constructed bottom-up. Concepts emerge from intuition, concepts form judgements, inferences are composite judgements and theories contain inferences. From a holistic point of view, logic is structured top-down. Theories consist of inferences, inferences can be used to derive judgements, concepts are reduced judgements, and intuition may be elements of concepts. What was interpreted by using the term 'compositionality' esp. in Dümig's study can be interpreted as reductionism regarding to Fischer's and Kewe's studies. The logica minor of WWR I is thus either an approach reduced to logic of concept or a compositionalist approach based on a logic of concept. In any case, it has not yet been shown why or how Schopenhauer's representationalism, when placed so close to authors like Fodor or Kotarbiński, can be understood as rational or non-naïve.

Regardless of whether one interprets the representationalism presented so far as causal or non-causal, comparisons with the two representationalists mentioned above show – and this is far more serious than other issues – that Schopenhauer's logica minor of the WWR I is incomplete: Either it does not explain what is supposed to be reducible and how exactly it is reducible (judgements, inferences), but only what it can be reduced to (five possible relations); or it does not explain how exactly judgement and especially inference compositions look like, which are to be composed of the conceptual spheres and relational figures. In both cases, the logic of judgement and especially the logic of inferences is very incomplete and can be the downfall of a system that claims to be complete.

This incompleteness becomes even more obvious when comparing the basic possibilities of relations of concepts in judgements in Schopenhauer's logica minor with those of Euler, Lambert or other geometric logicians of the 18th and 19th centuries. There is no need to make such a comparison in detail here because if one places only the four basic forms of judgement reproduced in Chapter 2.2.3 (Fig. 10) next to the five possible ratios of concepts in Schopenhauer's logic of judgements, it is striking that almost only affirmative judgements can be formed with § 9 of WWR I.[42] With Euler, Lambert and many other logicians in the 18th and 19th centuries, however, negative judgements can also be represented geometrically. As Chapter 1.3.3 will show, Schopenhauer recognised this shortcoming of WWR I, later on eliminated it in his *Berlin Lectures*, but did not correct it in the second and third editions of WWR I for the public audience.

---

[42] Lorenz Demey does not interpret the diagrams according to the five judgments of Schopenhauer quoted above, but also reads out negative or oppositional relations that the diagrams indicate. Schopenhauer does this only in the *Berlin Lectures*, WWR2.

1 The World and its Representationalist Interpretation

At least with the publication of the second edition of WWR I and the first edition of WWR II (1844) Schopenhauer seems to have tried to fill some other gaps in the doctrine of judgements and inferences. WWR II announces in Chapter 9 a treatise "On Logic in General" and in Chapter 10 a "Study of Syllogisms". Chapter 5, "On the intellect in the Abscence of Reason", already contained an introductory note in which Schopenhauer explained: "This chapter [sc. 5], together with the next, relates to §§ 8 and 9 of the First Volume [sc. WWR I] [...]."[43]

In what follows, I will outline the extent to which the above seven elements of § 9 of the WWR I are taken up again and supplemented in WWR II. The main focus here is not on a complete summary of all topics but on the modifications and additions of WWR II compared to WWR I.

(1) Philosophy of language: As in § 9 of WWR I, Schopenhauer addresses in WWR II the function of language from an anthropological perspective (Chapter 5), the relationship between intuition and concept (Chapter 7) as well as the abstraction of the concept from intuition and its relationship to images and words. Schopenhauer emphasises, above all, his connection with the Aristotelian and Lockean doctrine of concepts.[44]

(2) Reflection: The term 'reflection' is not as present in the second part of B I of WWR II as it is in § 9 of WWR I. Only at the beginning of Chapter 7, Schopenhauer contrasts reflection and intuition.

(3) Abstracta/ concreta: As mentioned in (1), Schopenhauer emphasises the abstraction of concepts from intuition. The most important passage on this pair of opposites is found at the beginning of Chapter 6, in which Schopenhauer claims

> that concepts form a stepwise progression, a hierarchy from the most specific to the most general, at the bottom of which scholastic realism could almost be correct, and at the top, nominalism. The most specific concept is practically individual, and thus practically real: and the most universal concept, e.g. being (i.e. the infinitive of the copula) is practically no more than a word.[45]

Schopenhauer uses this empirical logic of concepts at several points in his late work as a criterion for criticising philosophies that have a quantitatively strong proportion of abstract elements and derive *concreta* from them.[46]

---

[43] WWR II (1844), p. 65.
[44] Cf. WWR II (1844), p. 87ff. Schopenhauer's doctrine of concepts may have been particularly inspired by his teacher Gottlob Ernst Schulze, who saw the Locke tradition continued through Hume and followed it.
[45] WWR II (1844), p. 70.
[46] Cf. e.g. WWR II (1844), p. 89, pp. 90f. (Chapter 7). As Jacob Mühlethaler: Die Mystik bei Schopenhauer, p. 10, p. 38, p. 52f., p. 54 rightly states, in Schopenhauer's published work the mostly objective criticism of the early years only gave way from around the 1840s onwards to the well-known invectives and pejorations that were unfortunately all too often highlighted in the history of reception.

(4) Extension/ intension: The fact that in (3) Schopenhauer again introduces a conceptual scheme organized around spatial metaphors with a vertical connotation ("stepwise progression", Stufenfolge) is justified by the law of reciprocity, which is a fundamental element of the logic of concept: "Since, moreover, the content of concepts is inversely proportional to their scope, and the more that is thought under a concept, the less is thought in it".[47]

(5) Possibilities of composing two concepts: The text passages to be interpreted as a logic of judgement are found at the beginning of Chapter 9 and in the middle of Chapter 10:[48] Schopenhauer reflects in Chapter 9, on the one hand, the function of the copula 'is', 'is not' and, on the other hand, on logical connectives ("Logische Partikel" = 'logical particles'). He then discusses the differences of forms of judgement (general, particular, singular, ...) based on their supposed 'quantifiers' (all, some, ...), but reduces all quantitative forms of judgement to the general judgement.[49] Thus WWR II shows a similarly strong nominalism as Fischer had also attested for WWR I. In Chapter 10, Schopenhauer focuses on the central role of judgements regarding inferences and concepts. In contrast to WWR I, however, it is not the 'connection' or 'composition' but rather the 'comparison' that is emphasised as the central act of reasoning: The judgement is a comparison of concepts, the inference a comparison of judgements. Schopenhauer then outlines the role of subject and predicate in judgements.[50]

(6) Possibilities of composing three concepts: The syllogistic announced with the title of Chapter 10 is only interrupted by the last mentioned excursus on the doctrine of judgement discussed in (5).[51] Schopenhauer defines inferences at the beginning and classifies inferences into knowledge-expanding, knowledge-preserving, explicit, implicit, latent, liberated, and bound. After an excursus on the logic of judgement, he relativises his approach from WWR I, which is based on Eulerian diagrams: Inferences can be thought of as consisting of three concepts but should be better thought of as consisting of three judgements. Only the focus on judgment explains the typicality of the three syllogistic figures, which Schopenhauer explains on several pages. Although he uses two Euler diagrams to explain a contraposition in the third figure, he suggests that it would be better to symbolise inferences with sticks and hooks ("Stäbe und Haken").

---

[47] WWR II (1844), p. 70. Vide supra, Chapter 1.1.2. Here Schopenhauer takes up the law of reciprocity, which was known by Kant, cf. Jäsche Logic, § 7 (AA IX, pp. 95.31–33 = Lectures on Logic. Ed. by J. Michael Young. Cambridge 1992, p. 593), and later passed on by Bolzano and Frege. Cf. Rico Hauswald: Umfangslogik und analytisches Urteil bei Kant. In: Kant-Studien 101:3 (2010), pp. 283–308; Peter McLaughlin, Oliver Schlaudt: Kant's Antinomies of Pure Reason and the 'Hexagon of Predicate Negation'. In: Logica Universalis 14 (2020), pp. 51–67; Stefania Centrone: Der Reziprozitätskanon in den Beyträgen und in der Wissenschaftslehre. In: Zeitschrift für philosophische Forschung 64:3 (2010), pp. 310–330.
[48] Cf. WWR II (1844), pp. 111–114, pp. 117–125.
[49] Vide infra, Chapter 2.2.5.
[50] Cf.. WWR II (1844), p. 117.
[51] Cf. WWR II (1844), pp. 115–117, pp. 119f.

(7) Logical rules and laws: Before Schopenhauer moves on to the logic of judgement in Chapter 9, he discusses the reduction of all rules and laws of thoughts to the laws of the excluded middle and of sufficient reason, which he both explains, on the one hand, employing the metaphors of circumference and containment ("conceptual spheres") and from which he derives, on the other hand, a theory of conceivability and of truth.[52] Individual rules of deduction and contraposition are also discussed in the second part on syllogistic in Chapter 10 (6).

Even though this overview has only roughly presented the topics and only a few individual theses of the logica minor given in WWR II, it can be said that, above all, the relativization of Eulerian diagrams described in (6) and Schopenhauer's associated suggestion of preferring to use sticks and hooks is quite surprising. After all, by using the sticks and hooks for the inferences discussed in WWR II, Schopenhauer rejects the treatise on logic of WWR I, which was built up from the doctrine of concepts with the help of Eulerian diagrams. What diagrams or notations Schopenhauer exactly had in mind, however, remains a mystery to this day, as they are described in only a few sentences in Chapter 10 of WWR II, but not illustrated:

> When presenting the study of syllogisms using *conceptual spheres*, we picture them as circles. Similarly, when we use entire judgments, we picture them as sticks that for the sake of comparison are sometimes held together at the one end and sometimes at the other. The different ways in which this can take place result in the three figures. Now since each premise contains a subject and a predicate, these two concepts can be presented as located on the ends of each stick. [...]
>
> We can do this [sc. finding the terminus medius] if we think of the premises as two sticks and the concept as a hook that joins the sticks together: in fact sticks like this could be used during a lecture. By contrast, what distinguishes the three figures from each other is that[53]

Kewe has tried to interpret the allusions to sticks by combining them with Schopenhauer's parable of inferences as a Voltaic pile. The result of this interpretation is the diagram given in Fig. 2. Schopenhauer had given this parable at the end of Chapter 10: The "point of indifference in the middle represents the terminus medius, the two poles the disparate concepts; there the spark jumps out by connecting the wires, here the conclusion by focussing on the copula of the judgements".[54]

---

[52] Cf. WWR II (1844), pp. 111ff.
[53] Cf. WWR II (1844), p. 118, pp. 123f.
[54] Adolf Kewe: Schopenhauer als Logiker, p. 43.

1.3 The Status of Logic in WWR and WWR2

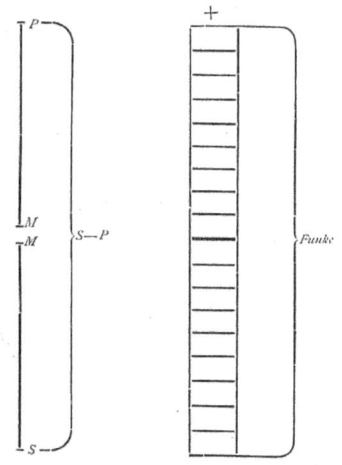

Fig. 2
Adolf Kewe: Schopenhauer als Logiker. Bonn 1907, p. 43.
(Funke = spark)

Kewe's interpretation may shed some light on this subject, but the Voltaic pile diagram also seems to me to be inadequate: on the one hand, it does not take up Schopenhauer's image of the hook, and on the other, Kewe's diagram fulfils no other function than that of the classical *pons asinorum*, whose sole task is to find the *terminus medius* in a syllogism.[55] Schopenhauer, as can be read from the few sentences given, is less concerned with the invention of the *terminus medius* than with the 'sensualization of the terminus medius in the premises' and, above all, with the representation of inferences including "entire judgments". An example of a sensual or diagrammatic representation of the syllogistic figures according to the arrangement of the terminus medius (the later so-called 'W-scheme' or 'syllogistic collar model')[56] would be Lange's 'Chirotecas'[57] published in 1712 (Fig. 3). What Schopenhauer has in mind, however, seems to be more in accordance with Krause's impressive legacy of logic manuscripts, in which 'sticks and hooks' are also used to

---

[55] Vide infra, Chapter 2.2.2.
[56] Cf. e.g. Alfred Swinbourne: Picture Logic. Or, The Grave Made Gay; An Attempt to Popularise the Science of Reasoning by the Combination of Humorous Pictures with Examples of Reasoning Taken from Daily Life. 2nd ed. London 1875, pp. 118f.: "The four figures may be remembered by the front of a collar. [...] The figures are thus easily remembered; \|||/, these lines being taken from the position of the middle term [...]."

The Front of a Collar.
Fig. 5
Alfred Swinbourne: Picture Logic. Or, The Grave Made Gay. 2. ed. London 1875, p. 118.

[57] Iohannes Christianus Langius: Nvclevs Logicae Weisianae. [...] illustrates [...] per varias schematicas [...] ad ocularem evidentiam deducta [...]. Editus antehac Avctore Christiano Weisio. Gissae-Hassorum 1712, p. 175.

represent valid inferences (Fig. 4).[58] This notation of Krause's was also brought into a close relationship with Frege's *Begriffsschrift* by numerous researchers.[59]

Fig. 3
Iohannes Christianus Langius: *Nvclevs Logicae Weisianae. Editus antehac Avctore Christiano Weisio.* Gissae-Hassorum 1712, p. 175.

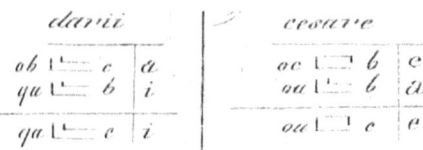

Fig. 4
Karl Christian Friedrich Krause: *Die Lehre vom Erkennen und von der Erkenntniss, als erste Einleitung in die Wissenschaft.* Ed. by Hermann Karl von Leonhardi. Göttingen 1835, Appendix, Table V, p. 128f.

Although it is known that Krause and Schopenhauer knew each other well in the 1810s,[60] this is not the place to pass judgement on the extent to which Krause may have influenced Schopenhauer's late logic (or vice versa). Many more preliminary works and studies are needed to judge Schopenhauer's late logic. Nevertheless, in my opinion, even at first glance Krause's diagrams fulfil more the function of a representation of inferences including "entire judgments" and contain more the elements of representation with sticks, lines or hooks of Schopenhauer's description than Kewe's diagram of the Voltaic pile. However, it is not at all decisive whether Schopenhauer's descriptions fit more on Kewe's or Krause's diagrams, but rather that Schopenhauer's descriptions, on the one hand, no longer harmonise with the circle diagrams of WWR I and that, on the other hand, according to the current state of knowledge, he has not even elaborated his new programme based on sticks and hooks.

On the basis of these non-complementary representations of logic in WWR I and WWR II, one can conclude that no beginner can become a master of common sense through these two logics. The logic in both volumes of WWR remains enigmatic and cryptic. This can also be seen in the fact that Schopenhauer's reductionist approaches

---

[58] Cf. Karl Christian Friedrich Krause: Die Lehre vom Erkennen und von der Erkenntniss, als erste Einleitung in die Wissenschaft. Ed. by Hermann Karl von Leonhardi. Göttingen 1835, pp. 199ff.
[59] Cf. Lothar Kreiser: Gottlob Frege. Leben – Werk – Zeit. Hamburg 2004, Chapter 3.2. Among others, Kreiser's judgment is followed by Claus Dierksmeier: Der absolute Grund des Rechts. Karl Christian Friedrich Krause in Auseinandersetzung mit Fichte und Schelling. Stuttgart-Bad Cannstatt 2003, p. 11; Danielle Macbeth: Frege's Logic. Cambridge 2009, p. 186, Fn. 3.
[60] Cf. Benedikt Paul Göcke: Karl Christian Friedrich Krause Einfluss auf Arthur Schopenhauers "Die Welt als Wille und Vorstellung". In: Archiv für Geschichte der Philosophie 103:1 (2021), pp. 148–168; Benedikt Paul Göcke: The Panentheism of Karl Christian Friedrich Krause (1781–1832). From Transcendental Philosophy to Metaphysics. Berlin 2018, Chapter 10. As one might see above, in Chapter 1.2, I do not agree with Göcke's interpretation of Schopenhauer in many parts, but I believe that there are nevertheless parallels between the two philosophers.

of (5) and (7) agree with the image of Fischer and Kewe that Schopenhauer is a simplifier – but his simplifications remain in need of explanation. Although the compositionalist image I have read out of Dümig's and Dobrzański's studies has also proved true in WWR II, why Schopenhauer replaces the conceptualist basis of § 9 of WWR I with a propositionalism in Chapter 10 or emphasises 'comparison' instead of 'connection' and 'composition' seems confusing. If, in addition, Schopenhauer uses logic metaphors of containment and circumference in (4), (6) and (7), does this mean that the logic of concepts and the logic of judgement based on them should be built up primarily by Eulerian diagrams, but the logic of inferences should be built up with sticks and hooks?

These open questions suggest that the logica minor of WWR I and WWR II do not provide a complementary picture. Rather, it seems that Schopenhauer wanted to revise his original approach of WWR I, but without implementing the revision in the second and third editions of WWR I, since § 9 of WWR I contains only slight modifications in the later editions. Similar to Kuno Fischer, Robert Schlüter, Arthur Lovejoy and others who had already argued against the fairy tale of a missing development of Schopenhauer and in favour of a revised doctrine in the whole oeuvre, especially regarding metaphysics and philosophy of nature, this can also be proven by logic: Conceptual spheres or Eulerian diagrams, to which the entire logic was reduced in § 9 of WWR I, would – as Schopenhauer then emphasises in Chapter 10 of WWR II – make syllogisms "easy to grasp [leicht faßlich]", but this "comprehensibility [Fasslichkeit] comes at the cost of thoroughness".[61] If one were even to apply strict standards of interpretation, one could speak not only of a revised doctrine but also of an incoherent picture and thus of a significant weakness of Schopenhauer's logic in his published works. Not only does Schopenhauer seem to have left behind a dark chapter in the German philosophy of language, as Coseriu had claimed, but its shadow also reaches into the depths of logic.

1.3.2 The System of WWR2

In Chapter 1.2.2, the trivial thesis was put forward that books such as WWR express an intention, that they name a goal, that they did not come into being out of nowhere and that they are written for a certain circle of addressees. From Chapter 1.2.3 onwards, I have argued that the purpose of WWR is not limited to the communication of a single thought, but that it is rather about the complete reflection of the concrete world in abstract concepts. Schopenhauer published the original system concept of WWR I twice during his lifetime after 1819, but only once did he actually revise it significantly. Although the second and third editions of WWR I announce in 1844 a

---

[61] WWR II (1844), pp. 117f.

"thoroughly improved and very increased" ("durchgängig verbesserte und sehr vermehrte") and in 1859 a "considerably increased" ("beträchtlich vermehrte Auflage") edition respectively, a substantial revision of the text of WWR I can only be found in the *Berlin Lectures* of the 1820s, published posthumously. They alone deserve to be regarded as a modified system design (WWR2).

The lectures have been edited several times: Parts were published by Frauenstädt and Grisebach as early as the 19th century; Franz Mockrauer presented the first complete edition in 1913 as part of the work edition edited by Paul Deussen, which was reproduced in modern print by Volker Spierling in the 1970s.[62] Apart from a few reviews, none of these editions has received closer attention from the research community. This is partly because the revision of the WWR in the *Berlin Lectures* has so far been evaluated very similarly in research. Among other things, these lectures have been perceived as inconsistent, since, on the one hand, they are purely textual adoptions of the WWR I system, but, on the other hand, they contain significant modifications that do not always correspond to the familiar system of WWR I. Heinrich Hasse, one of the first reviewers of the *Berlin Lectures*, captured this heterogeneous impression of WWR2 in words as follows: In WWR2 "we are, to a certain extent, facing a new work", but WWR2 follows WWR I from 1819 "in many cases to the letter, often down to the details of the wording and sentence structure".[63] The new character of the work, despite a similar choice of words and a similar systematic structure, is shown, above all, by the fact that WWR2 is a document of Schopenhauer's revised doctrine.

Hasse thus intensifies the positions of Schlüter, Fischer and others, who had already spoken out against the fairy tale of a missing development by Schopenhauer. Following Rudolf Lehmann, Hasse, too, had already argued in his book *Schopenhauers Erkenntnislehre als System* (*Schopenhauer's Epistemology as a System*) for the fact that many aporetic text passages can already be found in WWR I of 1819, but only become evident in his later work.[64] Hasse's review takes up this thesis again by showing both connections and modifications of WWR2 as opposed to WWR I. This results in two periods in Schopenhauer's oeuvre: An earlier phase, which includes in particular WWR I (1819) and the *Berlin Lectures* (WWR2), and a second, later phase of work starting from the second edition of WWR I and the publication of *Parerga and Paralipomena*.[65]

---

[62] For the history of the edition, see WWR2 I, pp. VI–XXXII [Mockrauer's preliminary remark]; Volker Spierling: Zur Neuausgabe. In: Arthur Schopenhauer: Theorie des gesamten Vorstellens, Denkens und Erkennens. Philosophische Vorlesungen Teil I. Aus dem handschriftlichen Nachlaß. Philosophical Lectures Part I. From the handwritten estate. Ed. by Volker Spierling. München et al. 1986, pp. 11–14.
[63] Heinrich Hasse: [Review of:] Schopenhauer, Arthur. Handschriftlicher Nachlass: "Philosophische Vorlesungen". Arthur Schopenhauers sämtliche Werke. Hrsg. v. Paul Deussen. München 1913. Vol. IX und X. In: Kant-Studien 19 (1914), pp. 270–272.
[64] Cf. Heinrich Hasse: Schopenhauers Erkenntnislehre als System einer Gemeinschaft des Rationalen und Irrationalen. Ein historisch-kritischer Versuch. Leipzig 1913, pp. 77ff.; Rudolf Lehmann: Schopenhauer. Ein Beitrag zur Psychologie der Metaphysik. Berlin 1894, pp. 107, Fn.
[65] Cf. Heinrich Hasse: [Review of:] Schopenhauer, Arthur. Handschriftlicher Nachlass.

1.3 The Status of Logic in WWR and WWR2

The two editors of the complete lecture editions, Deussen and Spierling, also acknowledged the "high value" that Hasse attested WWR2 had for Schopenhauer's intellectual development but relativised his impression of a "new work". Deussen explains that Schopenhauer reworked WWR to WWR2 in the winter of 1819/20 and added to it again about a year later. For this reason, too, one could not expect WWR2 "to have the same smoothness and perfection as the works Schopenhauer prepared for print"; rather, Schopenhauer had endeavoured with WWR2 to present WWR in "a more popular form calculated for the intellectual capacity of young students".[66] Similar to Deussen, Mockrauer says about WWR2 that it contains a system "reworked for beginners, often literal, generally broader repetition of well-known Schopenhauer sentences".[67] The last editor, Volker Spierling, has also confirmed this impression: "Schopenhauer's Berlin Lectures are a didactic version of the first volume of his [sc. Schopenhauer's] main work, the *World as Will and Representation*."[68] Thomas Regehly has recently confirmed these views in an article.[69]

I do not share this impression. On the contrary, one can read my Chapters 1 and 2 as a plea to see WWR I as a popular, didactically processed version of a philosophical system which has only grown in WWR2 to a certain academic maturity. This can be seen in the recognisable descriptive objectives and in the elaboration of significant topics for academic philosophy, for example in the theory of reason. By focussing on the WWR2 and especially on the revised logic therein, for example, Schopenhauer's suggestion that was still characterised as a triviality and as a joke in Section 1.2.2 can be reinterpreted as a serious statement: The WWR I was written for a broad and educated middle-class readership. In contrast, WWR2 limits the circle of addressees to an academic audience. WWR2 aims – following Gottsched's above-given formulation – to transform a new student into a master of common sense.

Before I go on to discuss this difference between WWR I and WWR2 in the further course of Chapter 1.3.2, which particularly concerns the elaboration of the logica minor discussed in Chapters 1.2.3 and 1.3.1 into a logica major, I would like to briefly discuss some system-related modifications that WWR2 offers in contrast to the system parts presented in Chapter 1.2. Since a detailed comparison between WWR I (1819) and WWR2 would lose sight of the logic too far, only those references that relate to the research questions discussed in Chapter 1.1 and the context of the logic part will be highlighted in the following. Those who are interested in the relationship between WWR and WWR2 beyond these two points should first refer to the survey article by Thomas Regehly and then to the dissertation by Salomon Levi.[70] As the

---

[66] WWR2 I, p. VI [Preface Deussen].
[67] WWR2 I, p. VII [Preface Mockrauer].
[68] Volker Spierling: Zur Neuausgabe, p. 11.
[69] Cf. Thomas Regehly: Die Berliner Vorlesungen.
[70] Cf. Thomas Regehly: Die Berliner Vorlesungen; Salomon Levi: Das Verhältnis der 'Vorlesungen' Schopenhauers (hrsg. von P. Deussen Bd IX u. X) zu der 'Welt als Wille und Vorstellung'. Gießen 1922. However, as Regehly correctly points out, a full comparison between WWR and WWR2 is still lacking.

following remarks, based primarily on epistemology, demonstrate, Schopenhauer relocates individual text sections in WWR2 from WWR I (1819) and adds many topics; nevertheless, the basic structure remains roughly the same, and in some cases, it even stands out more clearly than in the published main work.

The connection between WWR I (1819) and WWR2 is already evident in the title of the lecture "Lecture on the whole philosophy, i.e. the doctrine of the essence of the world and the human spirit. In four parts". As the concept of the world in the title of WWR suggests and as it was explained there in § 15,[71] the main title here, in WWR2, also has a quantitative connotation: it is about the *whole* of philosophy, and this consists – according to the subtitle announced with "i.e." – of two doctrines, namely the doctrine about the essence of the world (will) and the doctrine about the human mind (representation). The parallelism between the titles of WWR I and WWR2 reinforces the Blumenbergian thesis of the empirical-representationalist readability metaphor of the world discussed above, in Chapter 1.2.3: the entire philosophy should be a repetition of the world.

Similar to WWR I, WWR2 is divided into four parts according to its subtitle that are abbreviated below as 'P', analogous to the abbreviation 'B' (for 'books') in WWR I and II:

> (P1) "The Theory of the Whole Representation and Cognition",
> (P2) "Metaphysics of Nature",
> (P3) "Metaphysics of the Beautiful" and
> (P4) "Metaphysics of Morals".

As in many other parts of WWR2, a more system-related reference to Kant can already be seen in the choice of the title than in WWR I: P1 takes the place of the critical writings, whereas P2 and P4 are intended to occupy the systemic positions of "metaphysics both of nature and of morals",[72] which were derived from Kant in 1787, and these are also to be supplemented by a metaphysics of the beautiful in P3. From the titles, it is also clear that four parts of WWR2 correspond thematically to the four books of WWR I.

Schopenhauer reflected on this connection of his metaphysics to Kant as well as to Aristotle at the beginning of P2.[73] It can be seen that P1 is intended to replace Kant's CpR or represents a 'new organon' in the Baconian tradition, whereas P2–P4 build the canon of the system.[74] In the Kantian sense, Schopenhauer's metaphysics also coincides with transcendental philosophy, but shows that "we do have data for the

---

[71] Vide supra, Chapter 1.2.3.
[72] Cf. CpR, p. 26, p. 700f. (= B XLIII, B 878).
[73] Cf. WWR2 II, pp. 15–20. Cf. also WWR2 I, pp. 70ff.
[74] Cf. Ulrike Santozki: Die Bedeutung antiker Theorien für die Genese und Systematik von Kants Philosophie. Eine Analyse der drei Kritiken. Berlin 2006, p. 26, pp. 64ff.; Sonia Carboncini & Reinhard Finster: Das Begriffspaar Kanon-Organon. In: Archiv für Begriffsgeschichte 26 (1982), pp. 25–59, esp. pp. 55ff.

1.3 The Status of Logic in WWR and WWR2

cognition of the inner essence of the world", namely the 'double cognition'.[75] The questions connected with this, to what extent Schopenhauer now redefines the border between immanence and transcendence in comparison to Kant, or how he answers the Kantian question "What can I know?", I will only take up again later, especially in Chapter 2.3.

One major revision in contrast to WWR concerns the beginning and end of WWR2: Schopenhauer does not begin with the quasi-quotation "The world is my representation" in P1, nor does he end with the concept of 'nothing' in P4.[76] Although the two relevant expressions are also used in WWR2, Schopenhauer, interestingly enough, does not place them as prominently at the beginning or end of the book as he did in WWR I. Instead, the first sentence of Chapter 1 in P1 deals with "perfect philosophical prudence", which is the prerequisite for an adequate "cognition of the essence of the world".[77] The last chapter of P4, on the other hand, takes up the concluding question of freedom of will and thus reflects on the choice between the two basic ethical principles, affirmation and negation.

The restructuring of the beginning and the end, I believe, undermines the credibility of any strictly normative, linear or architectural reading of the system.[78] It does not seem justified to read the beginning of WWR2 as an indication of idealism and the end as a sign of normative nihilism or the like. Rather, the newly inserted or arranged sentences of WWR2 clarify the metaphilosophical and thus descriptive standpoint of representationalism, which was obscured in WWR I by the prominent occupation of text passages.

How decisive these transpositions of expressions or the restructuring of the beginning and end of WWR2 are for the different interpretations can be seen, for example, in the above-mentioned readings of the editors and reviewers. Deussen, who on the one hand, proclaimed the completion of Kant's idealism in Schopenhauer's philosophy,[79] on the other hand, sees in WWR2 only "a more popular form calculated for the intellectual capacity of young students". If, on the one hand, Schopenhauer in WWR2 adheres exactly to the wording and structure of WWR in many cases, how could he, on the other hand, have presented a popular form of it? What exactly makes WWR2 a popular introduction for students in contrast to WWR?

The reviewers and editors do not give a satisfactory answer to these questions. Hasse, for example, tries to justify the thesis of popularisation partly by the growth in "scope and detail" of the expanded parts of the system; however, as the main argument for his impression, he feels Schopenhauer's striving for "the greatest possible understanding between himself [sc. Schopenhauer] and his audience" and also refers to the "tone of personal warmth" that Schopenhauer displays towards students. However, as

---

[75] WWR2 II, p. 19.
[76] Vide supra, Chapters 1.2.3 and 1.2.6.
[77] WWR2 II, p. 113.
[78] Vide supra, Chapter 1.1.3.
[79] Vide supra, Chapters 1.1.3 and 1.3 (Introduction).

will be illustrated in Chapters 1.3.3 and 2.3 by the example of Schopenhauer's logic, the extended parts of the system, in particular, are devoted to academic topics rather than popular ones. As will be shown, for example, in Chapter 2.3, the complexity of proof theory itself goes beyond the sophisticated logics of the first half of the 19th century, such as Kant, Krause, Bolzano or Drobisch. (Which of course does not mean that Schopenhauer's logic is better or more understandable than those.) Finally, the thesis put forward by critics and editors that comprehensibility can be achieved through conceptual and argumentative precision does not seem to me to be popular, but rather the – albeit not always implemented – concern of academic philosophical writings, both today and then.

Due to the unsatisfactory answers of the editors and reviewers to the question of their impression of whether WWR2 is a popular introduction for beginners compared to WWR I, I believe that they have swapped the normative reading of WWR I for a didactic reading of WWR2. This may be because Schopenhauer actually makes some didactic remarks about the course of his lectures in the opening passages. Now that – from the point of view of the linear reading of WWR2 – didactics has been placed before a no longer prominent idealism, the work has been declassified as a work suitable only for beginners and not worth reading for supposed experts. Only Thomas Regehly has tried to read both the didactic thesis and the normative reading into the text of WWR2.

In fact, the editions of WWR2 published so far begin with an "exordium on my [sc. Schopenhauer's] lecture and its method", in which the metaphors of unity and the organic – as in the preface to WWR I –[80] also appear, but in which Schopenhauer reflects much more extensively than in WWR I on the structure of the lecture and its most important passages.[81] In the following "Introduction, on the study of philosophy" Schopenhauer explains that he could not assume "that most of you [sc. of the listeners present] have already been particularly interested in philosophy".[82] Although philosophy should not be confused with its history, which is a good prerequisite for it, Schopenhauer first presents a short history of philosophy before the actual beginning of P1. This structure, which does not announce any prerequisites in terms of content, may have been the reason why WWR2 has been declared a didactic variant of WWR I until now.

Schopenhauer divides P1 into five chapters. After an introductory Chapter 1, in which Schopenhauer begins, as discussed, with an explanation of philosophical prudence, Chapter 2 is devoted to the "intuitive representation" (space, time, causality) and Chapter 3 is devoted to "abstract representation, or reasoning" (especially logic). Chapter 4 discusses the "principle of reason and its four roots" and Chapter 5 "science in general".

---

[80] Vide supra, Chapter 1.2.2.
[81] Cf. WWR2 I, pp. 67–77.
[82] WWR2 I, p. 79.

1.3 The Status of Logic in WWR and WWR2

Apparently, the five-part division of P1 of WWR2 breaks through the three-part division of the text into an introduction (representation), sections on the understanding and reason of B I of WWR I, for which was argued in Chapter 1.2.3. Indeed, however, P1 is more closely related to B I of WWR than the simple division of the late Schopenhauer into two parts in WWR II, which is also described in Chapter 1.2.3. The division of Chapters 1, 2 and 3 in P1 corresponds to the textual tripartition into introduction, the understanding and reason part of WWR I: Chapter 1 defines the superordinate concepts of 'representation', 'subject' and 'object', Chapter 2 treats space,[83] time[84] and causality[85] as the faculties of the understanding and Chapter 3 deals with language,[86] science[87] and practical reason[88]. Schopenhauer reflects on this structure in three text passages in detail.[89]

With the end of Chapter 3, almost the entire content of B I of WWR I has thus been recapitulated. Therefore, Schopenhauer also explains directly at the beginning of Chapter 4, one has now gained a "generally complete overview of the essence of reason".[90] The chapter on science alone corresponds in the majority of cases literally to § 10 and ignores §§ 14 and 15 of the WWR I, which, according to the above-given Chapter 1.2.3, together make up the doctrine of science or knowledge. Schopenhauer explains in the chapter on science:

> Now let us first discuss knowledge in general: the consideration of methodological knowledge, i.e. science, would then follow suitably; but what I have to say about it presupposes many other considerations, which must therefore intervene, so that I will only later be able to teach what is necessary from science.[91]

The classification of the science of knowing announced here can also be read from §§ 10, 14 and 15 of WWR I: First, knowledge is defined in general terms (§ 10, Part 1), followed by methodology (§ 10, Part 2), which finally leads to the classification of various sciences (§§ 14, 15). Similar to WWR I, here in WWR2, too, three digressions between the two parts of the science of knowing can be found: (1.) the excursus on the relationship between the understanding and reason, (2.) the explanations on practical reason and (3.) the before mentioned Chapter 4.

Chapter 4 is, along with the additions in Chapters 1 and 2, one of the most interesting modifications of WWR2, since it provides two systematic answers to questions that

---

[83] Cf. WWR2 I, pp. 127ff.
[84] Cf. WWR2 I, pp. 136ff.
[85] Cf. WWR2 I, pp. 151ff.
[86] Cf. WWR2 I, pp. 242ff.
[87] Cf. WWR2 I, pp. 368ff.
[88] Cf. WWR2 I, pp. 399ff.
[89] Schopenhauer reflects this system structure especially in WWR2 I, p. 240ff., p. 366ff., p. 498.
[90] WWR2 I, p. 421.
[91] WWR2 I, p. 368.

1 The World and its Representationalist Interpretation

were repeatedly raised in research with the preface of WWR I: What does the reader need to know about Kant in order to understand WWR, and what role do Schopenhauer's writings, which were written up to 1819 but are not part of the corpus of WWR I, play in understanding WWR?

At the beginning of Chapter 2, Schopenhauer repeats particularly relevant definitions and arguments from the Introduction and the Transcendental Aesthetics of Kant's CpR, especially concerning the pairs of terms 'a priori/ a posteriori' as well as 'analytic/synthetic judgements'. In doing so, he repeatedly takes up points of criticism from the appendix of WWR I (*Critique of the Kantian Philosophy*), which had highlighted his adoptions of and differences from Kant. I will come back to these remarks in more detail in Chapter 2.2. After these remarks on Kant's philosophy, Schopenhauer also includes a sub-chapter in Chapter 2 entitled "Theory of Sensual and Empirical Intuition", which deals with optics and which goes far beyond the first chapter of the book *On Vision and Colours* published in 1816.[92] Chapter 4 finally offers a summary of the most important passages of *On the Fourfold Root of the Principle of Sufficient Reason*. This Chapter 4 also serves Schopenhauer as a superstructure for P1, since each of the four sentences of the principle of sufficient reason corresponds to one or more of the faculties of the understanding and reason.

The designated passages from Chapter 1, Chapter 2 and especially Chapter 4 of P1 require more detailed investigations in their relationship to WWR, which go beyond the insinuations I have made here. One can argue about whether it was advantageous for the structure of the system, for example, to separate the parts of the science of knowing even more than they were already separated in WWR. For the descriptive approach and our topic, however, two facts are decisive with regard to WWR2: *logic retains its primary location in the section on reason, and the interchange of large parts of §§ 14 and 15 with § 16 of WWR means that the representationalist and Baconian approach of § 15 is now prominently located and emphasised, namely at the end of T1*. This also serves to dismantle any normative bias.

As already announced, I will not go into the parallels between WWR I and WWR2 in more detail here. The explanations presented here can also be transferred to P2–P4: Schopenhauer shifts individual, sometimes larger text sections, expands certain topics of WWR I and makes text sections and thus also parts of the system and intentions altogether clearer. In spite of these modifications, as could already be seen on P1, the text seems to be closer to the first edition of WWR I than some interpretations of its system in the years from about 1840 onwards. This is particularly evident in the logica major.

---

[92] As far as I can see, this chapter also goes far beyond the later editions of *On the Fourfold Root of the Principle of Sufficient Reason* (1847) and *On Vision and Colour* (1847) as well as *Commentatio exponens Theoriam Colorum Physiologicam eandemque Primariam* published in 1830.

1.3.3 The Logica Major of WWR2

In Chapter 1.3.2, I joined the editors and reviewers of WWR2 in the formal evaluation of the *Berlin Lectures* in so far as I recognised in WWR2 a similar, albeit greatly expanded and clarified system design of WWR I (1819). In my opinion, the basic structure of the WWR I system described in Chapter 1.2 remains the same for WWR2, despite individual modifications and extensions. The most important changes of relevance to the system concern, above all, Schopenhauer's handling of the writings recommended for reading only in WWR I, especially his preparatory work (*On the Fourfold Root of the Principle of Sufficient Reason, On Vision and Colours*) and the appendix of WWR I. However, for P1 of WWR2 only the new arrangement of the last paragraphs of B I of WWR I is noteworthy, which gives the representationalist approach a more prominent position in the text. In WWR2, the logical, natural-philosophical (especially optics, biology and mechanics), objective-aesthetic and legal-philosophical parts of the system, some of which have not been dealt with sufficiently in WWR, are particularly well elaborated.[93]

However, also because of such enhancements, mentioned in Chapter 1.3.2, I did not join the evaluation of the content by the editors and reviewers of WWR2. On the contrary, I have argued that WWR2 is more fruitful for the academic readership than WWR. In this chapter, I will support this thesis regarding Schopenhauer's logic, first of all, and show how such an enhancement looks like in WWR2 as opposed to WWR. The presentation shall show that it is justified to call § 9 of the WWR discussed in Chapter 1.3.1 a 'logica minor' and to compare it with Chapter 3 of WWR2, which contains a 'logica major'. Finally, the argument introduced in Chapter 1.1 and explained for the first time in Chapter 1.2.2 is to be sharpened and brought to a point, namely that the history of the reception of Schopenhauer's philosophy is subject to the misunderstanding that up to now the popular version has been confused with the academic system.

In Chapter 1.3.2, evidence was presented that in WWR2 too, the primary location of logic lies in the chapter on reason and even almost completely fills it. While in WWR I Schopenhauer only published a 'logica minor', which consists almost exclusively of a conceptual logic and gives only insufficient key points for a geometric logic of judgement and inferences, the structure of the sub-chapters of Chapter 3 in WWR2 already shows that Schopenhauer has tried to remedy this shortcoming. In the main chapters of WWR2, Schopenhauer treats the doctrine of concept[94], judgement[95] and inference[96] in addition to an introduction and appendix on the philosophy of logic[97].

---

[93] Vide supra, Chapters 1.2.3 and 1.3 (Introduction).
[94] Cf. WWR2 I, pp. 242–260.
[95] Cf. WWR2 I, pp. 260–293.
[96] Cf. WWR2 I, pp. 293–340.
[97] Cf. WWR2 I, pp. 234–242, pp. 340–363.

1 The World and its Representationalist Interpretation

By elaborating on these three themes, Schopenhauer seems not only to want to overcome the qualitative lack of the logica minor of § 9 of WWR I and the lack of non-complementary additions in WWR II; the page numbers given above for the four main themes (concept, judgement, inference and philosophy of logic) also show that Schopenhauer has quantitatively expanded his logica minor of WWR I together with WWR II by well over a hundred pages. Similar to Hegel scholarship, the adaption and application of the old distinction between logia minor and major naturally only has the quantitative, but no longer the didactic sense that this distinction had in late scholasticism.[98] Nevertheless, the enhancement of logic by more than five times should justify the expression 'logica major'.

Before I come to the question of why Schopenhauer made this enhancement of logic, I will give a general overview of the individual themes of WWR2 regarding philosophy of logic, logic of concepts, of judgements and inferences. This overview will also show which of the topics (1)–(7) presented in Chapter 1.3.1 Schopenhauer has elaborated more intensively. Due to the many additions and the new chapter classification in WWR2, it is no longer useful to structure the overview of the logica major with the classification of the themes (1)–(7). For this reason, I have adopted Schopenhauer's structure (concept, judgement, inference and philosophy of logic) and reduced the topics contained therein to forty, i.e. from (1') to (40').

Logic of Concepts: This part begins with (1') semantics[99], which also includes sketches of (2') semiotics[100]. After an excursus (3') on reflection[101] and on the (4') distinctions between *abstracta* and *concreta*, the distinction between simple and compound concepts[102] is introduced and these distinctions are explained (5') through the process of abstraction from intuition. Schopenhauer attributes this process and these distinctions to (6') the empiricist tradition, particularly Locke,[103] and (7') criticizes the dogmatic distinction between clear and confused concepts that was developed in rationalism.[104] Following this, Schopenhauer introduces the metaphors of (8') subsumption[105] and (9') extension[106], (10') explains them with logic diagrams[107] and

---

[98] Vide supra, Chapter 1.3 (Introduction).
[99] Cf. WWR2 I, pp. 243–246. Vide infra, Chapter 2.1.
[100] Cf. WWR2 I, pp. 247ff.
[101] Cf. WWR2 I, pp. 249ff.
[102] Cf. WWR2 I, pp. 252ff.
[103] Cf. WWR2 I, pp. 252ff. Julian Young: Willing and Unwilling: A Study in the Philosophy of Arthur Schopenhauer. Dordrecht 1987, pp. 22–25 was one of the first to explain the influence of British empiricists and especially Locke on Schopenhauer's theory of concepts. According to the manuscripts remains, Schopenhauer twice read and commented on several of Locke's works, once in the summer of 1812 and once in January 1816.
[104] Cf. WWR2 I, pp. 253ff.
[105] Cf. WWR2 I, pp. 255f.
[106] Cf. WWR2 I, pp. 257.
[107] Cf. WWR2 I, pp. 257f.

1.3 The Status of Logic in WWR and WWR2

(11') clarifies the difference in content and scope by means of the rule of reciprocity,[108] with which he in turn (12') criticises philosophies that use very broad and thus abstract concepts.[109]

Logic of Judgements: After (13') a definition of the term 'judgement'[110], Schopenhauer (14') deals with the relationship between subject, predicate and copula in judgements.[111] With the subsequent definition of the central laws of thought, namely (15') the law of identity,[112] (16') the law of contradiction,[113] (17') the law of excluded middle[114] and (18') the principle of sufficient reason,[115] (19') Schopenhauer introduces four concepts of truth (logical, empirical, metaphysical, metalogical).[116]

These laws of thought concern the relations within a judgment (subsentential) and between several judgments (sentential). The relations (20') refer, on the one hand, to the question of identity or the difference between subject and predicate and are therefore – here Schopenhauer ties in with the beginning of Chapter 2 of P1 – either analytical or synthetical.[117] On the other hand, (21') Schopenhauer critically analyses, with the help of Euler diagrams, Gergonne relations and partition diagrams, whether the possible properties of these relations concern either the quantity, quality, relation or modality of judgements.[118] By examining the basic diagrams, (22') he concludes that quantity and quality express essential forms of judgements, that the Kantian title of relationality is only partially meaningful, and that modality is a form that does not depend on the judgement but on the one who judges.[119]

In the following chapter, (23') Schopenhauer derives four basic forms of judgement from the two essential properties of judgement, i.e. quantity and quality: a) universal affirmative judgments, e) universal negative judgments, i) particular affirmative judgments, o) particular negative judgments.[120] (24') These are in turn illustrated with Eulerian diagrams and mutually derived with the scholastic conversion rules.[121]

---

[108] Cf. WWR2 I, p. 258. Vide supra, Chapter 3.2.2.
[109] Cf. WWR2 I, pp. 258f. Cf. also Michel-Antoine Xhignesse: Schopenhauer's Perceptive Invective. In: Language, Logic, and Mathematics in Schopenhauer. Ed. by Jens Lemanski. Cham 2020, pp. 95–107.
[110] Cf. WWR2 I, pp. 260f.
[111] Cf. WWR2 I, pp. 261.
[112] Cf. WWR2 I, pp. 262.
[113] Cf. WWR2 I, pp. 262f.
[114] Cf. WWR2 I, pp. 263.
[115] Cf. WWR2 I, pp. 263f.
[116] Cf. WWR2 I, pp. 264–269. Cf. also Jean-Yves Béziau: Metalogic, Schopenhauer and Universal Logic. In: Language, Logic, and Mathematics in Schopenhauer. Ed. by Jens Lemanski. Cham 2020, pp. 207–257.
[117] Cf. WWR2 I, pp. 268f., also pp. 122ff., vide supra, Chapter 1.3.2.
[118] Cf. WWR2 I, pp. 270–282. Cf. also Jens Lemanski & Lorenz Demey: Schopenhauer's Partition Diagrams and Logical Geometry. In: Diagrammatic Representation and Inference. Diagrams 2021. Lecture Notes in Computer Science, vol 12909. Ed. by A. Basu, G. Stapleton, S. Linker, C. Legg, E. Manalo, P. Viana. Cham 2021, pp 149–165.
[119] Cf. WWR2 I, pp. 282ff. This result largely corresponds to the analysis of the table of categories and judgments given in the *Critique of the Kantian Philosophy*, i.e. WWR I (1819), pp. 480–500, but supports the thesis with the help of analytical diagrams.
[120] Cf. WWR2 I, pp. 284ff.
[121] Cf. WWR2 I, pp. 289ff. Cf. also Amirouche Moktefi: Schopenhauer's Eulerian Diagrams.

1 The World and its Representationalist Interpretation

Logic of Inferences: This section consists of the study of inferences (I) formed by three concepts,[122] (II) formed by more than three concepts,[123] and (III) formed by judgements.[124] (I) The first passage of the logic of inferences has a rough structure, but because of the high redundancy in this chapter, it seems to me to be more useful to structure the arguments.[125] Schopenhauer is less concerned with the correctness and validity of inferences, but more with the criteria of their naturalness.[126] One of the main theses is (25') that Kant's reduction of all four syllogistic figures to the first figure is possible via "indirect routes" ("Umwege"),[127] (26') but that Kant's reduction undermines the natural function of the terminus medius. (27') The terminus medius shows that the fourth figure represents only a particular function of the first figure. Schopenhauer illustrates this thesis (28'), on the one hand, by means of metaphors and, on the other hand, by means of Eulerian diagrams that correspond to them: In the first and fourth figures the medius has the function of "resolution" ("Entscheidung"),[128] in the second the function of "distinction" ("Unterscheidung")[129] and in the third the function of "exception" ("Ausscheidung")[130]. (29') The distinguishing criterion of the three figures was thus less the usual subject or predicate position of major, minor and medius[131] or (30') the rules associated with them,[132] but rather (31') the metaphorically expressed functioning of the medius, which was shown in the "schemes of the spheres": In the first figure, the medius is the middle sphere, in the second the widest and in the third figure the narrowest. (II) In the second part, Schopenhauer (32') also uses an Eulerian diagram to explain pro- and episyllogisms, those corresponding to Seneca and those to Goclenian sorites. While in the first two passages, (I) and (II), natural inferences were derived from the logic of concepts and were based on quantity and quality, Schopenhauer treats in the last section (III) a logic of inferences based on the logic of judgments of (I) and (II). (33') Complex inferences

---

[122] Cf. WWR2 I, pp. 293–331.
[123] Cf. WWR2 I, pp. 331–333.
[124] Cf. WWR2 I, pp. 333–356.
[125] For a classification of the sections in the logic of inferences vide infra, Chapters 2.3.5f.
[126] The topic of naturalness would have deserved a separate Chapter 2.4 in the further course of this book. However, it has already been dealt with in Hubert Martin Schüler & Jens Lemanski: Arthur Schopenhauer on Naturalness in Logic. In: Language, Logic, and Mathematics in Schopenhauer. Ed. by Jens Lemanski. Cham 2020, pp. 145–165.
[127] Cf. WWR2 I, pp. 302ff., pp. 318ff.
[128] Cf. WWR2 I, pp. 302ff., pp. 318ff., p. 323, p. 326. Schopenhauer also uses the metaphors of manipulator (1st figure) and receipt (4th figure).
[129] Cf. WWR2 I, p. 302, p. 316, p. 326, p. 329. Schopenhauer also uses the metaphors of septum and insulating stool here.
[130] Cf. WWR2 I, pp. 316ff., p. 327. Here Schopenhauer also uses the metaphors of inclusion and difference. Prima facie, these functions are reminiscent of Lambert's *dicta de diverso, de exemplo* and *de reciproco*. To what extent this assumption is correct, however, would have to be discussed in a separate study.
[131] Cf. WWR2 I, p. 324, pp. 327f.
[132] Cf. WWR2 I, pp. 324–327.

1.3 The Status of Logic in WWR and WWR2

with relational connectives are explained[133] and (34') the rules for modal logic are briefly discussed.[134]

Philosophy of Logic: The philosophical passage includes an introduction in which Schopenhauer, above all (35'), presents language and logic from an anthropological point of view as one of the "main faculties of reason" alongside knowledge (science) and practical reason.[135] The philosophical concluding passage consists of several parts: Before the actual philosophical passage in the final section, Schopenhauer deals with (36') various themes such as connectives,[136] enthymemes,[137] paralogisms and (37') especially sophisms.[138] The philosophical passage contains, first of all, (38') a short outline of the history of logic with special consideration of its respective additions,[139] (39') a treatise on the distinction between analytics (logic) and dialectics (eristics or also topics)[140] and some remarks on the value of logic.[141] Finally, as the last passage of this section on language and logic, one can consider Schopenhauer's remarks (40') on the persuasion (Eristic Dialectic), which take up some sophisms again and explain them with the help of Eulerian diagrams.[142]

If one compares the themes (1') to (40') of the logica major with the themes (1) to (7) of the logica minor – i.e. WWR I of 1819 and WWR II of 1844 – it becomes clear that the assertion that there is no development in Schopenhauer's work can certainly be called a fairy tale. In the logica minor of WWR I we already find the themes (3'), (4'), (8'), (9'), (10'), (13'), (21'), (22'), (32'), (35'); but it should be emphasised that many aspects, especially (22') and (32'), were only sketched out in WWR I. Also, the division of a logic of inferences into (I) and (II), which is only hinted at there, is known from WWR I and would be identified today as very small fragment of the logic of quantifiers. In the logica minor of WWR II, we then find the topics (4'), (5'), (11'), (12'), (13'), (14'), (17'), (18'), (25'), (27'), (35'), (36'). The approach of a logic of inferences including judgements (III) reminds of the logica minor from WWR II and would be identified today as a very small fragment of propositional logic (formerly Stoic logic).

However, as can be seen, for example, in (19'), many topics and arguments in the late work (WWR II) are only heavily abbreviated, sometimes to the point of incomprehensibility. It is also interesting to note that only the themes (4'), (5') and (10') are

---

[133] Cf. WWR2 I, pp. 333–339.
[134] Cf. WWR2 I, pp. 339f.
[135] Cf. WWR2 I, pp. 234–242.
[136] Cf. WWR2 I, p. 340.
[137] Cf. WWR2 I, pp. 340–343.
[138] Cf. WWR2 I, pp. 344–356.
[139] Cf. WWR2 I, pp. 356ff. Cf. Valentin Pluder: Schopenhauer's Logic in its Historical Context. In: Language, Logic, and Mathematics in Schopenhauer. Ed. by Jens Lemanski. Basel 2020, pp. 129–143.
[140] Cf. WWR2 I, pp. 358f.
[141] Cf. WWR2 I, pp. 359ff.
[142] Cf. WWR2 I, pp. 363–366. Cf. Jens Lemanski, Amirouche Moktefi: Making Sense of Schopenhauer's Diagram of Good and Evil. In: Diagrammatic Representation and Inference. 10th International Conference, Diagrams 2018, Edinburgh, UK, June 18–22, 2018, Proceedings. Ed. by Francesco Bellucci, Peter Chapman, Gem Stapleton, Amirouche Moktefi, Sarah Perez-Kriz. Cham 2018, pp. 721–724.

1 The World and its Representationalist Interpretation

actually found in all three works – WWR I, WWR2 and WWR II – whereby (4') and (10') are found in WWR I, but only (5') in WWR II in a considerably expanded form. It is also striking that Schopenhauer integrated (37'), (39') and (40') into the logic in WWR2, whereas in the logica minor they have been outsourced to dialectics. All this indicates that the logica major is, in terms of content, the document of a transition, but is at the same time the most mature form of logic we have of Schopenhauer.

If one wanted to assess the overall approach of the logica major, one would have to speak of a tragic document. Schopenhauer himself must not have been satisfied with the logica major. After all, he made additions to the document until about 1830, but fourteen years later he published an overall greatly changed approach in WWR II, which then only partially harmonized with the almost unchanged version of the logica minor in WWR I of 1844.[143] Why Schopenhauer considered the logica minor of WWR I untenable can also be explained, for example, by the transitional document on the topics (21') to (24'): Here, Schopenhauer expands the originally only five analytical diagrams to six and modifies them with negative judgments, which he omits in WWR I.[144] That the logica major of WWR2 as a whole is, however, a document of the early Schopenhauer, is shown, above all, in the fact that in WWR2, similar to WWR I, almost everything is explained based on analytical diagrams – an approach that Schopenhauer abandoned in WWR II and wanted to replace with an allegedly better, but not explicit notation of hooks and sticks.[145]

Although the late Schopenhauer himself was probably critical of his early logics (WWR I and WWR2), the logica major is an impressive philosophical-historical document containing several questions, topics and answers that are still being discussed today in individual historical and systematic fields of research on logic, philosophy of language, philosophy of mathematics, metaphysics, etc. This thesis is dealt with in more detail below, in Chapter 2. All in all, however, the great value of the logica major is almost inevitable due to the fact that Schopenhauer explicitly links several paradigms of logic independently with each other, some of which are still treated separately today: By this I mean, first of all, only Schopenhauer's investigation, continuation and critique of Kantian logic with the help of analytical diagrams, or in other words: the independent combination of those logics which form the starting point for today's algebraic and geometric logic. This connection will be shown in more detail in Chapter 2.2. In Chapter 2.3 it will also be shown that Schopenhauer still confronts this Kantian-Eulerian approach with Aristotelian and scholastic conceptions in proof theory. The result is such a complex and presupposition-rich text that it can hardly be evaluated as a didactic introduction for beginners.

---

[143] Vide supra, Chapter 1.3.2.
[144] Vide supra, Chapter 1.3.1. It is only through WWR2 that an Eulerian system can be established at all, since Schopenhauer does not give diagrammatic construction rules for negative judgements in the logica minor.
[145] Vide supra, Chapter 1.3.1.

1.3 The Status of Logic in WWR and WWR2

The tragedy of this document of transition, as mentioned above, can be seen, above all, in the fact that Schopenhauer's intensive elaborated logic ends at about the moment when he breaks with his academic career. All later allusions to logic, in WWR II and in *Parerga and Paralipomena* (= PP), show an interest in the topic because of their modifications, but a disinterest in its elaboration. What is the reason for this? What is the reason that must have moved Schopenhauer not to use the logica major of WWR2 extensively for the revision of § 9 of WWR I (1844) and the new chapters on logic published in WWR II? Or to put it even more radically: why did he withhold better knowledge from his readers?

I believe that the answer to this question, which is particularly evident in the section on logic, concerns Schopenhauer's self-understanding, i.e. his self-image as an academic philosopher, which has hardly been noticed in the reception history of his oeuvre and has therefore led to serious misjudgements. The thesis which I would like to present as the result of Chapter 1 is: WWR2 alone is the philosophical work that should be judged by academic philosophers, whereas WWR is only a didactic and popular form of it. If one wanted to express this in a metaphor of circumferential logic, one could probably say that the thematic content of WWR is smaller than that of WWR2, and this is because Schopenhauer intended that the readership of WWR should be larger than that of WWR2.

The reason for the restriction of the thematic content can be found in WWR I. After Schopenhauer has established the forms of judgement in § 9 on the basis of analytical diagrams ("Schematism of concepts"), he points out that the logic need not be dealt with further at this point. As justification for this we find two arguments:

> This *schematism of concepts*, which is *already explained quite well in many textbooks,* can be used to ground the doctrine of judgement as well as the whole of syllogistic logic and makes it very easy and uncomplicated to teach them both. The reason is that this schematism gives insight into the origin of all their rules and allows them to be derived and explained. *We do not have to burden our memory* with all the rules, since logic can only be of theoretical interest and never of practical use for philosophy.[146]

I read two quite different arguments from this quotation:

(A1) Analytical diagrams are already quite well developed in several textbooks.
(A2) It is not necessary to burden the memory with rules of logic.

---

[146] WWR I, p. 68. – My emphasis, J.L. The fact that one can find in the first sentence of this quote a "should" in the first edition of WWR I instead of a "can" is characteristic of the revised doctrine in logic.

1 The World and its Representationalist Interpretation

(A1) is mentioned in passing and seems to me to be only an external reason for not having to explain in more detail what was said before in § 9 about logic. To put it simply: Whoever asks for further explanations should look into special textbooks.[147] As mentioned in Section 1.2.6 in connection with the doctrine of the saints and ascetics (the imperative "Just read..."), here too in logic an external text is again used to supplement the incomplete system. (A2) is supported by the Hobbesian as well as the Cartesian phrase that, on the one hand, every person intuitively understands logic anyway[148] and that, on the other hand, logic has no practical use. However, this seems to me to be only an appeasement for those readers who – as was argued in Chapter 1.3.1 – might consider Schopenhauer's logica minor incomplete or dark and cryptic. In my opinion, however, neither of these reasons is sufficient justification for presenting the logic to its recipient in such an incomplete way – especially since logic, as discussed in Chapter 1.2.7, is not only a component of the system but also the instrument used for structuring the system.

In the further course, however, a further argument is added, following on from (A2), which can be found in one of the few text passages of § 9 that Schopenhauer revised considerably for the last edition of WWR I:

WWR I (1819), p. 68	WWR I (1859), pp. 69f.
For this reason logic must to an extent no longer be taught on its own and as a self-sufficient science, because as such it leads to nothing.	We are justified in treating logic on its own terms, independently of all other sciences (as well as in teaching it in universities) because it is a self-contained, self-subsistent, internally complete and perfected discipline that achieves absolute certainty.

Both quotes seem to be appeasement arguments to protect oneself from accusations of providing an incomplete logic: not only has logic already been better dealt with elsewhere (A1) and not useful to the reader anyway (A2), but it is also not a science in itself.[149] The quotation from the first edition seems contradictory in view of the logica major that Schopenhauer presented only two years later. If logic does not need to be taught alone and as a doctrine in its own right, why did Schopenhauer expand it so much for his lectures? This question can even be exaggerated when one considers

---

[147] For the textbooks that Schopenhauer may have had in mind, vide infra, Chapter 2.2.6.
[148] Cf. e.g. Thomas Hobbes: De Corpore I.1; René Descartes: Discours de la méthode I.1.
[149] On the assertion that logic is a self-contained and completed discipline, cf. Valentin Pluder: Schopenhauer's Logic in its Historical Context. The thesis is still known to us today mainly through Kant. Before Kant, however, it can also be found among numerous Leibnizians and Wolffians.

1.3 The Status of Logic in WWR and WWR2

that logic is the most extensive thematic expansion in WWR2;[150] or when one considers that Schopenhauer explicitly emphasised in many of his lecture announcements from the 1820s that he would treat logic in particular.[151]

The quote by the late Schopenhauer seems to respond to these exaggerations with two limitations: Of course, logic can be dealt with on its own and in its entirety, independently and in universities, i.e. in research and education. Is it possible to draw the reverse conclusion that § 9 of WWR I or the logic in the system of WWR I is not intended for teaching and research in the philosophy of the subject? I have given the answer using the concrete example in this Chapter (1.3): Schopenhauer may have used WWR I as a systematic basic framework for his *Berlin Lectures*, but in contrast to WWR I, he expanded on demanding topics such as logic or philosophy of law to a much greater extent.

Schopenhauer also explicitly draws attention to this elaboration of the philosophy of logic and the philosophy of law at the beginning of WWR2, in the "Exordium on my lecture": he tries to touch on all the subject areas of philosophy in one semester, even if, due to time constraints, these cannot always be as detailed and concrete as one might wish. Nevertheless, he explicitly emphasises two important topics of which he will at least present "the basis, the essence and the main teachings", namely logic and ethics.[152] The aspiring professional philosopher or academic in general is thus taught in WWR2 what the reader of WWR can only take note of as cryptic explanations. In my opinion, § 9 of WWR I, therefore, seems to provide only a makeweight function of a representationalist approach aiming at completeness – whereby 'completeness' here only means something like 'mentioning for the sake of completeness', but does not imply the complete explanation.

The reader of WWR may be a professional philosopher – Schopenhauer, as far as I know, never opposed the academic reading of his main work – but the much larger circle of readers of WWR, compared to WWR2, should also include – as the joke mentioned in Chapter 1.2.2 emphasises – those people who leave excessively exuberant books "in the dressing room or on the tea table of his educated lady friend". While the educated citizen is reassured that he does not need to take a closer look at § 9, the aspiring academic philosopher is presented with a logic in WWR2 that is almost complete according to the criteria of the zeitgeist. The history of reception that goes hand in hand with the confusion of the popular with the academic system has contributed to the tragedy that Schopenhauer has hardly been perceived as a logician to this day. However, Schopenhauer himself is to blame for this tragedy. His failed career as a Privatdozent (adjunct professor) is well known and has all too often been reinterpreted

---

[150] Cf. Heinrich Hasse: [Review of:] Schopenhauer, Arthur. Handschriftlicher Nachlass, pp. 270–272.
[151] Cf. WWR2 I, pp. XIIf. (Preface Mockrauer). However, it must also be remembered that in the 19th century logic was still part of the so-called 'Fuchskolleg' (verbal: collegium of foxes), i.e. the obligatory course of study for all students.
[152] WWR2 I, p. 72.

into heroic. We would certainly have a truly heroic and, above all, more positive picture of Schopenhauer today if he had further elaborated his syncretism of Aristotelian, Eulerian and Kantian logic academically or if he had drafted his modified theory of sticks and hook notation in more detail. In this way, however, the Schopenhauer picture of the contemporary world is at best sufficient to characterise an anti-hero. Nevertheless, in the following Chapter 2, arguments are put forward for the fact that even today's logician, philosopher, linguist and mathematician could still find interest in Schopenhauer's only academic system, even if it was never completed in its development.

2 Logic and its Geometrical Interpretation

In Chapters 1.1 and 1.2, I argued for a descriptive interpretation of Schopenhauer's system and showed that there are good reasons to classify it as a representationalist one: His system aimed to reflect and order the whole world in as few abstract concepts as possible. Within this representationalist approach, logic plays a dual role: it is part of the world and must therefore be represented in a complete conceptual system in the same way as all other sciences, phenomena, opinions and behaviours that are found in the world. Due to the aim of structuring the world conceptually, however, logic is much more than just a component of the system, for it determines the conditions of the possibilities of the theory of representation itself.

In Chapter 1.3 it was shown that Schopenhauer changed his logic several times around the 1820s, but also elaborated it. He elaborated the so-called logic minor of WWR I and, moreover, of WWR II only once for an academic audience. Schopenhauer's *Berlin Lectures* are a modified and, above all, greatly expanded version of WWR I, which deserves to be called 'WWR2' because of both its connection and elaboration. In these lectures, Schopenhauer developed the logica minor of WWR into a logica major by quintupling its scope quantitatively.

This logica major has remained largely unknown to research to this day and has never been intensively studied. It contains, besides many philosophical remarks, a unique doctrine of concepts, judgements and inferences. The uniqueness of this logica major is based, above all, on the fact that Schopenhauer formed in it a critical synthesis of the logics of Euler and Kant with his own reflections (Chapter 2.2), in order to establish a contextual and an abstraction-theoretical semantics (Chapter 2.1) and to bring about an improvement in scholastic and Aristotelian proof theory (Chapter 2.3). This synthesis is of interest today in so far as Kant's logic is a milestone to the paradigm of Fregean semantics, Euler's logic is a milestone to the paradigm of Venn's logic, and finally Aristotle's proof theory is an example of a calculus of natural inferences.

First, however, Chapter 2.1 will argue that Schopenhauer's logic of concepts introduced a unique semantics, which anticipates the methods of a use theory of meaning and a context principle often combined since Wittgenstein and explains them with the help of Eulerian diagrams (2.1.4–2.1.6). Schopenhauer's pragmatic semantics thus becomes the counter-evidence to the neologicist-inferentialist thesis of an oblivion of context that is said to prevail in the period between Kant and Frege (2.1.1–2.1.3). In Chapter 2.2, I will argue against critics of the definition of analyticity and the metaphor of containment that Kant's and especially Schopenhauer's theory of analytic judgements are justified primarily by so-called 'analytical' or 'Eulerian diagrams' and therefore do not require translation and reformulation (2.2.4–2.2.6). Rather, the aim is to show that Kant and, in particular, Schopenhauer's theory of analytic judgements are part of a long tradition of analytical diagrams that developed in the early modern period but dates back to antiquity (2.2.1–2.2.3). In Chapter 2.3 it is shown that Schopenhauer made a relevant contribution to a lively debate that was and still is held on the value and usefulness of intuitive figures in geometric proof theory (2.3.1–2.3.3). In addition, I will show that Schopenhauer also justifies the advantage of figures and diagrams in geometry and logic by the argument that a pure logicist proof theory itself is not justifiable (2.3.4–2.3.6).

2.1 Semantics – Context Principle, Use Theory and Representationalism

The statement that philosophy begins with Frege is an almost colloquial slogan in contemporary academic philosophy, expressing an unreserved commitment to so-called 'analytical philosophy'. Frege would probably only have agreed with this slogan to a limited extent, as he had pointed out in his *Begriffsschrift* or *Conceptual Notation* (= CN) that the idea of an exact sign language originally came from Leibniz.[1] Despite Frege's reference to possible predecessors, the prevailing opinion is that Frege is the originator of the context principle (= CTP) and the compositionality principle (= CPP) as two fundamental semantic principles, which on the one hand, were elaborated as a theory in the 20th century with the help of Wittgensteinian texts and which, on the other hand, can still be interpreted today as basic principles of conflicting philosophical schools:

(CTP$_F$) "it is only in the context of a proposition that words have any meaning."[2]

(CPP$_F$) "The possibility of our understanding sentences which we have never heard before rests evidently on this, that we construct the sense of a sentence out of parts that correspond to words."[3]

The situation is paradoxical: the two principles are usually considered to be incompatible, but they are both referred to as the 'Frege principle'. Differentiated studies such as those of Theo Janssen explain the attribution of both principles by the fact that (CTP) was first advocated by Frege in 1884 in the *Grundlagen der Arithmetik* or *Foundations of Arithmetic* (= FoA), whereas formulations reminiscent of (CPP) appear only around 1914 in Frege's letter to Jourdain.[4]

The fact that (CTP) and (CPP) can represent basic principles of semantics is rarely associated with Frege, but mostly with Wittgenstein. The standard interpretation of the *Philosophical Investigations* (= PI)[5] states that Wittgenstein agrees with (CTP) as the basic principle of a use theory of meaning (= UTM),[6] but rejects (CPP) as the basic principle of a representational or picture theory of meaning (= RTM):

---

[1] Cf. CN, p. 105.
[2] FoA, p. 73 (§ 62), also: p. xii.
[3] Frege to Jourdain (January 1914). In: Ibid.: Philosophical and Mathematical Correspondence. Ed. by G. Gabriel, H. Hermes, F. Kambartel, C. Thiel, A. Veraart; trans. by H. Kaal, Chicago, 1980, p.79.
[4] Theo M. V. Janssen: Frege, Contextuality and Compositionality. In: Journal of Logic, Language, and Information 10 (2001), pp. 115–136.
[5] Cf. e.g. Eike von Savigny: Die Philosophie der normalen Sprache. Eine kritische Einführung in die "ordinary language philosophy". 2. rev. ed. Frankfurt/ Main 1974, pp. 13–74.
[6] I am aware that esp. rationalists, for various reasons, will be disturbed by my use of the word 'theory' here as well as in other contexts, particularly in relation to Wittgenstein. While I do not share the aversion

2 Logic and its Geometrical Interpretation

(UTM$_W$)   "the meaning of a word is its use in the language."[7]
(CTP$_W$)   "a word had meaning only as part of a sentence [Satzzusammenhang]."[8]

(RTM$_W$)   "the individual words in language name objects–
(CPP$_W$)   sentences are combinations [Verbindungen] of such names."[9]

The role of (UTM) and (RTM) can be seen from the opposing formations of the philosophical schools in post-Wittgensteinian philosophy. (UTM), which originated in late Wittgenstein, had influence in game theoretical semantics (e.g. Hintikka), dialogical logics (e.g. Lorenzen) or pragmatic inferentialism (e.g. Brandom).[10] In contrast to this, (RTM), which originated in early Wittgenstein, is more related to categorical grammars (e.g. Ajdukiewicz), possible world semantics (e.g. Carnap) or Montague grammar, usually in connection with a truth theory of meaning.

Moreover, (UTM) and (RTM) are not only associated with semantics, but also with epistemology, philosophy of science, philosophy of mind and perhaps even with overall metaphysical connotations: (UTM) is holistic, (RTM) is atomistic; (UTM) is inferential or propositional, (RTM) is intensional or conceptual; (UTM) is top-down, (RTM) is bottom-up; (UTM) is expressive, (RTM) is impressive; (UTM) is syntactically recursive, (RTM) is syntactically progressive; (UTM) is idealistic, (RTM) is realistic, etc. Put simply, (UTM) is close to normal language philosophy and thus to the social theory of interaction, while (RTM) is linked to ideal language philosophy and mechanistic or artificial theories of argumentation.[11]

However, the systematic elaboration of the two theories of language also goes hand in hand with the historical attribution and evaluation of arguments from the history of philosophy, which serves in particular to sharpen the details.[12] Even the standard interpretation of PI states that Wittgenstein identified (CPP) and (RTM) with Plato, Augustine, Russell and with his early philosophy in the *Tractatus logico-philosophicus* (= Tlp). In addition, compositionalists such as Wilfrid Hodges see themselves as part of the great tradition that begins with the *Organon* of Aristotle, continued by the

---

to this word, especially since 'use theory of meaning' has become an established terminus technicus. Alternatively, however, one can read for (UTM), for example, 'Use-Thesis of Meaning' or, for Wittgenstein, use the term 'codex' proposed by Ingolf Max: Wittgensteins Philosophieren zwischen Kodex und Strategie. Logik, Schach und Farbausdrücke. In: Realism – Relativism – Constructivism. Proceedings of the 38th International Wittgenstein Symposium in Kirchberg. Ed. by Christian Kanzian, Sebastian Kletzl, Josef Mitterer, Katharina Neges. Berlin, New York 2017, pp. 409–424.

[7] PI I 43.
[8] PI I 49. Since Wittgenstein quotes the Fregean formulation in PI 49, is (CTP$_F$) = (CTP$_W$).
[9] PI I 1.
[10] Cf. Frédérick Tremblay: La rationalité d'un point de vue logique. Entre dialogique et inférentialisme, étude comparative de Lorenzen et Brandom. Nancy 2008.
[11] Cf. e.g. Richard Rorty: Philosophy and the Mirror of Nature. Princeton 1980, esp. Chapter IV.
[12] Consider here, for example, that Robert Brandom defines his closeness to Frege mostly in terms of Wittgenstein's rejection of the 'slab' language game.

2.1 Semantics – Context Principle, Use Theory and Representationalism

late antique and medieval Aristotelians and Leibniz, and transmitted by the late Frege, the early Wittgenstein and Russell to Tarski and Davidson.[13]

The current state of research on (UTM) is much more complicated: there is agreement that Wittgenstein formulated (UTM) in PI, which was then further developed by speech act theorists such as Austin and Searle. But whether Quine also interpreted (UTM) in *Ontological Relativity* in its full sense ("knowing how to use the word") and whether it, in turn, goes back to Dewey's 1925 behaviourist theory ("meaning [...] is primarily a property of behavior"), slightly anticipating Wittgenstein's theory,[14] is a question that, as far as I know, today's research has scarcely gone beyond.[15] To my knowledge, Michael Forster alone has argued in recent years that (UTM) was anticipated by several hermeneutical and 'quasi-empirical' approaches of the 18th century (Herder, Hamann, Ernesti, Wettstein),[16] which depend on Spinoza's dictum "Verba ex solo usu certam habent significationem".[17]

Especially by inferentialists, however, (CTP) is less associated with biblical hermeneutics than with transcendental philosophy. The relevant thesis on the origin of contextualism was put forward by Hans Sluga, popularly supported by Robert Brandom and substantiated by Theo Janssen: All three of them believe that contextualism was implicitly applied by Kant, but is explicitly shown for the first time by (CTP_F).

Brandom and Janssen disagree solely on the question of the role of philosophy between Kant and Frege: whereas Brandom claims that Kant's contextualist approach was forgotten in the 19th century and only taken up again by Frege, Janssen puts forward several 19th-century quotations that can be interpreted as variants of (CTP). Two further principles play an important role in the respective historical attribution of arguments from the history of philosophy to (CTP)/ (UTM) or (CPP)/ (RTM). I have mentioned both principles only incidentally above in the phrases "(UTM) is holistic, (RTM) is atomistic", "(UTM) is inferential or propositional, (RTM) is intensional or conceptual" and "(UTM) is top-down, (RTM) is bottom-up", as they have already been implicitly mentioned with (CTP_W) and (CPP_W). Both principles are simply called

---

[13] Cf. Wilfrid Hodges: Formalizing the Relationship between Meaning and Syntax. In: The Oxford Handbook of Compositionality. Ed. by Markus Werning, Wolfram Hinzen, Edouard Machery. Oxford 2012, pp. 245–261; some additions to the history of ideas and concepts can also be found in Wilfrid Hodges: Remarks on Compositionality. In: Dependence Logic. Theory and Applications. Ed. by Samson Abramsky, Juha Kontinen, Jouko Väänänen, Heribert Vollmer. Cham 2016, here: pp. 104–106.
[14] Cf. Willard Van Orman Quine: Ontological Relativity. In: The Journal of Philosophy 65:7 (1968), pp 185–212, here: p. 185, p. 187; ibid.: Use and its Place in Meaning. In: ibid.: Theories and Things. Cambridge/Mass., London 1981, pp. 43–55, here: p. 46.
[15] Cf. John V. Canfield: Wittgenstein versus Quine. The Passage into Language. In: Wittgenstein and Quine. Ed. by Hans-Johann Glock, Robert L. Arrington. London 1996, pp. 116–144. In Chapter 2.1, when comparing Wittgenstein and Quine, I refer to the terminology and content of the contributions in this volume.
[16] Cf. Michael Forster: Herder's Doctrine of Meaning as Use. In: Linguistic Content. New Essays on the History of Philosophy of Language. Ed. by Margaret Cameron, Robert J. Stainton. Oxford 2015, pp. 201–222.
[17] Baruch de Spinoza: Tractatus Theologico-Politicus. Continens Dissertationes aliquot, Quibus ostenditur Libertatem Philosophandi non tantum salva Pietate, & Reipublicæ Pace posse concedi: sed eandem nisi cum Pace Reipublicæ, ipsaque Pietate tolli non posse. Hamburgi [i.e. Amsterdam] 1670, p. 146 (= XII).

'priority of judgement' (= PJ) or 'priority of concept' (= PC). If one assumes, as discussed in Chapter 1.3, that a complete or comprehensive logic has to examine concepts, judgements and inferences, the order of logic depends on the choice between (CTP) and (CPP):

(PJ)   If (CTP) applies, a logical-semantic approach must begin with judgments from whose context it can only analyse concepts.

(PC)   If (CPP) applies, a logical-semantic investigation must begin with concepts with which it can compose judgements.

According to a vertical scheme, the order of logic can be roughly presented as follows:

Compositionality   *Contextualism*

Traditional Order	Traditional Designation	Doctrines	Moderne Designation	Modern Order
3	Syllogistic	Inferences	Inferentialism	1
2	Hermeneutics	Judgements	Propositionalism	2
1	Categories	Concepts	Conceptualism	3

bottom-up   *top-down*

The atomic multiplicity of concepts is found at the bottom, the holistic unity of inferences is found at the top of the diagram. But the scheme can be extended: Concepts can be composed of signs, ideas or objects; inferences can play a role in the context of theories, conceptual schemes or languages. (PJ) and (PC) thus express only a partial relationship; although, thanks to Sluga and Brandom, they are today mainly discussed at Kant's table of categories and at the so-called 'table of judgments', they are also only seen as derived principles from a larger picture that includes much more holistic or atomic variants of (CPP) or (CTP).[18]

Schopenhauer actually plays no role in this picture. On the contrary, if he is examined at all in the context of a semantic theory, he is usually classified as a representative of a standard variant of (RTM), as it is found almost exclusively in the paradigm of pre-Fregean philosophy. My remarks in Chapters 1.2 and 1.3, especially on the topic 'abstracta/ concreta', may have confirmed this impression. Some researchers even go so far as to claim that Schopenhauer had no sense for theories of meaning at all, because he, like many authors up to the 20th century, simply used language without reflecting it.[19] According to this view, Schopenhauer is a naïve representationalist who wants to reflect the world in concepts and also gives logic a

---

[18] Vide infra, Chapter 2.1.5.
[19] Cf. e.g. Gunnar Schumann: A Comment on Lemanski's 'Concept Diagrams and the Context Principle'. In: Language, Logic, and Mathematics in Schopenhauer. Ed. by Jens Lemanski. Cham 2020, pp. 73–85, esp. p. 76, p. 80.

place in his world system, but does not know that he must already have semantic and logical tools at his disposal to be able to realise this project.

In the following chapters, I would like to dispel these prejudices. Chapter 2.1 presents the main thesis, according to which Schopenhauer in his logica major in WWR2 represented a variant of (UTM), which he supported by (CTP). In order to build up this argumentation, in the first part of Chapter 2 (i.e. Chapters 2.1.1–2.1.3) I will first deal more intensively with the factual genesis of the context principle and formulate a critique of the current state of research. In Chapter 2.1.1, I will first introduce Frege research and especially the 'Kant/ Frege-Thesis' of Sluga, Brandom and Janssen in more detail. In Chapter 2.1.2, I will argue against Brandom's thesis of context oblivion in the 19th century and thus – to some extent – take Janssen's side and place his quote of Gruppe, Trendelenburg and Lotze, which are only presented as counter-evidence, in their historical context.[20] Chapter 2.1.3 will show why I cannot completely take Janssen's side. This is partly because I see the reception history of (CTP) and (PJ) with Frege and Wittgenstein differently from Janssen and partly because I consider the Kant/ Frege thesis to be problematic.

The second part of Chapter 2 (i.e. Chapters 2.1.4–2.1.6) will deal more intensively with a probable history of the reception and development of (CTP) and (UTM) and formulate a critique of the current state of research. Chapter 2.1.4 will begin by presenting Wittgenstein research and in particular the theses on Schopenhauer and Wittgenstein. The central Chapter 2.1.5 will then use Schopenhauer's critique of lexical semantics to present his variant of (UTM) and (CTP). Finally, I would like to discuss in Chapter 2.1.6 whether Schopenhauer's logica major represents a historical alternative to the contextualists *in spe* mentioned by Sluga, Brandom and Janssen – by which I mean Kant, Trendelenburg etc. – and to what extent Schopenhauer's (RTM) and (CPP) from Chapter 1 harmonise with the (UTM) and (CTP) developed in Section 2.1.

2.1.1 The Kant/ Frege-Thesis

Frege's historical contextualisation is still fraught with problems for Frege research today: the first question that will probably be asked is why so many researchers are even concerned with historical contextualisation: What do we gain from this knowledge, or why should we argue about whether Frege had certain predecessors or

---

[20] Janssen's paper (Theo M. V. Janssen: Compositionality. Its Historic Context. In: The Oxford Handbook of Compositionality. Ed. v. Markus Werning, Wolfram Hinzen, Edouard Machery. Oxford 2012, pp. 19–46) was published almost simultaneously with Jens Lemanski: Die neuaristotelischen Ursprünge des Kontextprinzips und die Fortführung in der fregeschen Begriffsschrift. In: Zeitschrift für philosophische Forschung 67:4 (2013), pp. 566–587 which in some respects serves as the basis for the following analysis. Although we have independently arrived at the same historical results, which to a certain extent invalidate Sluga's and, above all, Brandom's approaches, we are nevertheless divided on several detailed questions.

not? It can almost naturally be argued that knowledge of the historical position, especially of a school founder, can provide insights into the self-image of a particular discipline. In Frege's case, for example, it would be an insight into the self-understanding of the discipline called analytical philosophy. But the discussion about historical contextualisation goes even further: the outcome of this discussion says something about which methods and research results are scientifically relevant at all. If philosophy only begins with Frege, as it was said at the beginning of Chapter 2, the modern philosopher does not even have to deal with Aristotelism or Kantianism, for example. And what is true for the philosopher is even more true for those logicians, linguists, mathematicians etc. who are interested in the methods and results of philosophy.

As Christian Thiel had already pointed out in 1965, research on Frege's links with earlier thinkers has yet to be carried out, and research is trying to give the impression that "Frege created his logic 'out of nothing'".[21] However, as we have already seen in Chapter 1.2.2, such a creation of scientific work out of nothing is highly improbable. Therefore, research on Frege's historical contextualisation can at least record some preliminary work: Although Willard Van Orman Quine and John Wallace claim to have found variants of (CTP) in the history of philosophy, for example in Jeremy Bentham,[22] it was Hans Sluga who first put forward the historical and reception-historical theses that Leibniz's idea of a conceptual notation was brought via Trendelenburg to Frege[23] and that Kant founded (CTP), which was then received by Frege via Hermann Lotze.[24] Precursors of today's inferentialism have largely contradicted these theses: In 1981, Michael Dummett had not only argued against Sluga's interpretation but also fundamentally against the historicisation of Frege's development of thought:[25] Frege was the founder of the (CTP) and in the preceding history of philosophy there was general context amnesia.

To this day, Frege research seems to stand in the aporia between Sluga and Dummett: Whereas researchers such as Gordon Baker or Peter Hacker tend in principle more towards Sluga's historical view with their Frege criticism, Tyler Burge sees Kant

---

[21] Christian Thiel: Sinn und Bedeutung in der Logik Gottlob Freges. Meisenheim a. G. 1965, p. 9. (My transl., J.L.)
[22] Cf. Willard Van Orman Quine: Epistemology Naturalized. In: Ontological Relativity and Other Essays. New York 1969, pp. 69–90, here: p. 72; Id.: Five Milestones of Empiricism. In: Theories and Things. Cambridge, London 1981, pp. 67–72; John Wallace: Only in the Context of a Sentence do Words have any Meaning. In: Midwest Studies in Philosophy 2 (1977), pp. 144–164, esp. p. 145. Whether Jeremy Bentham really claims variants of (CTP) and (PJ), as Quine and Wallace suggest, is doubtful. For example, whereas Peter Michael Stephan Hacker: The Rise of Twentieth Century Analytic Philosophy. In: Ratio 9:3 (1996), p. 259 argues for this, recently Silver Bronzo: Bentham's Contextualism and Its Relation to Analytic Philosophy. In: Journal for the History of Analytic Philosophy 2:8 (2014), pp. 1–41, he has presented good counter-arguments.
[23] Hans D. Sluga: Gottlob Frege. The Arguments of the Philosopher. London 1980, pp. 48–52.
[24] Hans D. Sluga: Gottlob Frege., pp. 52–58, esp. p. 55.
[25] Cf. e.g. Michael Dummett: The Interpretation of Frege's Philosophy. Cambridge/Mass 1980, pp. XVff.

2.1 Semantics – Context Principle, Use Theory and Representationalism

as a forerunner of (CTP)[26] but is more concerned with modernising Frege's philosophy of language. In German-language research, on the one hand, Gottfried Gabriel and his students argue that Frege's formalism is related to traditional logic and epistemology of the New Kantians;[27] on the other hand, researchers such as Ulrike Kleemeier at least reject the historicisation of the (CTP) due to a lack of sources[28] and try to prove that (CTP) systematically does not necessarily have to have anything to do with the (PJ).[29]

All the authors mentioned above demand as a criterion for a plausible historicisation of Frege's thoughts not only the demonstration of a systematic parallel but also at least the naming of the respective historical author in Frege's oeuvre.[30] In the following, I will refer to this criterion as the 'naming criterion'. A further criterion, which can also occur independently of the first one, I call the 'citation criterion', according to which an influence of an author Y on an author X can be determined either by the adoption of whole sentences or of neighbouring lexemes, co-occurrences or other units that are semantically related.[31] Although admittedly both criteria depend heavily on the philological methods of (RTM),[32] they are generally accepted in Frege scholarship or even demanded as evidence.

Although Robert Brandom has been barely or insufficiently considered in the research dispute,[33] he claims to make Frege from the historical context (Sluga) fruitful for the current discussion (Dummett), as he believes "that ignoring the historical context in which Frege develops his theories, treating him we might say merely as a contemporary, leads to substantive misinterpretation of those theories".[34] In this sense, Brandom had early adopted the idea of a connection between Kant and Frege concerning the (CTP) or the (PJ) from Sluga[35] and modified it more and more until the writing of his major works.

---

[26] Cf. Tyler Burge: Truth, Thought, Reason. Essays on Frege. Oxford 2005, esp. p. 14.
[27] Gottfried Gabriel: Windelband und die Diskussion um die Kantischen Urteilsformen. In: Kant im Neukantianismus. Fortschritt oder Rückschritt?. Ed. by Marion Heinz, Christian Krijnen. Würzburg 2007, pp. 91–109, esp. p. 93.
[28] Cf. Ulrike Kleemeier: Gottlob Frege. Kontext-Prinzip und Ontologie. Freiburg i. Br. 1997, p. 22, p. 25, pp. 47ff., p. 142; cf. also Christian Thiel: Das Verhältnis von Syntax und Semantik bei Frege. In: Philosophie und Logik. Frege-Kolloquien, Jena, 1989/1991. Ed. by Werner Stelzner. Berlin 1993, pp. 3–16.
[29] Cf. Ulrike Kleemeier: Gottlob Frege. Kontext-Prinzip und Ontologie, pp. 35f., p. 58.
[30] Cf. e.g. Ulrike Kleemeier: Gottlob Frege, p. 141ff. and also Wolfgang Kienzler: Begriff und Gegenstand. Eine historische und systematische Studie zur Entwicklung von Gottlob Freges Denken. Frankfurt/ Main 2009, p. 15.
[31] Cf. John Lyons: Semantics. Cambridge 1977, Vol. 1, Chapter 8.4.
[32] Cf. Donald Davidson: Quotation. In: Theory and Decision 11:1 (1979), pp. 27–40. Also Willard Van Orman Quine: The Problem of Meaning in Linguistics. In: Ibid.: From a Logical Point of View. 2nd ed. Cambridge/Mass. 1963, pp. 47–65, here: p. 58.
[33] Delbert Reed: The Origins of Analytic Philosophy. Kant and Frege. London 2007 erwähnt ihn gar nicht; W. Kienzler: Begriff und Gegenstand, p. 22 zählt Brandom (esp. Robert Brandom: Tales of the Mighty Dead. Historical Essays in the Metaphysics of Intentionality. Cambridge/Mass 2002, p. 237) zu den ahistorischen Forschern der Dummett-Schule.
[34] Robert Brandom: Tales of the Mighty Dead, p. 237.
[35] Hans Dietrich Sluga: Gottlob Frege. The Arguments of the Philosopher, p. 60, p. 93; this is proven by Robert Brandom: Tales of the Mighty Dead, esp. p. 257f.

2 Logic and its Geometrical Interpretation

In his 1986 paper, *Frege's Technical Concepts*, Brandom critically reviews eight points raised by Sluga, which Frege shares with Lotze, but then concludes: "Sluga's most important and sustained argument, however, concerns the influence of Kant on Frege."[36] Brandom then discusses five of Sluga's arguments, the last of which relates to (CTP), which Frege may have taken over from Kant and which Frege is said to have represented even in his later years.[37] According to Brandom, although Sluga had pointed out that the late Frege also represented (CTP), he had not provided precise evidence or sufficient interpretation to support this. The crucial point for him, however, remains that the naming criterion concerning Kant is allegedly only met at a later period of in Frege's oeuvre.

In Sluga's study, which Brandom tries to follow up, Kant's influence on Frege could only be plausibly proven in the late oeuvre, but not in the early writings, which were so decisive for the (CTP). Michael Dummett had, however, claimed that for Frege's late phase (CTP) had made it superfluous due to the distinction between sense and meaning.[38] For him, it is also common knowledge that the late Frege had rather represented (CPP).

Here one can now recognise a significant difference between the two inferentialists in dealing with historical contextualisations. Brandom's strategy looks different from Dummett's one: if he can prove that Frege also represented in the late phase (CTP) and thus provides the decisive criterion for the unity of early and late phase, Brandom could – thanks to the fulfilment of the naming criterion in the late oeuvre – argue for a unity of both periods of the work. Therefore, there is no revised doctrine, but *only one* doctrine and this one then depends on Kant, although Frege mentions him only in later years.

In fact, Brandom also finds a relevant quote for (CTP) in Frege's late *Notes for Ludwig Darmstaedter* (*Aufzeichnungen für Ludwig Darmstädter*),[39] which Sluga is said to have referred to only rhapsodically.[40] For Brandom, this is evidence of Sluga's thesis of continuity in Frege's work[41] as well as Sluga's Kant/ Frege thesis, which Brandom can now connect with the Kant-reception of late works of Frege, in which

---

[36] Robert Brandom: Tales of the Mighty Dead, pp. 255f.
[37] Robert Brandom: Tales of the Mighty Dead, pp. 257f.
[38] Cf. e.g. Michael Dummett: Frege. Philosophy of Mathematics, Cambridge/Mass. 1991, p. 2; cf. also Ignacio Angelelli: Critical Remarks on Michael Dummett's *Frege and Other Philosophers*. In: Modern Logic 3 (1993), pp. 387–400. The context principle is still restricted by some researchers to early Frege, cf. e.g. Wolfgang Künne: Die philosophische Logik Gottlob Freges. Ein Kommentar. Frankfurt/ Main 2010, p. 595. For an assessment of this difference between Brandom and Dummett, cf. e.g. Ulrike Kleemeier, Christian Weidemann: Brandom and Frege. In: Robert Brandom. Analytic Pragmatist. Edited by Bernd Prien, David P. Schweikard. Heusenstamm 2008, pp. 116f.
[39] Cf. e.g. Robert Brandom: Making it Explicit. Reasoning, Representing and Discursive Commitment. 4th ed. Cambridge/Mass. 2001, p. 80 incl. Fn. 19.
[40] Brandom bezieht sich dabei wohl auf Hans Dietrich Sluga: Frege and the Rise of the Analytic Philosophy. In: Inquiry 18 (1975), p. 478.
[41] Cf. Robert Brandom: Tales of the Mighty Dead, p. 261.

2.1 Semantics – Context Principle, Use Theory and Representationalism

(CTP) is also present.[42] Brandom's interpretation is implicitly supported and even radicalised by Janssen's arguments. On the one hand, Janssen uncritically adopts Sluga's Kant/ Frege thesis[43] and, on the other hand, even argues against Dummett that (CPP) in Frege's late oeuvre is formulated explicitly, but instead applied (CTP) is mainly applied.[44]

But where is (CPT) to be found in Kant, according to Brandom? I am sure that a rudimentary satisfactory survey of the semantic themes, theories and principles of Kant mentioned so far would have to fill at least the rest of this book – if, as one might doubt, it is at all possible to present a satisfactory survey on this topic. In the chapters to follow, I will, therefore, only draw attention to a few difficulties of interpretation, which Sluga and other representatives of the Kant /Frege Thesis have, in my opinion, not yet considered critically enough.

Brandom first explains that in traditional logic there are three basic parts, the logic (1) of concepts, (2) of judgments and (3) inferences, which build on each other bottom-up – i.e. an ascending merging of simple concepts into more complex propositions and then into complete inferences with at least two judgments.[45] In *Making it Explicit*, Brandom argues at three text passages and in *Articulating Reasons* twice that it was Kant's "cardinal innovation"[46] to replace (1) the logic of concepts with (2) logic of judgments.

> "One of his [sc. Kant's] cardinal innovations is the claim that the fundamental unit of awareness or cognition, the minimum graspable, is the *judgment.* "As all acts of the understanding can be reduced to judgments, the understanding may be defined as the faculty of judging." [Fn. 13: *Critique of Pure Reason,* A69/B94] For him, interpretations of something as classified or classifier make sense only as remarks about its role in judgment. A concept just is a predicate of a possible judgment, [Fn. 14: Ibid.] which is why "the only use which the understanding can make of concepts is to form judgments by them." [Fn. 15: Ibid., A68/B93] Thus for Kant, any discussion of content must start with the contents of judgments, since anything

---

[42] Regardless of whether Brandom was able to improve Sluga's interpretation of the late Frege in terms of the context principle, Ulrike Kleemeier (Gottlob Frege. Kontext-Prinzip und Ontologie, p. 53, p. 59, pp. 106ff.) was able to interpret the context principle in a comprehensible way at another point in Frege's late writings.
[43] Cf. Theo M. V. Janssen: Compositionality. Its Historic Context. In: The Oxford Handbook of Compositionality. Ed. by Markus Werning, Wolfram Hinzen, Edouard Machery. Oxford 2012, pp. 19–46, here: p. 21.
[44] Cf. Theo M. V. Janssen: Frege, Contextuality and Compositionality, pp. 14ff.
[45] Vide supra, the scheme at the beginning of Chapter 2.1. To answer the question of why inferentialists claim that inferences including only two judgmentes can be considered complete, vide infra, Chapter 3.1.2.
[46] Cf. Robert Brandom: Making it Explicit. Reasoning, Representing and Discursive Commitment, p. 8, p. 79, p. 362f.; ibid.: Articulating Reasons. An Introduction to Inferentialism. 2nd ed. Cambridge/Mass. 2001, p. 125, p. 159.

2 Logic and its Geometrical Interpretation

else only has content insofar as it contributes to the contents of judgments. This is why his transcendental logic can investigate the presuppositions of contentfulness in terms of the categories, that is, the "functions of unity in judgment. [Fn. 16: Ibid., A69/B94]"[47]

I do not want to decide here whether or not we can read an anticipation of (CTP$_F$) in the Kantian quotes given. I find in the individual quotes given by Brandom, especially in their respective context, arguments for and against an anticipation of (CTP$_F$). Researchers on Kant are also divided into two positions concerning the priority question: Whereas (PC) is represented, for example, by Tonelli, Reich and Natterer,[48] other authors such as Krüger, Brandt and Wolff see a (PJ) in Kant.[49] As an alternative to both positions, some researchers such as Longuenesse, Goy or Pollok circumnavigate the priority question and claim, for example, that there is only a correspondence between the so-called 'tables of judgement' and table of categories.[50] Brandom's above-given quote is thus stirring up a hornet's nest.

Although it remains undecided how the quotes of Kant should be interpreted, Brandom seems to have good reasons for his thesis that Kant 'invented' (CTP), that it was then forgotten in post-Kantian philosophy and was only profitably taken up again by Frege: "This insight into the fundamental character of judgment and so of judgeable contents is lost sight of by Kant's successors [...]. It is next taken up by Frege."[51] Thus Brandom's thesis of contextual amnesia in the 19th century – more precisely: between Kant and Frege – has been clearly defined.

In the meantime, however, progress in Frege research has implicitly or unnoticedly both strengthened and criticised Brandom's thesis: Wolfgang Kienzler points out that Kant can already be seen as *"the main* opponent of the philosophical efforts" of the early Frege, as especially §§ 4 and 23 of CN are supposed to show.[52] Above all, § 4 of CN, which according to Kienzler is based on the Kantian table of judgement of CpR, would now strengthen Brandom's Kant/ Frege thesis, since Brandom places

---

[47] Robert Brandom: Making it Explicit. Reasoning, Representing and Discursive Commitment, pp. 79f.
[48] Cf. Giorgio Tonelli: Die Voraussetzungen zur Kantischen Urteilstafel der Logik des 18. Jahrhunderts. In: Kritik und Metaphysik. Studien. Heinz Heimsoeth zum achtzigsten Geburtstag. Ed. by Friedrich Kaulbach, Joachim Ritter. Berlin 1966, pp. 134–158, here: p. 147; Klaus Reich: Die Vollständigkeit der Kantischen Urteilstafel. Berlin 1948, p. 48; Paul Natterer: Systematischer Kommentar zur *Kritik der reinen Vernunft*. Interdisziplinäre Bilanz der Kantforschung seit 1945. Berlin 2003, p. 53.
[49] Cf. Lorenz Krüger: Wollte Kant die Vollständigkeit seiner Urteilstafel beweisen?. In: Kant-Studien 59:4 (1968), pp. 333–356, here: p. 337; Huaping Lu-Adler: Kant's Conception of Logical Extension and Its Implications. California 2012, p. 75; Reinhard Brandt: Die Urteilstafel. Kritik der reinen Vernunft A 67–76; B 92–101. Hamburg 1991, p. 8–43; Michael Wolff: Die Vollständigkeit der kantischen Urteilstafel. Mit einem Essay über Freges Begriffsschrift. Frankfurt/Main 1995, pp. 1–8.
[50] Cf. e.g. Béatrice Longuenesse: Kant and the Capacity to Judge. Sensibility and Discursivity in the Transcendental Analytic of the *Critique of Pure Reason*. Princeton 2001, p. 76ff.; Ina Goy: Architektonik oder die Kunst der Systeme. Paderborn 2007, p. 55; Konstantin Pollok: Kant's Theory of Normativity. Exploring the Space of Reason. Cambridge 2017, p. 69, pp. 86ff.
[51] Robert Brandom: Making it Explicit, p. 80.
[52] Wolfgang Kienzler: Begriff und Gegenstand, p. 251.

2.1 Semantics – Context Principle, Use Theory and Representationalism

(CTP_K) in the textual environment of the table of judgments in CpR (e.g. A68/B93), which the early Frege seems to have read.

However, Kienzler omits the fact that Kant is not expressly mentioned in CN § 4 and thus the naming criterion is not clearly fulfiled. This is particularly problematic in that the "distinction of judgments into categorical, hypothetical and disjunctive" mentioned in § 4, which Kienzler cites, is not only found in Kant's philosophy but is a common feature of many logics that circulated after Kant in the first and also still in the second half of the 19th century.[53] (We will see this, for example, in Schopenhauer's logica major.) The indirect criticism of Brandom, on the other hand, results from the consideration that these logics of the 19th century are not based on the CpR, but on Kant's so-called *Jäsche Logic*, which follows a traditional text structure and positions the logic of concepts before the logic of judgements. And neither in a further transcript of a Kantian logic nor in the logic of Georg Friedrich Meier[54] used by Kant is there any reference to the fact, asserted with Brandom's thesis, that Kant would prefer the logic of judgements. That Frege was familiar with Kant's *Jäsche Logic* is, incidentally, proven by a reference given in Frege's FoA, which thus fulfils the naming criterion.[55]

If one wanted to conclude by formulating the current state of knowledge on the historical origin of the context principle and reception in Frege's work in a somewhat exaggerated way for the sake of clarity, one could say: Although current research repeatedly attests Frege misunderstandings concerning his reading of Kant,[56] according to Brandom Frege is supposed to have been the first interpret of Kant who noticed the 'secret and forgotten doctrines' of (CTP) and (PJ) within the CpR – even though it could not be clearly clarified whether the early Frege (at the time of FoA) actually knew the CpR or only the *Jäsche Logic*.

What speaks against the Brandomian thesis of contextual amnesia are the quotes from the 19th century, compiled by Janssen in 2013, which prove in chronological order (CTP) or (PJ) and which, as he states, he took over partly from Sluga and partly from Oliver Scholz:[57] (1) Kant uses (PJ) in the CpR, (2) Schleiermacher uses a holistic (CTP), which he adds to the hermeneutical circle with (CPP), (3) Trendelenburg uses (CTP) in his *Logical Investigations* (*Logische Untersuchungen*, 1840) and refers to a philosopher named Gruppe, (4) Lotze was partly in favour of (CTP) and partly in favour of (CPP), (5) Wundt explicitly referred to Schleiermacher's hermeneutical circle in the sense of (CTP) and (CPP), (6) Frege would finally take up (CTP) again

---

[53] Vide infra, Chapter 3.12; vide supra, Chapter 1.3.3. Only § 23 of CN fulfils the naming criterion, although Frege must not have read the CpR for the statements made there, since the Kantian concepts and theses, especially the doctrine on synthetic judgements, could be found in countless textbooks before 1879.
[54] Cf. Georg Friedrich Meier: Auszug aus der Vernunftlehre. Halle 1752, esp. pp. 69–114 (§§ 249–414). In his logic, Meier, too, advocates the priority of the conceptual.
[55] FoA, p. 19 (§ 12).
[56] Cf. Joan Weiner: Frege in Perspective. Ithaca 2008, pp. 31–80, esp. pp. 35ff., p. 41.
[57] Cf. Theo M. V. Janssen: Compositionality. Its Historic Context, pp. 22f.

2.1.2 The Context Principle in 19th-Century Neo-Aristotelism

The research overview given in Chapter 2.1.1 leaves many questions open, especially about the origin, reception and developmental history of (CTP) and (PJ). Although the sustainability of Brandom's interpretation of Kant could only be discussed to a rudimentary extent, the difficult argumentation with which Brandom tries to support Sluga's Kant/ Frege thesis might lead to the suspicion that a history of reception and thus the tradition of the (CTP), which goes from Kant to Frege, is an interesting but historically improbable fact. However, since even the modern opponents of any historical contextualisation of Frege do not put forward any convincing reasons that would definitively rule out a historicisation of (CTP), it is reasonable to assume that the originator of the context principle may have been found before Lotze, but after Kant. However, Janssen's evidence, which can probably be considered an improved interpretation of Brandom and Sluga, leaves just as many questions unanswered, which can be bundled into one: Is there a continuous history of reception from Kant to Frege? The only indication of influence on Frege given by Janssen concerns the fact that Trendelenburg is mentioned in Frege's early phase (naming criterion) and that unusual metaphors there form co-occurrences of (CTP_T) und (CTP_F) (citation criterion).[58]

In the following, I will argue to some extent with Sluga and Janssen against Brandom that there was no context oblivion in the 19th century. With Sluga and Janssen, I will also focus on Trendelenburg but show that the context principle was virulently discussed in the Trendelenburg circle before 1840. Based on the basic conviction of the Trendelenburg district, it will quickly become clear that (CTP) and (PJ) can be traced far back to a time which, according to the prevailing opinion, did not form the cornerstones of (CTP) but of (CPP) in the Occident.

In fact, the first evidence of the early history of (CTP) is not to be found in Trendelenburg's ideas, but first in his reinterpretation of traditional Aristotelian logic. Five years after Hegel's death, Trendelenburg published the book *Elementa logices Aristotelicae* (= ElA), which can be interpreted as a neo-Aristotelian counter-draft to Kant's *Jäsche Logic* and Hegel's *Science of Logic*. According to Trendelenburg, both Kant and Hegel followed the traditional bottom-up structure in their (PC)-logics – (1) logic of concepts, (2) of judgments, (3) of inferences – which since Hellenism has been guaranteed by the corresponding arrangement of the Aristotelian oeuvre provided by the *Organon*, namely (1) *Categoriae*, (2) *De interpretatione*, (3) *Analytica*

---

[58] Cf. Theo M. V. Janssen: Compositionality. Its Historic Context, pp. 23.

2.1 Semantics – Context Principle, Use Theory and Representationalism

Priora.[59] Trendelenburg, however, breaks with this traditional, i.e. Aristotelian-Kantian-Hegelian arrangement right away in § 1 of ElA, as he compiles statements from the complete works of Aristotle which indicate a priority of the judgement, i.e. (PJ) instead of (PC):

Ἐν οἷς καὶ τὸ ψεῦδος καὶ τὸ ἀληθές, σύνθεσίς τις ἤδη νοημάτων ὥσπερ ἓν ὄντων· (de an. III 6 [430a27f.]) περὶ γὰρ σύνθεσιν καὶ διαίρεσίν ἐστι τὸ ψεῦδός καὶ τὸ ἀληθές. τὰ οὖν ὀνόματα αὐτὰ καὶ τὰ ῥήματα ἔοικε τῷ ἄνευ συνθέσεως καὶ διαιρέσεως νοήματι, οἷον τὸ ἄνθρωπος ἢ λευκόν, ὅταν μὴ προστεθῇ τι· οὔτε γὰρ ψεῦδος οὔτε ἀληθές πω. (de interpr. 1 [16a12–16]) Ὥστε ἀληθεύει μὲν ὁ τὸ διῃρημένον οἰόμενος διῃρῆσθαι καὶ τὸ συγκείμενον συγκεῖσθαι, ἔψευσται δὲ ὁ ἐναντίως ἔχων ἢ τὰ πράγματα. (metaphys. (Θ) IX. 10. [1051b3–5])[60]

Where the true and the false are found, there is already a synthesis of concepts as such, which are one. [de an. III 6, 430a27f.] For in the field of composition and division the false and the true take place. The names (of things) and the words (of operations) therefore resemble in themselves the concept without synthesis and division, e.g. 'man', or 'white', if nothing is added; for neither false nor true is anyhow. [de interpr. 1, 16a12–16] Whoever, therefore, thinks true, whoever thinks that what is divided is divided and that which is synthesised is synthesised; but whoever thinks false, whose thoughts are opposite to those of the things.[61]

The compilation and translation quoted here demonstrate quite explicitly that Trendelenburg, in his compilation of Aristotelian theorems, does not start with Aristotle's *Categories*, which corresponds to the logical doctrine of concepts. Rather, Trendelenburg uses a variant of the (CTP) from a quote given in *De anima*, which states that not concepts, but only compositions of concepts, i.e. judgements can express true and false. Only then does he turn to Aristotle's *De interpretatione*, which is the counterpart to the doctrine of judgement.[62]

More important than the *principium convenientiae* and the picture theory in judgments, however, is the discovered variant of (CTP), which for Trendelenburg was the systematic reason for interpreting (PJ) in Aristotle's writings, as can be seen from his

---

[59] Vide supra, Chapter 2.1 (Introduction). To what extent this does justice to Kant or Hegel I leave undecided.
[60] Friedrich Adolf Trendelenburg: Elementa logices Aristotelicae. In usum scholarium. Ex Aristotele excerpsit convertit illustravit. Berlin 1836, p. 1 (§ 1).
[61] My translation is guided by the German translation given in Friedrich Adolf Trendelenburg: Erläuterungen zu den Elementen der aristotelischen Logik. Zunächst für den Unterricht in Gymnasien. Berlin 1842, p. 1 (§§ 1.2).
[62] Friedrich Adolf Trendelenburg: Erläuterungen zu den Elementen der aristotelischen Logik., p. 2 (§§ 1.2).

2 Logic and its Geometrical Interpretation

commentary on the three quotes of Aristotle given in ElA § 1: "First of all, judgment is described as the starting point of logic and the area of judgment is defined."[63]

In § 2 of ElA, Trendelenburg confronts the judgment to be examined by logic with the sentence reserved for grammar, based on De int. 5, 17a8f., 6, 17a25f. and 9, 19a32f. Since judgments express truth or falsehood, or more generally – in Brandom's words – 'commitments' (in terms of truth or existence),[64] they are the most original part of logic, whereas truth-neutral sentences remain the field of grammar:

> Since the truth is the object of cognition, logic must begin where the claim to truth first arises. This happens in the *judgment*, which is aimed at representing reality in a mental way. By this reference, logical and grammatical considerations [...], which in its wider scope focus on the sentence as its object, are divided.[65]

Only in § 3 does Trendelenburg – similar to the early formal logicians of the 20th century such as Frege, Russell, Whitehead, Hilbert, Ackermann, Quine –[66] then derive the logic of concepts from the logic of judgments and continues to refer to the decisive statements from Aristotle's *De interpretatione* (2, 16a29–32 with the concluding addition 16b15 and de int. 10, 19b10–12).

Since so far only systematic reasons for this revolution in logic have been given, the question remains open as to how Trendelenburg comes to doubt the centuries-old authority of Aristotle since the Aristotelian *Organon* begins with the theory of concepts or with the book on *Categories*. The answer is not only systematic but also philological in nature. And this philological reason is best explained by Trendelenburg's scholar Carl Prantl, who points out that the composition of the Aristotelian Organ does not go back to Aristotle or his early compilers such as Hermippus of Smyrna and Andronicus of Rhodes, but is actually the product of later commentators.[67] Prantl then writes in the spirit of Neo-Aristotelism initiated by Trendelenburg:

> As is well known, this complete complex [sc. of the Aristotelian writings], which is called *Organon*, contains the books in the following order: Κατηγορίαι, Περὶ Ἑρμηνείας, Ἀναλυτικὰ

---

[63] Friedrich Adolf Trendelenburg: Erläuterungen zu den Elementen der aristotelischen Logik, p. 1 (§§ 1.2).
[64] Vide infra, Chapter 3.1.2 and also 3.2.1.
[65] Friedrich Adolf Trendelenburg: Erläuterungen zu den Elementen der aristotelischen Logik, pp. 1f. (§§ 1.2).
[66] Vide infra, Chapter 2.1.3. Cf. also Bertrand Russell, Alfred N. Whitehead: Principia Mathematica I. 2nd ed. Cambridge 1927, p. 190 (= 20); David Hilbert, Wilhelm Ackermann: The Principles of Mathematical Logic. Transl. by L.M. Hammond, G.G. Leckie, F. Steinhardt. New York 1950, pp. 3–40 (= Chapter 1); Willard Van Orman Quine: Methods of Logic. 4th revised ed. Cambridge/Mass. 1982, pp. 93–167 (§§ 14–26).
[67] Cf. Carl Prantl: Geschichte der Logik im Abendlande. Vol. 1. Leipzig 1855, p. 89 with the references given there.

> πρότερα two books, Ἀναλυτικὰ ὕστερα two books, Τοπικά eight books, Σοφιστικοὶ Ἔλεγχοι. Of course, there is no statement that this order originates from Aristotle himself, since it is a product of later school activity. Particularly as far as the first of the above-mentioned writings, the *Categories*, are concerned, we will later (already from the Stoics on [...]) become sufficiently acquainted with the bottomlessness and miserably low level of philosophical talent [...], from which the necessity of a priority of the categories is always pronounced and repeated to disgust; it was only from the trivial school manner that it was necessary to progress from the simplest to the composite that the categories were placed at the beginning of the *Organon*.[68]

What Prantl sketches here is the philological thesis that is decisive for the entire Neo-Aristotelianism of the 19th century: The late peripatetic and stoic school is solely responsible for the centuries-old misery of a "priority of categories" or (PC). Trendelenburg had already hinted at this misery in his *History of the Doctrine of Categories* (*Geschichte der Kategorienlehre*) and connected it with the systematic thesis from § 1 of the ElA that with Aristotle, the doctrine of judgement must actually represent the "starting point of logic" because – according to the mereological variant of (CTP) – truth cannot be expressed in parts (concepts) but only in the whole (judgements):[69]

> It [sc. the Aristotelian book on *Categories*] has been preceded from time immemorial in order, it seems, to progress from the simplest elements to the most developed forms, from the concepts to the judgment, from the judgment to the inference, from the inference to the proof and to science in the books that follow one another. Aristotle, however, hardly raised the logical consideration with isolated concepts as with dissected parts, since according to his characteristic expression the whole is earlier than the parts*. [Note *: "polit I, 2. p. 1253, a, 20."] Just as he begins with the whole, so he commands that what is put together be broken down into its simplest parts**. [Note **: "polit I, 1. p. 1252, a, 18 [...]."] It is probable that Aristotle started out from the investigation of the sentence or judgement as a logical whole, which first claims to be true. Thus, according to the

---

[68] Carl Prantl: Geschichte der Logik im Abendlande. Vol. 1, p. 90.
[69] Cf. Friedrich Adolf Trendelenburg: Logische Untersuchungen. 2 Vol., 2nd rev. ed. Leipzig 1862, Vol. II, pp. 298f.

system, the book περὶ ἑρμηνείας would have to come before the *Categories*;[70]

Hereby it is clearly stated by Trendelenburg, the main representative of Neo-Aristotelism in the 19th century, that (CTP) or (PJ) was first represented by Aristotle, but was undermined by his successors and therefore misrepresented for centuries. If Brandom thus believes that Kant was the originator of (CTP), but that this was then forgotten for decades until Frege, Trendelenburg with its reinterpretation of Aristotelian logic has definitely provided a strong counter-argument to this thesis.

However, one might now suspect that Trendelenburg's interpretation of Aristotle was inspired by Kant. However, this assumption could not be supported by the text.[71] Rather, Trendelenburg's *Logical Investigations* (*Logische Untersuchungen*) even suggest[72] that the priority of judgement was first introduced in 1834 by the thirty-year-old philosopher and philologist Otto Friedrich Gruppe.[73] In his 1834 book, *Turning points in Philosophy* (*Wendepunkt der Philosophie*), which Trendelenburg also quotes in his EIA,[74] Gruppe took a stand against the metaphysics and speculative philosophy of his time, i.e. the philosophy "which believes it can develop knowledge from mere concepts".[75]

In contrast to the metaphysicians, Gruppe's basic insight is based on the linguistic maxim that thinking and language are in an irreducible interrelation.[76] This prompts him to analyse the content of various propositions of natural philosophy, and he finally concludes that "concepts cannot be understood without judgments".[77] In this dictum one can already read out a semantic variant of (CTP), as it is later found similarly in Frege or Wittgenstein. But it is clearer that Gruppe derives the (PJ) from this dictum:

> Here too, I am in sharpest contradiction with all previous logic, which I cannot help but accuse of a major error. For it deals first

---

[70] Friedrich Adolf Trendelenburg: Geschichte der Kategorienlehre. Zwei Abhandlungen. Berlin 1846, p. 9.
[71] It is true that Trendelenburg (Geschichte der Kategorienlehre, p. 275; further: id.: Logische Untersuchungen, Vol. I, p. 360) claims that the Kantian table of judgements would point the way to the categories; but he also states (ibid., p. 352f.; id.: Geschichte der Kategorienlehre, p. 278f.) that from Kant's original-synthetic unity of apperception the categories arise and then the judgements, thus presenting a good argument against Brandom's interpretation of Kant.
[72] Friedrich Adolf Trendelenburg: Logische Untersuchungen II, p. 211: "Gruppe has shown that every concept is based on a judgement, and therefore the judgement is falsely treated on the basis of the concept and from the concept."
[73] On Gruppe's philosophy of language in general, see Guido Vanheeswijck: Otto Friedrich Gruppe. The Linguistic Turn and the End of Metaphysics. In: 1830–1848. The End of Metaphysics as a Transformation of Culture. Ed. by. Herbert de Vriese. Louvain 2003, pp. 261–313.
[74] EIA, pp. 85f., Fn. 2 (§ 35).
[75] Otto Friedrich Gruppe: Wendepunkt der Philosophie im neunzehnten Jahrhundert. Berlin 1834, p. 12.
[76] Otto Friedrich Gruppe: Wendepunkt der Philosophie, p. 28: "Thinking is not without language, as language is not without thinking; the two are interrelated. In its full significance and with all that follows from it, this has never yet been considered" Cf. also p. 72.
[77] Otto Friedrich Gruppe: Wendepunkt der Philosophie, p. 43.

2.1 Semantics – Context Principle, Use Theory and Representationalism

> with concepts and then with judgments, the latter it considers to be compositions of concepts, the concepts are therefore something finished before the judgments are; I, on the other hand, maintain that concepts are first the results of judgments, that they constantly expand with judgments and that they can only be explained by them;[78]

So even if we trace language back to the "last atoms", "*simplicia* and radical words", it remains undisputed for Gruppe that "there they are always grounded in thought and judgment".[79] However, Gruppe does not only appear as a co-founder of the (PJ), but his statements also make it clear that Neo-Aristotelism in the form of Trendelenburg would have found it difficult to follow Kant. The reason for this is that Gruppe accuses Kant of "suspiciously assuming what he [sc. Kant] should have criticised, e.g. the categories and logic of Aristotle", which traditionally, i.e. before Trendelenburg's Neo-Aristotelism (and again in today's interpretations), has indeed taken as its basis the theory of concepts and is thus regarded as a fundamental paradigm for (CPP) and (PC).[80]

The suspicion is obvious that Trendelenburg wanted to make the academically incendiary Gruppe[81] socially acceptable, or rather suitable for the academic auditorium, by proving philologically, on the one hand, that (PC) has been based on false Aristotelism for almost two millennia and, on the other hand, that the historical Aristotle must have propagated the (CTP) and also the (PJ) for systematic reasons.[82] The thesis of false Aristotelism is again not an invention of Trendelenburg, but rather goes back to Christian August Brandis, who argued against (PC) in Aristotle in his pertinent treatise *On the Order of the Aristotelian Organ* (*Über die Reihenfolge des Aristotelischen Organons*) against (PC) – albeit in milder terms than Trendelenburg, Prantl etc. later on:

> The sequence in which they [sc. "Andronicus of Rhodes and the following Peripatetics, Aspasius, Adrastus [of Aphrodisias], and others, or even earlier Alexandrians"] arrange these books on logic [sc. of the Aristotelian *Organon*] can be justified by at least very apparent reasons: Because the sequence represents a progress from the simple elements (concept and word) to judgment and proposition, and from these to inference. With the help of these inferences,

---

[78] Otto Friedrich Gruppe: Wendepunkt der Philosophie, p. 43.
[79] Otto Friedrich Gruppe: Wendepunkt der Philosophie, p. 79, p. 80; on the context principle and the priority of judgement, cf. also ibid, p. 49, p. 82.
[80] Otto Friedrich Gruppe: Wendepunkt der Philosophie, p. 22. Trendelenburg, too, seems to share this view, cf.. Geschichte der Kategorienlehre, pp. 268–297.
[81] Cf. on the reaction to Gruppe his own assessment in Otto. Friedrich Gruppe: Wendepunkt der Philosophie im neunzehnten Jahrhundert, pp. 1–13.
[82] To my knowledge, this is a thesis that Aristotelian research has not refuted to this day, although the reputation of the *Categories* could be partially restored at the beginning of the 20th century.

> it then proceeds to the form of knowledge in relation to truth and certainty, on the one hand, and probability, on the other. But that Aristotle put them together or even wrote them in this sequence was something that those exegetes hardly accepted themselves. It is also much more plausible that they came about in reverse order.[83]

The quote shows an explicit devaluation of the (PC) usually associated with the Aristotelian *Organon*. If Brandis' reasons for the inversion of the order indicated here are more of a philological nature, Trendelenburg later connected this with the systematic thesis of Gruppe.

Since a Gruppe/Frege thesis which claims that Frege has now (PJ) or also (CTP) inherited directly from Gruppe and which would thus replace the Kant/Frege thesis discussed in Chapter 2.1.1, fails on the naming and citation criterion, I think that an influence of Trendelenburg on Frege is more likely. In addition to the reasons already mentioned by Janssen above, there are other reasons for this. Trendelenburg's ElA were – as the subtitle suggests – written *in usum scholarium*,[84] and the influence of this work in the 19th century cannot be overestimated even outside of grammar school logic lessons. Klaus Christian Köhnke, for example, writes

> Between the summer semester of 1868 [...] and the summer semester of 1879, there are 16 courses in the course catalogues of German-speaking universities which announce that they are based on Trendelenburg's 'Elementa'. He himself [sc. Trendelenburg] also based them on exercises (e.g. summer semester 1872).[85]

If Frege did not come across Trendelenburg through his regular school or university education, it is still highly probable that he was made aware of Trendelenburg by the trendelenburgian (PC) either through the writings of a Trendelenburg scholar[86] or

---

[83] Cf. also the fundamental work by Christian A. Brandis: Über die Reihenfolge der Bücher des Aristotelischen Organons und ihre Griechischen Ausleger, nebst Beiträgen zur Geschichte des Textes jener Bücher des Aristoteles und ihre Ausgaben. In: Abhandlungen der Königlichen Akademie der Wissenschaften zu Berlin. Aus dem Jahre 1833. Berlin 1835, pp. 249–299, here p. 252. Brandis's treatise hints that the criticism mentioned in the quote had already been prepared by Immanuel Bekker, Barthold Georg Niebuhr, Adolf Wilhelm Theodor Stahr and others. Before Trendelenburg, however, most 'New Aristotelians' seemed to reject (PC) more on philological and less on systematic grounds.

[84] Cf. ElA, pp. V–XIV (Praef.).

[85] Klaus Christian Köhnke: Entstehung und Aufstieg des Neukantianismus. Die deutsche Universitätsphilosophie zwischen Idealismus und Positivismus. Frankfurt/ Main 1986, p. 447, Fn. 26 (my transl.; J.L.), cf. also pp. 23–58.

[86] According to Gottfried Gabriel: Preface. In: Hermann Lotze: Logik III. Vom Erkennen (Methodologie). Ed. by Gottfried Gabriel. Hamburg 1989, p. XIX, Hermann Lotze (Logic in Three Books, of Thought, of Investigation and of Knowledge. Transl. and ed. by Bernard Bosanquet. 2nd ed. in 2 Vols. Oxford 1888, Vol. II, p. 220 = § 321) also advocates the priority of judgement, although he also "against the better insight of his contemporaries (A. Trendelenburg, C. Sigwart and W. Wundt) [...] adheres to the traditional structure of concept – judgement – conclusion". However, one must bear in mind that Trendelenburg also defends the priority of judgement in the *Logischen Untersuchungen II*, although his stages of logic also begins with

2.1 Semantics – Context Principle, Use Theory and Representationalism

critic such as Rudolf Eucken.[87] In short, it was impossible to avoid the topics (CTP) and (PJ) if one only dealt with logic to a rudimentary extent during Frege's time, as these were discussed by the numerous Neo-Aristotelians in the German-speaking world, especially through the ElA.

It is therefore obvious to abandon Sluga or Brandom's unmediated Kant/Frege thesis for historiography and instead concentrate on the history of reception outlined here: This begins with Gruppe's (CTP), which Trendelenburg then reinterpreted as a Neo-Aristotelian variant and passed on to its students, which ultimately led to Frege. With Trendelenburg and Frege as *missing links*, it can also be explained how research has repeatedly produced comparisons between Gruppe and, for example, Wittgenstein, which, although systematically plausible, have had a rather irritating effect historically.[88]

2.1.3 The Context Principle in Early Analytic Philosophy

However, the historiography of reception presented here raises the question of the extent to which Neo-Aristotelism of the 19th century had a school-forming influence on early analytic philosophy. In order to somewhat limit the obviously broad field of this task, I will especially focus on the contents of the first three paragraphs of Trendelenburg's ElA or its Aristotelian contents and thus examine whether and – if so – how the early Frege and Wittgenstein refer to the fact that

> according to ElA § 1, the logic based on the (CTP) begins by the judgment,
> according to ElA § 2 there is a difference between judgments and sentences,
> according to ElA § 3 concepts must be derived as isolated elements from the judgment.

Whereas Frege scholarship to this day assumes that the (PJ) and the (CTP) only appear in the FoA in 1884, I believe I can show that already in 1879 at the beginning of Frege's CN all three paragraphs of Trendelenburg's ElA implicitly resonate. This, I can already anticipate, does not, however, suit modern rationalism, which builds its

---

the concept. The content (priority of judgement) and form (priority of concept) of the logics of the time thus do not necessarily seem to contradict each other. A perfect correspondence of content and form, however, is offered – as shown – by Trendelenburg's ElA.

[87] Cf. Lothar Kreiser: Gottlob Frege. Leben, Werk, Zeit. Hamburg 2004, p. 293. Precisely because Eucken returned to the classical structure of logic (ibid., p. 289ff.), this could have been decisive for Frege to deal more closely with the question of logical structure; cf. also Gottfried Gabriel: Vorwort. In: Hermann Lotze: Logik III, esp. p. 116f.

[88] To my knowledge, such comparisons are first found in Hans Sluga: Gottlob Frege. The Arguments of the Philosopher, esp. p. 19f., p. 186.

2 Logic and its Geometrical Interpretation

ontology based on a semantics that must clearly distinguish between subject and predicate (or several predicates) and assign proper names and definite descriptions a particular role in judgment.[89]

According to the set of definitions in CN § 1, which is to be regarded as a prolegomenon and in which Frege states that he wishes to adopt undetermined and determined symbols from the general theory of magnitudes (arithmetics) or mathematics, Frege turns at the beginning of CN § 2 to the doctrine of judgment, which states

> A judgement will always be expressed with the aid of the symbol ⊢ which stands to the left of the symbol or combination of symbols giving the content {Inhalt} of the judgement. If we *omit* the small vertical stroke at the left end of the horizontal one, then the judgement is to be transformed into a mere *combination of representations* of which the writer does not state whether or not he acknowledges its truth.[90]

From this initial quote from the CN, §§ 1 and 2 of the EIA can already be read: 1. Frege begins with the judgement and not with the concept, which at least proves that the JP has already been implemented in practice. 2. Frege makes a clear distinction between judgement and sentence in the Trendelenburg-Aristotelian sense since in Frege's opinion the "mere connection of representations" does not express an intentional "commitment" of the writer on a truth value. This implies that before the 'transformation' the judgement must express a commitment to truth values.

Following the quote given, Frege makes the omission of the vertical line clear employing an example and explains that the "representation" thus expressed can be circumscribed by the words *"the phrase that"*. Thus, Frege has integrated the difference between judgement and sentence in the Trendelenburg-Aristotelian sense and, in addition, continues the content of ElA § 3:

> Not every content can become a judgment by placing ⊢ before its symbol; for example, e.g. the representation 'house' cannot. We therefore distinguish *assertible* and *unassertible* contents.[91]

The house example shows that Frege subsumes both concepts and sentences under what he calls 'representation' (Vorstellung). But whereas representation such as 'there

---

[89] Vide infra, Chapters 3.1.2 and 3.2.1.
[90] CN, p. 111f. (§ 2). (My modification of the translation; J.L.)
[91] CN, p. 112 (§ 2). (My modification of the translation; J.L.)

2.1 Semantics – Context Principle, Use Theory and Representationalism

is a house' etc. are an assertible content or thought,[92] as Frege notes twice,[93] the concept or word 'house' cannot be judged in isolation. Thus CN presents *in concreto* the meaning of (CTP): Only in the context of judgments can concepts be examined for their assertible content.

Although Frege scholarship has so far only concentrated on the explicit definitions of the (CTP),[94] a comparison with Neo-Aristotelianism makes it clear that Frege – as previously Trendelenburg in ElA § 1– must presuppose (CTP) in order to legitimise the (PJ) practised in § 2 of CN. The fact that, in addition, ElA § 3 must also be implicitly presupposed in Frege's book is made clear by the fact that of a judgment such as 'There is a house' "the representation of a house is only part of it".[95] This expression refers to the mereological variant of the (CTP) in Neo-Aristotelism, which starts from the whole in order to make the parts assertible (polit. I 2, 1253a20).[96]

§ 3 of CN is also a reliable indication that Frege regards the logic of concepts at most as a derivative of the logic of judgments.[97] This paragraph is directly addressed to the representatives of Aristotelian conceptual logic, for whom the distinction between subject and predicate in categorical judgement plays an important role. Whereas in traditional logic it was largely the concepts which were decisive for the inference,[98] since it is only the *terminus medius* which mediates the *terminus major* and *minor*,[99] which are separated in the two premises (*propositiones major et minor*), and which could thus be brought together in a single judgment, the *conclusio*, yet "a distinction between subject and predicate does not occur".[100]

For Frege, this distinction has only rhetorical significance,[101] which is why he concentrates on the components of the judgment only on "that which influences its possible entailments".[102] Frege's focus on the entailments (Folgerungen) is often seen as the starting point for inferential semantics or modern inferentialism: The non-logical components of judgements acquire their meaning through what can be followed

---

[92] Cf. Gottlob Frege: Nachgelassene Schriften und wissenschaftlicher Briefwechsel. Ed. by Friedrich Kaulbach. 2 vols., 2nd rev. ed. Hamburg 1976ff, vol. 1, p. 120 (to P. E. B. Jourdain 1910): "Instead of 'assertible content' one can also say 'thought'." With this equation, Frege moves close to Gruppe as well as to Trendelenburg's Aristotle. The parallel is made clear by the fact that both the beginning of CN (e.g. Ulrike Kleemeier: Gottlob Frege. Kontext-Prinzip und Ontologie, p. 27ff., p. 106f.) as well as New Aristotelianism, whose priority of judgement is based on a quotation from *De anima*, were criticised at the time (e.g. by R. Eucken) as psychologism.
[93] CN, p. 112, Fn. ** (§ 2), p. 134 with Fn. * (§ 12).
[94] Robert Brandom: Making it Explicit, p. 94ff. sees only the priority of judgement in § 2 of CN.
[95] CN, p. 112, Fn. ** (§ 2).
[96] Vide supra, Chapter 2.1.2.
[97] Cf. also Volker Peckhaus: Logik, mathesis universalis und allgemeine Wissenschaft. Leibniz und die Wiederentdeckung der formalen Logik im 19. Jahrhundert. Berlin 1997, pp. 288ff.
[98] Vide supra, Chapters 1.3.1, 1.3.3 or infra, Chapter 2.2.2, 2.2.3.
[99] Vide supra, Chapter 1.3.3.
[100] CN, p. 112 (§ 3).
[101] CN, p. 113 (§ 3), pp. 128f. (§ 9), pp. 163f (§ 22).
[102] CN, p. 113 (§ 3). (My modification of the translation; J.L.)

from them.[103] In other words, the context of judgments within an inference makes explicit what was only implicitly contained as conceptual content in one single judgment. It should not matter where or at which position in the sentence the conceptual content occurs. This view of conceptual theory may have seemed radical to Aristotelian logicians.

In § 4 of CN, Frege also criticises the traditional doctrine of judgement in so far as he almost completely levels the "distinctions which people make with regard to judgements" established since Kant's *Jäsche Logic* for almost all logics of the 19th century.[104] By these distinctions, Frege means the fact that Kant and the Kantians of later periods divided the form of judgments into four main moments or titles, each with three moments.[105] The titles are (T1) quantity, (T2) quality, (T3) relation and (T4) modality. With § 4, Frege thus drives not only Aristotelian but also Kantian logic into a corner: the once relevant distinctions of judgements are now to play no role in logic. For Frege, they are only relevant to grammar or rhetoric.

From the main moment of (T1) *quantity*, Frege first takes up universal and particular judgements: These traditional moments or properties of the judgments are also applicable "to the content even when it is put forth, *not* as a judgment, but as a sentence".[106] The disposition between sentence and judgement again corresponds to § 2 of the ElA, and it should be clear why Frege abolishes the distinction between the moments of judgement as well as the distinction between subject and predicate: just as the latter are only relevant in rhetoric,[107] so they can only be significant for grammar.

(T2) From the *quality*, Frege takes up the negation and tries to show here too that the distinction does not concern the judgement but the content. It is irrelevant whether this occurs in the sentence or in the judgment. Frege rejects the judgments subsumed by Kant under the title (T3) of *relation* with the same argument: "The distinction of categorical, hypothetical, and disjunctive judgements appears to me to have only a grammatical significance."[108] Concerning the fourth principal moment (T4), Frege argues that the modality does not concern the judgement but is only an expression or an assessment of the judge.

While the argument against modality is a typical 19th-century criticism of the Kantian table of judgments, the other points of criticism may have seemed highly problematic or even incomprehensible from the point of view of a Kantian of that period. Let me give just a few examples: (1) The Kantian distinction was relevant in

---

[103] Cf. Jaroslav Peregrin: Inferentialism. Why Rules Matter. New York 2014, pp. 3ff.; Rudolf Carnap: Logical Syntax of Language. Translated by Amethe Smeaton. Reprint. London et al. 2010, pp. 175ff. (§ 49).
[104] CN, p. 114 (§ 4); one may think here of the logic textbooks of W. T. Krug, J. F. Fries, A. D. Ch. Twesten, F. Ueberweg, G. A. Lindner and many others.
[105] Cf. Jäsche-Logic, §§ 19–31 (AA IX, pp. 101–109; Lectures on Logic, pp. 598–605), also CpR, p. 206 (A70, B 95).
[106] CN, p. 114 (§ 4) (My modification of the translation; J.L.).
[107] Frege hints at this in § 3 of CN when he limits the distinction between subject and predicate to the relationship between speaker and recipient alone.
[108] CN, p. 114 (§ 4).

2.1 Semantics – Context Principle, Use Theory and Representationalism

deciding whether a judgment had to be treated according to Stoic or Aristotelian logic. This can be seen as a fragment of predicate logic, those as a fragment of propositional logic. (2) The actually leveled distinctions of the Kantian table of judgments can be applied again to Frege's definitions themselves: the hypothetical judgments can be found in § 5, the negative and disjunctive ones in § 7, the categorical ones in §§ 11 and 12 of CN, etc. (3) Whereas Kantians avoided modes such as Felapton because of their unnaturalness, Frege's artificial logic even insisted on equating them with natural modes such as Fesapo.[109]

Frege's levelling of the logic of judgements in § 4 was, on the one hand, a milestone on the way to modern inferentialism and, on the other hand, the reason of why (CTP) could not play a role in the CN. Frege's aim was the doctrine of inferences, its unification and its calculability: "Everything necessary for a correct inference is fully expressed; but what is not necessary usually is not indicated."[110] One sees in this quote a prioritisation of the inferential context. Judgments thus only play a role in the context of inferences and the conceptual content only in the context of these judgments.

The fact that Frege only mentions (CTP) explicitly in the FoA may be explained by these very objectives: Whereas CN completely levels out the theory of concepts in § 2 and also large parts of the theory of judgement in § 4 in order to focus on the doctrine of entailment and inference, the FoA focuses on a specific element that turns out to be part of the theory of concepts in the course of the investigation: namely the concept of number one.[111] In CN, Frege implicitly uses the (CTP) to precede the doctrine of judgement, which is the only way to introduce the new form of the doctrine of inferences. In the FoA, Frege also puts the doctrine of judgement first, but uses the (CTP) explicitly to demonstrate the development of the doctrine of concepts by means of an example. It is more difficult to explain, however, why Frege does not consistently continue the sentence/judgement distinction of CN (according to § 2 of the EIA) in the FoA, when he defines (CTP) at some points for the sentence[112] and at others for the judgement.[113] Does this mean that the Frege of the FoA also allows truth values for grammar or has the author – far from the unambiguity of his CN – once again been 'misled by the inaccuracy of the language'?

I believe I have shown, with the history of development mentioned so far, that essential distinctions and ideas of Neo-Aristotelism, such as (CTP) and (PJ), can already be identified in Frege's early work. It can certainly be argued that logical-semantic hybrids can be found between Frege's FoA and Wittgenstein's PI, which preserve remnants of the neo-Aristotelian (CTP) and (PJ) variants, but at the same time favour

---

[109] Cf. Jens Lemanski, Hubert Martin Schüler: Arthur Schopenhauer on Naturalness in Logic.
[110] The objective is shown in CN, p. 3 (§ 2): "Everything necessary for a correct inference is fully expressed; but what is not necessary usually is not indicates;"
[111] Cf. FoA, p. xiii.
[112] Cf. FoA, p. xxii, p. 71 (§ 60), p. 73 (§ 62), p. 116 (§ 106); cf. also Ulrike Kleemeier: Gottlob Frege. Kontext-Prinzip und Ontologie, p. 60.
[113] Cf. FoA, p. 59 (§ 46).

2 Logic and its Geometrical Interpretation

different approaches of (CPP) and (PC). However, since I see it as my task here in Chapter 2.1 to show what role Schopenhauer's major logic, especially in relation to the discussion of (UTM) and (CTP), played by the late Wittgenstein and the inferential-pragmatic rationalists, I will conclude by making only a few sketchy remarks on Russell, Whitehead and the early Wittgenstein, illustrating the role and changes in the above-mentioned semantic principles in the period before the writing of the PI.

Even before the variant of (CPP) in Frege's late writings mentioned in the introduction to Chapter 2.1, Bertrand Russell's *On Denoting* in 1905 contains an application of both (CTP) and (RTM), which depends on the type of words. In Russell's theory of indefinite description, x plays the role of the variable which, with the indefinite quantifiers $\forall, \neg\exists, \exists$ as descriptions, results in whole propositions such as '$C(x)$':

> C(everything) means '$C(x)$ is always true';
> C(nothing) means '"$C(x)$ is false" is always true';
> C(something) means 'It is false that "$C(x)$ is false" is always true.' [...]
>
> Here the notion '$C(x)$ is always true' is taken as ultimate and indefinable, and the others are defined by means of it. *Everything, nothing,* and *something* are not assumed to have any meaning in isolation, but a meaning is assigned to *every* proposition in which they occur. This is the principle of the theory of denoting I wish to advocate: that denoting phrases never have any meaning in themselves, but that every proposition in whose verbal expression they occur has a meaning.[114]

Which theories of truth, elimination and identity Russell presupposes here is of no interest to us here. Rather, I believe that we should focus on the combination of (CTP) with (RTM), which is expressed in this paragraph. The quote proves that indefinite descriptions, in general, and quantifiers, in particular, have no meaning in themselves or independent of judgements ("not assumed to have any meaning in isolation", "denoting phrases never have any meaning in themselves"): Expressions such as `a man` are meaningless in themselves, but have a meaning in context.

Since definite expressions – if they are not misused as a collective singular – and proper names reduce the existential ($\exists$) to the uniqueness quantification ($\exists!$), they refer to exactly one object and thus give meaning to the linguistic expression without context: `the father of Charles II` and `Charles II` mean something, even if we do not know the context.

(CTP) is thus necessary for indefinite descriptions, whereas definite descriptions are meaningful in a context-free manner due to an (RTM). Although this reading of the

---

[114] Bertrand Russell: On Denoting. In: Mind, New Series 14:56 (Oct. 1905), pp. 479–493, here: p. 480.

2.1 Semantics – Context Principle, Use Theory and Representationalism

previously given quote has been the subject of some interpretations in recent years,[115] many interpreters still read this quote primarily as evidence of an oblivion of context claimed by Peter Strawson for the first time in the post-Fregean period.[116] Wilfrid Hodges even sees the quote given as an example of an extension of traditional Aristotelian compositionality, as he describes it to Ammonius and Ibn Sina: Quantifiers, logical connectives etc. only have meaning when they are composed together with other expressions to form a proposition.

Although one might object that Hodges' interpretation is only a bottom-up formulation of my top-down interpretation, his description illustrates well the difference between traditional (CPP)-Aristotelism and 19th-century (CTP)-Neo Aristotelianism: For Trendelenburg and early Frege, (CTP) applies unreservedly, while Russell, in my opinion, with his two-part description theory, reopens the traditional Aristotelian distinction between (auto-)categoremata and syncategoremata.

The extent to which indefinite and definite descriptions differ in terms of their semantic principles is particularly evident in the *Principia Mathematica*. In Chapter III of the introduction, the (CTP) for indefinite descriptions is even associated with a (UTM), and both are then distinguished from the (RTM) for proper names:

> By an "incomplete" symbol we mean a symbol which is not supposed to have any meaning in isolation, but is only defined in certain contexts. In ordinary mathematics, for example, $\frac{d}{dx}$ and $\int_a^b$: are incomplete symbols: something has to be supplied before we have anything significant. Such symbols have what may be called a "definition in use." Thus if we put
>
> $$\nabla^2 = \frac{\partial^2}{\partial x^2} + \frac{\partial^2}{\partial y^2} + \frac{\partial^2}{\partial z^2} \text{ Df,}$$
>
> we define the use of ∇^2, but ∇^2 by itself remains without meaning. This distinguishes such symbols from what (in a generalised sense) we may call proper names: "Socrates," for example, stands for a certain man, and therefore has a meaning by itself, without the need of any context.[117]

Incomplete symbols are reminiscent of Frege's "unassertible content" in that they classify certain symbols such as 'house' or '$\frac{d}{dx}$' as semantically meaningless, provided they are "in isolation" and "by themselves" and not "in certain contexts". The theory, but also the choice of words seem to clearly favour (CTP), (PJ) and thus §§ 1, 3 of the

---

[115] Cf. what I consider the most convincing account by Robin Hörnig: Eigennamen referieren – Referieren mit Eigennamen. Zur Kontextinvarianz der namentlichen Bezugnahme. Wiesbaden 2003, Chapter 2.
[116] Peter F. Strawson: On Referring. In: Mind 59 (1950), pp. 320–344.
[117] Bertrand Russell/Alfred N. Whitehead: Principia Mathematica I, p. 66.

2 Logic and its Geometrical Interpretation

ElA: Not only expressions such as `context`, which refers to (CTP) but also mereological metaphors such as `isolation` and especially `incomplete` indicate that Russell and Whitehead base their assessment of the completeness of an expression on the judgement.[118] If there are no judgments, something must first be added – as in Frege's house example – in order for the content to be judgeable or the symbol to be complete.

But if incomplete symbols and unassessable contents have been integrated into a context, they gain meaning: the word becomes a concept in a judgment, the symbol becomes a quantity in a formula. This process described in the quote even shows a proximity to (UTM), which is suggested, above all, by expressions such as `definition in use`. Definitions of use can be understood as an exemplary tool for determining the meaning of incomplete symbols in context:[119] The symbols, the fraction, the integral and the operator given in the quote are meaningless in themselves, but they are defined in the use of a particular formula. Since even indefinite expressions in ordinary language are incomplete symbols, the same applies to them: "they have a meaning in use, but not in isolation."[120]. However, the above quote shows that proper names like `Socrates` do not need context to have meaning because they refer to a specific object.

As in *On Denoting*, there are two types of descriptions: (CTP) and a variant of (UTM) for incomplete symbols, (RTM) for proper names. But even if one puts the emphasis in the quote on the `supply`, one cannot get a compositional interpretation as Hodges has indicated. The incomplete symbols may form complex semantic molecules when taken together, but they are not themselves atomic semantic units. But since this only applies to incomplete symbols, the radical semantic holism of Neo-Aristotelianism has been relativised.

One can see the final separation of early analytic philosophy from Neo-Aristotelianism in Wittgenstein's Tlp. Wittgenstein never once speaks of judgements in the entire Tlp, but always only of sentences. (The only exception is the quote given below). Even the formulation of the (CTP), "Only the sentence has meaning; only in the context of the sentence does a name have meaning" (Tlp 3.3, further: 3.314), no longer seems to assign any weight to the difference between sentence and judgement in ElA § 2. This is explicitly confirmed in Tlp 4.442:

> Frege's 'judgement stroke' '⊢' is logically quite meaningless: in the works of Frege (and Russell) it simply indicates that these authors hold the sentence marked with this sign to be true. Thus '⊢' is no

---

[118] Vide supra, Chapter 2.1.2.
[119] Cf. Rudolf Carnap: Meaning and Necessity. A Study in Semantics and Modal Logic. Chicago 1947, p. 147.
[120] Bertrand Russell/Alfred N. Whitehead: Principia Mathematica I, p. 67.

2.1 Semantics – Context Principle, Use Theory and Representationalism

more a component part of a sentence than is, for instance, the number of the sentence. It is quite impossible for a sentence to state that it itself is true.

Wittgenstein seems to put forward two, albeit related, reasons against the concept or stroke of judgement here: Firstly, a judgement is only the psychologistic idealisation of a sentence (cf. also Tlp 4.063), insofar as the "authors" – Frege would say: the "writers" ("die Schreibenden", CN § 2, see above) – impose their intentional states, their own 'for-truth-keeping' on the sentence.[121] On the other hand, according to Russellian type theory (Tlp 3.332), a sentence cannot state its own truth, because it would then only ever convene with itself, but not with a "picture of reality" (Tlp 4.06, further: 2.173, 2.21, 4.462).

The schools of interpretation differ on the question of whether the priority of sentence and thus (CTP) or the image of reality and thus (PTM) must be emphasised: According to ElA § 3, the concepts arose as isolated elements or their content from the judgement. If Wittgenstein now levels the sentence/ judgement difference in favour of the sentence, the concept or the assertible content must now be derived from the sentence. At least this is what the above-mentioned (CTP) demands in Tlp 3.3 etc. If, on the one hand, one follows the deduction line of the Tlp, which leads from the world (Tlp 1.1) via the picture (Tlp 2.1) to the proposition (Tlp 3.1), then the (CTP) seems to be fulfiled in the form of the sentential priority – as a counterpart to the judgemental priority of ElA § 3. If, on the other hand, one emphasises the counterpart to the doctrine of concepts in traditional logic, which can already be read out before the (CTP) in the Critique of Names (Tlp 3.141ff.) and which is then developed in the theory of 'formal concepts' with the explicit confrontation with the "old logic" (Tlp 4.126), then the last link to ElA § 3 is also relativised.

I think that what has now been sketched makes it possible to see that from the perspective of a 19th-century logician who was socialised with ElA (even if only in the broadest sense such as Frege in the CN-period, for example) Tlp appears as if Wittgenstein transferred the logical doctrine of judgement into a grammatical theory, by the levelling of judgements in favour of sentences. In Tlp, although not the logic of inference, but the neo-Aristotelian logic of judgement and concepts was thus finally transformed into a philosophy of language. Moreover, contextualism could also be elevated to a principle that, regardless of its Aristotelian antecedents, did not have to be reserved for logic alone. As a result, it was then still possible for the late Wittgenstein to transfer (CTP) of Tlp into what was later called the 'ordinary language philosophy'.[122]

---

[121] Cf. Giorgio Lando: Assertion and Affirmation in the Early Wittgenstein. In: Wittgenstein-Studien 2 (2011), pp. 21–49, esp. pp. 29ff.
[122] Cf. PI I § 49; for a compilation of all text passages on the context principle in Wittgenstein, see Michael N. Forster: Wittgenstein on the Arbitrariness of Grammar. Princeton/N.J. 2004, p. 233, Fn 12.

2.1.4 The Schopenhauer/ Wittgenstein Theses

In the previous Chapters (2.1.1–2.1.3), I have shown how Robert Brandom tried to tie in with Hans Sluga's thesis according to which Frege took over the (CTP) forgotten in the early 19th century and the (PJ) from Kant. However, this Kant/ Frege thesis seemed problematic for several reasons, so I could not take up the cudgels for Brandom's claim of context oblivion in the 19th century. Rather, I have supplemented the historical references of Sluga and Janssen and attempted to show a historically more probable origin and reception history of the context principle and the priority of judgement, which started from Gruppe, which Trendelenburg transferred back to Aristotle a little later, and which was then advocated as a 19th-century neo-Aristotelian doctrine by his successors. Frege, whose FoA has so far been regarded in research as the founding document of contextualism, will most likely have received the (CTP) associated with the (PJ) in the course of academic New Aristotelianism of the 19th century.

In the end, I have shown that Frege already applied the (CTP) and the (PJ) before the FoA, namely in the beginning of CN. Although the (CTP) is only used here, but not reflected and explained, a greater proximity to New Aristotelianism is shown in the CN based on several criteria than then in the FoA and in the late writings. By making a separation between the (CTP) for indefinite statements and an (RTM) as well as a (CPP) for proper names and definite identifiers, Russell relativised the New-Aristotelian approach further than Frege had already done after CN. Wittgenstein's Tlp shows in the end only traces of the New-Aristotelian position, which favours (CTP) and (PJ) unreservedly.

Schopenhauer played no role in the history examined in Chapters 2.1.1–2.1.3. This should not be surprising insofar as I presented an interpretation of the system in Chapter 1 that should have already distanced Schopenhauer as a representationalist far from the supposed contextualists of his time – whether one thinks here of Kant, Gruppe, Trendelenburg or others. Like Gruppe,[123] Schopenhauer is only associated in the research literature with one of the early analytic philosophers, namely Wittgenstein. This is not surprising especially when, for example, interpreters take up one of the representationalist aspects of Schopenhauer and associate it with the standard interpretation of Tlp or with the critique of representationalism in the PI: Both Schopenhauer and the early Wittgenstein then appear as relevant representatives of an (RTM).

In the chapters that follow (2.1.4–2.1.6), I will present arguments that suggest that Schopenhauer offers counter-evidence to Brandom's thesis of context oblivion in the 19th century. Moreover, evidence will be presented to show that in Schopenhauer's representationalist system, which should naturally make use of a radicalised (RTM), the (CTP) and even a variant of the (UTM) are also to be found in prominent positions.

---

[123] Vide supra, Chapter 2.1.2.

2.1 Semantics – Context Principle, Use Theory and Representationalism

This will strengthen the impression that this (UTM) is closer to the late Wittgenstein of PI than Russell's and Whitehead's (UTM) in the *Principia Mathematica*. In this chapter, however, I would first like to present the research that deals with the relationship between Schopenhauer and Wittgenstein.

I have chosen the rather neutral terms 'relationship' and 'relationship research' here since Schopenhauer/ Wittgenstein scholarship does not always clearly distinguish between historically oriented influence or reception research and systematically oriented comparative research. The fact that such relationship research exists at all is motivated, on the one hand, by the fact that several of Wittgenstein's friends and students have claimed that he first studied Schopenhauer at the age of sixteen;[124] on the other hand, relationship research is legitimised by the fact that the late Wittgenstein explicitly admitted Schopenhauer's influence on his thinking.[125]

In order to be able to give a rough overview and a quantitative orientation on relationship research, I have compiled two tables below. In Table 1, the research literature (ordered by year) is listed in the first column and the main topics dealt with in the research are summarised in the header. The numbers in the table fields indicate the page numbers if a particular study deals with a corresponding topic in more detail. In Table 2, Schopenhauer's writings are listed in the first column and Wittgenstein's writings are listed in the header, which are discussed in the research literature (table fields). Marginalia, by-notes and short comments on individual topics, some of which are qualitatively interesting, could not be included in either table. Only English and German publications (without 'grey literature') were taken into account that aimed at a direct comparison between Schopenhauer and Wittgenstein. The complete bibliographical references can be found in the bibliography.

---

[124] To the best of my knowledge, the longest, if not most complete, summary and discussion of these friendship statements can be found in Allan S. Janik: Schopenhauer and the Early Wittgenstein. In: Ders: Essays on Wittgenstein and Weininger. Amsterdam 1985, pp. 26–48 [Orig.: Philosophical Studies 15 (1966), pp. 76–95]. It is not clear from any statement exactly which of Schopenhauer's works Wittgenstein received.

[125] Cf. e.g.. Ms. 154,15^v–16^r (= Ludwig Wittgenstein: Culture and Value. A Selection from the Posthumous Remains. Ed. by Georg Henrik von Wright et al. Rev. 2nd Ed. London et al. 1998, p. 16): "I [sc. L. W.] think I have never *invented* a line of thinking but that it was always provided for me by someone else & I have done no more than passionately take it up for my work of clarification. That is how Boltzmann Hertz Schopenhauer Frege, Russell, Kraus, Loos Weininger Spengler, Sraffa have influenced me."

2 Logic and its Geometrical Interpretation

Table 1: Research Topics

Topics / Literature	Aesthetics	Ethics	Logic/ Philosophy of Language	Mathematics	Metaphysik/ Philosophie	Philos. of Science/ Naturalism	Religion/ Mysticism	Solipsism/ Epistemoplogy	World/ Representationalism	Will/ Intentionality	Time
1963 Gardiner		275-282					281		280	279	
1966 Janik (Repr. 1985)	43-46	41-47	35, 38		36, 38-39	36-37, 39	29-30, 41, 44-47	32-35, 45-47	32-35, 43	39-43, 45-47	
1969 Engel			287-299		299-302						
1973 Griffiths	97, 115	97, 104-116	101-102			99	98, 112	100-103	102-103	105-108	99-100
1975 Hacker (Repr. 1986)	97-99				93-96			81-82, 87-93		91	
1976 Griffiths			4-10, 12-15	6-9					5-9, 12-16	15-19	11-12
1978 Clegg	29-30, 42-44	29-30	29-30, 33-36, 41-43		32, 39		32-35, 43-46	30-36	40-44	40-43	
1979 Goodman		437-447			439-440			440-441, 444-445	440	438, 443	445-447
1981 Worthington			481-489	482-483, 494-496	489-496		481, 487, 490-492, 494		482, 490-492	484	484, 486-489
1982 Janik			272-273, 275-278					276			
1983 Churchill	499	499	492-493, 497-499		491-500		497-499	490-492	491-500	491, 493-496	497, 499
1983 Magee (Repr. 2002)		319-320	321-322					322-324, 338	316, 322-324	311, 314-316	
1988 Clegg			83-100						82-84	90	
1989 Janaway		333-342	331-332				318-321	321-331	321-328	336-342	
1989 Lange		29	32-52, 53-88		1-31, 32-41, 109-110		10, 109-112	41, 53-54, 69-134	26-28, 32-52, 64-88, 104-106	7-9, 97-103	

2.1 Semantics – Context Principle, Use Theory and Representationalism

Topics / Literature	Aesthetics	Ethics	Logic/ Philosophy of Language	Mathematics	Metaphysik/ Philosophie	Philos. of Science/ Naturalism	Religion/ Mysticism	Solipsism/ Epistemoplogy	World/ Representationalism	Will/ Intentionality	Time
1992 Janik			73-77		70-73	69-77					
1992 Weiner	84-88, 105	94-111	24-45	50	9-27	50-51	92, 94-98, 104-108	46-79	46-79	67-72	80-111
1995 Weimer		29-32	17-20, 43-45		27-29	26-27		18, 20-22, 35-42		23-32	
1999 Glock	437-441	437-443	435-437		427-435		441-443	443-449		449-455	
2002 Han	117-118	116-119						112-115		115-116	116-119
2003 Cakmak	124	121-124	118-119				119-120, 124		117-118	116-117	
2011 Millet		76-81			67-70			65-73	69-70, 73-76	73-79	78-79
2011 Tejedor		93-102			97-98			89-102		94-96	
2012 Schröder			368		371-372, 378-379		379-380	380	368-375	368-369	375-377	

2 Logic and its Geometrical Interpretation

Table 2: Work comparisons

Schopenhauer \ Wittgenstein	Tractatus/Notebooks 1914-16	PI	Blue Book	Philosophical Remarks	Big Typoscript/ Philosophical Grammar	On Certainty
WWR	Gardiner 1963, 275-282; Janik 1966, 26-47; Engel 1969, 287-301; Griffiths 1973, 96-116; Hacker 1975, 81-100 ; Clegg 1978, 29-46; Goodman 1979, 437-445; Worthington 1981, 481-496; Churchill 1983, 489-501; Clegg 1988, 82-94; Janaway 1989, 318-342; Magee 1989, 313-315; Lange 1989, 1-134; Weiner 1992, 9-111; Weimer 1995, 23, 32-33; Glock 1999, 427-452; Han 2002, 112-119; Cakmak 2003, 115-125; Millet 2011, 63-81; Tejedor 2011, 85-102; Schröder 2012, 367-375	Engel 1969, 287, 297; Clegg 1988, 94-100; Magee 1989, 313, 326; Lange 1989, 110-134; Janik 1992, 69-77; Weimer 1995, 13-25, 40-45; Glock 1999, 452-455; Schröder 2012, 375-380	Lange 1989, 123-134; Weimer 1995, 23-24	Lange 1989, 9, 96	Lange 1989, 96, 117-134; Schröder 2012, 373-375	Janik 1992, 69-77
On the Fourfold Root of the Principle of Sufficient Reason	Griffiths 1976, 4-19; Janik 1982, 275	Janik 1992, 69-77	Engel 1969, 295-299			Janik 1992, 69-77
Parerga and Paralipomena	Magee 1989, 312-313			Goodman 1979, 445-447		
Manuscript Remains	Weimer 1995, 23	Engel 1969, 287-294				

2.1 Semantics – Context Principle, Use Theory and Representationalism

As both tables show, the thematic and work-related scope of research is so extensive today that a discussion and qualitative-critical evaluation of the individual theses and arguments cannot even begin here. At this point, I would therefore like to pick out only a few research results and problems that are of importance for the present study.

First of all, one can see from Table 2 that although most studies deal with the relationship between the WWR (esp. I) and the Tlp, there are already research efforts that contrast Schopenhauer's Nachlass with Wittgenstein's late work. Reasons for why research is not exclusively limited to the relationship between Schopenhauer's main work and Wittgenstein's Tlp can be drawn on: Bryan Magee and Hans-Johann Glock in particular repeatedly emphasise that Schopenhauer's influence on Wittgenstein did not cease after the Tlp and that there are also important allusions to Schopenhauer in the late Wittgenstein.[1]

Garth Hallett has also indexed a list of over 20 lemmas with about 40 Schopenhauer allusions and parallels in the PI,[2] and according to Jerry Clegg, this list could even be extended almost indefinitely.[3] One of the best-known allusions was first researched by Morris Engel and states that Wittgenstein borrowed the concept of 'family resemblance' (Familienähnlichkeit) from Schopenhauer;[4] but it is also known that Magee stated a Wittgensteinian borrowing of Schopenhauer's concept of 'life form' (Lebensform);[5] Janik also differentiated the similarity between the concept of 'training' (Abrichten) in both authors;[6] and Glock emphasised that the concept of 'metalogical' was also of Schopenhauerian origin and that the late Wittgenstein had critically examined Schopenhauer's concepts of the will and intentionality.[7]

Severin Schroeder, on the other hand, questioned Schopenhauer's influence on Wittgenstein's late work, especially by examining the concepts of 'life form' and 'family resemblance',[8] and cited rather negative statements by the late Wittgenstein about

---

[1] Cf. Hans-Johann Glock: Schopenhauer and Wittgenstein. Representation as Language and Will. In: The Cambridge Companion to Schopenhauer. Ed. by. Christopher Janaway. Cambridge 1999, pp. 422–458, here: pp. 423f., p. 426; Bryan Magee: The Philosophy of Schopenhauer. Oxford 1983, p. 311.
[2] Cf. Gareth Hallett: A Companion to Wittgenstein's Philosophical Investigations. Ithaca 1977, p. 799. A similarly long but unrelated list of allusions can be found in David Pears: The False Prison. A Study of the Development of Wittgenstein's Philosophy. Vol. 1. Oxford 1987, esp. pp. 166ff.
[3] Cf. Jerry S. Clegg: Schopenhauer and Wittgenstein on Lonely Languages and Criterialess Claims. In: Schopenhauer. New Essays in Honor of His 200th Birthday. Ed. by Eric v. Luft. Lewiston et al. 1988, pp. 82–100, here: p. 94f.
[4] Cf. S. Morris Engel: Schopenhauer's Impact on Wittgenstein. In: Journal of the History of Philosophy 7:3 (1969), pp. 285–302, here: p. 287, Fn. 8. [Repr.: Schopenhauer. His Philosophical Achievement. Ed. by Michael Fox. Brighton 1980, pp. 236–254]
[5] Cf. Bryan Magee: The Philosophy of Schopenhauer, p. 326.
[6] Cf. Allan S. Janik: Wie hat Schopenhauer Wittgenstein beeinflußt?. In: Schopenhauer-Jahrbuch 73 (1992), pp. 69–78, here: p. 72f.
[7] Cf. Hans-Johann Glock: Schopenhauer and Wittgenstein, pp. 456f., Fn. 15. Vide supra, Chapter 1.3.
[8] Cf. Severin Schroeder: Schopenhauer's Influence on Wittgenstein. In: A Companion to Schopenhauer. Ed. by Bart Vandenabeele. Chichester et al. 2012, pp. 367–385, here: p. 378. Schroeder's judgement, however, should not be final. For example, he rightly questions Engel's thesis that Wittgenstein took the concept

Schopenhauer as evidence of the difference. In their writings, Glock, Janik and Ernst Michael Lange relativise these comments with positive statements by the late Wittgenstein about Schopenhauer.[9] Even before Schroeder, Linhe Han had already defined the relationship between the two thinkers critically and Wolfgang Weimer even negatively.[10]

With regard to the Quinean problem of translatability presented e.g. in *Word and Object*, Weimer remarks that Schopenhauer would probably take the standpoint "that the concepts are the same in all cultures, but the words are different – a view whose dubiousness immediately strikes me [sc. W. Weimer]".[11] Weimer's main argument emphasises Schopenhauer's linguistic-philosophical representationalism and is based, above all, on the (RTM) in WWR I, § 9 and WWR II, Chapter 6f. that we addressed in Chapter 1.3.1; compared to the (UTM) of the late Wittgenstein, this philosophy of language is naïve and therefore deficient. Weimer thus attests to Schopenhauer's overall "inadequacy of his philosophy of language".[12] In doing so, he has even extended Eugenio Coseriu's judgement – also discussed above in the introduction to Chapter 1.3 – that the case of Schopenhauer is a dark chapter in German philosophy of language – from e(xternal)- to i(nternal)-language-theory.[13]

For these reasons, not only the influence of Schopenhauer on the late Wittgenstein, but of course also an influence of one on the other in relation to the subject areas of 'logic' and 'philosophy of language' remains controversial. It should be noted that the recently published studies by Sascha Dümig and Daniel Schmicking do indeed present Schopenhauer as an i-language linguist in the sense of cognitive representationalism according to Jerry Fodor, Jerrold Katz or Zenon Pylyshyn and thus present good arguments against Coseriu's e-language interpretation;[14] however, it must also be noted

---

'form of life' (Lebensform) from WWR I § 54, since the original only says "Form des Lebens" (not Lebensform); however, Schroeder does not mention that "Lebensform" appears at least once in the main work (WWR II, chap. 25), albeit in a context that is not very relevant.

[9] Cf. Hans-Johann Glock: Schopenhauer and Wittgenstein, p. 424; Allan S. Janik: On Schopenhauer's Relationship to Wittgenstein. In: Zeit der Ernte. Studien zum Stand der Schopenhauer-Forschung. Ed. by Wolfgang Schirrmacher. Stuttgart-Bad Cannstatt 1982, pp. 271–279, here: p. 275; ibid.: Schopenhauer and the Early Wittgenstein, p. 31; ibid.: Wie hat Schopenhauer Wittgenstein beeinflußt?, p. 76; Ernst Michael Lange: Wittgenstein und Schopenhauer. Logisch-philosophische Abhandlung und Kritik des Solipsismus. Cuxhaven 1989, p. 2.

[10] Cf. Linhe Han: Wittgenstein and Schopenhauer. In: Wittgenstein and the Future of Philosophy. A Reassessment after 50 Years / Wittgenstein und die Zukunft der Philosophie. Eine Neubewertung nach 50 Jahren. Ed. by R. Halle, K. Puhl. Wien 2002, pp. 112–121.

[11] Wolfgang Weimer: Ist eine Deutung der Welt als Wille und Vorstellung heute noch möglich? Schopenhauer nach der Sprachanalytischen Philosophie. In: Schopenhauer-Jahrbuch 76 (1995), pp. 11–53, here: p. 21.

[12] David Avraham Weiner: Genius and Talent. Schopenhauer's Influence on Wittgenstein's Early Philosophy. Rutherford 1992, p. 29. The justification can be found on pp. 19ff.

[13] Cf. Noam Chomsky: Knowledge of Language. Its Nature, Origin, and Use. New York et al. 1986, esp. pp. 19–40.

[14] Sascha Dümig: Lebendiges Wort?, pp. 161ff. Cf. also the similar thesis by Daniel Schmicking: Zu Schopenhauers Theorie der Kognition bei Mensch und Tier – Betrachtungen im Lichte aktueller kognitionswissenschaftlicher Entwicklungen. In: 86. Schopenhauer Jahrbuch (2005), pp. 149–176, here: p. 154.

2.1 Semantics – Context Principle, Use Theory and Representationalism

that this result only plays into the hands of a supposed contextualist like Weimer again, who generally regards mentalist-functionalist (RTM)/ (CPP) approaches before and after Wittgenstein's PI as a flawed approach to normal language theory. For rationalists, Schopenhauer's representationalism remains simply naïve.

As controversial as Schopenhauer's philosophy of language is, especially in relation to the late Wittgenstein, it is remarkable that a large quantitative part of Engel's paper also falls into the two problem areas 'logic/ philosophy of language' and 'Schopenhauer/ late Wittgenstein'. Engel justifies the choice of examining these two subject areas, which are now designated as problem areas, by the fact that Wittgenstein had been interested in logical topics and that the latter had therefore followed the logical traces of Schopenhauer's main work to the latter's estate. Similar to Glock, Lange, Magee and many others, Engel therefore also believes he has found evidence of Wittgenstein's multiple readings of Schopenhauer, which can be seen, among other things, in the fact that both Schopenhauer in the *Manuscript Remains* and Wittgenstein in the later writings (RTM) have to reject in each case plausibly pointing to conceptual confusions in argumentative (normal language) discourses.[15] It is also interesting that Schroeder also emphasises – and this before naming his arguments that are supposed to speak against an influence of Schopenhauer on late Wittgenstein – that Schopenhauer rejects variants of (RTM) in WWR I, § 9, which results in a parallel to late Wittgenstein.[16]

Whereas Engel and Schroeder – in contrast to Weimer – do not reduce Schopenhauer's philosophy of language to (RTM) alone, Hacker, Clegg and Churchill even emphasise that the originality of the relationship between the two thinkers lies in the fact that Wittgenstein tried to transform Schopenhauer's 'metaphysics' and 'mysticism' into a theory of meaning.[17] Although these three approaches come close, in the broadest sense, to my thesis of a common ground between Schopenhauer's and Wittgenstein's theories of meaning, in Churchill's view Hacker and Clegg fail because of too strong limitations of content and argumentative gaps. But since Churchill himself also reduces the theory of meaning only to a theory of the language boundary (in the sense of Tlp 5.6), in my opinion, none of the three approaches is convincing.

Nevertheless, Hacker, Clegg and Churchill have credibly suggested that there are interesting aspects to Wittgensteinian theories of meaning in Schopenhauer's work.

---

[15] Cf. S. Morris Engel: Schopenhauer's Impact on Wittgenstein, pp. 290f., p. 295. David Avraham Weiner: Genius and Talent, pp. 32ff. also sees a clear similarity between Schopenhauer's and Wittgenstein's critiques of language. That Schopenhauer stands in this tradition of language criticism is shown by Dieter Birnbacher: Schopenhauer und die Tradition der Sprachkritik. In: Schopenhauer-Jahrbuch 99 (2018), pp. 37–56.

[16] Cf. Severin Schroeder: Schopenhauer's Influence on Wittgenstein, p. 378: "A more interesting point of contact is that Schopenhauer rejects the idea that understanding words is a process of translating them into mental images (WWR I § 9)."

[17] Cf. Peter Michael Stephan Hacker: Insight and Illusion. Themes in the Philosophy of Wittgenstein. Rev. Edition. Oxford 1986, Chapter IV.2; Jerry S. Clegg: Schopenhauer and Wittgenstein on Lonely Languages and Criterialess Claims, esp. Chapter III; John Churchill: Wittgenstein's Adaption of Schopenhauer. In: The Southern Journal of Philosophy 21 (1983), pp. 489–502.

2 Logic and its Geometrical Interpretation

Schroeder's claim that Schopenhauer is completely unaware of the problem of justification of linguistic meaning that has existed since Locke[18] thus appears to be outdated in light of Hacker's, Clegg's and Churchill's research.[19] Moreover, Schroeder's assertion seems far too premature insofar as Schopenhauer explicitly traces his theory of concepts back to Book III of Locke's *Essay Concerning Human Understanding* in both the logica major of 1820 and the logica minor of 1844, the contents of which Schroeder claims Schopenhauer did not know at all.[20]

In a paper that has unfortunately hardly been received, Janik goes one step further than Hacker, Clegg and Churchill and claims "that, despite many differences of opinion regarding details, Schopenhauer's linguistically immanent conception of logic completely anticipates Wittgenstein's position".[21] This seemingly generous thesis, however, remains limited in its justification to Janik's central theme, namely that our human way of thinking is situated in nature or is an extension of nature – a thesis reminiscent of the main argument of John McDowell's *Mind and World*.[22] This brings Schopenhauer's supposedly naïve representationalism close to an inferentialist position that focuses primarily on the material of the concept.

Janik sees this famous 'naturalism of second nature' not only as the central theme in Wittgenstein's PI and *On Certainty*, but also in a passage from Schopenhauer's dissertation (*On the Fourfold Root of the Principle of Sufficient Reason*), in which Schopenhauer describes that in the natural learning of language, every child intuitively and playfully learns to practically apply all the rules of logic that philosophical logic can only formulate theoretically as rules with great difficulty.[23] According to Janik, Wittgenstein and Schopenhauer thus meet in the "idea that the logic of thought consists not of rules but of an appropriate application of expression".[24] Although Janik briefly mentions the application of expression here, and elsewhere also the theory of meaning, it remains to be explored whether Schopenhauer's 'naturalism of second nature' also includes (UTM) or (CTP). In Janik, at least, we find no satisfactory indication of this, similar to Hacker, Clegg and Churchill.

---

[18] Cf. Severin Schroeder: Schopenhauer's Influence on Wittgenstein, p. 379.

[19] Cf. ibid., p. 379. – Moreover, it should be noted that the quotation from WWR I, § 9, which Schroeder gives as evidence for his negative judgement, is a quotation of which S. Morris Engel: Schopenhauer's Impact on Wittgenstein, pp. 295f. thinks may have been the inspiration for a passage in Wittgenstein's *Blue Book*.

[20] Vide supra, Chapter 1.3.3. For details cf. also David E. Cartwright: Locke as Schopenhauer's (Kantian) Philosophical Ancestor. In: 84. Schopenhauer Jahrbuch (2003), pp. 147–156.

[21] Allan S. Janik: Wie hat Schopenhauer Wittgenstein beeinflußt?, pp. 73f.

[22] Cf. esp. John H. McDowell: Mind and World: With a New Introduction. 5th ed. Cambridge/ Mass. et al. 2000, p. XX (Intr., § 8), pp. 84ff. (IV, § 7).

[23] Citing other texts by the two authors, David Avraham Weiner: Genius and Talent, pp. 40ff. also sees a similarity between Schopenhauer and Wittgenstein due to this topic.

[24] Allan S. Janik: Wie hat Schopenhauer Wittgenstein beeinflußt?, p. 74.

2.1 Semantics – Context Principle, Use Theory and Representationalism
2.1.5 Schopenhauer's Use Theory of Meaning and the Context Principle

I briefly summarise the results so far. In the first part of Chapter 2.1, it was first pointed out that the currently prevailing opinion on the history of contextualism approves of an influence of Kant on Frege and thus rejects Dummett's vote of a general context amnesia before Frege. The only dispute is whether there was contextualism in the period between Kant and Frege (Brandom) or whether one can still find traces of (CTP) in the 19th century (Sluga, Janssen). I have argued that Gruppe plays a crucial role in this history since Trendelenburg combined his variant of (CTP) with Brandis' critique of the traditional (PC) to form a New Aristotelian approach, which can then be found in Frege's CN. With the help of some sketches, I have also tried to show that Frege from the FoA onwards, Russell and Whitehead in the *Principia Mathematica* and Wittgenstein in the Tlp weaken the radical approach of New Aristotelianism and develop mixed forms of semantic theories and principles, the assessment of which, however, depends strongly on the mode of interpretation.

Although Sluga and Janssen have argued that with Schleiermacher there is at least one contextualist between Kant and New Aristotelianism, other authors in this era have not been discussed so far, to my knowledge. Although the approach of Russell and Whitehead, as indicated above with Carnap, shows closeness to (UTM), in research only Dewey and especially Wittgenstein are regarded as founders of (UTM). Although Engel, Janik, Hacker, Clegg and Churchill have made suggestions that Schopenhauer may have played a role in the history of early analytic philosophy because of his possible influence on Wittgenstein's theories of meaning, perhaps even on (UTM), no relevant argument is yet available to make Schopenhauer worth mentioning in the discussion between Dummett, Brandom, Sluga, Janssen et al.

Consequently, on the one hand, there are doubts that Schopenhauer is a pure representative of a traditional (RTM); on the other hand, a possible anticipation of a (UTM) has at most been hinted at by Schopenhauer, and the question of whether he plays a role in the history of (CTP) investigated by Sluga and Janssen has not been examined at all by so-called 'relationship research'; on the contrary, there have even been attempts to label Schopenhauer as a representative of a naïve representationalism: However, only Weimer sees Schopenhauer as a represent of (RTM), and only Schroeder believes that Schopenhauer did not even know the problem of meaning that could have led him to either (UTM) or (RTM). Likewise, Weimer believes that even in the case of semantic problems such as the Quinean translation problem, Schopenhauer would still advocate a naïve variant of (RTM). Consequently, from a perspective close to inferentialism, Schopenhauer's representationalism is naïve and deficient. Only Janik's thesis seems to be a glimmer of hope for reconciling Schopenhauer with modern rationalism.

2 Logic and its Geometrical Interpretation

The present chapter sets out to refute Weimer's and Schroeder's claims based on a Schopenhauer quote from the logica major. In doing so, I will argue against Dummett's thesis of a general context amnesia and Brandom's accusation of a partial context oblivion in the 19th century. In addition, I will cite further evidence both for Sluga's and Janssen's history of the (CTP) and for the parallels between Wittgenstein's and Schopenhauer's (UTM) suggested by Janik, Hacker et al.

The quotation I announced as relevant is found right at the beginning of the logic of concept within WWR2. The entire context of the quotation has been given by me in Chapter 1.3.1 in the list of topics of the logica major as no. (1'). In the context of the quotation, Schopenhauer is concerned with the acquisition of conceptual meaning and of linguistic competence, whereby the topic can first be understood quite generally as a *tertium comparationis* between him and Wittgenstein. For on a general level, Schopenhauer's lectures already reveal a systematic parallel to the opening passages of PI, in which (UTM$_W$) and then later also (CTP$_W$) are closely linked to the theme of language learning (keywords: "training", "ostensive teaching of words", "teaching of language", etc.). On a more specific level, however, slight differences emerge: While Wittgenstein shows us the process of language learning, especially in children (especially starting at § 5 of the PI), Schopenhauer comes to the question of how words as a simple sequence of signs become semantically charged concepts in the course of considering the possibility of foreign language learning and the associated problem of translation. Schopenhauer thus discusses precisely the topic that Schroeder is convinced the latter did not know, and which Weimer even thinks Schopenhauer could only have answered with (RTM). The Schopenhauer quote, on the other hand, which I believe contains a variant of (UTM) together with (CTP), now reads in the context of this theme of foreign language acquisition:

> ^{5}That is why one learns not the true value of the words of a ^{6}foreign language with the help of a lexicon, but only *ex usu* [by using], through ^{7}reading, if it is an old language, and through speaking, staying in the ^{8}country, if it is a new language: namely it is only from the various ^{9}contexts in which the word is found that one abstracts ^{10}its true meaning, finds the concept that the ^{11}word designates.[25]

For a more detailed discussion of the quotation, I have taken the line numbers (= L.) of the quoted German Mockrauer/ Deussen edition of 1913 and prefixed them in superscript numbers to the respective line in the quotation. According to my thesis, the quotation contains a variant of (UTM) as well as of (CTP): the variant of (UTM) is found in L.5–8; the variant of (CTP) is found in L.8–11 and begins after the colon.

---

[25] WWR2 I, p. 246.

2.1 Semantics – Context Principle, Use Theory and Representationalism

This colon together with the following "namely" (nämlich) are revealing in that they indicate that (CTP$_S$) is a justification for (UTM$_S$). Through this network of theory and justification, Schopenhauer seems to distinguish even more differentially between (UTM) and (CTP) than some Wittgensteinians do, as I found the thesis several times in the research literature that the relevant (UTM$_W$) and (CTP$_W$) sentences in PI §§ 43 and 49 are equal. So far, however, I have only talked about the rough structure and an interesting aspect of the Schopenhauer quote. What actually entitles me now to really read (CTP$_S$) and (UTM$_S$) out of this quotation?

I first come to (UTM$_S$) in L.5–8 and initially ignore the "therefore" (Darum) in line 5 and thus also the context of the quotation. The actual thesis, i.e. (UTM$_S$), is: "One learns the true value of the words of a foreign language *ex usu* [by using]" That this is clearly about the use of "language" is indicated by the "ex usu". Exactly how Schopenhauer imagines the use of language in foreign language learning can be made concrete: With old languages, the use consists in passive reading (l.6f.); with new languages, it consists in active speaking (L.7f.).

Now, one might think that it would be a little more difficult to prove why this is really a semantic theory and not simply, for example, the trivial statement that people use words or that the "value of words" is an aesthetic one. However, I believe that at least four arguments can be made for a semantic theory, the last of which places the quotation in the context of the whole passage, which I believe contains within it an approach that is revolutionary for the early 19th century:

(1) Although the context of the quotation shows this even more clearly, I think it is sufficient to point out that the expressions "true value of the words" (wahren Werth der Wörter) in L.5 and "true meaning" (wahre Bedeutung) in l.10 are mutually salva veritate et significatione substitutable.

(2) The use of the metaphor of value instead of the concept of meaning is also common in semantic theories in the German-speaking world to this day. Lorenz Puntel's paraphrase of the context principle, namely that "linguistic expressions only have a semantic *value* in the context of a sentence", can be cited as evidence ("sprachliche Ausdrücke [haben] nur im Zusammenhang eines Satzes einen semantischen *Wert*").[26]

(3) The third piece of evidence becomes obvious if one consults the original manuscript, in which Schopenhauer first wrote "concept" (Begriff), but then crossed it out and replaced it with "wahrer Werth" (true value).[27] For this, however, it is necessary to recall Chapter 1.3, in which it was explained that Schopenhauer makes a strict distinction between word and concept: Concepts

---

[26] Cf. Lorenz B. Puntel: Grundlagen einer Theorie der Wahrheit. Berlin et al. 1990, p. 146 (My emphasis, J.L.), also p. 147 et al. Furthermore, a good example is Tlp 3.313f.
[27] Cf. SBB-IIIA, NL Schopenhauer, Fasc. 24, 102^{r}f.

have meaning, the word is "the sensuous sign of the concept", just as the cypher is the sign of the number.[28] The correction in the manuscript thus shows that the expression "true value of the words" originally meant the concept of the word and thus its semantics.

(4) Furthermore, one can also pursue a different interpretative strategy and show that Schopenhauer in L.5f. even tries to distinguish (UTM) from the semantics of (RTM) – a thesis that should completely refute Weimer's claim. Schopenhauer, in fact, denies a variant of (RTM), especially through the following emphasised phrase: "we do not learn the true value of the words of a foreign language through the lexicon". With this, Schopenhauer refers to the context in which the relevant quotation was given. I will present this quotation context in more detail in the following.

In my opinion, Schopenhauer's expression "through the lexicon" refers to an interlinear translation, i.e. a word-for-word translation in which a word of a foreign language that is not yet semantically known by the translator is represented by a concept of the mastered language that is already full of meaning. Such a translation would have to presuppose a purely lexical semantics that assumes a bijective function between the source and target languages (*total equivalence*):[29] every word of one language corresponds in its meaning to a word of another language. However, this is precisely a too naïve semantic view, which Schopenhauer only allows to apply to concrete concepts such as 'Baum' and their translations 'arbor', 'tree' etc. For example, there are words in one language that have no conceptual equivalent in another language (*zero equivalence*). Words like 'chaos', 'affect', 'naïve' were therefore introduced as loan words. Most often, however, translations require a surjective representation between the two languages (*facultative equivalence*): "That is why", Schopenhauer says, "in the lexicon the word of one language is usually explained by several words of the other".[30]

What is revolutionary about this thesis is not even the assertion itself, but Schopenhauer's explanation and consequence. The consequence has already been illustrated above with the relevant quotation: Since there is no total equivalence between two languages and thus interlinear translations are problematic, Schopenhauer holds that languages are best understood from usage, i.e. he argues for (UTM). However, the explanation for the facultative equivalence of two languages is equally original. Unlike much of post-Fregean philosophy and linguistics, Schopenhauer explains the semantic problem not in terms of an algebraic formalisation, but in terms of geometrical logic.[31] That is, he applies the geometrical view to illustrate the facultative

---

[28] WWR2 I, p. 243. One can also relate this distinction to EIA § 2.
[29] On this typology of equivalence in translation studies introduced by Otto Kade, see Holger Siever: Übersetzen und Interpretation. Die Herausbildung der Übersetzungswissenschaft als eigenständige wissenschaftliche Disziplin im deutschen Sprachraum von 1960 bis 2000. Frankfurt/ Main 2008, pp. 52ff.
[30] WWR2 I, p. 245.
[31] Vide infra, Chapter 2.2.1.

2.1 Semantics – Context Principle, Use Theory and Representationalism

equivalence on the basis of the intersection between the source and target language. The fact that concepts of different languages are never completely congruent semantically (in the sense of a symbolic logic of containment) is shown, for example, by the Latin word `honestum`, whose scope of meaning and conceptual sphere

> is never hit concentrically by that of the word which any German word designates, such as Tugendhaft, Ehrenvoll, anständig, ehrbar, geziemend [i.e. virtuousness, honourable, decent, appropriate, glorious et al]: they all do not hit concentrically: but as shown:

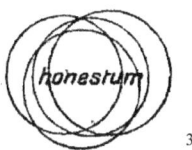

32

The analytical diagram or Eulerian diagram here shows the facultative or 1:4 equivalence between the source language and the target language:[33] The range of meaning of four words of a source language is necessary to 'encircle' the meaning of a word in the target language. Schopenhauer thus illustrates here with the diagram only the optional equivalence, which he also favours; but he does not symbolise the total equivalence, which he does address, but also rejects at the same time ("never hit concentrically"). The fact that Schopenhauer considered it nonsensical to illustrate a 1:1 equivalence in which two circles overlap concentrically in such a way that only one circle can be seen is already clear from the first of the five forms of connection shown in Chapter 1.3.1.

As tempting as it is here to place the above-given quotation in relation to Quine's, Davidson's, Putnam's or Chomsky's discussion of the translation problem, I would prefer to focus on the result, which is the combination of the (UTM) and the (CTP) given earlier. First of all, the quotation should have made it clear that Schopenhauer here completely identifies lexical learning with a naïve form of the (RTM), which amounts to a simple (CPP): If one could translate the conceptual content of a word from a source language 1:1 into a target language, then a judgement in a foreign language could emerge compositionally from the concepts thus translated. However, since experience shows that such an interlinear version does not work, the translator must acquire linguistic competencies. How these competencies are acquired is described by (UTM$_S$): One understands the meaning of a word through its use and not by learning 1:1 equivalents by heart or looking them up in a dictionary.

---

[32] Ibid.

[33] One may be surprised that the text, unlike the diagram, indicates a 1:6 equivalence. But anyone who sees Schopenhauer's spidery handwriting and especially his shaky circles in the original will understand that any additional circle here would have led to illegibility.

2 Logic and its Geometrical Interpretation

Before I turn to the (CTPs), the question arises as to what extent Schopenhauer's critique of the (RTM) in the problem of translation and his (UTM) inferred from it are consistent with the other use theories addressed in Chapter 2.1 (Russell, Dewey, Wittgenstein, etc.). The question is delicate in that in answering it I am forced to make a general classification of use theories, which will, in any case, be open to discussion. Nevertheless, I think it is advisable to commit myself to certain forms of (UTM) and (CTP) in order to show that Schopenhauer at the beginning of the logica major is indeed close to semantics that modern philosophers of language also hold.

For the time being, I will stay with Wittgenstein and use some comparative criteria that Engel and Forster have established. Engel highlights, for example, from PI 27f. the ambiguity of the ostensive teaching as a critique of an (RTM). One can certainly also see such a problem in Schopenhauer's critique of a naïve lexical semantics that propagates a total equivalence, although there is always only a facultative equivalence. Engel also highlights § 39 of the PI, in which the logatome Nothung is used to discuss whether words must always correspond to something at all. I think I also see this aspect in Schopenhauer's reference to the fact that concepts such as the Greek banausos or chaos actually have no lexical equivalent in other languages, i.e. in these cases there is a zero equivalence (1:0).

Forster sets up four quite different criteria by which he measures Herder's with Wittgenstein's (UTM), and which can be partially transferred to a comparison with Schopenhauer: (1) holism, (2) rule-following, (3) sociality, (4) psychologism. Since (1) and (3) concern the relationship between (CPP) and (CTP), which will be discussed further below, the question of the extent to which Schopenhauer fulfils these criteria will be answered later. (2) and (4) can be examined together based on the following passage from the logica major, which follows on almost directly from the relevant quotation containing (UTMs) and (CTPs):

> Thus, in learning a new language, one must cut off [abstechen] entirely new spheres of concepts in one's mind: conceptual spheres must arise in us where there were none before: thus we do not merely learn words, but acquire concepts.[34]

(2) The concept of rule or a description of rule-following is not found in the context of Schopenhauer's quote. However, as the cited quote shows and as was illustrated in Chapters 1.2 and esp. 1.3, for Schopenhauer, linguistic competencies are not directly conceptual, but either only indirectly and intuitively demonstrable employing a circumferential metaphor ("conceptual spheres") or directly by means of intuitive geometry. Therefore, both Wittgensteinian rule-following (in the non-behaviourist sense) and definitions of use in the sense of Russell and Whitehead would probably

---

[34] WWR2 I, p. 246.

not be satisfactory for him. (4) Although affective or perceptual impressions do not play a role in the whole context of the translation problem with (UTM$_S$) and (CTP$_S$), Schopenhauer's point of view is not purely anti-psychological, since language acquisition presupposes certain cognitive abilities (e.g. "cut off entirely new spheres of concepts in one's mind").

However, all these interpretations, which were developed primarily based on Forster's and Engel's criteria of comparison, depend on readings of Wittgenstein that are certainly worthy of discussion. On the other hand, it is safe to assume that Weimer's assertion that according to Schopenhauer "the concepts are the same in all cultures, but the words are different" must be considered just as false as Schroeder's opinion that Schopenhauer was completely unaware of problems and theories of meaning. In fact, the opposite is the case: Schopenhauer explains semantic problems in a hitherto novel way, namely with the help of Eulerian diagrams.

I will now turn to (CTP) in L.8–11. Schopenhauer's actual thesis, i.e. (CTP$_S$), is: "only from the various contexts in which one finds the word does one abstract its true meaning". That a variant of contextualism is clearly described here is indicated, above all, by the expression "context" (Zusammenhang), with the help of which one is supposed to find the "true meaning" of a "word". However, L.9 does not offer an explicit answer to the question of exactly which variant of contextualism Schopenhauer intends. Several variants can be ruled out, however:[35]

(1) That this is not a holistic (CTP) in the sense of Quine, Davidson or, further, Sellars, according to which a sentence only becomes meaningful in connection with a theory or with a language,[36] can clearly be seen in the fact that Schopenhauer limits contextualism here only to the word (L.9).

(2) Now, based on the phrase "staying in the country" (L.7f.) preceding the quote, one could assume that Schopenhauer intended a purely situational linguistic context in the form of social field research. However, such a behaviourist (CTP), as some researchers have interpreted from the Quinean Gavagai example, is probably based on a grammatical misinterpretation of L.5–8 in relation to Schopenhauer. The expression "staying in the country" is in the German version quite clearly an apposition to "speaking" (l.7) due to the comma placement ("Sprechen, Aufenthalt im Lande,...") and thus serves solely to draw attention to the fact that new languages (compared to old ones) can or even should be learned communicatively with native speakers.

---

[35] In presenting the variants, I follow Lorenz B. Puntel: Grundlagen einer Theorie der Wahrheit, p. 156ff.
[36] Cf. Willard Van Orman Quine: Two Dogmas of Empiricism. In: From a Logical Point of View. 2. rev. Ed. New York et al. 1963, pp. 20–47; Wilfrid Sellars: Truth and 'Correspondence'. In: Journal of Philosophy 59:2 (1962), pp. 29–56, here: p. 35 (Repr. in Science, Perception and Reality. Atascadero 1991, pp. 197–224); Donald Davidson: Truth and Meaning. In: Synthese 17:1 (1967), pp. 304–323, here esp. pp. 306–310. (Repr. in Inquiries into Truth and Interpretation. Oxford 1967, pp. 17–42). Wittgenstein's statement in PI 43 is not usually understood holistically, since it is not about the meaning of a sentence but about the meaning of a word in language.

2 Logic and its Geometrical Interpretation

(CTP$_S$) is thus always to be understood as social practice, as in the New Hegelian interpretation of Wittgenstein, but Schopenhauer differentiates more strongly than some modern philosophers of language do:[37] Speaking is a direct social practice that can include, above all, a 'behaviourist' moment (staying in the country). Reading is an indirect social practice that must almost exclude a behaviourist approach. Nevertheless, in language acquisition, reading also has the social component of being a form of communication between author and interpreter.[38]

Not only because of the exclusion of the holistic and the purely behaviourist approach, but also because of the reference to apposition, the suspicion that Schopenhauer represents a *linguistic-logical contextualism* is strengthened, since (CTP$_S$) can be meaningfully specified as follows: "only from the various contexts in which one finds the word *when reading or speaking* does one abstract its true meaning".

In contrast to Quine, Wittgenstein or Frege, the propositional context (theory, proposition, part of a proposition) in which the word must be embedded in order to unambiguously infer its semantics is not precisely determined – which, in my opinion, can be interpreted as a deficiency of (CTP$_S$); but the fact that a larger context is broken down to a smaller semantic unit here is attested to by the expression "abstracts" (L.9).[39] Since the next higher unit, according to the structure of the logica major discussed in Chapter 1.3.3, is the judgement, however, one can assume that the process of understanding associated with (CTP$_S$) implies a (PJ): From judgements, the meaning of words can be abstracted and thus understood.

It remains to be noted lastly that Schopenhauer does not oppose (PC) at this point; rather – and this is not unusual in the 19th century[40] – he restricts (CTP) to semantics and to certain processes of understanding (Verstehen), and uses (PC) together with (CPP) didactically for the construction of logic.[41] As outlined at the beginning of Chapter 2, most contextualists and compositionalists contrast (CPP) and (CTP) as contradictory principles, believing that traditional logics and their modern mentalistic and mechanistic offshoots start with (RTM) at the level of intuition, ideas or concepts and then transfer this semantic theory to judgements and inferences by means of (CPP). In contrast, however, Schopenhauer can be regarded as an example of the fact that neither the semantic theories, (UTM) and (RTM), nor their principles, (CTP) and (CPP), must necessarily be brought into opposition. Rather, there seem to be good

---

[37] Here I largely agree with the criticism of Michael Forster: Herder's Doctrine of Meaning as Use, pp. 214ff. but have doubts about his Crusoe thought experiment.
[38] Imitation of a writing style can be seen as a form of behaviourism.
[39] Schopenhauer's description of (CTP) in the form of a process of abstraction is still common today, cf. for example the relevant text by Donald Davidson: Truth and Meaning, esp. p. 308.
[40] Vide supra, Chapter 2.1.2.
[41] Cf. WWR2 I, pp. 234ff.

reasons to understand them subcontrarily. I will discuss why this is possible in the last Chapter of 2.1.

2.1.6 Representationalism and Contextualism

I hope that I have convincingly shown in the previous chapter that Schopenhauer, in his logica major, which can be found in his *Berlin Lectures*, advocated a use theory of meaning (UTM), which he justified with a variant of the context principle (CTP). Firstly, I believe that I have at least supported Janik's seemingly generous thesis that Schopenhauer's language-immanent conception of logic completely anticipates Wittgenstein's position. Secondly, I believe I have presented two relevant arguments against the criticisms of Schopenhauer's philosophy of language formulated by Schroeder and Weimer.

Of the two critics, Weimer had most strongly emphasised the "inadequacy" of Schopenhauer's philosophy of language. Weimer had claimed that a problem of translation, such as one finds in Quine, does not exist in Schopenhauer: If Schopenhauer had discussed it, he could only have advocated the (RTM) and furthermore the (CPP). In Chapter 2.1.5 it was explained that Schopenhauer was aware of the translation problem and treated it in exactly the opposite way to Weimer's assumption.

Even more generally than Weimer, Schroeder had argued that Schopenhauer had no sense for the Lockean problem of semantics. This premature thesis overlooks, on the one hand, that Schopenhauer's theory of concepts was already explicitly oriented towards Locke in the 1820s and, on the other hand, that those semantic theories do not necessarily have to depend on Locke. Contrary to Dummett's assumption of a general context amnesia before Frege or Brandom's weakened thesis of a context oblivion between Kant and Frege, Sluga, Janssen and Forster named numerous authors who advocated (UTM) or (CTP) and furthermore (PJ) long before Frege and Kant or even between Kant and Frege.

Regardless of Locke, Schopenhauer's semantics could also have had numerous other influences: For example, Schleiermacher, with whom Schopenhauer studied between 1811 and 1813, or one of the numerous authors from the biblical hermeneutical circles mentioned by Forster. Perhaps, but I can only speculate here, (UTM$_S$) and (CTP$_S$) were also inspired by traditional Indian logicians to whom Schopenhauer refers in his philosophy of logic and who aroused the interest of modern contextualists a few years ago;[42] or even Aristotle, whose contextual approach I drew attention to in Chapter 2.1.2.

---

[42] Cf. WWR2 I, p. 357. Cf. for example the volume Roy W. Perrett (ed.): Indian Philosophy: Logic and philosophy of language. New York 2001.

That Schopenhauer could then also have had an influence on Wittgenstein, especially on (UTM$_W$) and (CTP$_W$), remains pure conjecture.[43] It is true that both the naming criterion and the citation criterion are fulfiled, which form the foundation for research into the relationship between Schopenhauer and Wittgenstein in the first place;[44] however, I am too far from being able to take up an argument as a mortgage in order to secure credit for the thesis of a necessary dependence between (UTM$_W$) and a historical precursor.[45] As unresolved as the question must remain whether Schopenhauer could have influenced Wittgenstein, so too does the question remain whether a thinker before the 1810s or -20s influenced Schopenhauer.

What is certain, however, is that Schopenhauer's semantics not only closes another gap between Kant and Frege, but that the logica major also reveals several deviations from the previously known pre-Fregean history of semantics: Schopenhauer explains the (UTM) with the (CTP), he develops both on the basis of the discussion of the problem of translation, he uses forms of geometrical logic for an explanation, he discusses semantic problems even before the New Aristotelians in and based on logic, and he integrates (UTM) and (CTP) into an overall logical approach that was explicitly developed on the basis of the Lockean countermovement, namely on (RTM) and (PC).

I would like to conclude Chapter 2.1 with some rather free remarks concerning the relationship between (UTM$_S$) and (RTM$_S$), which, according to the prevailing opinion, have so far been assessed as either contradictory or even contrary. In Chapter 1, I drew attention to the fact that Schopenhauer represents a conceptual logic that is particularly influenced by Locke's empiricist theory of abstraction, according to which concreta refer to intuition and abstracta are only obtained from concreta by means of abstraction. Schopenhauer uses this kind of conceptual logic, above all, to structure his systems, WWR and WWR2, which are subject to the objective of Baconian representationalism to reflect the intuitive world in the abstract.

Such an (RTM) is usually perceived today as a naïve and conservative representationalism and opposed to a more progressive rationalism that starts from (UTM) in order to determine the role of concepts in judgements and inferences. Chapter 2.1 has shown, however, that in Schopenhauer's guileless representationalism we find a (UTM) as well as a (CTP) that run counter to the actual (RTM) concern and in this respect can lead to an aporia.[46] So what role can (UTM) and (CTP) play in a system of (RTM)? Why should an (RTM) include a (UTM) when meanings are based on the correspondences between words and ideas?

---

[43] I have addressed the question of a possible influence of Schopenhauer on Wittgenstein more intensively in Jens Lemanski: Schopenhauers Gebrauchstheorie der Bedeutung und das Kontextprinzip. Eine Parallele zu Wittgensteins *Philosophischen Untersuchungen*. In: 97. Schopenhauer-Jahrbuch 2016, pp. 171–196 discussed, but have come to the same conclusion, only hinted at here.

[44] Vide supra, Chapter 2.1.4.

[45] I also see Michael Forster's thesis of Herder's influence on Wittgenstein via Mauthner as similarly problematic.

[46] Vide supra, Chapter 1.1.4.

2.1 Semantics – Context Principle, Use Theory and Representationalism

By highlighting the systemic and substantive possibilities and limitations as well as the relationship between (UTM$_S$) and (RTM$_S$) in the following, I would like to make the case, first and foremost, that Schopenhauer did not advocate a naïve or causal representationalism, as one might think based on Chapter 1. To support this thesis, four arguments will be made, with both the first and the last two arguments being closely linked. Points (3) and (4) represent, in my view, the two main arguments that make it possible to build representationalism without neglecting rational semantic principles. Arguments (3) and (4) will be further asserted primarily in Chapter 3; moreover, they already lead over to Chapter 2.2 here:

(1) The apparent aporia between (RTM) and (UTM) can be straightened out by distinguishing between method and content or even form and matter in Schopenhauer's work.[47] Only through such a distinction have the differences in readings in contemporary Schopenhauer research become clear:[48] Whereas the descriptive reading of Schopenhauer's system refers primarily to passages in the text in which Schopenhauer describes his own method, his objectives and his results, the normative reading, which still prevails, focuses on the content discussed in Schopenhauer's work. However, as was shown in Chapter 1.3 using the example of logic, these contents are subject to a process of cognition, which is in part even based on the progress of logical research during the years in which Schopenhauer was occupied with this part of the system. If Schopenhauer's systems, WWR or WWR2, now claim, based on an (RTM), to represent all real and ideal facts of the world, i.e. the 'world as will' (realism) and the 'world as representation' (idealism), and if language as well as theories of language incl. (RTM) and (UTM) are subsumed under such facts, then Schopenhauer's WWR with its (RTM) method must also represent contents that concern the question of the validity of (UTM). To put it simply: a representationalist like Schopenhauer can describe a (UTM) in his system as an adequate position for a system domain without having to apply it as a philosophical method.

(2) The fact that (UTM) is a content of the system, but does not describe the method that brings this system about, can be traced back to the subject-centred paradigm of (early) modern philosophy. This paradigm does not see language as the centre of philosophical thought, but rather the subject or consciousness. Dummett's assertion of a paradigm shift that only began with Frege – away from Cartesian epistemology towards a logico-mathematically oriented philosophy of language[49] – fails to recognise Kuhn's result that paradigm shifts do not occur abruptly and

---

[47] Similarly, some Wittgenstein scholars explain how it is possible that Wittgenstein in Tlp, on the one hand, advocates (RTM) (e.g. Tlp 2.1–2.225) and, on the other hand, discusses also alternative elements such as (CTP) (Tlp 3.3, 3.314), which usually only occur with (UTM).

[48] Vide supra, Chapter 1.1.3.

[49] Cf. Michael Dummett: Frege. Philosophy of Language. New York et al. 1973, pp. 665ff. Cf. also Richard Rorty: Philosophy and the Mirror of Nature.

2 Logic and its Geometrical Interpretation

that they are prepared by crises. Such a crisis can be seen in the representationalist and empiricist approaches of Schopenhauer or Gruppes – or further also Herder – who develop their philosophy in critical confrontation with the – in their opinion – abstract language use of idealism, without yet renouncing the subject-centred standpoint. For representatives of this period of crisis, it is generally true that they can represent semantically modern-looking theoretical elements, but without ascribing to them a methodological or substantive centre of philosophising.

(3) The relationship between (RTM$_S$) and (UTM$_S$), however, primarily concerns the distinction between language emergence or development, on the one hand, and language use or learning, on the other. Both sides can also be equated with the difference between the *explanation* of meaning and *understanding* of meaning. After discussing the problem of translation, Schopenhauer explains possible scenarios of the emergence of concepts, which clarify essential relations of the logic of concepts. The concept itself remains a pure metaphor if one does not schematically connect it with intuition: 'concept', 'conceptus', 'termini', 'horoi' are expressions taken from geometry that refer to its circumference and sphere.[50] Schopenhauer first states, that

> Each concept (as a general, not particular representation) has a *sphere*, a *comprehension*, i.e. several other, certain concepts, or at least many real objects, which therefore lie within its sphere, can be thought of with its help. The concept *conceives* [der Begriff *begreift*] several things: this is without doubt the origin of the name 'concept'. So the name is appropriate; it says as much as a sum-total [Inbegriff]: We say, for example, "pack animal" contains all horses, camels, donkeys, and so on, or "countryman" conceives more than just the peasants. This is why such a general representation is called *concept*, in contrast to the individual representation which is intuition.[51]

For Schopenhauer, there are at least two forms of abstraction that could have been used in the original formation of concepts and which he reflects on using examples with analytical diagrams: From the already "formed concept (e.g. `bird = Vogel`)" all "determinations and differences" could have been abstracted except for one (e.g. `animal = Tier`) so that only one essential determination remained (see Fig. 1). Alternatively, several determinations could have been taken from a concrete concept such as `tree (Baum)`, e.g. `green (grün)`, `flower-bearing (blüthetragend)`, which formed subsets with each other but had a common intersection in the concrete concept (see Fig. 2).

---

[50] Cf. WWR2 I, p. 257, p. 297. Vide infra, Chapters 2.2.2 and 2.2.3.
[51] WWR2 I, p. 257.

2.1 Semantics – Context Principle, Use Theory and Representationalism

Fig. 1
WWR2 I, p. 258.

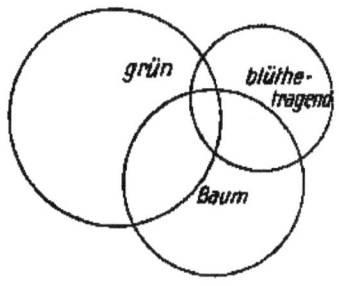

Fig. 2
WWR2 I, p. 257.

Let us take up here only the last procedure in conceptualisation since it is of interest in that the given example already approximates the functioning of Venn diagrams with three variables. In contrast to ordinary Venn diagrams, however, Fig. 2 describes semantic concept formation:[52] To clarify the syntax of the example, we use Venn's variables in Fig. 3, which correspond to the Schopenhauerian semantics given in Fig. 2, so that x = tree, y = green, z = flower-bearing. We interpret the process of abstraction as negation and symbolise it with an overline. From the three concepts and the negation, we can now create nine different combinations that correspond to the nine segments or regions in Fig. 3 (or Fig. 2). Here, for example, xyz denotes the intersection of all three circles, $\overline{xy}z$ the region in the z-circle that is not intersected by x and y, and $\overline{xyz}$ the entire region outside the three circles.

The concrete perception of an object in the outside world that has the properties of being a tree, bearing flowers and being green can best be described with xyz. Now, however, during the formation of the concept that finally led to the concept flower-bearing (z), more and more semantically occupied regions of z, which formed subsets with x and y, were omitted. This leads to abstraction steps such as (1) first xyz, (2) then $\overline{x}yz$, (3) finally $x\overline{y}z$. What remains is (4) the pure concept $\overline{xy}z$ = flower-bearing. If one draws this progression of abstraction steps in the diagram, one obtains what today would be called a directed graph.

Schopenhauer thus developed a method that is probably unique for the 19th century, since it applies the analytical diagrams for the logic of judgment and inference to semantics. As inspiring as this method is, it was described casually by Schopenhauer. Overall, I am therefore not very satisfied with Schopenhauer's examples using

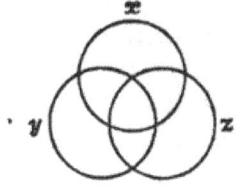

Fig. 3
John Venn: Symbolic Logic. 2nd rev. ed. London 1894, p. 115.

---

[52] Cf. John Venn: Symbolic Logic. 2nd ed. London et al. 1894, p. 115.

everyday language concepts and will therefore argue in Chapter 3.2.2 for the use of a more concrete data basis when investigating semantic relations. However, the above-mentioned examples already show that the procedure for describing semantics with the help of analytical diagrams opens up many possibilities and has potentials that, to my knowledge, have not yet been extensively elaborated and applied to semantics and didactics.[53]

What can be shown with the second method, however, is that (RTMs) and (UTMs) are not contradictory theories: In a representationalist theory, as found in WWR, we want to conceptually explain concrete objects such as a green, flower-bearing tree, but mostly use concepts such as x to understand them, and then have to find out their concrete-implicit meanings from the usage (xyz). Analytical diagrams thus represent relations that we use in language.

This difference between (RTMs) and (UTMs) becomes even clearer with another example. Schopenhauer defines that intuitions are usually to be called 'clear' (klar) if they do not appear 'obscure' (dunkel) due to sensory distractions etc. A concept, on the other hand, is 'distinct' (deutlich)

> when one can not only divide it into its characteristics, analyse it, define it; but when one can also analyse these characteristics again, if they are again abstracts, and [so on] down to the concretes, and then have clear perceptions corresponding to these and can substantiate them with them. [...] Concepts are confused (verworren) if one does not know their sphere correctly, that is, if one cannot break them down into their characteristics by specifying the other conceptual spheres that intersect or fill them, or surround them, that is, by definition; consequently, one either omits essential characteristics [or] adds false or unessential ones.[54]

Even if concepts can be formed representationally and used to represent intuitions, there are still different forms of concept formation and use that prevent the unambiguous representations of meanings between communication partners from different socialisation and language communities. Thus, although language formation and language use may have a representationalist origin and an equally communicative objective, language acquisition and understanding always require

---

[53] In computer science, researchers at the University of Brighton have developed a similar semantics in recent years, which is oriented towards Peirce diagrams, see for example Jim Burton, Gem Stapelton, Aidan Delaney, Jon Howse, Peter Chapman: Visualizing Concepts with Euler Diagrams. In: Diagrammatic Representation and Inference. Diagrams 2014. Lecture Notes in Computer Science, vol 8578. Ed. by T. Dwyer, H. Purchase, A. Delaney. Berlin, Heidelberg 2014, pp. 54–56. A similar method was apparently used successfully in translation studies in the late 20th century, although it was only insufficiently elaborated theoretically, see Ross Vander Meulen: Using Venn Diagrams to Represent Meaning. In: Die Unterrichtspraxis / Teaching German 23:1 (1990), pp. 61–63.
[54] WWR2 I, p. 254f.

2.1 Semantics – Context Principle, Use Theory and Representationalism

the context of the concepts. This context of concepts can be expressed in a judgement, which, however, is a reflection of several conceptual spheres.

(4) Schopenhauer's reference to semantics in the theory of concepts has the preparatory function for the theory of judgement of introducing a set-theoretical semantics of regions. It is precisely at the last quote given in (3) that the terms 'intersecting', 'filling' and 'surrounding' (schneidend, füllend, umgebenden) are introduced, which have been explained by the analytical diagrams given so far in Chapter 2.1. These and similar descriptions are used by Schopenhauer to determine relations, which must first distinguish between the connection or separation (of conceptual spheres) and then clarify to what extent the conceptual relations are connected or separated. Schopenhauer's critique of the assumption that processes of understanding or translation can generally always be one-to-one enables him to illustrate connection and separation using the 'honestum diagram' given in Chapter 2.1.5. Words in the target language such as `virtue`, `honourable`, `decent`, `respectable` etc. have a semantic intersection that is approximately identical to the concept in the source language, namely `honestum`. At the same time, however, the diagram also shows a symmetrical difference for the set of words specified for the target language: without, however, specifying the exact ratio of the target language words to each other – Schopenhauer's analytical diagram is not that precise – each of the target language words nevertheless shows a residual set that is not identical in meaning to the concept of the source language. Schopenhauer has thus introduced the essential function of logic diagrams based on semantics, in order to be able to describe judgements on the basis of relations.

Before I turn to Schopenhauer's logic of judgement and its relation to the analyticity debate, I would like to conclude by explaining why all four of the arguments mentioned should relativise the impression of a naïve representationalism that may have been created for the reader in Chapter 1 by my interpretation or by one of the prejudices of Weimer, Schroeder or Coseriu that have been refuted here.

All four arguments given show, in my opinion, that (UTM) does not have to assume a contrary or contradictory relationship to (RTM), but that the two semantic theories can each occupy their own position in the space of reasons in a rational representationalism: (UTM), as a rational method, is crucial for understanding the meaning of words, whereas (RTM), as a representationalist method, explains how it comes about that words have a meaning that we can understand in the first place. I will explicate this in more detail in Chapter 3.2.1 and then rely primarily on geometric logic to explain meaningful expressions in Chapter 3.2.2.

However, I do not want to get too far ahead of myself here, so I will take a final look at the four arguments developed on Schopenhauer's text and their significance for the philosophies of language presented in Chapter 2.1: While arguments (3) and (4) represent reasons for the harmonious coexistence of both approaches to language philosophy due to different uses and functions, arguments (1) and (2) have raised the question of the evaluation of content in a broader context. Arguments (3) and (4) already indicate, in a tentative way, the grounding of the logic of concepts for the logic of judgement to be dealt with in Chapter 2.2. Arguments (1) and (2), on the other hand, indicate the revisionism announced at the beginning of this book. This revisionism concerns, on the one hand, the content of Schopenhauer's philosophy on the basis of a system-related interpretation and, on the other hand, the neologicist reinterpretation of the pre-Fregean paradigm. For even if the linguistic-philosophical paradigm took off with Frege's elaboration of Neo-Aristotelian methods, this does not mean that – contrary to the programme sentence stated at the beginning of Chapter 2.1 – philosophy also begins with Frege.

2.2 Analyticity – Analytic judgements, Containment Metaphors and Logic Diagrams

Empiricist and representationalist theories, even if they have elements of progressive or rational semantics, are known to be subject to dogmas. Prominently, Willard Van Orman Quine debunked two of these dogmas in the year 1951: analyticity and reductionism. Quine sees both the distinction between analytic judgements (AJ) and synthetic judgements (SJ) and reductionism, which attributes the immanent logic of statements to intuitive experiences, as unfounded.

Both dogmas of empiricism, as one can already surmise based on Chapter 2.1, are closely related to each other. Empiricists claim that the classification of judgements is based on conceptual-logical relations, which in turn are based on intuitions and the relations of these intuitions. Quine famously finds it difficult to argue against the dogmas of empiricism, not only because he first attempts to criticise analyticity and reductionism as two independently posited dogmas, but also because he already fails on Kant's central definition of both (AJ) and (SJ) for explicitly stated reasons of understanding.

This failure was preceded by many attempts at understanding in the pre-Kantian history of philosophy. Undoubtedly, the critique of (AJ) and (SJ) is almost as old as Kant's distinction itself: Kant's contemporaries such as Johann August Eberhard, Johann Gebhard Ehrenreich Maaß or even Aenesidemus-Schulze had already questioned either the strict separation of (AJ) and (SJ) or analyticity itself.[1] And this criticism can also be found in the 'founding documents' of analytic philosophy: Frege criticises in § 88 of the FoA that even an illustration of (AJ) by a "geometrical illustration [...] [n]othing essentially new, however, emerges in the process" does not bring anything essentially new to light".[2] This criticism was taken up again and again until the 1950s and continued by Quine, Nelson Goodman and Clarence Irving Lewis,[3] so that Morton G. White, in a lecture two years before the publication of *Two Dogmas of Empiricism*, spoke of a revolution directed against the dualism given in the logic of judgment and this revolution had been instigated by the leading formal logicians.[4] The revolution against Kant's unacceptable dualism was directed primarily against the following introductory passage from the CpR:

---

[1] Cf. Henry E. Allison: The Kant-Eberhard-Controversy. Baltimore et al. 1973; James Van Cleve: Problems from Kant. Oxford et al. 1999, esp. pp. 18–21.
[2] FoA, p. 100 (§ 88).
[3] Cf. the overview provided by William H. Walsh: Reason and Experience. London 1947, pp. 30–51.
[4] Cf. Morton White: The Analytic and the Synthetic. An Untenable Dualism. In: Semantics and the Philosophy of Language. Ed. by Leonard Linsky. Urbana 1952, pp. 272–286.

2 Logic and its Geometrical Interpretation

> In all judgments in which the relation of a subject to the predicate is thought (if I consider only affirmative judgments, since the application to negative ones is easy), this relation is possible in two different ways. Either the predicate B belongs to the subject A as something that is (covertly) *contained in* this concept A; or B *lies* entirely *outside* the concept A, though to be sure it stands in connection with it. In the first case I call judgment analytic, in the second synthetic.[5]

Quine explains that his disagreement with this quotation is due to the Kantian metaphor. He sees the linguistic problem primarily in the fact that there is no generally valid definition of (AJ) and (SJ) in the quote, but rather a kind of a definition in use.[6] Nevertheless, in order to clarify and expose the dogma raised, he seeks refuge in a translation and reformulation that raises a strict dualism between meanings and facts, with which probably neither empiricists nor rationalists can be comfortable. According to Quine, the deficit of Kant's distinction between (AJ) and (SJ) is shown by two justifications: On the one hand, Kant's reduction of the doctrine of judgement to judgements with subject-predicate structures is insufficient;[7] on the other hand, Kant's definition in use remains inadequate because of the containment vocabulary (here: 'to be contained in', 'lying outside') remains unclear:

> Kant conceived of an analytic statement as one that attributes to its subject no more than is already conceptually *contained* in the subject. This formulation has two shortcomings: it limits itself to statements of subject-predicate form, and it appeals to *a notion of containment which is left at a metaphorical level*. But Kant's intent, evident more from the use he makes of the notion of analyticity than from his definition of it, can be restated thus: a statement is analytic when it is true by virtue of meanings and independently of fact. Pursuing this line, let us examine the concept of meaning which is presupposed.[8]

Quine's seminal paper *Two Dogmas of Empiricism* has itself been widely criticised. Nevertheless, the critique of the Kantian containment metaphor is still relevant today, both in post-Quinean philosophy and in Kant research.[9] If one wonders why metaphor has a negative connotation for Quine, to my knowledge only in Quine's *A Postscript*

---

[5] Kant: CpR, p. 130 (A6f., B10; My emphasis, J.L.).
[6] Vide supra, Chapter 2.1.3.
[7] Cf. also Rudolf Carnap: The Old and the New Logic. In: A.J. Ayer (ed.): Logical Positivism. Repr. Westport 1978, pp. 133–146. pp. 137ff.
[8] Willard Van Orman Quine: Two Dogmas of Empiricism, p. 20f. (My emphasis, J.L.).
[9] Vide infra.

2.2 Analyticity – Analytic judgements, Containment Metaphors and Logic Diagrams

on Metaphor can one find a reference that accuses the foundations of transcendental logic of inauthenticity. In his brief theory of metaphor, Quine has called for a conceptualization of technical language that goes hand in hand with scientific expansion. Max Black later assigned this form of metaphor theory to the "substitution view of metaphor".[10] For Quine, this means that if metaphors are transfers of words from a sensual-pictorial context into a technical linguistic one, then in the course of scientification the imagery and sensuality inherent in them should give way to the dry literalness and clarity of the concepts. Metaphors are to be translated into concepts or substituted by them.[11] As the last two sentences of the quote suggest, Quine's attempt at a 'restatement' of the containment metaphor can be found in an investigation of the 'concept of meaning'.

It remains questionable, however, whether Kant's containment vocabulary was really meant 'only' metaphorically and, above all, whether the genuine figurativeness of vocabulary such as 'to be contained', 'to lie outside', furthermore 'to include', 'to encompass' does not show a strong correspondence with the Eulerian diagrams introduced in Chapters 1.3, 2.1.5 and 2.1.6. Starting from Quine, this question has been the subject of controversy in four different fields of research in particular: (1) Kantian research, (2) cognitive linguistics, the inter- and partly transdisciplinary research communities called (3) 'diagrammatology' and (4) the so-called 'geometric logic'. Only a few milestones of these research areas are mentioned here, which, above all, should point the way to the approach envisaged here in Chapter 2.2.

(1) About four years after the publication of *Two Dogmas of Empiricism*, Stephen Körner reiterated Quine's two main criticisms of Kant, stating that the metaphorical formulation of (AJ) is not serious; while one cannot say that a subject contains its predicate as, say, one box contains another, nevertheless (AJ) can be extended not only to conceptual but also to relational propositions:[12] "A judgment is analytic if, and only if, its denial would be a contradiction in terms, or what amounts to the same thing, if it is logically necessary."[13] Körner's definition of (AJ) is based primarily on § 2 of Kant's *Prolegomena*.[14] Picking up on this definition, Richard Robinson argues that (AJ_K) and (SJ_K) were inspired by Leibniz' distinction of necessary and accidental propositions, which in turn depended on Aristotle.[15] Kant's detrimental reformulation

---

[10] Cf. Max Black: Metaphor. In: Proceedings of the Aristotelian Society, New Series 55 (1954/55), pp. 273–294, here: pp. 278ff.

[11] Willard Van Orman Quine: A Postscript on Metaphor. In: Critical Inquiry 5:1 (1978), pp. 161–162. The representationalist critique in Chapter 2.1.6, according to which thought and concepts cannot have metaphorical properties such as "neatly" and "clear", already indicates that Quine approaches empiricism with ideas that deviate strongly from his object of investigation.

[12] Cf. Stephen Körner: Kant. Baltimore/Maryland 1955, p. 22.

[13] Stephen Körner: Kant, p. 23.

[14] Cf. AA IV, p. 267.6–15 (= Kant's Critical Philosophy, Vol. II. The Prolegomena. Ed. and Transl. by J. P. Mahaffy, J. H. Bernard. London 1989, p. 15).

[15] Cf. Richard Robinson: Necessary Propositions. In: Mind 67 (1958), pp. 289–304, esp. p. 289–286. Cf. also Sybille Krämer: Tatsachenwahrheiten und Vernunftwahrheiten. In: Gottfried Wilhelm Leibniz: Monadologie. Ed. by Hubertus Busche. Berlin 2009, pp. 95–111, esp. pp. 108f. Vide infra, Chapter 2.3.1.

2 Logic and its Geometrical Interpretation

of Leibniz's criterion of contradiction into a containment metaphor was probably mainly based on the attempt to show that mathematics was synthetic.[16]

The fact that analyticity was not only a – as Quine summed up: non-empiricist and metaphysical–[17] dogma of logical empiricism,[18] but was also in the tradition of rationalism and empiricism in Kant's time, is nowadays explained, above all, from the history of tradition:[19] In 1958, Arthur Pap filled the historical gap left by Robinson between Aristotle and Leibniz as well as Leibniz and Kant with evidence of (AJ)-like criteria in Locke and Hume, which is still recognised by most researchers today.[20]

Paul Grice and Peter Strawson had already claimed in 1956 that the distinction between (AJ) and (SJ) had its justification in philosophy, but that rather the reasons for the distinction were semantically circular and therefore problematic.[21] After *In Defense of a Dogma* by Grice and Strawson, the debate developed in different directions: While, on the one hand, the question of (AJ) was in part only implicitly discussed based on the questions of meaning and translatability,[22] on the other hand, authors such as Jerrold Katz criticised that the intensional and holistic arguments of Quine and Davidson were themselves holistic and thus circular because of their criticism of the intensional distinction between (AJ) and (SJ).[23] The alternative to this, he argues, is a compositional approach.

Jonathan Bennett had already prepared this critique of Quine and Davidson; he had also criticised that Kant's definitions of (AJ) and (SJ) presuppose psychologistic assumptions:[24] Depending on the speaker, judgements can sometimes be used analytically, sometimes synthetically, just as tennis rackets can sometimes be used right-handed, sometimes left-handed.[25] This view was also held by Arthur Pap, but at the same time he pointed out that there were numerous psychologistic arguments in traditional and analytical philosophy, which were, however, not a disadvantage.[26] Hilary Putnam also criticised Quine, on the one hand, since he considered the distinction

---

[16] Cf. Richard Robinson: Necessary Propositions, pp. 297f.

[17] Cf. Willard Van Orman Quine: Two Dogmas of Empiricism, p. 37.

[18] Cf. e.g. Alfred Ayer: Language, Truth and Logic. London 1936, pp. 64–83 (= Chapter 4); Cory Juhl, Eric Loomis: Analyticity. New York et al. 2010.

[19] Cf. e.g. Albert Newen, Joachim Horvath: Apriorität und Analytizität: Zwei Grundbegriffe der Philosophie und ihre Entwicklung – Eine Einleitung. In: Apriorität und Analytizität. Ed. by Albert Newen, Joachim Horvath. Paderborn 2007, pp. 9–33, here: pp. 10ff. Vide infra, Chapter 2.3.1.

[20] Arthur Pap: Semantics and Necessary Truth. An Inquiry into the Foundations of Analytic Philosophy. New Haven 1958, Chapters 3 and 4, esp, pp. 59ff. and pp. 69ff.

[21] Cf. Herbert Paul Grice, Peter Frederick Strawson: In Defense of a Dogma. In: The Philosophical Review 65:2 (1956), pp. 141–158.

[22] Vide supra, Chapter 2.1.

[23] Cf. Jerrold J. Katz: Analyticity and Contradiction in Natural Language. In: The Structure of Language. Ed. by Jerry A. Fodor, Jerrold J. Katz. Prentice-Hall 1964, pp. 519–543; ibid.: Some Remarks on Quine on Analyticity. In: The Journal of Philosophy 64:2 (1967), pp. 36–52.

[24] Cf. Jonathan Bennett: Analytic–Synthetic. In: Proceedings of the Aristotelian Society 59 (1958/59), pp. 163–88; ibid.: On Being Forced to a Conclusion. In: Aristotelian Society Supplementary Volume 35 (1961), pp. 15–34.

[25] Cf. Jonathan Bennett: Kant's Analytic. Cambridge 1966, pp. 4ff., pp. 53f.

[26] Cf. Arthur Pap: Semantics and Necessary Truth, pp. 30ff., pp. 84ff., pp. 394ff.

2.2 Analyticity – Analytic judgements, Containment Metaphors and Logic Diagrams

between (AJ) and (SJ) to be justified, albeit exaggerated; on the other hand, he also criticised Strawson and Grice for not justifying the tenability of the distinction by giving a more precise definition of (AJ) and (SJ).[27]

Putnam's paper shows very well that the debate had lost its original object of reference: Kant is not even mentioned or quoted in the paper. This alone is evidence for Robert Hanna's thesis that the peak of the debate on (AJ) and (SJ) had been reached by the end of the 1950s and that the discussion declined sharply between the 1970s and -90s.[28] Reduced to the essential criticisms of the 1950s debate, four points can be named (I will add two more later):

(AJ) and (SJ)
- (a) are each defined differently by Kant,
- (b) apply only to judgements with subject-predicate form,
- (c) are neither conceptually nor logically defined by the metaphorical expression 'containment',
- (d) are only psychologistic assumptions.

I will argue in this chapter that of these criticisms, especially (c) plays a crucial role in understanding (AJ) and (SJ). This is shown, among other things, by the fact that (c) in particular has been discussed since the 1980s in other fields of research that were inspired by Quine's critique of Kant but do not refer to it in any significant way.

(2) Between the early and mid-1980s, a new debate ignited in linguistics, fuelled primarily by George Lakoff and Jerrold Katz. Initially, Lakoff and Mark Johnson argued in their book *Metaphors We Live By* that thinking in almost all languages takes place in three guiding metaphors: "The speaker puts ideas (objects) into words (containers) and sends them (along a conduit) to a hearer who takes the idea/ objects out of the word/ containers".[29] In the mid-1980s, Lakoff concretised that cognitions, but also emotions, are essentially described in metaphors and that central metaphors of thinking and feeling such as 'containment' (inside/ outside), 'boundaries' (interior/ exterior), 'verticality' (up/ down), among others, represent kinaesthetic image schemata.[30] The cognitive metaphors would be used to describe Boolean logic, which in turn would correspond to schematic representations, namely Venn diagrams.[31]

In his book *Cogitations*, Katz had attempted to question Descartes' analytical proposition "ego cogito, ego existo". To this end, he distinguished between two forms of containment metaphors, both of which go back to Locke: 1. sentence-containment,

---

[27] Cf. Hilary Putnam: The Analytic and the Synthetic. In: Minnesota Studies in the Philosophy of Science 3 (1962), pp. 358–397.
[28] Robert Hanna: Kant and the Foundations of Analytic Philosophy. Oxford et al. 2001, pp. 123f.; Cory Juhl, Eric Loomis: Analyticity, pp. 6ff.
[29] George Lakoff, Mark Johnson: Metaphors We Lived By. Chicago 1980, p. 10.
[30] George Lakoff: Women, Fire, and Dangerous Things. What Categories Reveal About the Mind. Chicago 1987, pp. 271ff.
[31] George Lakoff: Women, Fire, and Dangerous Things, pp. 456ff.

2 Logic and its Geometrical Interpretation

insofar as the consequent is contained in the antecedent of a conditional, 2. concept-containment, insofar as the predicate is contained in the subject. Concept-containment is particularly problematic because "Kant's account of containment is informal, highly metaphorical, and expressively weak".[32]

Although Katz, with his quote of Locke, provides stronger evidence than Körner that analyticity is not limited to the subject-object structure and that circumference metaphors are also useful in relational judgments, Lakoff's approach has prevailed overall.[33] Especially in the 1990s, Lakoff's theses were discussed controversially but productively in various disciplines such as psychology, linguistics, semiotics, cultural studies and phenomenology, computer science and logic, some of which work interdisciplinarily and also transdisciplinarily, but some of which apply their own research standards and methods.[34] Already in Lakoff's cognitive semantics, only a marginal engagement with Quine's metaphor critique was noticeable.[35]

Even in the two interdisciplinary research communities of (3) 'diagrammatology' and (4) so-called 'geometric logic', only traces of Lakoff's and Katz' argument about the containment metaphor can be seen from the 1990s onwards. Due to interdisciplinarity, both research communities, (3) and (4), have very heterogeneous methods, goals and viewpoints. Although scholars from both disciplines maintain an exchange with each other nowadays, clear differences in the scientific handling and evaluation of diagrams can be seen.

Whereas the field of (3) 'diagrammatics' or 'diagrammatology' debates the function and application of diagrams in general in connection with the so-called 'spatial turn' of the late 1980s and is strongly influenced by semiotic, structuralist and cultural studies approaches, today's (4) research on 'geometric logic' or 'logic diagrams' emerged from the 'New Mathematics', the 'Diagrammatic Reasoning' movements of the 1960s and 1970s and on the paradigmatic works of the Barwise school published in the 1990s, especially Sun-Joo Shins *The Logical Status of Diagrams*.[36]

However, the differences between (3) and (4) are not only historical but, above all, systematic. I summarise them in an abbreviated and exaggerated way as follows:

---

[32] Jerrold J. Katz: Cogitations. A Study of the Cogito in Relation to the Philosophy of Logic and Language and a Study of Them in Relation to the Cogito. Oxford et al. 1988, pp. 55.
[33] Sentence-containment wird aber auch in der analytischen Philosophie bes. seit Dummetts *Justification of Deduction* diskutiert, siehe unten, Kap. 2.3.
[34] Exemplary in the German-speaking world is the volume Diagrammatik und Philosophie. Akten des 1. Interdisziplinären Kolloquiums der Forschungsgruppe Philosophische Diagrammatik, 15./16.12.1988 an der FernUniversität/ Gesamthochschule Hagen. Ed. by. Thomas Keutner, Petra Gehring. Amsterdam 1992.
[35] Cf. George Lakoff: Women, Fire, and Dangerous Things, pp. 208ff.
[36] Vide infra, Chapter 2.3.

2.2 Analyticity – Analytic judgements, Containment Metaphors and Logic Diagrams

(3) The cultural studies approach[37]
- is primarily interested in the design of diagrams and the linguistic reflection on diagrams in general,
- makes a strict distinction between literacy and diagrams including the logocentric dogma that diagrams can only be understood through written explanation,
- starts from a subject-specific point of view of a thinker or school (e.g. Peirce, Cassirer, Serres) in order to explain the nature or commonality of diagrams,
- uses expressions such as 'logic' or 'epistemics' partly metaphorically, namely as if there were an independent 'logic of diagrams' or an 'epistemics of the diagrammatic'.

(4) Geometric logic[38]
- is primarily interested in the application and validity of specifically geometrical figures in logic,
- assumes that logic diagrams can be understood through their (contextual) use and defined by rules,
- assumes a specific logic (or its inherent modes of expression, rules, axioms, calculi or the like),
- does not use the term 'logic' metaphorically, but uses fundamentals of the discipline of 'logic' to inquire into the function of diagrams.

Certainly, this distinction is not shared by every researcher who deals with diagrams; but this distinction may help to recognise that there are different interests and goals in research to deal with logic diagrams. Anyway, to my knowledge, neither of the latter two research communities has revisited Quine's genuine Kantian critique and especially the meaningfulness of the metaphors of containment and circumference. Nevertheless, in recent years there have been a few (1) Kant scholars who have tried to bring together several of the above-mentioned research areas.[39]

(1) As the research groups in areas (3) and (4) expanded, the question of the validity of (AJ) and (SJ) was revisited in Kant research with the help of modal logic. With reference to Rudolf Carnap and Jaakko Hintikka,[40] researchers had increasingly attempted to determine Kant's logic as extensional or intensional since around the

---

[37] Cf. e.g. Martina Heßler, Dieter Mersch: Bildlogik oder Was heißt visuelles Denken?. In: Logik des Bildlichen. Zur Kritik der ikonischen Vernunft. Ed. by Martina Heßler, Dieter Mersch. Bielefeld 2009, pp. 8–62.
[38] Cf. e. g. Jon Barwise, John Etchemendy: Heterogeneous Logic. In: Diagrammatic Reasoning. Cognitive and Computational Perspectives. Ed. by J. Glasgow, N. Hari Narayanan, B. Chandrasekaran. Cambridge/Mass. 1995, pp. 209–232, here: p. 214.
[39] In the English-speaking world, one would probably use the term 'crossdisciplinarity'.
[40] Rudolf Carnap: Meaning and Necessity, Chapter V; Jaakko Hintikka: On the Logic of Perception. In: Models for Modalities. Selected Essays IV. Ed. by. Jaakko Hintikka. Dordrecht et al. 1969, pp. 151–183.

1980s. I first understand the term 'extensional' here as the set of objects and 'intensional' as the set of properties. In the broadest sense, many Kant scholars also accept this definition, but on the one hand, they discuss whether the term 'objects' only connotes concepts or also objects, and on the other hand, they like to replace the term 'set' with 'extension', which in my opinion hardly avoids confusion of terms.[41]

Due to this discussion stimulated by modal logic, two further points of criticism arise concerning the four points of criticism mentioned above around (AJ) and (SJ). Both criticisms are:

(AJ) and (SJ)
(e) cannot be unambiguously determined extensional or intensional,
(f) cannot be determined unambiguously object- or concept-related.

For Kant scholars, modal logic represents a suitable means of revising the debate from the 1950s insofar as, on the one hand, the reformulations of (AJ_K) and (SJ_K) as in § 2 of the *Prolegomena* include contradictory modal operators such as 'necessary' and 'not necessary', but, on the other hand, their application is often considered problematic. Nevertheless, Kant scholars such as Robert Hanna, James Van Cleve and others repeat that the use of a containment metaphor is not a sufficient criterion for determining analyticity.[42] Decades after Robert Körner's relevant text, Robert Hanna currently argues similarly, namely that there are judgements that can be classified as (AJ) according to the criterion of contradiction, but as (SJ) according to the containment metaphor.[43] Rico Hauswald has convincingly shown that Hanna made several exegetical errors in his attempt to support this thesis.[44] Nevertheless, the containment metaphor remains the most controversial point in the definition of (AJ) and (SJ), even in the modal-logical paradigm of Kant research.[45]

I had defined above that by 'extensional' I mean the set of objects. However, this view has become controversial. From the debate about the validity of a modal logical interpretation of (AJ), a new debate has arisen about the meaning of the containment metaphor, which mainly discusses the question (f) whether by metaphors such as 'circumference', 'containment' or 'extension' is meant either a set of objects[46] or of 'non-

---

[41] Cf. Rico Hauswald: Umfangslogik und analytisches Urteil bei Kant, pp. 284f., pp. 287f.
[42] Cf. James Van Cleve: Problems from Kant, pp. 18ff.
[43] Cf. Robert Hanna: Kant and the Foundations of Analytic Philosophy, pp. 123ff.
[44] Cf. Rico Hauswald: Umfangslogik und analytisches Urteil bei Kant, pp. 297ff.
[45] Cf. Rico Hauswald: Umfangslogik und analytisches Urteil bei Kant, p. 298.
[46] This is the view of e.g. Peter Schulthess: Eine systematische und entwicklungsgeschichtliche Untersuchung zur theoretischen Philosophie Kants. Berlin et al. 1981, pp. 103ff.; Bernd Prien: Kants Logik der Begriffe. Die Begriffslehre der formalen und transzendentalen Logik Kants. Berlin et al. 2006, p. 76, p. 83; John MacFarlane: Frege, Kant, and the Logic in Logicism. In: The Philosophical Review 111:1 (2002), pp. 25–65, here: p. 51.

2.2 Analyticity – Analytic judgements, Containment Metaphors and Logic Diagrams

real entities' such as truth values, concepts or ideas[47] or even both.[48] In addition, it is disputed – and here modal-logical arguments emerge again – whether the encompassed set contains only actualia or only possibilia or both.[49] That these questions are not unimportant in the clarification of (AJ) becomes clear, above all, from the question of the status of example judgements with possible entities without concrete reference: "A triangle has three corners" or the like.

Although Peter Schulthess had already pointed out in 1981, albeit rather incidentally, that Kant himself illustrated (AJ) and (SJ) in his lectures on the basis of analytical or Euler diagrams, only a few years ago Robert Lanier Anderson drew attention to the fact that Kant probably had Euler diagrams in mind in his containment metaphors, especially with regard to the definition of (AJ) and (SJ).[50] This would, however, make Quine's careless classification of the containment expression as a metaphor as well as the accusation of a missing conceptualization in the sense of this more differentiated description problematic in today's logic.[51] An interpretation of (AJ_K) and (SJ_K) as expressions of a set-theoretical scheme could bring the groping and repeatedly stuck research back upon the secure course of a science.

A first intensive comparison between historical Euler diagrams and Kant's circumference diagrams can be found in a study by Huaping Lu-Adler from 2012, which was inspired by Robert Lanier Anderson.[52] As profitable as this approach is in my opinion, it must be noted, in addition to several problems of detail,[53] that the study, on the one hand, unfortunately only refers to the three historical approaches of Euler, Lambert and Leibniz' and, on the other hand, primarily focuses on the question of which properties can be attributed to the encompassed set (objects/ ideas, actual/ possible etc.). Lu-Adler comes to the intermediate conclusion that in the *Logic of Port-Royal*, Wolff and Wolffians such as Martin Knutzen, Karl Daniel Reusch and Georg Friedrich Meier, etc., there is no certain criterion as to what exactly conceptual circumference or containment refer to, whereas in the geometric figures of Leibniz, Lambert and

---

[47] This view is held, for example, by Rainer Stuhlmann-Laeisz: Eine Interpretation auf der Grundlage von Vorlesungen, veröffentlichten Werken und Nachlaß. Berlin et al. 1976, pp. 87f.; Lanier Anderson: It Adds up After All. Kant's Philosophy of Arithmetic in Light of the Traditional Logic. In: Philosophy and Phenomenological Research 69:3 (2004), pp. 501–540, here: p. 507f.; ibid.: Containment Analyticity and Kant's Problem of Synthetic Judgment. In: Graduate Faculty Philosophy Journal 25:2 (2004), pp. 161–204, here: pp. 186ff.; Clinton Tolley: Kant's Conception of Logic. Chicago (Diss.) 2007, pp. 429ff.; Timothy Rosenkoetter: Are Kantian Analytic Judgments About Objects?. In: Recht und Frieden in der Philosophie Kants, Vol. 5. Ed. by Valerio Rohden, Ricardo R. Terra, Guido A. Almeida, Margit Ruffing. Berlin et al. 2008, pp. 191–202, esp. p. 199.
[48] This view is held, for example, by Robert Hanna: Kant and the Foundations of Analytic Philosophy, pp. 130ff.; H. Lu-Adler: Kant's Conception of Logical Extension and Its Implications, p. 18.
[49] Cf. ibid.
[50] Cf. Robert Lanier Anderson: The Poverty of Conceptual Truth. Kant's Analytic/Synthetic Distinction and the Limits of Metaphysics. New York 2015, pp. 100ff.; ibid.: Containment Analyticity and Kant's Problem of Synthetic Judgment, pp. 161–204.
[51] Cf. Peter Bernhard: Euler-Diagramme. Zur Morphologie einer Repräsentationsform in der Logik, Paderborn 2001, pp. 63f. The topic will be considered several times in the following chapters.
[52] H. Lu-Adler: Kant's Conception of Logical Extension and Its Implications, Chapter 2.
[53] Some problems are discussed in Chapter 2.2.5.

2 Logic and its Geometrical Interpretation

Euler it is clear that an infinite number of possible objects and concepts are meant.[54] As with the Wolff disciples, however, Kant's diagrams and quotations would not lead to any unambiguous result; moreover, Lu-Adler argues, one could understand (AJ) in terms of a geometric logic – according to the scheme '$\forall x\ bx \rightarrow ax$'.[55]

The results of the study depend on many successively introduced interpretive premises, which in my opinion are often open to attack. Nevertheless, the comparison between the three geometrical logics (Leibniz, Lambert, Euler) and Kant seems to me to be a meritorious pioneering achievement, which – even if Lu-Adler does not explicitly point it out – calls into question an entire tradition of analytic philosophy starting from Quine, who might have misunderstood Kant's 'notion of containment' as a metaphor and not as a technical term of geometrical logic: 'circumference', 'being contained', 'lying outside' are verbal descriptions of logical schemata; they refer to intuitions that cannot simply be substituted, reformulated, translated or conceptualized.

Despite the progress in knowledge made by Lu-Adler's research, many questions still arise: Erich Adickes, for example, had already noted over 100 years ago that Kant drew logic diagrams in his lectures that were not only dependent on Euler or Lambert,[56] but probably on the *Nucleus Logicae Weisianae* (1712).[57] Lu-Adler did not consider the diagrams found in this work or their possible predecessors. It also seems strange that one of the first critics of (AJ) and (SJ), namely the above-mentioned Maaß, himself plays an important role in the history of logic diagrams. If Maaß did know about the functioning of supposed circumferential metaphors, why did he criticise them and not identify them as such?

In the subsequent chapters, I will follow Lu-Adler and Robert Lanier Anderson in defending against Quine and many of his successors the thesis that Kant and early Kant followers had Euler diagrams in mind when they formulated (AJ) and (SJ) in an apparently metaphorical way. Moreover, it is argued for that Kant must have been influenced by more sources in his use of logic diagrams than Lu-Adler and Adickes have assumed. Furthermore, it is also necessary to clarify the question of what reasons must have moved the geometrical logician Maaß to degrade (AJ) and (SJ) as insufficient. Subsequently, however, I will argue that the only logic in the paradigm of the transcendental philosophy until the end of the 1820s in which (AJ) and (SJ) are completely explicated with Euler diagrams is Schopenhauer's logica major.

The last point is central. Although I am convinced that the research questions and results of the Kant research presented here also promote the understanding of Schopenhauer's logic and vice versa, I do not believe that the interpretations presented in

---

[54] H. Lu-Adler: Kant's Conception of Logical Extension and Its Implications, Chapter 2.
[55] Ibid., Chapter 4.III. This refers to the diagram in Jäsche-Logic, § 29 (AA IX, p. 108; Lectures on Logic, p. 604); vide infra, Chap 2.2.4.
[56] Huaping Lu-Adler does not point out that it is very unlikely that Kant was aware of Leibniz's diagrams, which were not published until the 20th century.
[57] Vide infra, Chapter 2.2.4.

2.2 Analyticity – Analytic judgements, Containment Metaphors and Logic Diagrams

Chapter 2.2 will solve and answer profound problems and questions of Kant research. Nevertheless, there is a justified hope that the remarks on Kant's and Schopenhauer's diagrams and their descriptions will contribute to the question of how (AJ) and (SJ) differ from each other and how we can better understand both from their history of origin. Finally, I assume that most researchers who deal with analyticity are more interested in what (AJ) and (SJ) mean, what functions, properties and, above all, benefit they can have, and they are only secondarily interested in whether these are explored in a hero like Kant or in an anti-hero like Schopenhauer.

In order to illustrate the use of the logic diagrams that Kant and especially Schopenhauer use to explain (AJ) and (SJ), Chapters 2.2.2 and 2.2.3 will roughly sketch the history of the development of these diagrams up to Kant's time. This history of development will be presented by means of a few selected quotations and examples of so-called 'analytical diagrams' of geometrical logic. The selection of geometrical logicians, which will be dealt with in Chapters 2.2.2 and 2.2.3, will be justified in Chapter 2.2.1 in the course of a review of the state of research on geometrical logic up to the 19th century. I would also like to point out that the history of the development or ideas of these logic diagrams presented in Chapters 2.2.2 and 2.2.3 is supplemented by Chapter 2.3.4; in this I have taken a closer look at those text passages of the history of logic and philosophy that reflect and evaluate the functioning of the diagrams that were used. One could thus say that Chapter 2.2 argues for dismissing the genuine definition of analyticity not as an empty metaphor, but as a precise construction rule for analytical diagrams of geometric logic.

2.2.1 The State of Research on the History of Analytical Diagrams

It was already known to influential 19th-century geometric logicians such as John Venn or Charles Sanders Peirce that Leonhard Euler was not the inventor of the diagrams named after him today. Venn did claim that his diagrams, which visualise all semantic combination possibilities between elements of a set (or class), were inspired by Euler; but he also shows that Euler himself was part a much older tradition of logicians who had already used similar geometric figures in logic.[58] Peirce, who made a fundamental extension of Euler's diagrams with his existential graphs, also makes an explicit effort to provide a complete prehistory of Euler's diagrams.[59]

The historical studies of Venn and Peirce, in particular, show how much meticulousness even innovative and pioneering logicians of the 19th century had already dealt

---

[58] Vide infra, Chapter 2.
[59] Cf. e.g. Charles Sanders Peirce: Book II. Existential Graphs: In: Collected Papers of Charles Sanders Peirce. Vol. 4. The Simplest Mathematics. Ed. by Charles Hartshorne, Paul Weiss. 5th ed. Cambridge/MA 1980 (Repr. 1933), pp. 293–470 (4.347–4.584), here: pp. 298ff. (4.353ff.).

with their supposed predecessors and how even the energetic drive for historical completeness reached its limits in the pre-digital age.[60] Venn and Peirce, for example, had made intensive, if futile, efforts to obtain books on geometrical logic, such as the fabulous *Nucleus Logicae Weisianae* of 1712. Although they suspected that the prehistory of Euler's diagrams was richer than they could prove, their claim to completeness failed due to the unavailability of those works whose contents they either knew only second- or third-hand or not at all, but whose titles and contents seemed promising for their research.

It is astonishing, however, that even in the age of increasingly digital availability, i.e. even today, many ambiguities and prejudices still exist outside the special logical literature: For example, even in the field of diagrammatics, one often finds references to the fact that the history of the so-called 'Euler diagrams' begins with the year 1768, i.e. with Leonhard Euler's publication of his diagrams in *Letters to a German Princess* (*Lettres à une Princesse d'Allemagne*); or in numerous current textbooks on mathematics or philosophy, there is a reference to Euler-Venn diagrams, although most Euler diagrams cannot be Venn diagrams from a mere syntactical point of view. Nevertheless, since John Venn's relevant work *Symbolic Logic*, specialists from individual disciplines on geometric logic and diagrammatology have provided a great deal of valuable information on the prehistory and systematics of Euler diagrams.

Unfortunately, however, especially in the historical studies, the results and findings of one's own source research were not always compared with predecessor studies.[61] In the following, only those historical reference works on the prehistory of Euler diagrams in German, English and French have been listed in chronological order that deal with geometric logics up to the time of Kant and Schopenhauer:[62]

1. John Venn: *Symbolic Logic*. 2 red. ed. London 1894, esp. pp. 504–527 (= Chapter XX.II).[63]
2. Theodor Ziehen: *Lehrbuch der Logik auf positivistischer Grundlage mit Berücksichtigung der Geschichte der Logik*. Bonn 1920, esp. pp. 227–236 (= § 54).

---

[60] This is not to say, of course, that there are no boundaries in the age of digital availability of texts, but only that these boundaries have shifted considerably.

[61] Thus, unfortunately, the judgement of Christian Thiel is still valid today: Die Quantität des Inhalts. Zu Leibnizens Erfassung des Intensionsbegriffs durch Kalküle und Diagramme. In: Die intensionale Logik bei Leibniz und in der Gegenwart. Ed. by Albert Heinekamp, Franz Schupp. Wiesbaden 1979, p. 22: "It is regrettable that due to the neglect of diagrammatic procedures in the historiography of logic, a systematic as well as historical overview is still lacking today [...]." (My transl.; J.L.)

[62] Detailed studies on individual historical authors were not included, but have been considered and supplemented in the presentation of the following chapters. Studies that are not primarily historical but deal with the history of analytical diagrams in a few paragraphs have also been omitted, e.g. Jesse H. Shera, Conrad H. Rawski: The Diagram is the Message. In: Journal of Typographic Research 2:2 (1968), pp. 171–188, here: pp. 178ff. (Here, however, Schopenhauer's diagrams are particularly emphasised).

[63] This chapter is a revised version of John Venn: On the Employment of Geometrical Diagrams for the Sensible Representation of Logical Propositions. In: *Proceedings of the Cambridge Philosophical Society* IV (Oct. 25, 1880 – May 23, 1883), pp. 47–59.

2.2 Analyticity – Analytic judgements, Containment Metaphors and Logic Diagrams

3. Martin Gardner: *Logic Machines and Diagrams*. New York, Toronto et al. 1958.
4. Margaret E. Baron: A Note on the Historical Development of Logic Diagrams. Leibniz, Euler and Venn. In: *The Mathematical Gazette* 53:384 (May 1969), pp. 113–125.
5. E[rnest] Coumet: Sur l'histoire des diagrammes logiques, 'figures géométriques'. In: *Mathematiques et Sciences Humaines* 60 (1977), pp. 31–62.
6. Peter Bernhard: *Euler-Diagramme. Zur Morphologie einer Repräsentationsform in der Logik*. Paderborn 2001, esp. pp. 69–80 (cf. index).
7. Amirouche Moktefi, Sun-Joo Shin: A History of Logic Diagrams. In: *Logic. A History of its Central Concepts*. Ed. by Dov M. Gabbay, John Woods. Oxford et al. 2012, pp. 611–682
8. Deborah Bennett: Origins of the Venn Diagram. In: *Research in History and Philosophy of Mathematics*: *The CSHPM 2014 Annual Meeting in St. Catharines, Ontario*. Ed. by M. Zack, E. Landry. Heidelberg et al. 2015, pp. 105–119.

Although all historical reference works have commonalities with regard to their conceptual scheme and some references to authors from the history of geometrical logic, there are nevertheless numerous differences concerning taxonomy and historiography.

As an example of the taxonomic differences, I will take the conceptual schemes of Ziehen and Venn, some of whose concepts I have already used in the previous chapters and some of which I will further differentiate. Ziehen, for example, speaks in 1920 as an umbrella term of a 'mathematical' or 'symbolic logic', which in turn is divided into an 'algebraic' and 'geometric logic',[64] of which the latter includes line, triangle, cube, sphere or circle diagrams. Such symbolic logics thus imitate either algebra (or even arithmetic) or geometry. The accompanying tree diagram (Fig. 1) shows an excerpt of this subdivision.[65]

---

[64] Theodor Ziehen: Lehrbuch der Logik auf positivistischer Grundlage mit Berücksichtigung der Geschichte der Logik. Bonn 1920, pp. 227ff., p. 409.
[65] Ziehen gives an example for each type of geometric logic (e.g. line diagrams: Lambert). Since one example is given in each case, I have refrained from showing these examples as individuals in Fig. 1. For the interpretation of tree diagrams vide infra, Chapter 2.2.2.

2 Logic and its Geometrical Interpretation

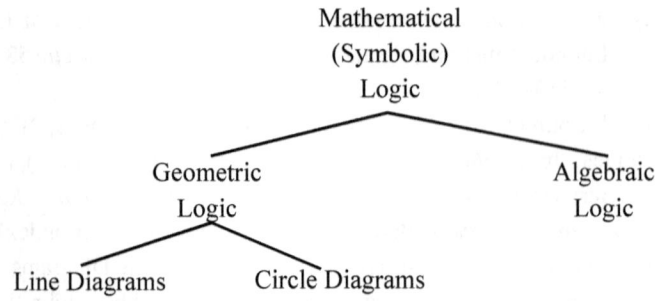

Fig. 1
Excerpt of Ziehen's Taxonomy

In 1881, Venn had already written a chapter in his *Symbolic Logic* entitled 'On the employment of geometrical diagrams for the sensible representation of logical propositions', in which he spoke of 'analytical diagrams', i.e. diagrams "which deal directly with propositions, and which *analyse* them".[66] In the second edition of *Symbolic Logic* of 1894, Venn takes up this definition again and speaks in the same sense of analytical diagrams, "which are meant to distinguish between subject and predicate and between the different kinds of propositions".[67]

According to Venn, the term 'analytical diagram' generally encompasses circular diagrams and other geometric figures (squares, triangles, lines) with similar logical functions; it does not, however, encompass the arbores porphyrianae, pontes asinorum, quadrata formula or even some semicircular diagrams included in the term 'logic diagram'. An excerpt of this division is illustrated by the accompanying tree (Fig. 2).[68]

---

[66] John Venn: Symbolic Logic. 1st ed., London 1881. p. 504.
[67] John Venn: Symbolic Logic. 2nd ed., London 1894, p. 504, p. 506.
[68] One should note in this diagram that the entities on the bottom left (Vives diagram, Euler diagram, ...) express individuals, whereas those on the bottom right express species whose individuals are not specified. Moreover, the individuals on the left are only examples and not listed completely.

2.2 Analyticity – Analytic judgements, Containment Metaphors and Logic Diagrams

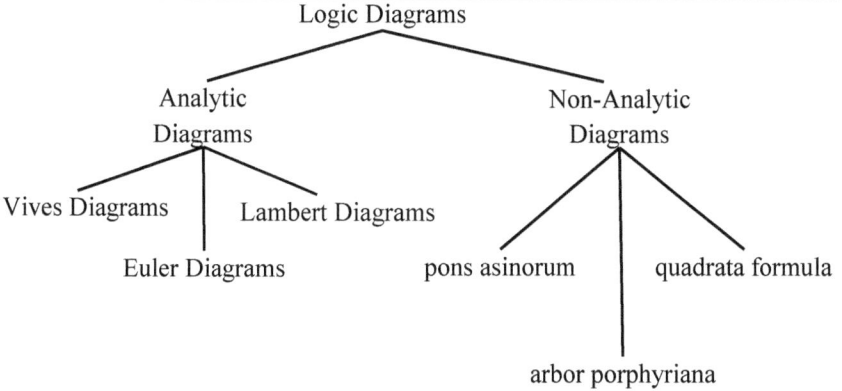

Fig. 2
Excerpt of Venn's Taxonomy

As one can observe from the two taxonomies, Ziehen is more concerned with the form of the diagram, Venn more with the function. Both taxonomies have several advantages and disadvantages, and I would like to take up just a few problematic points here for clarification, which concern the context: In Ziehen's taxonomy, it must be known to some extent whether a particular diagram is in the context of a mathematical or logical treatise, for example, otherwise one would quickly classify a typical Euclidean figure as in *Elementa* 3.11 as a logic circle diagram or a geometric figure as often found in *Elementa* 5.1 as a line diagram of geometric logic. In Venn's taxonomy, moreover, it must be clear not only whether it is a diagram of logic, but also exactly how this diagram works in order to be able to classify it at all. The following applies to both taxonomies: not only the diagram but also its context play an important role.

In today's research, neither taxonomy has fully established itself. Although most researchers tend towards a functional taxonomy, a uniform way of speaking has not yet been established. What is striking is that the terms once understood as individuals, such as 'Euler's diagrams' or 'Venn's diagrams', are now used as generic or species terms: Today, researchers speak of 'Euler diagrams' or 'Venn diagrams', and since the 1990s at the latest, they have come to mean certain formal logic systems with their own syntax and semantics.[69] In addition, expressions such as 'Euler diagrams' or 'Euler-type diagrams' are used to refer to diagrams that are closest to Euler diagrams but do not correspond to their syntax or semantics.[70]

The established expression 'Euler diagrams' only slowly became established from the 1910s onwards, even though geometric logicians (especially followers of Kant)

---

[69] Cf. e.g. Eric M. Hammer: Logic and Visual Information. Stanford 1995.
[70] Cf. e.g. Amirouche Moktefi, Sun-Joo Shin: A History of Logic Diagrams. In: Logic. A History of its Central Concepts. Ed. by Dov M. Gabbay, John Woods. Oxford et al. 2012, pp. 611–682. I thank Amirouche Moktefi for numerous and invaluable references related to the history of logic diagrams.

repeatedly spoke of 'Euler's diagrams', 'Eulerian diagrams', 'diagrams of Euler' in the form of a reference point from the 1790s onwards.[71] Euler diagrams were understood by Venn as an ideal type of analytical diagrams, by Ziehen as an exemplary form of circular diagrams within geometrical logic. For both Venn and Ziehen, Euler's diagrams form a central point of reference for comparison, which should make it possible to discuss family resemblances between analytical diagrams (Venn) or diagrams of geometric logicians (Ziehen). Even in histories of the development of geometric logic in the 20th century, Euler diagrams remain not only the historical but also the systematic point of reference for a multitude of modern diagrams in logic, e.g. Venn, Lewis, Randolph, KV, Spider or also *CL* diagrams.[72]

By the term 'Euler diagram' one can understand, in a very generalised and systematic sense, geometric figures that show the relationship between sets of objects, concepts, classes, etc. (extensional) or the relationship between sets of properties, characteristics etc. (intensional).[73] Although the designation 'Euler diagrams' is exceedingly well-established in a systematic sense, I use it in the following solely in a historical sense when authors in Euler's succession are either explicitly of the opinion or give indications that their use of diagrams in logic resembles that of Euler. Otherwise, I tend to use neutral concepts such as 'logic diagram', 'analytical diagram', 'circle diagram' etc. in the sense of Ziehen or Venn, since especially in the field of geometric logic and diagrammatics there are more specific connotations associated with 'Euler diagram' than can be discussed or need to be discussed in the following.[74]

The above-mentioned works of research provide a heterogeneous picture not only with regard to their taxonomy but also with regard to the historical logicians treated or mentioned in each case, who used geometric diagrams up to Kant and Schopenhauer. The following table shows, on the one hand, all the historical logicians mentioned in the reference works with the dates of their relevant works and, on the other hand, in which reference work (header) they are treated. The sign ✓ in the table field indicates that the corresponding work of research deals more intensively with the respective historical logician and assigns him or her a positive role in the

---

[71] Cf. Jens Lemanski: Periods in the Use of Euler-Type Diagrams. In: Acta Baltica Historiae et Philosophiae Scientiarum 5:1 (2017), pp. 50–69.
[72] For the history of development in the 20th century cf. Amirouche Moktefi, Sun-Joo Shin: A History of Logic Diagrams; Amirouche Moktefi, Francesco Bellucci, Ahti-Veikko Pietarinen: Continuity, Connectivity and Regularity in Spatial Diagrams for N Terms. In: Diagrams, Logic and Cognition. Ed. by J. Burton, L. Choudhury. CEUR Workshop Proceedings 1132 (2013), pp. 23–30; Jens Lemanski: Logic Diagrams, Sacred Geometry and Neural Networks. Logica Universalis 13 (2019), pp. 495–513.
[73] Cf. Catherine Legg: What is a Logical Diagram?. In: Visual Reasoning with Diagrams. Ed. by Sun-Joo Shin, Amirouche Moktefi. Basel 2013, pp. 1–18.
[74] The semantics of expressions such as 'Euler diagram', 'analytical diagram', 'logic diagram' and others are exceedingly dificile, especially in historical studies. I am aware through numerous debates that the conceptual scheme I have chosen here will not satisfy all recipients, but I believe I can at least give good reasons why I have chosen some expressions and what concepts are associated with them.

2.2 Analyticity – Analytic judgements, Containment Metaphors and Logic Diagrams

historical development of Euler diagrams; the sign ✗ was used for a negative role.[75] The brackets – (✓) or (✗) – indicate that the respective historical logicians are mentioned only incidentally in a positive or negative sense in the respective work of research.

Works of Research Logicians/ Dates	1. Venn	2. Ziehen	3. Gardner	4. Baron	5. Coumet	6. Bernhard	7. Moktefi/ Shin	8. Bennett
Aristotle (~4th c. BCE)	(✓)		✓	(✗)	(✓)	(✓)		
Porphyry (~ 3rd c. CE)			✓	✓			✓	
Ammonius (~ 5th c. CE)				(✓)				
J. Philoponus (~ 6th c. CE)		✓		(✓)		(✗)		
R. Lull (~1305)			✓	✓				(✓)
J.L. Vives (1551)	✓				(✓)	✓		(✓)
P. Tartaretus (1581)	(✓)							
J. Pacius (1584)					✓			
N. Reimers (1589)					✓	✓		
J.H. Alsted (1611)	✓			✓				
J.C. Sturm (1661)			✓		(✓)	✓		(✓)
A. Geulincx (1662)		✓						
R. Sanderson (1680)	(✓)							
C. Weise (1691)	✗	✓		✓	(✓)	✗		(✓)
G.W.F. Leibniz (~ 1690)			✓	✓	(✓)	✓	✓	✓
E. Weigel (1693)						(✓)		(✓)
J.C. Lange (1712)	(✓)		✓	✓	✓	(✓)		
G. Ploucquet (1759)	✓	✓		✓	(✓)	✓		
J.H. Lambert (1764)	✓		✓	✓	(✓)	✓	✓	(✓)
L. Euler (1768)	✓	✓	✓	✓	✓	✓	✓	✓
J.A.H. Ulrich (1792)	(✓)							
J.G.E. Maaß (1793)	✓	✓				✓		
I. Kant (1800)	✓				(✓)			
K.C.F. Krause (1803)	✓							
A. Schopenhauer (1819)		(✗)			(✓)			

---

[75] What exactly is meant by 'positive' and 'negative role' must be taken from the respective reference works. As a rule, 'positive role' means the historical anticipation of Euler diagrams or a systematic similarity to them. The 'negative role' argues against such historical or systematic references.

2 Logic and its Geometrical Interpretation

On the one hand, the table illustrates that all reference works taken together provide a valuable historical panorama with certain focal points. On the other hand, it also illustrates the heterogeneous historiography of geometrical logic up to Schopenhauer, which is partly based on the fact that the results of the respective preceding studies were not always taken into account or critically re-examined by the subsequent works of research.

Because of the inconsistencies that can be seen by the table, I first compiled the writings of the historical actors mentioned, reviewed them, checked the judgements of the reference works and outlined the results below. I was also able to compile almost all the historical books and also research papers in a freely accessible digital repository.[76]

Since the repository allows for quick verifiability of all relevant diagrams, only exemplary evidence for the individual geometric logics will be cited in the following. Due to the above-given table, I also take the liberty of not further mentioning the listed works of research in the following chapters, since I assume that researchers on the respective authors, epochs and geometric logics will compare my now following historical overview with the respective works of research or are familiar with them. In Chapters 2.2.2 and 2.2.3 in particular, I therefore only refer to special studies or to the respective reference works if the evidence is not provided by the table in this chapter. In Chapter 2.2.2, I will first give an overview of the most common logic diagrams between antiquity and the early modern period, discussing the extent to which these diagrams can be regarded as analytical in Venn's sense or as part of geometric logic in Ziehen's sense. In Chapter 2.2.3 I will then present three periods of analytical diagrams in early modern geometrical logic, ending with Schopenhauer.

2.2.2 Logic Diagrams from Antiquity to Early Modern Times

In order to understand the tradition in which Kant and his successors, such as Schopenhauer, placed themselves when they used the metaphor of containment in defining analytical diagrams, it is useful to look deeply into the history of logic and its related sciences. In doing so, this chapter and the one to follow cannot claim to be exhaustive, but it can at least compile some of the known historical research findings and also systematisations. First of all, however, the question arises as to how far a look into the history of logic should go.

Whether an explicit picture-related representation of logic *more geometrico* already began in antiquity is disputed in research. Researchers agree that no ancient logic papyri with geometric diagrams have yet been found as secure textual witnesses; but there are nevertheless numerous interpreters who try to work out an implicit image-relatedness in ancient texts on logic and the philosophy of language. Already in the

---

[76] http://blog.fernuni-hagen.de/euler-venn-diagrams

2.2 Analyticity – Analytic judgements, Containment Metaphors and Logic Diagrams

early modern period, numerous geometrical logicians pointed out a similarity between their logic diagrams and the geometrical allusions in Pythagorean, Platonic and Aristotelian writings.[77] In post-Eulerian logic, Friedrich Ueberweg in particular had compiled quite a few allusions to logic diagrams in Plato's *Sophistes* and Aristotle's *Prior Analytics*.[78] Pirmin Stekeler-Weithofer also sees, for example, in the sailcloth analogy that the Platonic Parmenides utters to Socrates (Plat. Parm. 131b f.), evidence that Plato was of the opinion that concepts behave in a judgement in the same way as models of surfaces in geometry.[79] As Peter Bernhard, in particular, points out many modern interpreters assume use of geometric diagrams in Aristotle, since he explicitly speaks of the 'scheme' and of '*middle* terms', uses logical metaphors of the circumference or also makes analogies between logicians and mathematicians (e.g. An. pr. 49b).[80] This is supported by Marian Wesoły, who sees in the Aristotelian terminology descriptions of 'lost diagrams' that later reappear in a similar form in the Byzantine tradition.[81] Marko Malink has argued that all valid inferences of Aristotelian logic can be used like a one-dimensional diagram, which Aristoxenus of Tarentum used to represent musical intervals.[82]

Also in Theophrastus of Eresos and later in Alexander of Aphrodisias, for example, in their commentaries on Arist. an. pr. 43b36-39 (in APr.), one finds mention of the choice of diagram and syllogisms ("ἐκλογὰς καὶ τὸ διάγραμμα ὅλον καὶ τοὺς συλλογισμούς").[83] More explicit references to logical schemes and diagrams are then found in Augustine: he reports (Conf. IV 16) that in his time teachers made the *Categories* of Aristotle perceptible not only by oral speech but also by many illustrations painted in the dust ("non loquentibus tantum, sed multa in pulvere depingentibus intellexisse"). The comments of Theophrastus, Alexander and Augustine, however, do not allow any conclusions to be drawn as to exactly what kind of drawings are involved.

---

[77] Vide infra, Chapter 2.3.4.
[78] Cf. Friedrich Ueberweg: System of Logic. Transl. by Thomas M. Lindsay. London 1871, pp. 134 (§ 53).
[79] Cf. e. g. Pirmin Stekeler-Weithofer: Grundprobleme der Logik. Elemente einer Kritik der formalen Vernunft. Berlin et al. 1986, pp. 27–88.
[80] Cf. Peter Bernhard: Euler-Diagramme, pp. 69f.
[81] Cf. Marian Wesoły: Αναλυσις περι τα σχηματα. Restoring Aristotle's Lost Diagrams of the Syllogistic Figures. In: Peitho. Examina Antiqua 1:3 (2012), pp. 83–114 with further references.
[82] Cf. Marko Malink: Aristotle on Principles as Elements. In: Oxford Studies in Ancient Philosophy 53 (2017), pp. 163–214. I also thank Marko Malink for pointing me to the following writings, which deal with diagrams in ancient logic, but which are not discussed in detail here: Benedict Einarson: On Certain Mathematical Terms in Aristotle's Logic: Part II. In: The American Journal of Philology 57:2 (1936), pp. 151–172; Lynn E. Rose: Aristotle's Syllogistic. Springfield 1968, pp. 22–24, 133–137.
[83] Kevin L. Flannery: Ways into the Logic of Alexander of Aphrodisias. Leiden et al.1995 interprets these diagrams in a logical of circumference (p. 136ff.) and refers to an Aristotelian tradition (p. 1ff., p. 41). Cf. also William Hamilton: Discussions on Philosophy and Literature, Education and University Reform. Chiefly from the Edinburgh Review; Corrected, Vindicated, Enlarges in Notes and Appendices. 2nd ed. London 1853, p. 670. Hamilton gives numerous reasons why one should be sceptical about the late antique diagrams.

2 Logic and its Geometrical Interpretation

For the early Middle Ages, three types of diagrams can be attested, which Venn describes as non-analytical:[84] 'squares of oppositions' (quadrata formula/ schema oppositionum),[85] 'bridge of asses' (pons asinorum)[86] or also 'tree diagrams' (arbor porphyriana/ scientia etc.).[87] Based on such evidence, some historians of logic argue that these graphic illustrations of logic were already introduced by the Middle and Neoplatonists such as (Ps.-)Apuleius, Ammonius, Philoponus or Porphyry. Such assumptions are problematic in several respects, however, since the reference of modern historians to copies, incunabula or editions made later does not rule out the possibility that the figures mentioned were added to the late antique works only afterwards.[88]

Nevertheless, the three non-analytical diagram types mentioned can be clearly traced between the 9th and 13th centuries. I only draw on exemplary results (1) for so-called *bridges of asses* in Byzantine-Slavic ecclesiastical areas, (2) for *tree diagrams* in Central European commentaries and glosses, (3) for the *square of opposition* on archaeological finds from the Scandinavian region. (1) The logical 'diagrams' (διάγραμμα) and 'schemata' (σχῆμα)[89] so called by Michael Psellos in the 11th century have been found numerous times in the writings of Gregory Palamas' and in slides of his Serbian-Christian-Slavonic translations in the monastery of Dečani and contain, among other things, bridges of asses.[90] (2) The oldest tree diagrams, to my knowledge, can be found in a gloss on the *Isagoge* written by an author called 'Jepa' in the 9th or 10th century, and later in Boethius translations of Porphyry's *Isagoge* of the 11th century.[91] (3) Archaeological evidence of drawings in logic classes can be found, for example, from the 13th century on the tower walls of the Gothic church of

---

[84] Vide infra, Chapter 2.2.1.
[85] According to prevailing opinion, the squares of opposition are found for the first time in the writing *Peri hermeneias* attributed to (Ps-)Apuleius of Madaura, cf. Heinrich Schepers: Logisches Quadrat (Art.). In: HWPh, vol. 7, pp. 1733–1736. However, due to known doubts about the authenticity, dating and transmission history of the writing, the attribution is problematic.
[86] Cf. Heinrich Schepers: Eselsbrücke (Art.). In: HWPh, vol. 2, pp. 743–745; Charles Leonhard Hamblin: An Improved Pons Asinorum?. In: Journal of the History of Philosophy 14:2 (1976), pp. 131–136.
[87] Cf. e.g. William Kneale, Martha Kneale: The Development of Logic. 2nd ed. Oxford et al. 1971 (Repr.), pp. 71f.
[88] The assessment is supported, among others, by Heinrich Scholz: Abriß der Geschichte der Logik. 3. ed. Freiburg et al. 1967, pp. 43f., Fn. 25; Michael Krewet: Zum Wissenstransfer in Ammonios' Kommentierung des neunten Kapitels von Aristoteles' *De Interpretatione*. (Working Paper des SFB 980 Episteme in Bewegung). Berlin 2019, pp. 50f., Fn. 161.
[89] Cf. Katerina Ierodiakonou: Psellos' Paraphrasis on *De interpretatione*. In: Byzantine Philosophy and its Ancient Sources. Ed. by Katerina Ierodiakonou. Oxford 2004, pp. 157–183.
[90] Cf. Ioannis Kakridis: Codex 88 des Klosters Dečani und seine griechischen Vorlagen. Ein Kapitel der serbisch-byzantinischen Literaturbeziehungen im 14. Jahrhundert. Munich et al. 1988, esp. pp. 150ff. – Numerous photographs of the diagrams discussed by Kakridis can be found in Slobodan Žunjić: Logički dijagrami u srpskim srednjovekovnim rukopisima. In: *Theoria* 54:4 (2011), pp. 127–160.
[91] Cf. Annemieke Rosalinde Verboon: Lines of Thought. Diagrammatic Representation and the Scientific Texts of the Arts Faculty, 1200–1500. S.l. 2010 (http://hdl.handle.net/1887/16029), esp. pp. 35–57 (including numerous figures). However, I do not share one of the main theses of the chapter: Verboon claims that tree diagrams can only be called such if the diagrams also have an iconic resemblance to a tree. This, she claims, can first be demonstrated in Petrus Hispanus (Paris, BN, ms. lat. 16611). Unfortunately, however, she omits that almost all logic diagrams (e.g., also squares of opposition) in this manuscript are drawn in a tree-like manner.

2.2 Analyticity – Analytic judgements, Containment Metaphors and Logic Diagrams

Bro.[92] Squares of oppositions have been preserved in the Bro church and reconstructed by archaeologists using modern methods. For each of these three types of diagrams, I will take an example from medieval literature in order to provide historical evidence and to outline the logical function in a rudimentary way.[93]

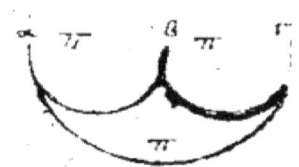

(1) Fig. 1 shows a crescent-shaped *bridge of the asses* that depicts the subject-predicate structure of the judgements in the syllogism: From top left to top right, the three peaks are occupied by three terms (α = terminus maior, β = medius und Γ = minor); between the three peaks are the respective quantifiers – here (gr: π[άντως] = All; τ[ίς] = Some; o[ὐ πᾶς] = No, resp. lat.: O[mne]; Q[uoddam]; N[ullam]) – drawn on three curved connecting lines, which in turn map the two premises of equal length and located above and the conclusion drawn below by the longer line (α_β = prop. maior; β_Γ = prop. minor; α_Γ = concl.). Thus, one can read from Fig. 1: All α are β, all β are Γ, thus all α are Γ.

Fig. 1
Aristotle: *Organon*, Bibliotheca Augusta, catalogue no. 4211, Cod. Guelf. 24 Gud. graec., fol. 32ʳ.

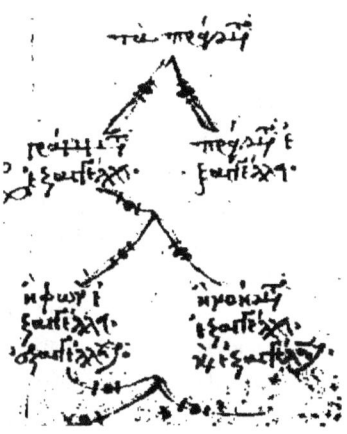

(2) Fig. 2 shows a typical *tree diagram*, with a term at the top (L_0), which is differentiated top-down dichotomously over three levels (L_{-1}, L_{-2}, L_{-3}) in each case. Apart from the highest term (G_0) and possibly the lowest term, each term of a level (L_{-1}, L_{-2}, L_{-3}) can be interpreted both as subject/ generic term/ superset (= A) in relation to the terms below it and as predicate/ species term/ subset (= B) in relation to the degrees above it. Since tree diagrams also illustrate transitive rules such as 'What is true of A is also

Fig. 2
Aristoteles: *De interpretatione*, with a commentary by Michael Psellus, Terra d'Otranto, 13ᵗʰ–15ᵗʰ c., Magdalen College, P. Magdalen Gr. 15, fol. 1ʳ.

---

[92] Cf. Uaininn O'Meadhra: Medieval Logic Diagrams in Bro Church, Gotland, Sweden. In: Acta Archaeologica 83 (2012), pp. 287–316 incl. a dating of the earliest squares of opposition to the ninth century.
[93] The diagrams used below are from two manuscripts, namely (1) a 12ᵗʰ-century Central European manuscript of the *Organon* (Herzog August Bibliothek Wolfenbüttel Guelf. Gud. gr. 24) and (2) a late 13ᵗʰ-century collected manuscript from Terra d'Otranto containing excerpts from *De interpretatione* with a commentary by Psellos (Magdalen College, gr. 15).

true of B',[94] predicate-logical/ ontological/ set-theoretical inferences can be drawn,[95] such as: All L_{-1} are L_0, all L_{-2} are L_{-1}, thus all L_{-2} are L_0.[96]

(3) Fig. 3 shows a *square of opposition* which indicates by the lines the relations between four relata such as symbols, concepts, judgements etc., which are found at the respective corners. Usually, the categorical judgements of assertoric syllogism are used as relata and shown in the corners: *A*-judgments top left (All S are P), *E*-judgments top right (No S is P), *I*-judgments bottom left (Some S are P), *O*-judgments bottom right (Some S are not P). The following relations apply, indicating the lines between the four corners: The upper horizontal line A^-E stands for contrariety, the diagonal lines, $A\backslash O$ and I/E

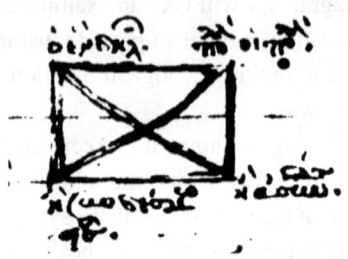

Fig. 3
Aristoteles: De interpretatione (Commentary by Michael Psellos). Otranto 13th c., Magdalen College MS Gr 15, fol. 11r.

indicate contradiction, the vertical lines $A|I$ and $E|O$ stand for subalternation and the lower horizontal line I_O shows subcontrarity. The following applies to these relations: *Contrary* relata cannot be true at the same time, but they can be false at the same time. In the case of *subcontrary*, both relata cannot be false at the same time, but they can be true at the same time. In the case of *subalternation*, the more general relatum implies the particular one. A relation is *contradictory* if one of the two relata is true iff the other is false.

Although the three diagram types depict subject-predicate structure, the circumference of concepts or subordination, and they can also be combined with or transformed into analytical diagrams, it is questionable whether they already possess the complete functioning of analytical diagrams that Venn had in mind.[97] I would like to leave out a more detailed treatment of this question here and follow Venn and Ziehen in their assessments. Before I turn to the analytical diagrams of the early modern period, several medieval candidates for analytical diagrams will first be considered. In particular, four theses by scholars are worth discussing, according to which there are medieval

---

[94] Quidquid de subiecto/ genere/ omni dicitur, etiam de praedicato/ specie/ quibusdam dicitur.
[95] The expressions 'predicate-logical/ ontological/ set-theoretical' here refer to the respective *S/U/P* or *S/U/B* interpretation.
[96] For a detailed interpretation and application of tree diagrams see e.g. John F. Sowa: Knowledge Representation, Chapter 1.1 and 2.
[97] Venn did not attribute the functions of analytical diagrams to the types of diagrams discussed so far (see above, Chapter 2.2.1). For the thesis that tree diagrams do have analytic functions, argue Margaret E. Baron: A Note on the Historical Development of Logic Diagrams. Leibniz, Euler and Venn and Lu-Adler: Kant's Conception of Logical Extension, chap. 2.1; for the thesis that squares of opposition have analytic functions, authors in the field of oppositional geometry argue, e.g. Alessio Moretti: Arrow-Hexagons. In: The Road to Universal Logic. FS for the 50th Birthday of Jean-Yves Béziau. Vol. 2. ed. by A. Koslow, A. Buchsbaum. Cham 2015, pp. 417–489. bridges of asses also developed such complexity in the 16th century that Venn's thesis would also have to be scrutinized more closely than can be done here.

2.2 Analyticity – Analytic judgements, Containment Metaphors and Logic Diagrams

diagrams that meet Venn's criterion for analyticity and Ziehen's criterion for geometric logic: (1) Gardner's thesis on Ramon Lull's combinatorial circles; (2) Frampton's thesis on Borromean rings found in a Calcidius' manuscript; (3) Nolan's thesis on Afflighem's wheel diagram; (4) Hodges' thesis on the line diagrams of Abu'l-Barakāt al-Baghdādī.

(1) Gardner and – following him – Baron have claimed that a diagram (Fig. 4) attributed to Ramon Lull showing four intersecting circles and the concepts 'One', ' Being', 'Truth', and 'Good' ('Vnum', 'Esse', 'Verum' 'Bonum') would represent the historically first anticipation of plane analytical diagrams. Gardner found this scheme on a portable sundial with Lullian motifs from 1593, and he refers solely to a study by Ormonde Maddock Dalton for proof.

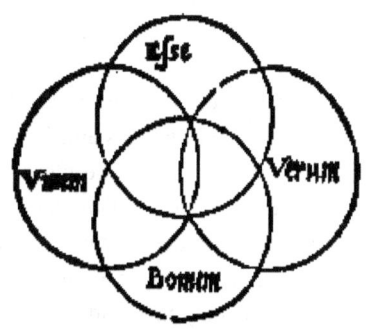

Fig. 4
Ps.-Lull: De Audito Kabbalistico seu Kabbala. In: *Raymundi Lulli Opera ea quae ad adinventam ab ipso artem universalem* […], Argentinae 1598, p. 109.

In my perusal of Lull's manuscripts of the so-called 'first generation',[98] I did discover numerous diagrams with combinatorial circles,[99] many tree diagrams[100] and also illustrations that are also on the corresponding clock,[101] but I could not find the diagram cited by Gardner and Baron or even a scheme that would obviously indicate geometrically plane analytical diagrams.

It is true that a three-dimensional representation of Borromean rings[102] can be

---

[98] Cf. Albert Soler: Els manuscrits lul·lians de primera generació als inicis de la primera generacio. In: Estudis Romànics 32 (2010), pp. 179–214.
[99] Cf. Arras, Bibliothèque Municipale, Ms. 78, fol. 1ᵛ; Città del Vaticano, Bibliotheca Apostolica Vaticana, Ottob. lat. 832, fol. 3ᵛ–4ʳ; Città del Vaticano, Bibliotheca Apostolica. Ottob. lat. 2347, fol. 1ᵛ–2ʳ; Città del Vaticano, Bibliotheca Apostolica., Vat. lat. 3858, fol. 1ᵛ–2ʳ; Città del Vaticano, Bibliotheca Apostolica Vaticana, Vat. lat. 5112, fol. 3ᵛ–8ʳ; Milano, Biblioteca Ambrosiana, I 121 Inf., fol. 1ʳ; Milan, Biblioteca Ambrosiana, P 198 Sup., fol. 1ᵛ–2ʳ, fol. 137ᵛ–139ʳ; Munich, Bayerische Staatsbibliothek, Clm. 10496, fol. 1ᵛ, fol. 2ᵛ; Munich, Bayerische Staatsbibliothek, Clm. 10495, fol. 171ᵛ; Munich, Bayerische Staatsbibliothek, Clm. 18446, fol. 1ᵛ–2ʳ; Oxford, Bodleian Library, Canon. Misc. 141, fol. 69ʳ, fol. 70ᵛ–73ʳ; Paris, Bibliothèque Nationale, NL Petrus von Limoges, MS lat. 16113, fol. 51ᵛ–52ᵛ, fol. 61ʳ; Paris, Bibliothèque Nationale, MS lat. 16116, fol. 84ʳ; Sevilla, Biblioteca Capitular y Colombina, 5-6-35, fol. Iᵛ, 1ʳ; Venice, Biblioteca Nazionale Marciana, Lat.VI.200 (2757), fol. 2ᵛ–4ʳ, fol. 157ᵛ–160ʳ.
[100] Cf. Bologna, Biblioteca Universitaria, Ms. 1732, fol. 2ᵛ–4ʳ; Dún Mhuire, Killiney, Franciscan Library, B 95, fol. Iᵛ–IIʳ; Milan, Biblioteca Ambrosiana, D 549 Inf, fol. 260ʳ, fol. 265ᵛ, fol. 304ᵇⁱˢ·ᵛ, fol. 318ʳ; Milano, Biblioteca Ambrosiana, I 121 Inf., fol. 11ʳ; Padua, Biblioteca Capitolare, C 79, fol. 1ʳ; París, Bibliothèque Nationale, lat. 15385, fol. 1ʳ; Paris, Bibliothèque Nationale, NL Petrus von Limoges, lat. 16114, fol. 15ᵛ–17ᵛ; Paris, Bibliothèque Nationale, MS lat. 16116, fol. 84ʳ; París, Bibliothèque Nationale, fr. 22933, fol. 61ᵛ–64ʳ; Palma de Mallorca, Collegi de la Sapiència, Biblioteca Diocesana de Mallorca, F-129, fol. 1ᵛ, fol. 52ʳ–55ʳ; Palma de Mallorca, Collegi de la Sapiència, Biblioteca Diocesana de Mallorca, F-143, fol. 153ʳ, fol. 156ᵛ, fol. 160ʳ, fol. 180ʳ, fol. 188ʳ.
[101] Cf. Paris, Bibliothèque Nationale, lat. 16115, fol. 84ᵛ.
[102] Cf. Peter Cromwell, Elisabetta Beltrami, Marta Rampichini: The Borromean Rings. In: Mathematical Intelligencer 20:1 (1998), pp. 53–62.

2 Logic and its Geometrical Interpretation

found in a pseudo-Lullian treatise from the early 16th century; but the two-dimensional representation of circles (Fig. 4) relevant for Dalton, Gardner and Baron I could only prove in a new edition from 1598.[103] Thus, although Lull's merit for combinatorics remains unaffected, the assertion by Gardner and Baron that Lull was a predecessor of analytical diagrams or even Euler diagrams is, in my view, untenable given the thin evidence.[104] As will be shown in the following, Fig. 4 probably goes back to a Byzantine tradition, which, however, does not have much to do with analytical diagrams.

Whether one should generally interpret Borromean rings, which are well known to the Middle Ages (e.g. Aug. Trin. IX.4.7), as precursors of analytical diagrams or even Euler diagrams cannot be discussed in more detail here.[105] The fact is, however, that such diagrams can be found even before Ps.-Lull, for example in versions of the *Liber Figurarum* from the early 13th century.[106]

2) However, in order to establish a direct link between the Middle Ages and Venn's analytical diagrams, it might be more obvious to address the thesis of the medical historian Michael Frampton, who claimed a few years ago that there is a connection between Venn's diagrams and a natural philosophical scheme found in a 12th-century manuscript containing a Calcidius translation of the *Timaeus* and the *Song of Roland*.[107] This diagram (Fig. 5) shows a combinatorial inner ring with the seasons and four outer incomplete rings, each of which is both differentiated and connected by further (quarter) rings in three areas (quality, element and age). Although Frampton claims an analogy with Venn, he does not substantiate it, but rather suggests a relationship with diagrams designed to highlight the terminus medius in a syllogism.

Fig. 5
Calcidius: Transl. of Plato's Timaios, France, 1st half of the 12th c., Bodelain Lib. MS Digby 23, fol. 54v.

---

[103] Cf. S.a. [possibly Pietro Mainardi]: Opvscvlvm Raymvndinvum de avditv Kabbalistico Sive ad omnes scientias introdvctorivm. S.l., s.a. [1518], s.p. [ca. p. 90]; S.a.: De Audito Kabbalistico seu Kabbala. In: Raymundi Lulli Opera ea quae ad adinventam ab ipso artem universalem [...]. Argentinae 1598, p. 109.
[104] Lambert and Ploucquet reached a similar conclusion in their texts mentioned in Section 2.2.3.
[105] Numerous similar diagrams are illustrated and discussed in the anthology by Alexander Patschovsky: Die Bildwelt der Diagramme Joachims von Fiore. Zur Medialität religiös-politischer Programme im Mittelalter. Ostfildern 2003 and in Stephan Meier-Oeser: Die Präsenz des Vergessenen. Zur Rezeption der Philosophie des Nicolaus Cusanus vom 15. bis zum 18. Jahrhundert. Münster 1989. The Werner Oechslin library is also a treasure chest of this diagrammatic culture; even if I thank the founder for the many references, which I was allowed to pursue in Einsiedeln during a conference in December 2016 organized by Petra Lohmann, all of the representations I sifted through do not exhibit the function of analytical diagrams.
[106] Cf. e. g. Corpus Christi College, Ms. 255A, fol. 7v.
[107] Cf. Michael Frampton: Embodiments of Will. Anatomical and Physiological Theories of Voluntary Animal Motion from Greek Antiquity to the Latin Middle Ages, 400 B.C.–A.D. 1300. Saarbrücken 2008, p. 307.

2.2 Analyticity – Analytic judgements, Containment Metaphors and Logic Diagrams

Thus, neither Frampton nor Gardner and Baron could convincingly prove a clear logical function of diagrams in the Middle Ages, e.g. illustrating containment. It should also be noted that the diagram discussed by Frampton, in my view, goes back to illustrations of the so-called *Liber Rotarum* (Isidore of Seville: *De naturum rerum*),[108] which opens up a whole new history of tradition, and this concerns diagrams of cosmology, not logic.

Fig. 6
John of Afflighem: *De musica cum tonario,* ca. 1100, StB Mainz, Hs II 375, fol. 9ᵛ.

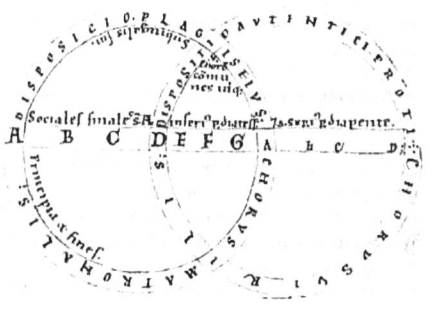

Fig. 7
Aribo: Dialogus de Musica, before 1100, Sibley Music Library, Rochester US-R MS Vault ML 96, fol 19ᵛ.

(3) A medieval variant of Euler or Venn diagrams has been presented by Catherine Nolan in a way that is plausible in itself.[109] These diagrams are found in the treatise *De musica cum tonario*, written around 1100 by John of Afflighem (Ioannis Cottonis); they illustrate differences and similarities between plagal and authentic cadences (Fig. 6). Afflighem's diagrams, as Anthony William Fairbank Edwards in particular points out, were even anticipated some 30 years earlier by Aribo Scholasticus (Archiepiscopus Moguntinus) in his work *De musica* (see Fig. 7). Even though the two rings in Fig. 7 already point to the semantic function of regions in Euler and Venn diagrams,[110] respectively, a closer look at the interfaces still shows the representation of Borromean rings. Afflighem's representation in Fig. 8 is indeed a clear two-dimensional figure representing a circumference of circles and not an entanglement of rings, but Afflighem explicitly speaks of "rotae" and not of 'circuli' in the text – only Edwards consistently translates 'rotae' as 'circles'.

---

[108] Cf. e.g. Zofingen, Stadtbibliothek, Pa 32, fol. 62ʳ.
[109] Cf. Catherine Nolan: Music Theory and Mathematics. In: The Cambridge History of Western Music Theory. Ed. by T. Christensen. Cambridge 2002, pp. 272–304, here: p. 282; Anthony William Fairbank Edwards: An Eleventh-Century Venn Diagram. In: BSHM Bulletin: Journal of the British Society for the History of Mathematics 21:2 (2006), pp. 119–121.
[110] In Fig. 9 we see three compartments: 1. on the left, namely plagal *and not* authentic (A, B, C, D), 2. the vesica piscis, namely plagal *and* authentic (E, F, G, a), and 3. on the right, namely authentic *and not* plagal (h, c, d).

2 Logic and its Geometrical Interpretation

The fact that in the tradition of music-theoretical diagrams of the Middle Ages not only were 'wheels' explicitly named but were usually also clearly drawn as such, is shown in the study by Anna Maria Busse Berger.[111] Berger ultimately does not refer to a logical tradition of these diagrams, but with Mary Carruthers' mnemonic studies on the Middle Ages to a universal 'learning by rote'.[112] Thus, while Afflighem's diagrams appear to be analytical circle diagrams in appearance, the explicit description is that the two-dimensional circles are simplified three-dimensional (Borromean) wheels. In order to clarify to what extent these diagrams could belong to the history of the development of analytical diagrams, I believe that much research into the historical context is still necessary.

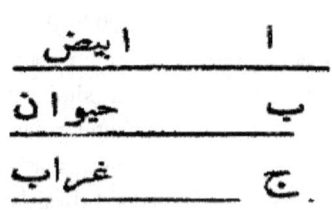

Fig. 8
Abu'l-Barakāt al-Baghdādī's: al-Kitāb al-Muʿtabar. Hyderabad 1938, p. 139.

(4) The most convincing thesis on analytical diagrams in medieval geometrical logic has been put forward by Wilfrid Hodges.[113] He argues that line diagrams were already used by Abu'l-Barakāt al-Baghdādī in the 12th century. In his work *al-Kitāb al-Muʿtabar*, about 25 pages are devoted to syllogistic, whereby Barakāt's logic deviates greatly from Aristotle. According to Hodges, Barakāt uses his line diagrams to find out when true premises also lead to a true conclusion, without the aid of Aristotelian proof techniques such as conversions. Barakāt uses 86 line diagrams for this purpose, which on the one hand, are reminiscent of the interval diagrams that Malink used to interpret Aristotelian metaphors; on the other hand, they also have a similarity to the line diagrams found in Keckermann and other authors from the 17th century onwards. Fig. 8 shows the term 'white' on the top line, 'animal' on the middle line and 'crow' on the bottom line. As Hodges argues, Barakāt's line diagrams have been transcribed incorrectly in several editions and are therefore in great need of interpretation.[114] It is currently unknown from which period the oldest manuscript we have come from. Since research on this diagram tradition is only in its infancy, let us leave it at the fact that it is currently one of the most promising contributions to analytical logic diagrams in the Middle Ages and approach the early modern period.

---

[111] Cf. Anna Maria Busse Berger: Medieval Music and the Art of Memory. Berkeley 2005, pp. 105ff.
[112] Cf. Anna Maria Busse Berger: Medieval Music and the Art of Memory, pp. 105ff.
[113] Cf. Wilfrid Hodges: Two Early Arabic Applications of Model-Theoretic Consequence. In: Logica Universalis 12 (2018), pp. 37–54; ibid.: Medieval Arabic Notions of Algorithm. Some Further Raw Evidence. In: Fields of Logic and Computation III. Ed. by A. Blass, P. Cégielski, N. Dershowitz, M. Droste, B. Finkbeiner (Lecture Notes in Computer Science, vol 12180). Cham 2020, pp. 133–146.
[114] Cf. esp. W. Hodges: A Correctness Proof for al-Barakāt's Logical Diagrams. In: The Review of Symbolic Logic (forthc.). I thank Wilfrid Hodges for sending me his unpublished manuscripts and the annotations of an earlier version of this section.

2.2 Analyticity – Analytic judgements, Containment Metaphors and Logic Diagrams

From the late 15th century onwards, geometric diagrams are found, which – thanks to the first printing machines – are also clearly recognisable as an integral part of textbooks of logics: Numerous different logic diagrams (square of opposition, hexagon of opposition, pons asinorum, tree diagram, and many others) were used in incunables and printed editions between the late 15th and early 18th centuries. It can be seen that different scholarly cultures favoured different diagrammatic systems: (1) Byzantine scholars and, in some cases, Aristotle editions with Neoplatonic or Arabic commentaries use increasingly complex forms of pontes asinorum. (2) Squares of Oppositions and their extensions are used almost exclusively by Aristotelians of the Roman Catholic denomination. (3) Thomists in particular (mostly Dominicans) concentrate on the development of tree diagrams, (4) Scotists (mostly Franciscans) mainly on so-called Phoebifer Axis diagrams. And as I will show in Chapter 2.2.3, it is critics of Aristotelianism and especially Protestant logicians who develop increasingly complex forms of analytical diagrams, especially line and circle diagrams. In the following, I will give a few selected examples of the non-analytical strands of tradition (1)–(4) and then try to present the analytical strand of tradition (Chapter 2.2.3) in more detail in its chronological development.

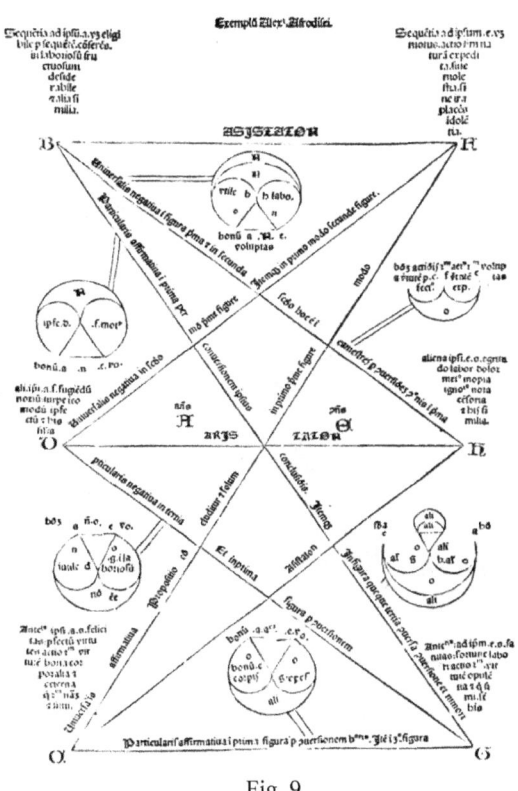

Fig. 9
Liber i priorum analecticorum, cap. 29. In: Omnia Aristotelis opera cum commento Averrois. Venetiis 1489.

(1) In the printed works of Byzantine scholars and editions of Aristotle with Arab commentators, the bridges of asses already appear prominently in the late 15th century, often together with tree diagrams and squares of opposition and their various extensions. A good example of an early document in this tradition is an edition published in Venice in 1489 with texts by Aristotle, Averroes and Pseudo-Aristotle, in which one finds combinations

of the types of diagrams already discussed, along with many other diagrams. Extensions of the square of opposition to the logical pentagon or octagon were known to narrower circles of scholars of the 14th century through Buridan or Nicholas of Oresme.[115] In Fig. 9, however, one can see how in Aristotle's edition of 1489 the opposites in an asystaton hexagon are supported with complex representations of pontes asinorum.[116] Jacques Lefèvre d'Étaples made this method prominent from 1492 onwards, among other things through the Aristotle edition by Ermolao Barbaro. Just as d'Étaples made the method of using bridges of asses known in the French region, Byzantine humanists such as Georgios Trapezuntios (*Dialectica*, 1509), Johannes Argyropulos (*Aristotelis Stagyrite Dialectica*, 1517) and Leo Magentinus (*Philoponi Commentaria*, 1536) handed it down for the Italian scholarly circle of the 16th century. More prominent examples in this reception history are the logics, commentaries and editions by Agostino Nifo (*Dialectica ludrica*, 1521), Giacomo Zabarella (*Tabulae logicae*, 1583), Giulio Pace (*Principis Organon*, 1584) and Giordano Bruno (*De progressu et lampade venatoria logicorum*, 1587).[117]

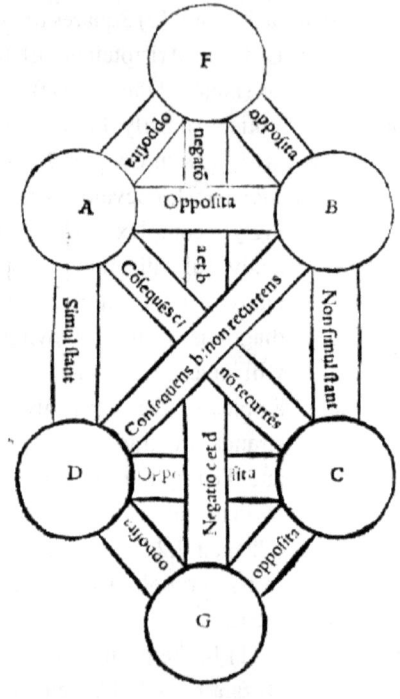

Fig. 10
Jacobus Faber Stapulensis: Libri logicorum, ad archetypos recogniti [...]. Parisius 1503, fol. 142ʳ.

(2) The line of tradition relating to the square of opposition and its extensions has also been shaped

---

[115] Cf. bspw. Stephen Read: John Buridan's Theory of Consequence and His Octagons of Opposition. In: Around and Beyond the Square of Opposition. Ed. by Jean-Yves Béziau, Dale Jacquette. Basel 2012, pp. 93–110; Lorenz Demey: Between Square and Hexagon in Oresme's *Livre du Ciel et du Monde*. In: History and Philosophy of Logic 41:1 (2020), pp. 36–47.
[116] For some other works of this tradition not mentioned here, see also Ivo Thomas: The Later History of the Pons Asinorum. In: Contributions to Methodology and Logic in Honour of J.M. Bochenski. Ed. by Anna-Teresa Tymieniecka. Amsterdam 1965, pp. 142–150.
[117] More detailed information on this line of tradition can be found in Letizia Panizza: Learning the Syllogisms. Byzantine Visual Aids in Renaissance Italy – Ermolao Barbaro (1454–93) and others. In: Philosophy in the Sixteenth and Seventeenth Centuries. Conversations with Aristotle. Ed. by Constance Blackwell, Sachiko Kusukawa. London, New York 1999, pp. 22–48. However, research on this type of diagram has not yet progressed beyond individual works.

2.2 Analyticity – Analytic judgements, Containment Metaphors and Logic Diagrams

by d'Étaples.[118] The Aristotle commentary *Libri logicorum ad archetypos recogniti* by d'Étaples, published in 1503, in which numerous diagram types were united and combined, also brings the first logical hexagon, which resembles the modern extensions by Robert Blanché and Augustin Sesmat.[119] Here, on the one hand, concepts are placed in relation instead of judgements and, on the other hand, the corners that are contrary and subcontrary in the square are connected to form two further corners:[120] In Fig. 10, F indicates the relation 'Neither A nor B' ($\neg(A \lor B)$) and G the negation of F. Similar extensions are then found only occasionally in the 16th and 17th centuries, for example in the logics of Fabio Glissenti (*In Priora Analytica Aristotelis*, 1594) or Juan Caramuel y Lobkowitz (*Logica vocalis, scripta, mentalis, obliqua*, 1680).[121] Although from the 17th century onwards the square of opposition can be found in almost every logic textbook by a Catholic clergyman, the medieval and early modern extensions are only rediscovered by August de Morgan and then by Blanché and Sesmat.[122]

---

[118] William Hamilton: Lectures on Metaphysics and Logic. Vol. II, p. 420 even claims that there are no diagrams in d'Étaples, but revises his judgment in ibid.: Discussions on Philosophy and Literature, Education and University Reform, pp. 669ff.
[119] Cf. e.g. Alessio Moretti: The Geometry of Logical Opposition. Neuchâtel 2009, Chapter 8.
[120] Cf. Jacobus Faber Stapulensis: Libri logicorum, ad archetypos recogniti [...]. Parisius 1503, fol. 27ʳ (to Cat. 4b–5b), fol. 141ᵛ–142ʳ. I thank Werner Oechslin for his insights into the diversity of d'Étaples' diagrammatic works.
[121] For Caramuel's Pentagon cf. Wolfgang Lenzen: Caramuel's Pentagon of Opposition and his Vindication of the Principle Ex contradictorio quodlibet. In: History of Logic and its modern Interpretation. Ed. by Ingolf Max, Jens Lemanski. London 2022 (forthc.).
[122] Cf. Anna-Sophie Heinemann: 'Horrent with Mysterious Spiculæ'. Augustus De Morgan's Logic Notation of 1850 as a 'Calculus of Opposite Relations'. In: History and Philosophy of Logic 39:1 (2018), pp. 29–52. Heinemann has found many more diagrammatic representations of De Morgan, but they have not yet been published.

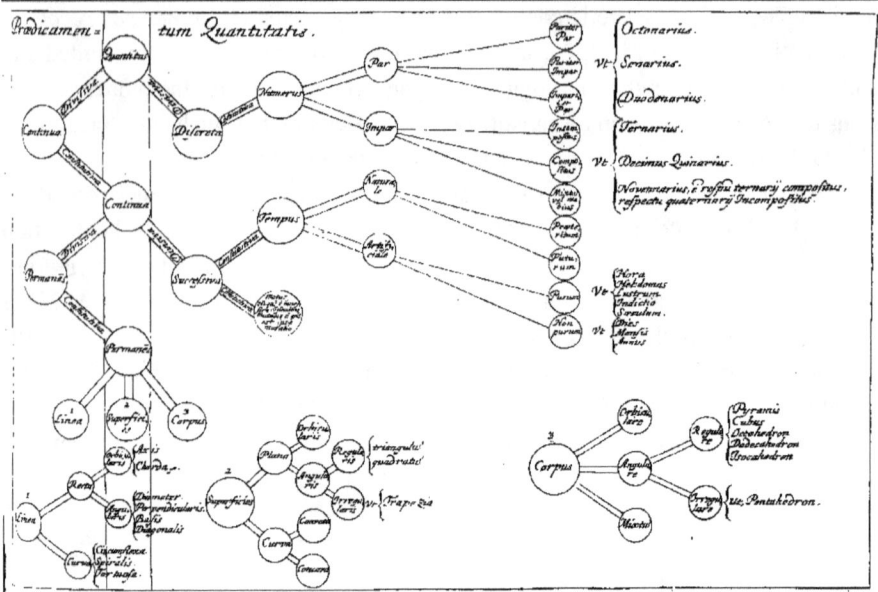

Fig. 11
Celestino Sfondrati: Cursus Philosophicus I. Logica Major. S. Galli 1696, fol. 361.

(3) Tree diagrams were printed in numerous variations in the early modern period. In particular, in the table works of the 16th and 17th centuries, they served to structure almost all possible subject areas. Besides the well-known ramistic tables,[123] two types of tree diagrams in logic are particularly noteworthy: On the one hand, tree diagrams exemplify certain philosophical or logical theories, such as the *arbor purchotiana* in the Cartesian school; on the other hand, tree diagrams in the Thomistic tradition, which subdivided not only the Aristotelian category of substance but also the further categories into genera, species and individuals.[124] The latter category trees were already used in High Scholasticism,[125] then became widespread in the early 16th century, for example, through the logic textbooks of Magnus Hundt (*Compendium totius logicae*, 1507), Johannes Murmellius (*In Aristotelis decem praedicamenta isagoge*, 1513) or Johannes Eck (*In summulas Petri Hispani extemporaria et succincta*, 1516). Finally, large-scale category trees were used especially in Thomistic textbooks. For example, the category trees of Celestino Sfondrati (*Logica*, 1696), who uses the 10 Aristotelian categories to develop all subject areas of science and everyday life employing a diaíresis, have a special expressiveness. In Fig. 11, one can see one of these category trees, which combines a Porphyrian tree on the left with a Senecean tree

---

[123] Cf. on tree diagrams in general and on ramistic tables in particular Siegel: Tabula. Figuren der Ordnung um 1600. Berlin 2009, esp. Chapter III.3.
[124] Cf. e.g. Paul Richard Blum: Studies on Early Modern Aristotelianism. Leiden et al. 2012, Chapter 15.
[125] See, for example, the first pages of the above-mentioned manuscript of the Summula of Petrus Hispanus, Paris, BN, ms. lat. 16611.

2.2 Analyticity – Analytic judgements, Containment Metaphors and Logic Diagrams

diagram on the right side.[126] This combination allows the category of quantity (top left) to be transformed into the areas of arithmetic (top right), time (middle right), plane (bottom left) and spatial Euclidean geometry (bottom right).

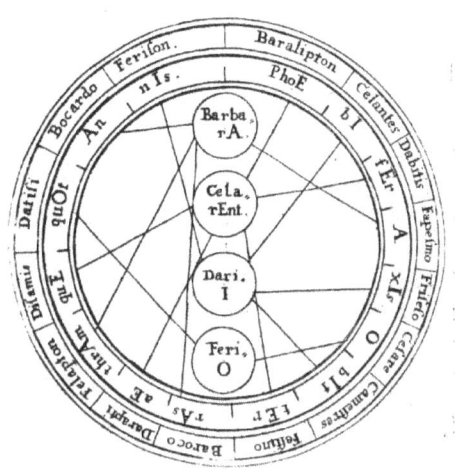

Fig. 12
Amand Hermann: Sol Triplex […]. Sultzbaci [Sulzbach] 1676, p. 46.

(4) In humanism and then especially in Baroque Scotism, a method developed of apagogically testing the validity of syllogisms: For this purpose, one hypothetically took the contrary or contradictory judgement of the conclusion of the syllogism to be tested as the premise of a new syllogism. This premise of the new syllogism was supplemented with one of the two premises of the syllogism under test so that the new syllogism now corresponded to one of the four perfect modes (Barbara, Celarent, Darii, Ferio). If the conclusion of the new syllogism is now a contrary or contradictory judgement to the second premise of the syllogism under test, which was not used in the test procedure, then it could be considered valid. However, if the new conclusion did not contradict the non-used judgement of the syllogism under test, the latter was considered invalid.

The procedure was particularly appreciated by Scotists, since a uniform method for proving all syllogisms had been found in just a few steps. In order to memorise the procedure, however, at least for all modes known to be valid, one first developed mnemonics such as "Phebifer axis obit terras aethramque quotannis" and later diagrams that visualised this mnemonic. If one knows the order of the imperfect modes (Baraliption, Celantes, Dabitis usw.), one can assign each mode in turn to a vowel in the mnemonic, whereby the vowel points to the conclusion of the perfect mode to be constructed in the apagogic proof: for example, "PhEbIfEr" helps to know that the validity of Baralipton can be apagogically tested via CelarEnt, Celantes via DariI, and Dabitis via CElarent.[127] In order not to have to do the rather complex steps of proof

---

[126] For the distinction between trees of Porphyry and Seneca cf. Jonathan Barnes: Commentary. In: Porphyry's Introduction, Translated with a Commentary. Oxford 2003, pp. 21–312, here: pp. 108–112; Jaap Mansfeld: Heresiography in Context. Hippolytus' Elenchos as a Source for Greek Philosophy. Leiden et al. 1992, pp. 78–109.

[127] For example, one forms the contradiction of the conclusion of Baralipton for the first premise of the new syllogism. The first premise of Baralipton is then taken as the second premise of the new syllogism. Now one can form the conclusion of the new syllogism, which is in sum equivalent to Celarent. However, the

in one's mind, Phoebifer Axis diagrams were printed in Scotist textbooks of the southeastern region of Central Europe, especially from the middle of the 17th century onwards, and were also copied into textbooks by pupils and students. Such diagrams can be found, among others, in the Jesuit Melchior Cornäus in Würzburg (1657), in the Franciscan Antonín Brouček in Prague (1663), in the Franciscan Bernhard Sannig in Prague (1684), and in the Benedictine Cölestin Pley in Salzburg (1693). Fig. 12 shows a Phoebifer Axis diagram printed in 1676 in *Sol Triplex* by the Olomouc Franciscan Amand Hermann.

After the Thirty Years' War, the Phoebifer Axis diagrams first indicated the confessional affiliation in the countries of the Roman Catholic Church, but at the same time also the difference that was reflected on completely different levels between Protestants and Catholics: Thus, the mnemonic was an expression of a bitter and conservative anti-Copernicanism, which can be seen in the literal translation of the mnemonic: "The solar axis revolves around the Earth and the Aether every year". Moreover, the mnemonic was the product of a highly contested logical problem of the late 17th century between Catholics and Protestants, namely the question of how to prove the whole syllogistic by only one method: Whereas Catholics propagated indirect proofs by the Phoebifer Axis method, Protestants developed increasingly complex forms of analytical diagrams in geometric logic for a direct proof procedure.[128]

2.2.3 Analytical Diagrams of Geometric Logic

As was shown in Chapter 2.2.1, Venn had defined analytical diagrams as those that can analyse judgements and concepts, and Ziehen had described geometric logic as one that mimics the intuitive figures of geometry. In the search for analytical diagrams of geometrical logic, I first listed several theses in the previous Chapter 2.2.2 that indicate that logic diagrams may have already existed in ancient logic. For medieval logic, four theses were discussed. The 12th-century syllogistic of Barakāt al-Baghdādī came closest to the line diagrams that correspond to Venn and Ziehen as precursors of the tradition of analytical diagrams in geometrical logic beginning with Euler. Then, with the advent of the printing press, five diagrammatic traditions in logic emerged: (1) the Byzantine tradition with pontes asinorum in the 16th century, (2) the Roman Catholic tradition of squares of opposition throughout the early modern period, (3) the Thomistic tradition with category trees, and (4) the Scotist tradition with phoebifer axis diagrams, both in the 17th century.

---

conclusion of Celar*E*nt is in contrariety with the second premise of Bar*A*lipton. Therefore, Baralipton must be valid.

[128] I gave a more detailed description of this type of diagram at the 2019 Inaugural Pan-American Symposium on the History of Logic at UCLA.

2.2 Analyticity – Analytic judgements, Containment Metaphors and Logic Diagrams

Whether these lines of tradition meet Venn's and Ziehen's criteria for analytical diagrams of geometric logic is often questionable. In any case, all the diagrams mentioned so far do not seem to give us any clue as to what Kant might have had in mind when he used metaphors of circumference and containment to define analytic judgements. Or at least, the types of diagrams mentioned so far cannot invalidate the criticisms that Quine and his predecessors and successors have made of the metaphors of Kant's definition.

In what follows, I will argue for the existence of yet another line of tradition that is crucial for the development of analytic logic diagrams up to Kantianism: First, the anti-Aristotelian Protestant tradition with analytical diagrams consisting primarily of lines or circles, which can be divided into three periods between the early 16th and late 19th centuries: (1) a first period is confined to the 16th century; (2) after the Thirty Years' War, a second period then develops, which is forgotten with the rise of rationalism at the beginning of the 18th century; (3) finally, the period still known today begins in the middle of the 18th century, which only gains influence with the decline of rationalism and the rise of Kantianism from the 1790s onwards.

(1) From the 16th century onwards, the line of tradition that must be attributed to the analytical diagrams of geometrical logic developed. The beginnings of this line of tradition are difficult to grasp. A recourse to one of the forms of diagrams discussed in the Middle Ages is not yet known. It has been assumed several times that analytical diagrams originate from one of the works mentioned above: From a geometrical point of view, similar diagrams can be found, for example, in the works of Lull, d'Étaples or Bruno. These come very close to the analytical circle or line diagrams, but do not have a directly recognisable analytical function in the sense of Venn. Such diagrams are also further elaborated in Charles de Bouelle's *De mathematica Rosa* (1509), but without reference to logic. An indication that analytical diagrams might already have been known in logic at the beginning of the 16th century is also provided by some quotations. For example, in Erasmus of Rotterdam's *Moriae encomium*, it is said that philosophers felt elevated above the rabble when they superimposed their triangles, quadrilaterals, circles and such mathematical pictures ("triquetris, & tetragonis, circulis, atque huiusmodi picturis mathematicis aliis super alias inductis").[129]

But locating the first early modern representation corresponding to Euler's logic diagrams is not possible through these references. John Venn had claimed that the first analytical diagrams originated with Juan Luis Vives; Charles S. Peirce, on the other hand, had put forward the thesis in 1903 that these diagrams had already been traced back to Lorenzo Valla. Peirce's thesis is puzzling, however, because I could not find any references to logic diagrams in any of the early editions of Valla's writings that I

---

[129] Cf. Erasmus Roterodamus: Moriae encomium. S.l.: s.n., s.a. [1511], fol. Fii^v. It should be noted, however, that contemporaries of Erasmus interpreted this passage not necessarily in the sense of geometrical logic, but in the sense of contradictory statements of scholasticism (cf. for example the translation by Sebastian Franck: Das Theür vnd künstlich Büchlin Morie Encomion. S.l.: s.n., s.a. [ca. 1543], fol. 49r).

2 Logic and its Geometrical Interpretation

have looked through (1499, 1509, 1531, 1540). Nor is it possible to tell whether Peirce makes this claim himself or whether he has taken it unchecked from other sources on the history of logic. Peirce's statement thus remains nebulous.[130]

The first schematic representation, which John Venn himself interpreted as a precursor of his and Euler's diagrams, dates from 1531 by Juan Luis Vives.[131] In the chapter on syllogistic of the second book of *De censura veri et falsi*, Vives first describes the classical syllogism, then the difference between the terminus maior and minor, and finally outlines the Aristotelian *dictum de omni* (*et nullo*) with the help of the following inference:[132]

If three triangles are drawn, one of which *b* is the largest and another comprises *a*, but a third is the smallest in *a*, which is *c*, we say that if all *b* are now *a* and all *c* are *b*, [then] all *c* are [also] *a*.

vt si tres trianguli pingantur, quorum vnus *b* sit maximus, & capiet alterum *a*, tertius sit minimus intra *a*, qui sit *c*: ita dicimus si omne *b* est *a*, & omne *c* est *b*, omne *c* est *a*.[133]

The quote is by no means to be understood metaphorically, since Vives explicitly speaks of triangles (trianguli), but not of concepts or judgments that are comprehended (capiet).[134] Nevertheless, the conclusion shown is not for a geometrical exercise, but to illustrate logical comparisons (comparationes). The whole quote is embedded in a text passage that is meant to show an analogy between logic and geometry. Thus, the direct conclusion of the terminus minor from the maior is to be exemplified by the example of the diagram.

Coumet had already referred to the independent work *Metamorphosis Logicae* by Nicolaus Reimarus Ursus from 1589. Reimers is of great importance not only as a

---

[130] I thank Ahti Pietarinen for pointing out this thesis. The thesis is found in Houghton Library, MS 530, c.1902, and this manuscript is published as ch. 30 in Charles Sanders Peirce: Logic of the Future. Peirce's Writings on Existential Graphs. Ed. by Ahti Pietarinen. Berlin, Boston.

[131] Venn takes this reference from Friedrich Albert Lange: Logische Studien. Ein Beitrag zur Neubegründung der formalen Logik und der Erkenntnistheorie. Iserlohn 1877, p. 10. Since in Vives "instead of the triangles mentioned in the text, only angles" are found, Lange assumes a "typographical convenience" and also believes that the illustration is hardly "an invention of the astute Spaniard", but rather a school tradition. As a hint for further research, however, it should be noted here that one can find numerous logical illustrations and also angle illustrations in the Spanish edition of Thomas Bradwardine: Preclarissimum mathematicarum opus [...]. S.l. [Valencia] 1503, which was made for Vives' teacher Hieronymus Amiguetus. The first reference to analytical diagrams in Vives is found in Ignatius Denzinger: Institutiones logicæ. Vol. II. Leodii 1824, p. 66.

[132] Cf. Arist. Cat. III. 1b10–16, V. 3b4f.; An. pr., I. 24b26–30, IV. 25b39–26a2, 26a23–26, IX. 30a17–23, XIV. 32b38–33a38; Top. D. I, 121a25f.; Porph. eisag. VIII.2–3 (Aristoteles Stagaritae Peripateticorum: Principis Organon. Ed. by Iulius Pacius. Morgia 1584).

[133] Ioannes Ludovicus Vives: De censura veri et falsi. In: ibid.: De disciplinis Libri XX, Tertio tomo de artibus libri octo. Antverpia 1531, fol. 57ᵛ. Other schemes in tom. III are found on fol. 27ᵛ, fol. 37ʳ.

[134] An imposing comparison between Aristotle, Vives and the early modern editions of Euclid's *Elements* (esp. lib. V) would be a separate research topic and cannot be done here.

2.2 Analyticity – Analytic judgements, Containment Metaphors and Logic Diagrams

Copernicus translator but also as a geometric logician, since he is the first to use a circle diagram to sketch and prove inferences.[135] In *Metamorphosis Logicae* he first introduces a table (deductionis tabulâ), which, similar to the vertical tree diagram, now horizontally divides generic terms into two subordinate terms each according to the Platonic principle of διαίρεσις or divisio. Similar to the famous examples of divisions by Seneca (Ad Luc. 58) or Porphyry (Eisag. I), the generic term 'animal' (animal) is broken down into 'irrational, such as animals' (irrationale, vt brutum), or 'rational, such as humans' (rationale, vt Homo), and the latter instance is in turn distributed into 'male or man' (Mas seu vir) and 'female or woman' (Fæmina seu mulier). This conceptual scheme now enables Reimers to set up the modus Barbara (universally quantified by O[mnia]) and to indicate to the containment metaphor (*inest*):

$$\text{Animal} \begin{cases} \textit{Irrationale, vt brutum} \\ \textit{Rationale, vt Homo} \begin{cases} \textit{Mas seu vir,} \\ \textit{Fæmina seu} \\ \textit{Mulier.} \end{cases} \end{cases}$$

All humans are animals (because human is contained in animal)
All women are human beings (because woman is also contained in human)
Therefore all women are animals.
O. Homo est Animal: (quia Homo inest Animali)
O. Mulier est Homo: (quia et Mulier inest homini)
O. Ergo Mulier est Animal.[136]

The sense of the logical expressions of containing and not containing ("illa Vocabula Logica, Inesse, &, non inesse") now refers to the division presented.[137] Reimers explains that the inference given above can be traced back to the "Principle through inner insight" (Principium per intellectum internum), namely the *dictum de omni* (*et nullo*), which he illustrates abstractly with a circle diagram, which in turn proves the validity of the inference by means of the logical expression of being contained ("per Inesse Demonstremus"):[138]

---

[135] Nicolaus Raymarvs Vrsvs Dithmarsivs: Metamorphosis Logicae [...]. Argentorati 1589, p. 32.
[136] Nicolaus Raymarvs Vrsvs Dithmarsivs: Metamorphosis Logicae, p. 31. (My transl. – J.L.)
[137] Nicolaus Raymarvs Vrsvs Dithmarsivs: Metamorphosis Logicae, p. 30.
[138] Raymarvs: Metamorphosis Logicae, p. 31.

2 Logic and its Geometrical Interpretation

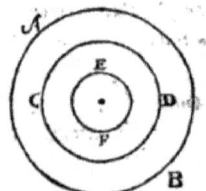
Let *EF* be the innermost circle, *CD* the middle circle and *AB* the outermost circle. If therefore the innermost *EF* is contained in the middle *CD*, and again the middle in the outermost *AB*, then the innermost *EF* will also be contained in the outermost *AB*.

Sit enim Circulus *EF*. intimus: *CD*. intermedius: *AB*. verò extremus. Cùm itaq[ue] intimus *EF*. sit comprahensus in intermedio *CD*. rusumquè intermedius in extreme *AB*. necessariò erit etiam intimus *EF*. contentus in Extremo *AB*.[139]

With this example, all the basic uses of sphere or circle diagrams in geometric logic have been touched upon: (1) Concepts such as 'animal' have a sphere or a circumference in which sub-concepts such as 'brutum', 'homo' and further mediated sub-concepts are contained (*comprahensus in*). (2) The conceptual scheme makes it possible to distinguish true statements (such as 'Homo inest Animali') from false ones (such as 'Mulier inest Brutum') and to explain them ('quia...'), so that (3) these judgements taken together can yield a conclusion ('...Ergo Mulier est Animal'). Reimer's diagram, like Vives', thus fulfils the basic functions of analytical diagrams of geometric logic according to Venn's and Ziehen's definition given in Chapter 2.2.1.

Analytical diagrams in the form of lines can be found before the 18th century in Bartholomäus Keckermann: Whereas in the *Systema Logicae* of 1601, conceptual circumferences in the justification of judgements were still represented by lines of equal length (Fig. 1), later editions, such as that of 1611, illustrate the same ratios by lines of different lengths (Fig. 2). Keckermann explains that the first of the three canonical

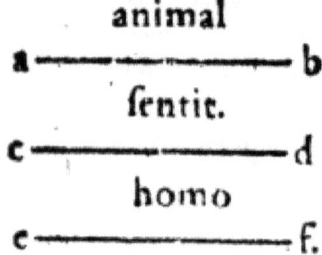
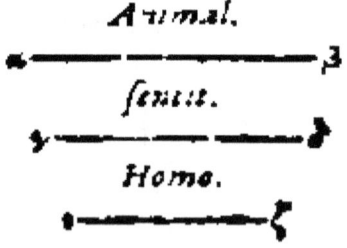

Fig. 1
Bartholomæus Keckermannus: *Systema Logicæ. Compendiosa methodo adornatum* [...]. Hanoviae 1601, p. 92 (= III, I 3)

Fig. 2
Bartholomæus Keckermannus: *Systema Logicæ. Tribus Libris Adornatvm*, [...]. Hanoviae 1611, p. 426 (= III, I 6).

---

[139] Raymarvs: Metamorphosis Logicae, p. 33. (My transl. – J.L.)

2.2 Analyticity – Analytic judgements, Containment Metaphors and Logic Diagrams

figures is evident, especially the modus Barbara, since it coincides with the natural principle of *dictum de omni (et nullo)*. This naturalness is shown by the enthymemic inference in Fig. 1:

> Man is an animal because he feels: here *animal* is represented by the line ab, *man* by the line ef, *he feels* by the line cd.
> Homo est animal, quia sentit: hîc *animal*, est instar lineæ a, b, *homo* instar linæ e, f: sentit, *instat* linæ c, d.[140]

Johann Heinrich Alsted adopts Keckermann's earlier diagrammatic version in 1614.[141] Although Venn argued against Hamilton that Alsted's line diagram was not yet an analytical one similar to Lambert's, I see the better arguments rather on Hamilton's side, since Fig. 2 (and even Fig. 1) is supposed to represent the dictum de omni by means of the modus Barbara. Whether Keckermann and Alsted are in a tradition of line diagrams to which Barakāt also belongs remains questionable.

After Keckermann and Alsted, however, there is a break in the history of analytical diagrams in geometric logic. One can say that Vives, Reimers, Keckermann and Alsted form a first period in the traditional line of analytical diagrams; and one can furthermore also assume that they wanted to set themselves apart from the Byzantine tradition (Argyropulos, Trapezuntios, d'Étaples etc.) with their method. Analytical diagrams in Central Europe, surprisingly, ended before the Thirty Years' War and were only used afterwards by a new generation of logicians.

(2) Several times Arnold Geulincx and his 'cubus logicus' are mentioned by Ziehen and other historians of logic[142] in this second period of analytical diagrams. In Geulincx's works, I was able to find both longer logical descriptions with circumferential metaphors that may describe analytical diagrams and drawings that might be of interest to the currently fast-growing field of research called logical geometry or oppositional geometry;[143] however, I did not find any analytical diagrams.[144]

After the Thirty Years' War, however, a period of intensive use of logic diagrams and especially analytical diagrams began in Central Europe that lasted until the early 18th century. Almost all logicians remained faithful to the diagrams of their own school and sometimes polemicised against diagrammatic procedures of other lines of tradition and especially of other denominations. The only logicians who, as far as I

---

[140] Cf. Bartholomæus Keckermannus: Systema Logicæ. Compendiosa methodo […]. Hanoviae 1601, p. 91 (= III, I 3).
[141] Cf. Johanne-Henrico Alstedio: Logicæ Systema Harmonicum […]. Herbornæ Nassoviorum 1614, p. 395 (= VII, IV 1).
[142] Cf. Gabriel Nuchelmans: Geulincx Containment Theory of Logic. Amsterdam 1988. A negative verdict falls Carl Friedrich Bachmann: System der Logik. Ein Handbuch zum Selbststudium. Leipzig 1828, pp. 148f.
[143] Cf. Alessio Moretti: Arrow-Hexagons; Lorenz Demey: From Euler Diagrams in Schopenhauer to Aristotelian Diagrams in Logical Geometry.
[144] A more detailed study of diagrams in Geulincx can be found in Chapter 2 of Jens Lemanski: Calculus CL – From Baroque Logic to Artificial Intelligence. In: Logique et Analyse 249 (2020), pp. 111–129.

2 Logic and its Geometrical Interpretation

know, used several diagram types of different schools were the Cistercian monk Caramuel at the beginning of this period and the Pietist Johann Christian Lange at the end.[145] However, both also break with numerous other logical conventions of their time. In Caramuel's explicitly Thomistic project of a *Logica Vocalis* in 1654, he classifies various forms of naturally occurring syllogisms, also introducing a sympathetic syllogism (*De Syllogismo Sympathetico*).[146] The distinction between sympathy and antipathy, which had been common since Neoplatonic natural philosophy, could also be applied to the relationship of judgements in syllogisms and thus see the similarity between the *dictum de omni et nullo* and the first Euclidean axiom. This similarity is the starting point for understanding syllogisms with the help of the Aristotelian category of quantity and for transferring them to other categories such as action (actio) or where (ubi):

Line AB is smaller than Line CD:	
Line CD is smaller than Line EF	Quantity
Thus, line AB is smaller than line EF.	
To acquire wealth is daring.	
Daring is disobedience.	Action
Therefore, wealth is disobedience.	
I went from Prague to Vienna.	
From Vienna also to Linz,	local motion, where
so I went from Prague also to Linz.	
Linea AB est minor, quàm linea CD:	
Atqui linea CD est minor, quàm linea EF:	Quantitas
Ergo linea AB est minor, quàm linea EF.	
Divitiae pepererunt audaciam	
Audacia inobedientiam.	Actio
Ergo tandem divitiae inobedientiam.	
Ivi Pragâ Viennam:	
Vienna verò Linzium:	motus localis, Ubi
Ergo Pragâ ivi Linzium.[147]	

Whether such figures really have to be understood as logic diagrams is certainly debatable. After all, one can also argue that logic is not imitating geometry here (Ziehen's definition), but that analogies are being drawn between geometry and logic. Either way, this example shows the analogy that several early modern logicians saw

---

[145] In Leibniz, there are several descriptions reminiscent of the square of opposition (e.g., the title page of *De Arte Combinatoria*). Großer lists various types of diagrams in his Logic, but explicitly distances himself from the scholastic methods.
[146] Ioannes Caramuel: Theologia Rationalis, Sive In Auream Angelici Doctoris Svmmam [...] Praecursor Logicvs [...]. Francofurti [Frankfurt] 1654, p. 354, cf. also p. 235.
[147] Cf. I. Caramuel: Theologia Rationalis, p. 354.

2.2 Analyticity – Analytic judgements, Containment Metaphors and Logic Diagrams

between geometry and logic, and which characterises the multiple meanings of the term 'geometric logic'. Moreover, it can be assumed that such parallels between logic and geometry motivated Euler's diagrams: In § 14 of Jakob Bernoulli's 1685 *Parallelismus Ratiocinii Logici et Algebraici*, one finds the same analogy between the *dictum de omni et nullo* and the first Euclidean axiom as given in Caramuel.[148] Bernoulli, however, translates the proportionality expressed by the two fundamental principles into an algebraic notation. The fact that Euler was already very familiar with these theses as a student is proven by the dissertations submitted in 1722 for the vacant professorship of logic in Basel. In these dissertations, for which Euler was a respondent, Bernoulli's approach was discussed intensively.

In the period before Euler, especially the diagrams found in the so-called Weigel and Weise circles between the 1660s and 1710s are of great relevance to geometric logic.[149] I will first discuss Sturm and Leibniz, the two students of Erhard Weigel. In the 1661 treatise *Novi Modi Syllogizandi*, Johann Christoph Sturm not only uses five diagrams ("diagrammate") to illustrate new logical inferences, but also, for the first time, the circular schemes without reference to the *dictum de omni* (*et nullo*) or the modus Barbara corresponding to it. As an example, let us take Sturm's first geometrical diagram, in which a particular affirmation with an indeterminate subject is to be inferred from a universal affirmation in the *propositio maior* and a universal negation in the *minor*:

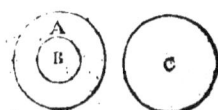

If all B are A and no C is B, then it follows formally and by necessity that some not-C are A.

Si omne B est A, & nullum C est B, sequitur formaliter & ἐξ ἀναγ´κης [sic] haec: Quodam non-C est A.[150]

---

[148] Jakob Bernoulli: Parallelismus ratiocinii logici et algebraici. Basileae 1685, p. 4.
[149] For a more detailed analysis of these diagrams, cf. Jens Lemanski: Logic Diagrams in the Weise and Weigel Circles. In: History and Philosophy of Logic 39:1 (2018), pp. 3–28.
[150] Cf. Johann Christopherus Sturmius: Universalia Euclidea [...]. Accedunt ejusdem XII. Novi Syllogizandi Modi in propositionibus absolutis, cum XX. aliis in exclusivis, eâdem methodo Geometricâ demonstrates. Hagæ-Comitis 1661, p. 84, Fig. p. 86.

2 Logic and its Geometrical Interpretation

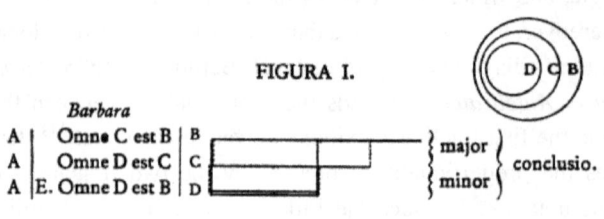

Fig. 3
G. W. Leibniz: De formæ logicæ per linearum ductus. In: *Opuscules et fragments inédits de Leibniz. Extraits des manuscrits de la Bibliothegue royale de Hanovre*. Ed by Louis Couturat. Paris 1903, pp. 292–321, here: p. 294.

Weigel's former student Gottfried Wilhelm Leibniz was also familiar with Sturm's work[151] and had used combinatorial and logic diagrams early on.[152] But it was not until around 1690 that a series of fragments – not published until 1903 by Louis Couturat – emerged in which both line and circle diagrams and arithmetical representations of logic culminated in order to perfect the Aristotelian syllogism. For example, right at the beginning of *De formæ logicæ comprobatione per linearum ductus* (Fig. 3), Leibniz sets up a schematic representation of the traditional four types of categorical judgement (*A, E, I, O*) and then illustrates the modus Barbara on the *dictum de omni* (*et nullo*). And in the famous *Generales Inquisitiones*, line diagrams are also used to represent intensional and extensional conceptual relations.[153]

Researchers have repeatedly discussed whether Sturm's and Leibniz's diagrams were influenced by their teacher Weigel. Whereas Bernhard, for example, considers this hypothesis to be unconfirmed, Maarten Bullynck has recently argued for an influence based on a geometriclogical analogy of the *dictum de omni et nullo* in Weigel's early work.[154] It

Fig. 4
Erhardus VVeigelus: *Philosophia Mathematica. Archimetria*. Jenæ 1693, I p. 122, II p. 105, Appendix.

---

[151] Cf. Stefan Kratochwil: Johann Christoph Sturm und Gottfried Wilhelm Leibniz. In: Johann Christoph Sturm (1635 – 1703). Ed. by Hans Gaab, Pierre Leich, Günter Löffladt. Frankfurt/Main 2004, pp. 104–119, esp. pp. 107f.
[152] On Leibniz's combinatorial diagrams and their tradition, cf. Hubertus Busche: Leibniz' Weg ins perspektivische Universum. Eine Harmonie im Zeitalter der Berechnung. Hamburg 1997, pp. 135ff. Leibniz's circular diagram, which is considered by researchers to be the earliest and can be interpreted as an analytical diagram, is found on sheet N. 493$_2$ in Gottfried Wilhelm Leibniz: Sämtliche Schriften und Briefe. Ed. by Preußische/ Deutsche/ Göttinger/ Berlin-Brandenburgische Akademie der Wissenschaften. Darmstadt et al. 1923ff., vol. VI 4 A, p. 2773. (I thank Hubertus Busche for this reference).
[153] Cf. ibid., pp. 772ff.
[154] Cf. e. g. Maarten Bullynck: Erhard Weigel's Contributions to the Formation of Symbolic Logic. In: History and Philosophy of Logic 34 (2013), pp. 25–34.

2.2 Analyticity – Analytic judgements, Containment Metaphors and Logic Diagrams

is known to some researchers that in 1693 one can find illustrations in the appendix of Weigel's *Philosophia Mathematica* visualizing relations of containment and non-containment. As shown in Fig. 4 with the example of the modus Barbara (All C are B, all a are C, thus all a are B) and Celarent (No C is B, all a are C, thus no a is C), Weigel visualised the termini minor, maior and medius in different modes of syllogistic by means of the nested and diverged initials A, B and C respectively. Weigel had already discussed this method in 1669 in his *Idea Matheseos universae* and had given his analytical diagrams the name "Logometrum" or "Inference measure" (Schluss-Maaß).[155] According to him, he invented the Logometrum around 1660, which his student Sturm ("Sturmium meum")[156] then made known to the Belgians according to his instructions ("à me compendium"). In this respect, Weigel claimed to be the inventor of the Logometrum, which could be illustrated not only by circles or initials but also by lines.[157] Indeed, one finds such an idea outlined in Weigel's *De Definitione Diagrammatica* of 1658.[158]

John Venn had already rightly stated that the claim that Christian Weise's books on logic – the *Doctrina Logica* of 1686, *Nucleus Logicæ* of 1691 and *Curieuse Fragen über die Logica* (*Curious Questions on Logic*) of 1696 – contained logic diagrams, which is still frequently found in the literature today, was false. Nevertheless, it is clear from the writings of Weise's students that Weise employed numerous analytical diagrams in his classes on logic in Zittau, using them to explain the content of his textbooks. Some students then published these techniques in their own books.

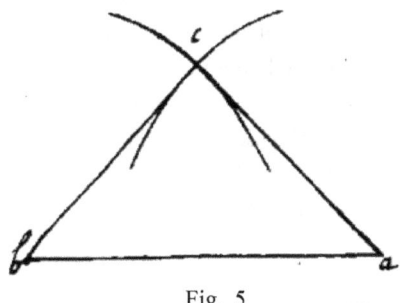

Fig. 5
Samuelus Grosserus: *Pharus Intellectus, sive Logica Electiva*. Lipsiae 1697, Supplement between p. 110 and 111.

Samuel Grosser, a student of Weise, published two works on logic in 1697 with similar content, *Pharus Intellectus, sive Logica Electiva* and *Gründliche Anweisung zur Logica* (*Thorough Instruction on Logica*). It is not yet known in research that, in addition to several triangular diagrams, one also finds a kind of analytical diagram in the third chapter of the *Gründliche Anweisung* or as an insert also in *Pharus Intellectus*.

---

[155] For a more detailed description of Weigel's method, vide infra, Chapter 2.3.4.
[156] Erhardus VVeigelus: Idea Matheseos universæ cum speciminibus Inventionum Mathematicarum. Jenae 1669, pp. 46f. (= VIII, § 18).
[157] Erhardus VVeigelus: Idea Matheseos universæ, p. 46.
[158] Erhardus Weigelus: Analysis Aristotelica ex Euclide restituta. Jena 1658, pp. 60ff.

2 Logic and its Geometrical Interpretation

Similar to Caramuel and Bernoulli, Grosser explains in this diagram (Fig. 5) that the relationship between subject and predicate has an "analogy with mathematics [Mathesi]". He justifies this by saying that two semicircles drawn with a compass can be used to show "at which point the two extremes of the imaginary line come together".[159] What is innovative, above all, is that the semantic function of the intersection $(A \land B)$ is reduced to the point of intersection, whereby the analytical diagrams can be indicated by semicircles, which moreover also correspond to the first Euclidean figure in the *Elementa*. I found a similar reduction only in the 19th century again in Krause and Lindner and then in the 20th century in Peirce, McCulloch and Randolph.[160]

A milestone of geometric logic is represented by the *Additamenta* to the *Nvclevs Logicae Weisianae* of 1712, which the Weise student Johann Christian Lange produced and which contain numerous logic diagrams on approx. 700 pages. Lange is the first logician in the tradition presented so far to express an interest in historical antecedents: he explicitly states that with his triangular bridges of asses ("Schema Triangvlare", "Schematibus semi-circvlaribvs") he was following authors such as Georgios Trapezuntios, Johann Heinrich Schellenbauer, Jakob Martini or also Samuel Grosser and had taken over his "*circles* or *spheres*" (*Circulos* aut *Sphœras*) from Sturm.[161] In addition, many of the diagrams that Lange discusses in his *Additamenta* were made known by Weigel.[162]

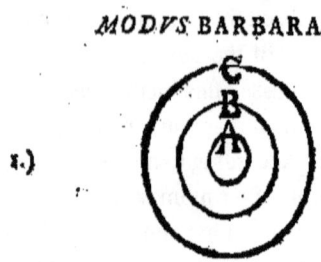

Fig. 6
Iohannes Christianus Langius: *Nvclevs Logicae Weisianae*. [...] illustrates [...] per varias schematicas [...] ad ocularem evidentiam deducta [...]. Editus antehac Avctore Christiano Weisio. Gissae-Hassorum 1712, p. 250.

---

[159] Samuel Großer: Gründliche Anweisung zur Logica [...]. Budißin, Görlitz 1697, pp. 117–118. Grosserus: Pharus Intellectus, pp. 208f.

[160] Cf. Karl Christian Friedrich Krause: Lehre vom Erkennen und von der Erkenntnis; Gustav Adolph Lidner: Lehrbuch der formalen Logik. 2nd ed. Wien 1867; Warren Sturgis McCulloch: Machines that Think and Want. In: Brain and Behavior. A Symposium. Ed. by W.C. Halstead. Berkeley/CA 1950, pp. 39–50; John F. Randolph: Cross-Examining Propositional Calculus and Set Operations. In: The American Mathematical Monthly 72 (1965), pp. 117–127. Vide infra, Chapter 2.3 and my interpretation in Chapter 3.1.

[161] Cf. Johannes Christianus Langius: Nvclevs Logicae Weisianae. [...] illustrates [...] per varias schematicas [...] ad ocularem evidentiam deducta [...]. Editus antehac Avctore Christiano Weisio. Gissae-Hassorum 1712, esp. p. 248, also: p. 160, p. 205, pp. 398ff., p. 603; cf. also Jacobus Martini: Institutionum Logicarum Libri VII. Wittebergae 1610, pp. 359ff., p. 432, pp. 491ff. (bridges of asses), p. 472 ("circuli probatio"); Joannes Henricus Schellenbauerus: Compendium logices. Stuttgardiae 1715, pp. 163ff. [the editions of 1682 and 1704 often mentioned in the literature cannot be found, but an edition of 1702 with similar diagrams]; Samuelus Grosserus: Pharus Intellectus, sive Logica Electiva. Lipsiae 1697, Supplementary sheets according to p. 110, p. 132.

[162] Cf. Langius: Nvclevs Logicae Weisianae, p. 707, p. 757, p. 827 und pp. 28–29 (Dissertationis Apologeticae).

2.2 Analyticity – Analytic judgements, Containment Metaphors and Logic Diagrams

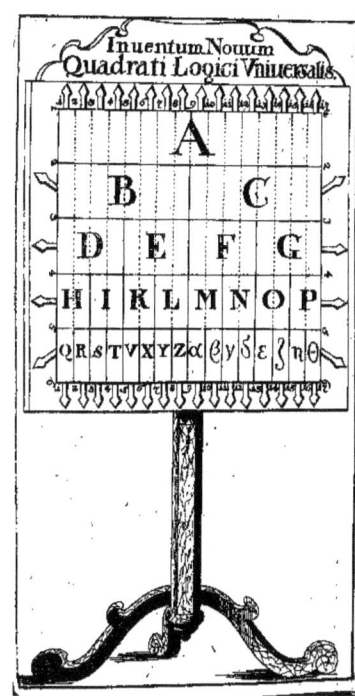

Fig. 7
Iohannes Christianus Langius:
Inuentum Nouum Quadrati Logici Vniversalis. Gissae Hassorum [Gießen] 1714.

Lange himself had used regular inferences – beginning with the modus Barbara (Fig. 6) – to illustrate not only basic classes and levels of conceptual relations but also logical and methodological special cases such as conditionals, enthymemes, sorites and induction with the help of analytical diagrams. Moreover, he uses analytical diagrams to go beyond traditional syllogistic.[163]

Venn had pointed out that the numerous 'quadrati' or 'cvbi logici' in Lange's *Inuentum Nouum Quadrati Logici Vniversalis* of 1714 were only a plane and solid geometric representations of tree diagrams:[164] A divides into B and C; B into D and E, etc. This, however, is a fundamental misconception that was adopted from Lambert several times in the English-language literature of the 19th century. Lange describes the design seen in Fig. 7 as a universal scheme ('schematis universalis'),[165] intended to combine the advantages of his line and circle diagrams from the *Nvclevs* with the square of opposition. Inspired by Leibniz and John Napier, his aim was to use the diagram as the plan for building a logic machine ('machina', 'apparatus').[166] The stakes and sticks ("Clauis vel Bacillis") shown in Fig. 7 indicate vertically a complete subordination/ inclusion/ implication (e.g. All B is some A), horizontally a complete exclusion (e.g. No D is any E) and transversally a partial or complete exclusion (e.g. Some B is not some E) by using numerically quantified terms.[167] According to Lange, not only the *dictum de omni* (e.g. in the form of the modus Barbara: All B is some A, All D is some B, thus all D is

---

[163] Cf. Jens Lemanski: Euler-type Diagrams and the Quantification of the Predicate. In: Journal of Philosophical Logic 49 (2020), pp. 401–416.
[164] Venn's judgment, however, stands in opposition to Baron and Johann Heinrich Lambert: Anlage zur Architectonic, oder Theorie des Einfachen und des Ersten in der philosophischen und mathematischen Erkenntniß. 2 vols. Riga 1771, here: Vol. 1, p. XIII, p. XXI, p. 128, who assume analytical diagrams there. In the likewise debatable work of Johann Andreas Segner, *Specimen Logicae vniversaliter* (1740), I could not find any analytical diagrams.
[165] Cf. Langius: Nvclevs Logicae Weisianae, 1712, p. 830.
[166] Cf. Langius: Inuentum Nouum Quadrati Logici Vniversalis. Gissae Hassorvm [Gießen] 1714, pp. 66ff.
[167] On Lange's numerically exact syllogistic cf. Jens Lemanski: Extended Syllogistics in Calculus *CL*. In: Journal of Applied Logics 8:2 (2021), pp. 557–577.

2 Logic and its Geometrical Interpretation

some *A*), but also the entire logic can be explained on only one diagram, since it has over 30 interpretation functions.[168]

Lange sent Leibniz the design of this machine and at the same time put forward the idea of establishing a scientific revision academy (Societas recognoscentium) in Gießen. Leibniz wrote to Lange shortly before his death that he was deeply impressed by the book. Lange had pursued the plan to create a universal logic ("logica universalis"). But Leibniz saw even more in Lange's machine: for him, it was a true universal algebra ("algebra universalis"),[169] i.e. a step towards the perfection of logicism. All arithmetic operations were to be broken down to logical functions, which could then be calculated by Lange's machine. On the one hand, Leibniz took up Lange's idea and tried to continue it. At the same time, he supported him in the academy plan.

But Leibniz's death put an end to all plans. Both Lange's design of the machine and the plan for an academy in Gießen soon fell into oblivion. Formerly professor of logic and metaphysics in Giessen, Lange became superintendent in Idstein in 1716 and from then on devoted himself almost exclusively to theology. For the following generations of logicians, Lange's designs were incomprehensible, if not ridiculous. Lambert and Ploucquet no longer understood Lange's ideas ad hoc, and August de Morgan was no longer concerned with the content, but only made fun of Lange's baroque linguistic style.[170]

Nowadays, it is known that Lange's machine can indeed be used as a diagrammatic calculus in formal logic.[171] Since it was Lange's idea to imitate the cognitive abilities of humans by machine and thus bring ontology and logic together, Lange's diagrammatic design of a logic machine not only has enormous potential for the field of artificial intelligence but is also understood as artificial intelligence *avant le lettre*.[172] Thus Lange's writings from the 1710s represent the culmination, but also the conclusion, in the history of analytical diagrams before Euler. Unlike his predecessors, he

---

[168] Cf. Langius: Inuentum Nouum Quadrati Logici Vniversalis, p. 151.
[169] Cf. [Johann Christian Lange:] Ausführliche Vorstellung von einer neuen und gemein-ersprießlichen zu beßtem Behuf und Auffnahm Aller wahren und rechtschaffenen Gelehrtheit gereichenden Anstalt [...]. Idstein: Lyce, 1720, pp. 209f.
[170] A detailed portrayal of the Weise School can be found in Jens Lemanski: Logikdiagramme und Logikmaschinen aus der Zittauer Schule um Christian Weise. In: Neues Lausitzische Magazin 141:1 (2019), pp. 39–57.
[171] Cf. the modern interpretations and applications e.g. in Jens Lemanski, Ludger Jansen: Calculus *CL* as a Formal System. In: Diagrammatic Representation and Inference. Diagrams 2020. Lecture Notes in Computer Science, vol 12169. Ed by. A.-V. Pietarinen, P. Chapman, L. Bosveld-de Smet, V. Giardino, J. Corter, S. Linker. Cham 2020, pp. 445–460; Jens Lemanski: Calculus *CL* as Ontology Editor and Inference Engine. In: Diagrammatic Representation and Inference. 10th International Conference, Diagrams 2018, Edinburgh, UK, June 18–22, 2018, Proceedings. Ed. by P. Chapman, G. Stapleton, A. Moktefi, S. Perez-Kriz & F. Bellucci, Cham 2018, pp. 752–756; Jens Lemanski: Extended Syllogistics in Calculus *CL*.
[172] Cf. Henry Prade, Pierre Marquis, Odile Papini: Elements for a History of Artificial Intelligence. In: Guided Tour of Artificial Intelligence Research. Vol. 1: Knowledge Representation, Reasoning and Learning. Heidelberg et al. 2020, pp. 1–45.

2.2 Analyticity – Analytic judgements, Containment Metaphors and Logic Diagrams

does not attempt to polemicise against certain types of diagrams, but to harmonise their modes of operation.

Lange's *Inventum* marks the end of the second period of analytical logic diagrams in the early modern period. Just as there is a break between the first and second periods of authors who used analytical diagrams in the period of the Thirty Years' War, there is also a break between the second and third periods. The second period is dominated by the Weigel and Weise schools. And similarly, as the representatives of analytical diagrams of the first period set themselves apart from the Byzantine tradition with the mnemonic diagrams, some of the representatives of the second period polemicised themselves especially against so-called Phoebifer Axis diagrams of their contemporaries.

The choice of the respective logic diagram becomes an expression of denominational affiliation after the Thirty Years' War: Catholic logicians use the square of opposition, tree diagrams (Thomists) or Phoebifer Axis diagrams (Scotists); Protestant logicians, on the other hand, use analytical diagrams. This seems curious at first, but regardless of this striking coincidence, which I attest here to the second period of analytical diagrams, other researchers see a continuation of these traditions well into the 20th century.[173] With Lange's *Inventum*, however, this rivalry ends in Central Europe and once again logic diagrams are forgotten.

This is because in the first half of the 18th century a new influential tradition spread in Central Europe: Rationalist logic, which was influenced by the few known writings of Leibniz and especially Christian Wolff at the time, and which started from the ideal of no longer using diagrams or geometric forms in logic at all. Diagrammatic thinking was replaced by the ideal of the pure concept, which emphasised logic as a prerequisite for a mathematical proof.[174] One can see in this school, shaped by Leibniz and Wolff, the birth of modern logicism. Moreover, the metaphysical-logical concentration on innate, eternal truths and concepts also seems to suppress the use of sensualistic means such as diagrams.

(3) It was not until the end of the 1750s that diagrams again appeared prominently in the history of logic and opened up a third period of analytical logic diagrams in the early modern period.[175] In these years, a dispute – which seems anachronistic in view of the history of logic known today – broke out between Johann Heinrich Lambert and Gottfried Ploucquet about the authorship of analytical logic diagrams and their function as logical calculus.[176] It is certain that Ploucquet drew three nested squares

---

[173] Cf. Dany Jaspers, Peter A. M. Seuren: The Square of Opposition in Catholic Hands. A Chapter in the History of 20th-century Logic. In: Logique et Analyse 59 (2016), pp. 1–35. (I thank Dany Jaspers for this and some other references on this topic.)

[174] Cf. Wilhelm Risse: Die Logik der Neuzeit. 2 Vols. Stuttgart-Bad Cannstatt 1964/1970, Vol. 2, pp. 259ff.

[175] Between 1715 and 1760 diagrams are so rare in writings on logic that e.g. a triangular diagram as in Hermann Samuel Reimarus: Vernunftlehre, als eine Anweisung zum richtigen Gebrauche der Vernunft [...]. Hamburg 1756, p. 196 (= § 136) already seems worth noting.

[176] Some illustrations of diagrams of the early modern period can also be found in Wilhelm Risse: Die Logik der Neuzeit, Vol. 1, p. 221 (Square of Opposition), p. 225 a. p. 542 (bridges of asses); Vol. 2, pp.

2 Logic and its Geometrical Interpretation

in his *Fvndamanta Philosophiæ Speculativaæ* in 1759 to demonstrate the following intensional inference:

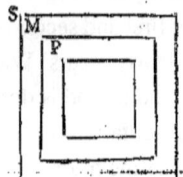 By intuition, it is open to the eye that *P* is the predicate of all *M*, & *M* the predicate of all *S*. But the predicate of the predicate is the predicate of the subject. *P* is therefore the predicate of all *S*, so that descriptively: All *S* is *P*.

Ex intuitione patet, *P* esse prædicatum omnis *M*, & *M* esse prædicatum omnis *S*. Sed prædicatum prædicati est prædicatum Subjecti. *P* itaque est prædicatum omnis *S*, id quod ita exprimitur: Omne *S* est *P*.[177]

What is remarkable about this quotation is first of all the resumption, in the second sentence and typical of the 18th century, of a lemma developed between Neoplatonism and late scholasticism, known as the *regula de quocunque*, which I consider being an intensional variant of the Aristotelian *dictum de omni*.[178] The three nested squares show that the judgement of the conclusion "All S is P" is identical with the judgement "P is in all S" (*P esse in omni S, seu, quod idem est, omne S esse P*).

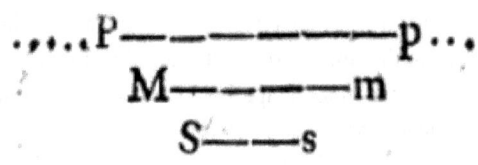

Fig. 8
J. H. Lambert: Neues Organon. 2 Vol. Leipzig 1764, Vol. 1, p. 124 (= § 201).

In 1762, Johann Heinrich Lambert had found an "old scholastic logic, or [...] a commentary on the logic of Aristotle" with logical "figures in woodcuts" illustrating "many concepts and relationships" in the Bürgerbibliothek of the Zurich Wasserkirche.[179] The fact that he still sent a letter to Zurich six years later with the request to be allowed to inspect the book again may be taken as an indication that the logical line and dot diagrams developed in 1764 in the work *Neues Organon* were inspired by this "scholastic logic". For Lambert, the development of such an intuitive

---

90f. (Geulincx), p. 128 (tree diagrams), p. 145 (Weigel), p. 168 (Sturm), p. 202 (Leibniz), pp. 280f. (Ploucquet), pp. 286–289 (Euler), pp. 562–564 (Lange), pp. 656f. (Reimarus).
[177] Gottfredus Ploucquet: Fvndamenta Philosophiæ Speculativæ. Tübingae 1759, p. 25 (= § 71).
[178] For example, one finds a strongly exaggerated scholastic variant of this phrase in [Ps.-]Joslenus Suessionensis: De generibus et speciebus. In: Ouvrages inédits d'Abélard. Ed. by Victor Cousin. Paris 1836, p. 520: "Si enim aliquid praedicatur de aliquo et aliud subiciatur subiecto, subiectum subiecti subicitur praedicato praedicati." Ploucquet's quote probably goes back to Leibniz's *De casibus perplexis* II, XXI ("praedicatum praedicati est praedicatum subjecti").
[179] Johann Heinrich Lambert to Johann Jakob Steinbrüchel, 14th April 1768. In: Johann Heinrich Lamberts deutscher gelehrter Briefwechsel. Ed. by Johann II. Bernoulli. 2 Vols. Berlin s.a. [1782], Vol. 1, pp. 403–408.

2.2 Analyticity – Analytic judgements, Containment Metaphors and Logic Diagrams

logic was, above all, a didactic improvement over arithmetic and purely algebraic calculi.[180] In the *Neues Organon*, the "extension" (Ausdehnung) of an abstractum is represented using a drawn line, the length of which illustrates the number of all individua, which are also represented by dots if they are not intended to indicate an extension or if their number is indeterminate. The diagram of Fig. 8, therefore, allows an explanation of inferences that depend on the *dictum de omni* (*et nullo*), such as: All M are P, all S are M, thus all S are P.[181]

In a newspaper article in January 1765, Lambert had praised the *Abhandlung über die Mathematik* by Georg Jonathan von Holland, which tended to favour Ploucquet's method, at least for attempting to "fix the epochs of such types of calculation, so that once they have reached their true perfection and usefulness, one does not quarrel so bitterly about their invention, as has happened with the differential calculus".[182] Lambert affirmed that he had developed his geometric method at least one year before writing the *Neues Organon* – apparently, he assumed that Ploucquet's method had only been developed in 1763/64.

Of course, Ploucquet did not miss the opportunity to refer to his writing from the 1750s, which he had already mentioned: According to Ploucquet, it was "not unhelpful [...] to satisfy the request of Prof. Lambert and to establish the epochs of this type of account". As early as 1758, he had the idea of "drawing inferences and presenting them in figures" (Schlüsse zu zeichnen, und in Figuren vorzustellen).[183] From a historical point of view, it is remarkable that Ploucquet also resisted Heinrich Wilhelm Clemm's historical attribution[184] that his logic had similarities with the *characteristica universalis* of Ramon Lull, Richard Suiseth or Gottfried Leibniz.[185] Lambert could see in all these historical rebukes of Ploucquet only a "more circumstantial narrative" (umständlichere Erzählung) and quickly swung to criticism of the content.[186]

It can be seen as a confirmation of the old saying "duobus litigantibus tertius gaudet" that Leonhard Euler, of all people, was declared the namesake of circumferential logic diagrams in geometric logic[187] and that of all the diagrammatic logics mentioned so

---

[180] For a more detailed description of the Lambertian method vide infra, Chapter 2.3.4.
[181] J. H. Lambert: Neues Organon. oder Gedanken über die Erforschung und Bezeichnung des Wahren und dessen Unterscheidung vom Irrthum und Schein. 2 Vols. Leipzig 1764, here: Vol. 1, pp. 109–125 (= §§ 173–202).
[182] Cf. J. H. Lambert: Neue Zeitung von gelehrten Sachen 1765:1 (3rd January). In: Sammlung der Schriften, welche den logischen Calcul Herrn Prof. Ploucquets betreffen, mit neuen Zusåzen. Ed. by August Friedrich Bŏk. Frankfurt, Leipzig 1766, p. 152.
[183] Gottfried Ploucquet: Untersuchung und Abänderung der logikalischen Constructionen des Hrn. Prof. Lambert. In: Bŏk (ed.): Sammlung der Schriften, pp. 157–202, here: p. 157. For a more detailed description of Ploucquet's method vide infra, Chapter 2.3.4.
[184] Henricus Gvilielmus Clemmius: Novae amoenitates literariae. Fascicvlvs Qvartvs. Stvtgardiae 1764, pp. 549–556, here: p. 554.
[185] G. Ploucquet: Untersuchung und Abänderung der logikalischen Constructionen des Hrn. Prof. Lambert. In: Bŏk (ed.): Sammlung der Schriften, pp. 157–160.
[186] Cf. J. H. Lambert: Neue Zeitungen von gelehrten Sachen. 1765:58 (22th July). In: Bŏk (ed.): Sammlung der Schriften, pp. 207–215, here: p. 207.
[187] Vide supra, Chapter 2.2.1.

far, it was his that was first elaborated into the diagrammatic calculus in the 1990s.[188] Euler had written his famous circle diagrams almost at the same time as Ploucquet and Lambert, namely between 1760 and 1762; however, he had not published them until 1768 in volume 2 of his *Lettres à une princesse d'Allemagne sur divers sujets de physique et de philosophie*. Unlike Lambert and Ploucquet, Euler's analytical diagrams were not connected with the logician's idea of building a logical calculus. This is not surprising, since Euler was, after all, known as a sharp critic of rationalist philosophy, especially of the so-called 'Leibniz-Wolff school'. Euler was much more concerned with using analytical diagrams to test the validity of syllogisms.[189]

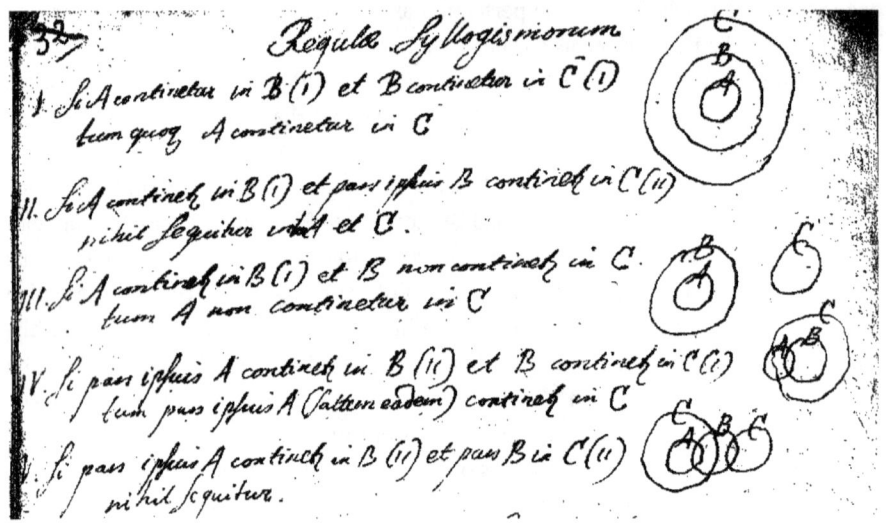

Fig. 9

Leonhard Euler: Theses Logicae (Manuscript, ca. 1740). Санкт-Петербургский филиал Архива ПФА РАН Ф. 136. Оп. 1. № 134, p. 32.

As can be seen from Euler's much earlier manuscripts, however, there are not the 'Euler diagrams', but also analytical diagrams that can be understood as preliminary work to the *Lettres*. Vladimir Ivanovich Kobzar, who, as far as I know, published these Euler diagrams from the Nachlass for the first time, argues that the diagrams were created in the late 1730s in St. Petersburg for teaching purposes since logic diagrams were disreputable in Central Europe at that time.[190] Since the Euler diagrams

---

[188] Cf. Eric M. Hammer: Logic and Visual Information. Stanford, 1995, pp. 69–83; Eric M. Hammer, Sun-Joo Shin: Euler's Visual Logic. In: History and Philosophy of Logic 19:1 (1998), pp. 1–29. Shin had shortly before developed a sound and complete calculus for Venn (resp. Peirce) diagrams, vide infra, Chapter 2.3.
[189] Cf. Peter Bernhard: Euler-Diagramme, pp. 45ff.
[190] For dating cf. Владимир Иванович Кобзарь: Элементарная логика Л. Эйлера. In: Логико-философские штудии [Logiko-filosofskie studii] 3 (2005), pp. 130–152, here: p. 134. A more detailed discussion with illustration of the relevant manuscripts is offered by ibid: Гносеология и логика Л. Эйлера

2.2 Analyticity – Analytic judgements, Containment Metaphors and Logic Diagrams

are not known outside the Russian-speaking world, I have taken the liberty of illustrating in Fig. 9 a longer section of *Theses Logicae* entitled *Regulae Syllogismorum*, which shows five of eleven syllogistic rules. The first rule of the manuscript page shows the modus Barbara in the metaphor of being contained according to the *dictum de omni* (*et nullo*): If A is contained in B, and B is contained in C, then A is also contained in C ("Si A contineatur in B (I) et B contineatur in C (I) tum quoq[ue] A contineatur in C.").[191]

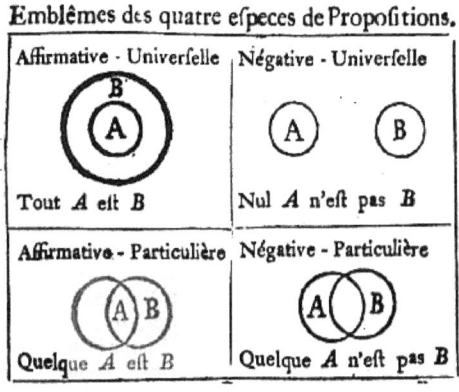 

Fig. 10
Leonhard Euler: Lettres à une princesse d'Allemagne sur divers sujets de physique & de philosophie. 2 vols. Saint Petersbourg 1768, here: Vol. 2, p. 101 (= L. CIII).

Fig. 11
L. Euler: Lettres. Vol. 2, p. 101 (= L. CIII).

In the famous *Lettres*, Euler wanted to explain – in the Lockean manner – the difference between the individuals or proper names (as *concreta*) and the abstracta abstracted from them, in order to then build up a logic of judgement that presents the relation of subject and predicate on the basis of circular analytical diagrams ("figures rondes").[192]

Euler begins by representing the traditional four types of categorical judgement (A, E, I, O) by means of a cross-classification of quantity and quality (Fig. 10). With the help of the circle diagrams, conditionals such as "If the notion C is entirely contained in the notion A, it will be so likewise in the notion B" can then be illustrated, which then justify an inference (Fig. 11) in the manner of modus Barbara: "Every A

в "Письмах к немецкой принцессе о разных физических и философских материях". In: Логико-философские штудии [Logiko-filosofskie studii] 8 (2010), pp. 98–120.

[191] I thank especially Larissa Tonoyan for making the copies of the manuscript and point out that all rights to the manuscripts belong to the Russian Academy of Sciences and Prof. Kobzar. My deep gratitude goes to the many Russian colleagues who supported my research, especially Ivan Mikirtumov and Yuri Chernoskutov.

[192] Leonhard Euler: Lettres à une princesse d'Allemagne sur divers sujets de physique & de philosophie. 2 Vols. Saint Petersbourg 1768, here: Vol. 2, pp. 96–101 (= L. CIIf.). For a more detailed description of the Eulerian method vide infra, Chapter 2.3.4.

is B: but every C is A: Therefore Every C is B."[193] The expression of being contained, dismissed today as a mere metaphor, becomes here the foundation of the entire syllogistic. As Euler explicitly states, his logic is based on the principles of containment, which correspond to the *dictum de omni et nullo*:

> The foundation of all these forms is reduced to two principles, respecting the nature of *containing* and *contained*.
> I. Whatever is in the thing *contained* must likewise be in the thing *containing*:
> II. Whatever is out of the *containing* must likewise be out of the *contained*.[194]

Euler's logic and metaphysics were initially little noted in the German-speaking world and even intensely opposed by Leibnizians.[195] Nor were Ploucquet's and Lambert's diagrams elaborated further in the 1770s and -80s.[196] Even the attempt to develop a calculus that would be limited to logic alone was considered a failure – regardless of whether the calculus was based on an algebraic or geometric system of signs. The logician's dream of creating a universal calculus that could do far more than just calculate logic had receded into the distant future. As early as 1766, August Friedrich Bök summarised the judgement of his contemporaries as follows:

> An invention of this kind, which could be called a real calculus, and with which Leibniz' insatiable inquisitiveness was busy for many years without success, does not seem to belong to the sphere of mortals, and will probably have the same fate as the search for a philosopher's stone, the squaring the circle and the construction of a perpetuum mobile.[197]

The logic diagrams of Lambert, Ploucquet and Euler were not taken note of again until the dispute between Leibnizians and Kantians arose around 1790.[198] It can be said that it was to the credit of Kant's philosophy, which had an affinity with intuition, that a third period of analytical logic diagrams was given any continuity at all until the so-called 'crisis of intuition' around the year 1880. To put it simply: without Kant and

---

[193] Ibid., Vol. 2, p. 104 (= L. CIII): "Si la notion C est contenuë tout entiere dans la notion A, elle sera aussi contenuë toute entière dans l'espace B." "Tout A est B: Or Tout C est A: Donc Tout C est B." (Translation taken from Letters of Euler. Ed. by David Brewster, John Griscom. New York 1833, Vol. 1, p. 342)
[194] Ibid., Vol. 2, p. 118 (= L. CIV). "Le fondement de toutes ces formes se réduit à ces deux principes sur la nature du *contenant* & du *contenu*." (Translation taken from Letters of Euler, 1833, Vol. 1, p. 350)
[195] The theses still following in this chapter were treated in more detail in the article Jens Lemanski: Periods in the Use of Euler-Type Diagrams.
[196] Already rare is a more detailed explanation of Lambert's diagrams, such as in Johann Carl Christoph Ferber: Vernunftlehre. Helmstädt, Magdeburg 1770, pp. 429ff.
[197] Bök: Vorrede. In: ibid. (ed.): Sammlung der Schriften, [s.p.]. Vide infra, Chapter 2.3.
[198] On this dispute vide infra, Chapter 2.3.1.

2.2 Analyticity – Analytic judgements, Containment Metaphors and Logic Diagrams

the Kantians of the first generation, we would not call analytical diagrams Euler diagrams today. Had the rationalism of the Leibnizians and Wolffians prevailed, 19th-century logicians would probably have had as little idea of the logic diagrams of Euler's time as they had of the logic diagrams of the first and second periods. In the dispute between Kantians and Leibnizians, i.e. around the year 1790, the controversy centred on the questions of what role diagrams possessed for cognition and science and which analytical diagrams should be used.

Gotthelf Samuel Steinbart and Johann August Heinrich Ulrich took up Lambert's line diagrams again in a rather neutral way around 1790.[199] However, the fact that Lambert was appropriated by both Leibnizians and Kantians during these years can be seen in several examples: Kantians such as Johann Gottfried Kiesewetter and Georg Samuel Albert Mellin endeavoured to harmonise Euler's diagrams with Lambert's; Leibnizians such as Wilhelm Ludwig Gottlob von Eberstein and Johann Gebhardt Ehrenreich Maaß, on the other hand, wanted to harmonise the Kant critic Ploucquet with Lambert. As an example, I will only present one diagram each by Kiesewetter and Maaß in the following.

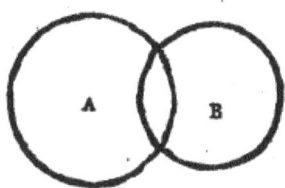

Fig. 12
Johann Gottfried Kiesewetter: Grundriß einer reinen allgemeinen Logik nach kantischen Grundsätzen [...]. Frankfurt 1793, p. 125.

In 1793, Kiesewetter used two circle diagrams in his *Grundriß einer reinen allgemeinen Logik* to illustrate conversion rules. In Fig. 12, Kiesewetter uses the circle diagrams to indicate a simple inversion of affirmative judgements: "if some A are B, then some B are also A."[200]

In Maaß' *Grundriß der Logik*, published in 1793, there is a semiotics in which the sign is represented in the form of geometrical triangles in order to simplify both the discussion of term extensions and subordinations and the heuristics of conclusions from existing premises.[201] As an example of this, his first two logic diagrams are shown in Fig. 13, on which it is discussed whether the "sphere of a concept" extends to α, κ or μ.[202] Maaß did not use triangles at random. The decision in favour of triangles – apart from the reasons given above at the beginning of Chapter 2.2 by Henry

---

[199] Cf. Gotthelf Samuel Steinbart: Gemeinnützige Anleitung des Verstandes zum regelmäßigen Selbstdenken. 2nd ed. Züllichau 1787, pp. 14ff.; Ioannes Avgvstvs Henricus Vlrich: Institvtiones logicae et metaphysicae. Scholae svae scripsit perpetva Kantianae disciplinae ratione habita. Ienae 1792, p. 171.
[200] Johann Gottfried Kiesewetter: Grundriß einer reinen allgemeinen Logik nach kantischen Grundsätzen [...]. Frankfurt 1793, p. 126.
[201] For a more detailed description of the method of Maaß, vide infra, Chapter 2.3.4.
[202] Johann Gebhard Ehrenreich Maaß: Grundriß der Logik, zum Gebrauche bei Vorlesungen. Halle 1793, p. 294 (= § 365).

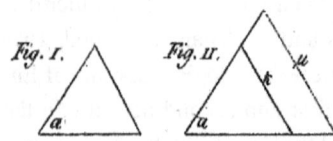

Fig. 13
Johann Gebhard Ehrenreich Maaß: Grundriß der Logik, zum Gebrauche bei Vorlesungen. Halle 1793, Supplement p. III.

E. Allison – also provides an explanation why Maaß questions (AJ) and (SJ) in Kant: The extension metaphors, which are useless on their own, and which in Euler and then also in Kant are represented primarily by two-dimensional circular or quadrilateral diagrams, are vague, he argues, because they compete with the much more perfect subordination metaphors in Lambert, which are represented by lines.[203] In his triangular diagrams, Maaß, therefore, focuses on the subordinating sides, which taken together, however, also represent an extensional surface.

From the early 19th century, the Kantians dominated logic and with them Euler's analytical diagrams of geometrical logic as well as the metaphor of circumference and containment. All other lines of the tradition of geometrical logic fell more and more into oblivion, only Hegelian rationalism adopted the diagram hostility of the Leibniz-Wolff school.[204] In the German-speaking world, Euler's diagrams were used in the most influential logic textbooks between Kant and Schopenhauer: Krause published his *Grundriss der historischen Logik* in 1803, Wilhelm Traugott Krug his *Logik oder Denklehre* in 1806, Jakob Friedrich Fries his *System der Logik* in 1811. All contained analytical diagrams in the sense of Euler and Kant. In the English-speaking world, too, the influence of Kant and his school can be seen very well in Thomas Wirgman's article on logic in the *Enyclopædia Londinensis*, in which he attempts to represent large parts of the Kantian system with the help of logic diagrams (such as Fig. 14).

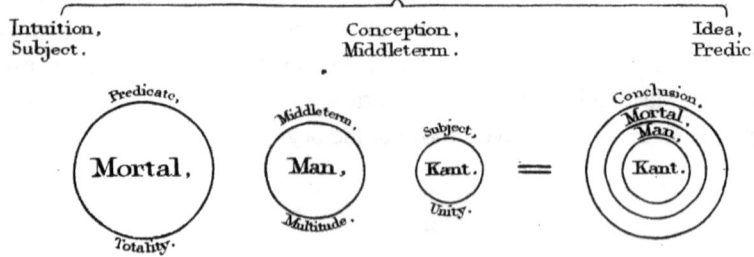

Fig. 14
Thomas Wirgman: Logic (art.). In: *Enyclopædia Londinensis*, Vol. XIII. London 1815, pp. 1–51, here: p. 25 (plate III).

---

[203] Ibid., pp. IXff. (Vorrede).
[204] Cf. Valentin Pluder: The Limits of the Square. Hegel's Opposition to Diagrams in its Historical Context. In: The Exoteric Square of Opposition. Ed. by Jean-Yves Beziau and Ioannis Vandoulakis. Basel 2021.

2.2 Analyticity – Analytic judgements, Containment Metaphors and Logic Diagrams

2.2.4 Kant's "Notion of Containment"

In the previous chapters, it was shown that already in antiquity one can find evidence for understanding many technical expressions of logic as descriptions of geometric relations. It was not until the Middle Ages that these relations were not only described but also increasingly visualised through diagrams of geometric logic (Chapter 2.2.2). At the latest in the early modern period, increasingly differentiated logic diagrams developed, which possessed analytical properties in Venn's sense (Chapter 2.2.1) and at the same time also visualised the technical expressions of circumference and containment (Chapter 2.2.3), which were dismissed as incomprehensible metaphors in the early 20th century (Chapter 2.2).

Kant usually plays only a marginal role in historical reference works on logic diagrams. Very cautiously, several historians refer to the so-called *Jäsche Logic*, whose publication falls precisely on the year 1800. This cautious reference is probably because Kant did not write his logic lectures himself, but had them written by Gottlob Benjamin Jäsche, and also because logic diagrams appear only sporadically in them. As I have shown in Chapter 2.2.3, analytical diagrams of geometric logic, especially Euler diagrams, which are based on the principle of containment, only became popular through Kant and his followers. So when Ernst Schröder writes in 1890 that since Euler logic diagrams have been used or at least referred to in all works on logic,[205] it must be improved that Euler's logic diagrams only found their way into the logic textbooks of the 19th century through Kantianism.

The logic of judgement contained in *Jäsche Logic* is structured in the usual Kantian manner by the so-called 'table of judgments' ("Logische Formen der Urtheile"). That is, Kant[206] divides the logic of judgement into the four main moments: quantity, quality, relation and modality (§ 20). In note 5 of § 21, he combines the geometric figure of a circle with the figure of a square with regard to the quantity of judgements in order to be able to represent particular judgements:

> Of *particular judgments* it is to be noted that if it is to be possible to have insight into them through reason, and hence for them to have a rational, not merely intellectual (abstracted) form, then the subject must be a broader concept (*conceptus latior*) than the predicate. Let the predicate always = ○, the subject ☐, then

---

[205] Cf. Ernst Schröder: Vorlesungen über die Algebra der Logik (Exakte Logik). Vol. 1, Leipzig 1890, p. 155.
[206] For the sake of simplicity, in the following I will refer to Kant as the author of the *Jäsche Logic*.

2 Logic and its Geometrical Interpretation

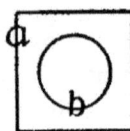

is a particular judgment, for some of what belongs under a is b, some not b – that follows from reason. But let it be

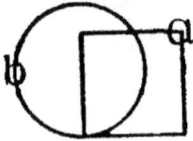

so then at least all a can be contained under b, if it is smaller, but not if it is greater, hence it is particular only by accident.[207]

After the presentation of quality (§ 22), Kant explains the relation of the judgements from § 23 onwards and sketches the conceptual "spheres" (Sphären) by means of a "scheme" (Schema) in order to illustrate the "peculiar character of disjunctive judgments". Strictly speaking, § 29 shows two schemes, since Kant contrasts the disjunctive judgements with the categorical ones, citing a variant of the modus Barbara in the wording of the *dictum de omni et nullo*, which he also discusses in §§ 14 and 63:

> The following *schema* of comparison between categorical and disjunctive judgments may make it more intuitive that in disjunctive judgments the sphere of the divided concept is not considered as *contained* in the sphere of the divisions, but rather that which is *contained under* the divided concept is considered as contained under one of the members of the division.
>
> In categorical judgments x, which is contained under b, is also under a:

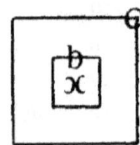

---

[207] AA IX, p. 103.14–22 (= Kant: Lectures on Logic, p. 599f.)

2.2 Analyticity – Analytic judgements, Containment Metaphors and Logic Diagrams

In disjunctive ones x, which is contained under a, is contained either under b or c, etc.:

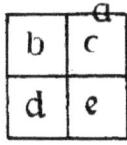

208

The two reasons conjectured above for the restrained engagement of historians of logic with the schemata of the *Jäsche Logic* are, in my opinion, goundless. On the one hand, in Kant's interleaved edition of Georg Friedrich Meier's *Auszüge aus der Vernunftlehre*, which Kant based his logic lectures on from 1765 onwards, one finds several and, moreover, very different schemata and diagrams, as I will prove below. On the other hand, some of these notes and reflections by Kant correspond strongly with the text and diagrams in the *Jäsche Logic*, since Jäsche compiled Kant's manuscripts and notes when compiling the published textbook on logic.[209] For example, Note 5 to § 21 (*Jäsche Logic*) above corresponds very closely to Kant's own commentary on Meier's § 292; and the quoted paragraph from § 29 is also textually and especially schematically close to Kant's commentary on §§ 307ff. of Meier's *Auszüge aus der Vernunftlehre*.

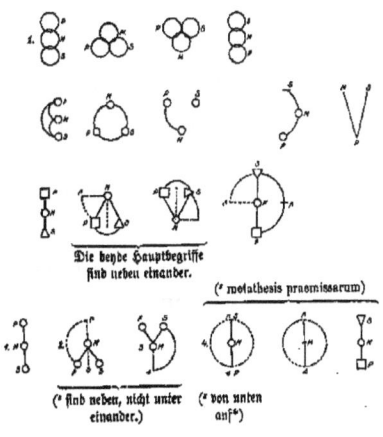

Fig. 1 (a)
Immanuel Kant: Handschriftlicher Nachlaß, Logik. AA XVI, p. 726 (= No. 3235).

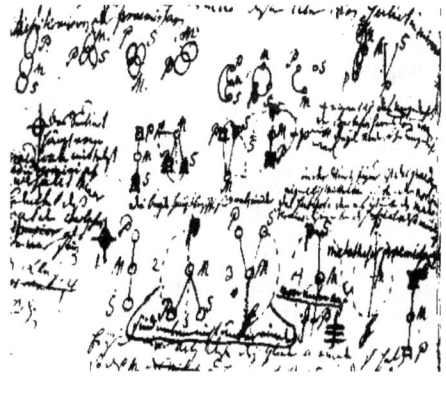

Fig. 1 (b)
Immanuel Kant: Vorlesungen über Logik. In: G. F. Meier: Auszüge aus der Vernunftlehre. TÜR, Mscr 92, p. 95.

---

[208] AA IX, p. 108.1–8 (= Kant: Lectures on Logic, p. 603f.); My emph., J.L. The first diagram illustrates the $\forall x\ bx \rightarrow ax$ scheme mentioned at the beginning of Chapter 2.2.
[209] Cf. AA IX, p. 3f. (= Kant: Lectures on Logic, p. 521).

2 Logic and its Geometrical Interpretation

In addition to the *Jäsche Logic*, the volume XVI of Kant's Academy Edition edited by Erich Adickes (with the notes on Meier) contains at least seven more reflections with diagrams or schemes: Nos. 3063, 3215, 3216, 3229, 3235, 3236, 3239–3240.[210] Adickes had pointed out in his commentary on No. 3215 that Kant's circle diagrams might have been taken from Euler.[211] Schulthess has generalised Adickes' assumption by the thesis that Kant's diagrams may have been influenced by Euler, since, on the one hand, Kant himself refers to Euler's diagrams in the so-called *Logik Philippi* and, on the other hand, the diagrams in the reflections on logic only appear after the publication of Euler's *Letters* around 1768.[212] From a historical point of view, therefore, there is no reason why Kant's analytical diagrams should not be called 'Euler diagrams'. Moreover, not much depends on the date itself, although one can note that according to Adickes' dating, some of the seven reflections with diagrams mentioned may have been written before 1768.

The fact that not much depends on the date can be explained as follows. It is not the dating but the drawings themselves that show that Kant knew more about logic diagrams than was contained in Euler's *Letters* (and also in the manuscripts published by Kobzar). Especially reflection no. 3235 shows many different diagrams (with S = subject, M = medius, P = predicate) in many variants, several of which form Borromean Rings, lunar horns or semicircles and triangles (esp. second line of Fig. 1 (a)) – i.e. those schemes that have already been demonstrated in Chapters 2.2.2 and 2.2.3 in the Middle Ages and early modern period. However, since such diagrams do not appear in Euler's *Lettres* or manuscripts, they cannot have been Kant's only source of information on logic diagrams.

This is the reason why Adickes in his commentaries (on Nos. 3215 and 3235) used Lange's *Additamenta* to the *Nucleus* as a means of comparison in addition to Euler, although he found no indication that Kant knew this edition. However, the majority of the diagrams in Fig. 1 are found neither in Euler nor in Lange (e.g. lines 3 and 4 of Fig. 1 (a)).[213] The comparisons made by Adickes, Lu-Adler, Anderson and Schulthess fall short when they compare Kant's diagrams only with Euler or with one or two other geometric logicians from the early modern period. Although the history of logic diagrams outlined in Chapters 2.2.2 and 2.2.3 is certainly incomplete, it is extremely unlikely that Kant could have taken his knowledge of geometric logic from only one source, since one finds, especially in the transcripts of Meier, very many different diagrams with different functions, which even far-reaching compendia on geometric logic – such as Lange's additions to the *Nucleus* – cannot cover all.

---

[210] I thank Margit Ruffing for pointing me to the original manuscripts in Tartu.
[211] Adickes refers to the following sentences of Kant taken from the *Logic Philippi*: "The more the concepts are from reason, the more they contain among themselves, but the less we can make this sensual by means of figures." (AA XXIV.1, p. 454; My transl. J.L.)
[212] Cf. Peter Schulthess: Relation und Funktion, p. 101, Fn. 28.
[213] The uncommon diagrams in the lower two rows of Fig. 1 are somewhat reminiscent of the diagrams in Heinrich Ernst Seebach: Introductio in iuris et polities utrium per viam logices. Wittebergae 1697, Pars III.

2.2 Analyticity – Analytic judgements, Containment Metaphors and Logic Diagrams

The lack of a 'Q source' for the various diagrams does not, in my opinion, justify the *argumentum e contrario* that Kant developed all the diagrams himself. For it would be a very great coincidence if Kant, without knowledge of their history, had constructed precisely those counter-intuitive lunar horns, triangular, circle and line diagrams and provided them with the same functions, as they can also be found in isolated instances in many different works of the Middle Ages and the early modern period. Since here, too, no conclusive proof can be provided as to where Kant knew the many logic diagrams, it is obvious to first agree with the judgement of Gardner, Baron and many other historians that especially the analytical diagrams and bridges of asses were handed down orally over the centuries through the teaching of logic at secondary schools. Kant's marginal notes on Meier's logic are themselves evidence of this thesis.

However, special consideration of the geometric forms mentioned so far must be given to the circle and square diagrams, as they are in the analytical tradition. At the beginning of Chapter 2.2, I argued that the use of logical expressions such 'circumference', 'being contained', 'lying outside' that Kant uses in defining (AJ) and (SJ) is not a "shortcoming", as Quine has claimed and as has long been discussed in Kant scholarship. Rather, following Anderson and Lu-Adler, I suggested that these logical expressions imply a pictoriality that has grown out of the logical tradition and is nowadays discussed under the heading of 'Euler diagrams' or, more generally, 'analytical diagrams'.

Kant himself explicitly depicted the imagery underlying (AJ) and (SJ) in one place in his hand copy of Meier's *Auszüge aus der Vernunftlehre*. This account is found in reflection no. 3216 and, together with reflections 3214–3219, to which Adickes attested a dependence on Euler diagrams, comments on § 363 of Meier's logic. This reference to § 363 of Meier is by no means useless, but rather shows distinctly the tradition in which Kant's analytical diagrams stand in these reflections: Meier first introduces the theorem of contradiction in § 362 and uses it to justify the *dictum de omni et nullo* in § 363. Like Vives, Reimers, Keckermann, Alsted, Leibniz, Ploucquet, Lambert and also Euler himself, Kant uses analytical diagrams to comment on, present and evaluate the *dictum de omni et nullo*.

Due to various aspects, Kant's evaluation of the *dictum de omni et nullo* appears both progressive and problematic as well as conservative and tradition-conscious. From the *Jäsche Logic* (§ 63) and many other logic manuscripts by Kant's students, one can see that Kant progressively wanted to derive the *dictum de omni (et nullo)* from the supreme principle *nota notae est nota rei ipsius (et repugnans notae, repugnat rei ipsi)* – i.e., what belongs to the mark of a thing belongs also to the thing itself, and so on.[214] From the syntax alone, it is easy to see that the *nota notae* principle, which Kant had already publicised in 1762 in *Die Falsche Spitzfindigkeit der vier syllogistischen Figuren*, is a paraphrase – possibly adapted to the wording of Meier (§

---

[214] Cf. Peter Schulthess: Relation und Funktion, p. 43.

115–123) – of the Neoplatonic-Scholastic lemma *praedicatum praedicati est praedicatum subjecti* already cited above in Chapter 2.2.3 in the quote of Ploucquet.[215]

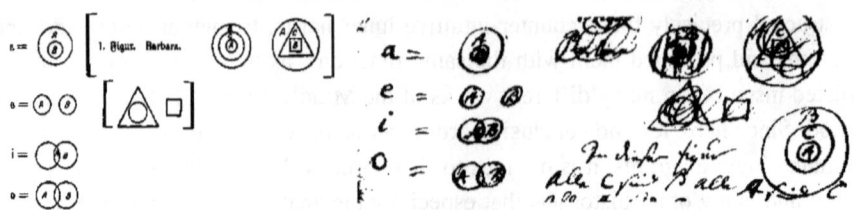

Fig. 2 (a)
Immanuel Kant: Handschriftlicher Nachlaß, Logik. AA XVI, p. 715 (= Nr. 3215).

Fig. 2 (b)
Immanuel Kant: Vorlesungen über Logik. In: G. F. Meier: Auszüge aus der Vernunftlehre. TÜR, Mscr 92, p. 94.

On the one hand, the dependence of the *dictum de omni* on the *nota notae* principle is problematic, since variants of the principle and the *dictum* in their use in the history of ideas were mostly used synonymously from the Neoplatonists to Kant (see Ploucquet) and, moreover, Kant himself even translates the *dictum de omni* (*et nullo*) with *nota notae*, for example in Reflection no. 3218. On the other hand, however, the explanation of the dictum or principle through the supposed 'Euler diagrams' shows that Kant did not choose the diagrams at this point either because they had become fashionable through Euler, or because they could demonstrate any problematic inferences, for example in Sturm.[216] Thus there is one aspect of conservatism and one of progression: Kant's logic diagrams not only tie in with a tradition – but they also consciously continue it.[217]

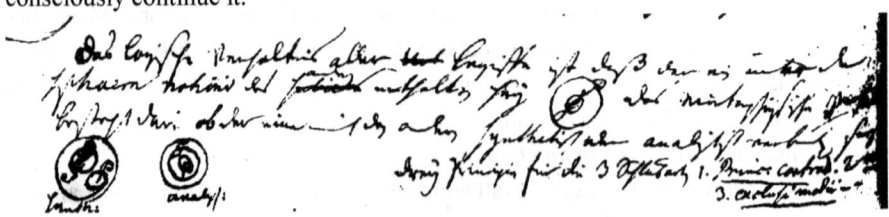

Fig. 3
Immanuel Kant: Vorlesungen über Logik. In: G. F. Meier: Auszüge aus der Vernunftlehre. TÜR, Mscr 92, p. 94.

---

[215] Cf. also the detailed explanation based on a triangular diagram in Wilhelm Traugott Krug: Logik oder Denklehre (System der theoretischen Philosophie I). Königsberg 1806, pp. 306ff. (= § 79).
[216] Vide supra, Chapter 2.2.3.
[217] Already Georg Samuel Albert Mellin: Figur, logische (art.). In: ibid.: Encyclopädisches Wörterbuch der kritischen Philosophie, vol. 2:2. Jena, Leipzig 1799, pp. 581–611 connects Kant's *nota notae* principle as the highest rule of all inferences of reason with the *dictum de omni* (*et nullo*) and derives from it the Lambertian dicta, which he illustrates with line and circle diagrams.

2.2 Analyticity – Analytic judgements, Containment Metaphors and Logic Diagrams

To substantiate this thesis, I now draw on the diagrams of the Kantian containment metaphors as found in Kant's commentaries on Meier's *Auszüge aus der Vernunftlehre* § 363, especially in Reflexion No. 3216. In Reflexion No. 3214, Kant first gives an example of the dictum or principle; in Reflexion No. 3215, one finds the Eulerian diagram for four categorical judgments (see Fig. 2), as well as the representation of three modi (Barbara, Bamalip, Darii) by means of two diagrammatic combinations of the four types of judgments.[218]

Finally, in reflection no. 3216, (AJ$_K$) and (SJ$_K$) are clarified using analytical diagrams:

> The logical relation of all ~~judgm~~ concepts is that the one is contained under the sphaera notionis of the ~~subject~~ other: 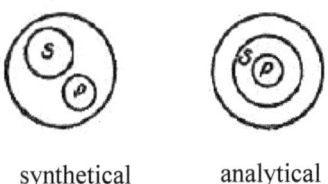 . The metaphysical relation consists in whether the one is synthetically or analytically connected with the other:
>
> synthetical   analytical

Kant's quote and the corresponding Fig. 3 are likely to provide as many answers as they raise new questions. At this point, I do not want to discuss the difference between the 'logical' and the 'metaphysical relation' that has been mentioned,[219] nor do I want to discuss the question that arises as to whether the illustration of the 'logical relation' fully corresponds with the Euler diagram for categorical *A*-judgments of reflection no. 3215 and with Euler's own diagram for affirmative universal judgments ('Tout A est B').[220] Nor will the obvious conjecture about the function of reflection no. 3216 in the commentaries on Meier's *Auszüge aus der Vernunftlehre* § 363 – namely, that Kant inserts the distinction between (AJ) and (SJ) here in order to clarify the status of the nota notae proposition itself – be dealt with here.

I would like to note, however, that the schemata representing analyticity and syntheticity, i.e. the two 'metaphysical relations', are inspired by Euler diagrams, but are

---

[218] Adickes has already pointed out several difficulties in interpreting these diagrams in comparison with Euler: Compare, for example, the *A*-judgments in Fig. 2(a) with Fig. 4(a).

[219] Cf. Schulthess: Relation und Funktion, pp. 118–121. Schulthess interprets the 'logical relation' of *P* (larger circle) and *S* (smaller circle) extensional, the metaphysical relation of *S* (larger circle) and *P* (smaller circle) in analytic judgments intensional. Even if the interpretation of Peter Bernhard: Euler-Diagramme, p. 42f., p. 55–69 is added, only Kant's 'logical relation' corresponds to the extensional Euler diagrams.

[220] Vide supra, Chapter 2.2.3.

2 Logic and its Geometrical Interpretation

not themselves Euler diagrams. The Eulerian basis of these diagrams consists primarily in the fact that Kant resorts, on the one hand, for (AJ) to the nested circle relationship for categorical *A*-judgments (Reflection No. 3215) or to Euler's 'Affirmative – Universelle' judgments and, on the other hand, for (SJ) to two non-intersecting circles for categorical *E*-judgments or Euler's 'Négative – Universelle' judgments.[221] The difference between Kant's schemes and Euler's diagrams, however, arises from the outer circle in each case, since in Kant's diagram this has no clearly assigned function in the form of a variable, a constant, or the like. Since all diagrammatic elements in Euler's *Lettres à une princesse* have such a clear function – e.g. *A*, *B* as a variable, an asterisk ✶ for indicating an intersection, etc. – the last two diagrams in the quote cannot be Euler diagrams in the strict sense. One can also make these two essential differences between Kant's and Euler's diagrams clear by interpreting and contrasting them using a square of opposition (Fig. 4).[222]

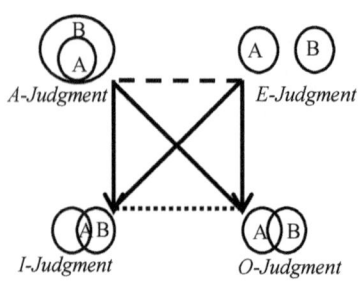
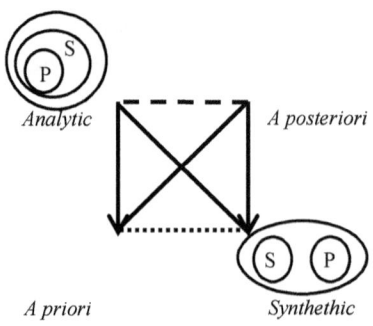

Fig. 4 (a)
Square of Opposition incl. Euler Diagrams for categorical judgments

Fig. 4 (b)
Square of Opposition incl. Kant diagrams for (AJ) und (SJ)

The fact that Kant is dealing with judgements at all here and that 'synthetical' and 'analytical' mean judgements is not only clear from the context (Meier), but also from the fact that the quote deals with the "relation" (Verhältnis) of two concepts. Since in the second sentence of the quote Kant establishes the difference between (AJ) and (SJ) by the way in which the concepts are connected, I assume that the outer circle in each case represents the connection (copula) or the judgement itself. If the two inner circles represent the concepts and *S* and *P* stand for 'subject' and 'predicate' respectively, then in (SJ_K) the subject and predicate are only connected by the copula,

---

[221] Vide supra, Chapter 2.2.3.
[222] For the function of the square of opposition vide supra, Chapter 2.2.2. For the interpretation of fig. 4(b) cf Jean-Yves Béziau: The Power of the Hexagon. In: Logica Universalis 6 (2012), 1–43, here: pp. 27f.; Peter McLaughlin, Oliver Schlaudt: Kant's Antinomies of Pure Reason and the 'Hexagon of Predicate Negation', p. 60f.

2.2 Analyticity – Analytic judgements, Containment Metaphors and Logic Diagrams

whereas in (AJ$_K$) the connection of subject and predicate already seems to be completely given by the subject, since the predicate is already 'contained' in the subject. In other words, in (SJ$_K$) subject and predicate can – due to their spatial difference ('Négative – Universelle') – only be connected by the copula, whereas in (AJ$_K$) the copula can either replace the predicate since the predicate already implies the subject or can only function as an aid to make the contained predicate explicit.[223]

Since the two schemes discussed illustrate figuratively what the definitions of (AJ$_K$) and (SJ$_K$) from the CpR given at the beginning of Chapter 2.2 represent conceptually, the expressions 'to be contained in' and 'lying outside' can be understood neither as metaphorical "shortcomings" (Quine) nor as non-authenticity that requires conceptualization. For 'being contained in' and 'lying outside' belong in the CpR to that technical vocabulary that verbalises what Kant only exemplified figuratively in reflection no. 3216. What seems to be implied 'metaphorically' alone in the CpR is schematised visually in Reflexion No. 3216 as excluding alternatives: In (AJ$_K$) the predicate *is contained* in the subject (and does not lie outside the subject); in (SJ$_K$) the predicate *lies outside* the subject (and is not contained in the subject).

One could now debate whether the expressions Kant uses in defining (AJ) and (SJ) in the CpR, which Quine subsumes under the "notion of containment", actually have a metaphorical meaning.[224] Certainly, a decision would depend on what one again defines as metaphor oneself and which metaphor theory one uses. And of course, in CpR, one could use the lack of such figurativity as in reflection no. 3216 as an argument that the expressions 'contained in' and 'lying outside' remain metaphorical, even though they are meant schematically.

Of course, the question also arises as to which of the two is the image: Is the (AJ)/(SJ) metaphoric in the CpR a verbalisation of diagrammatic relations or are analytical diagrams, such as one finds in Reflection No. 3216, merely a visualisation of the transferences already at home in the logical conceptual scheme? To my knowledge, Kant does not discuss this question, although one could of course answer it by recourse to individual theoretical elements of transcendentalism (e.g. schematism, the idea of transcendental logic, etc.). How a rational representationalism should deal with these questions about the explanation of intuition and concept or world and logic, however, will be the subject of Chapter 3. Until then, the clue given in this chapter, which is still to be discussed, should suffice us to understand the relationship.

If one considers that the teaching of logic, at the latest since the time of Augustine, has kept alive those images and concepts that find a verbal equivalent in expressions such as 'contained in' and 'lying outside', is it not fair to ask whether Kant and his contemporaries actually felt "that the notion of containment is left at a metaphorical

---

[223] This is the meaning of the scholastic mnemonic: Omne subjectum est praedicatum sui. Quod in subjecto implicite est, in praedicato est explicite.
[224] For example, Peter Bernhard: Euler-Diagramme, Chapter 4.2.1, drawing on many studies of logic diagrammatics, argues that there is less of a negatively connoted metaphorical relationship between the logical vocabulary of containment or circumstance and Euler diagrams than an equivalent isomorphism.

level"? Is not the person who knows how to assign expressions such as 'linking', 'adding', 'dividing'. 'splitting', 'thinking in something.', 'extracting', etc. (Verknüpfung, hinzutun, Zergliederung, zerfällen, in etwas denken, herausziehen) – i.e. the expressions Kant uses in further definitions of (AJ) and (SJ) – to the corresponding logic diagrams at a kind of interpretive advantage over the person who understands these expressions only as such metaphors that have to be reformulated and translated?

2.2.5 Schopenhauer's Geometrical Doctrine of Judgement

I have argued since the beginning of Chapter 2.2 against Quine and contrary to the view of many other philosophers that the definitions of (AJ) and (SJ), including metaphors of containment and circumference in Kantian and post-Kantian philosophy, are not deficient and do not need reformulation. Rather, my thesis is that Kant's definitions and verbal descriptions of (AJ) and (SJ) correspond to a visual and geometrical logic, which, however, he only explicated in the notes to Meier. With this thesis, I have tried to continue the new research approach of Anderson and Lu-Adler, who have interpreted Kant's definitions of (AJ) and (SJ) with geometric logic.

However, as can be seen from Kant's logic manuscripts on Meier, the previous comparisons between Kantian geometric logic and individual geometric logics of the early modern period fall short. This is because Kant used many different logic diagrams, not all of which can be traced back to Euler, Lambert or Lange alone. For this reason, with the help of the research results on the history of the development of geometric logic (Chapter 2.2.1), an attempt was made to show an overview of the variety of geometric logics up to the beginning of the 19th century (Chapters 2.2.2 and 2.2.3).

As announced at the beginning of Chapter 2.2, however, it was not crucial for me to solve the exegetical problems of Kant research, but to relativise Quine's problematisation of the concept of containment and circumference in the definition of (AJ) and furthermore (SJ). I was less concerned with an understanding of Kant than with an understanding of (AJ) and (SJ) themselves. With the evidence that Kant already attempted to represent these definitions geometrically-visually, Quine's demand for a reformulation of the definitions seems to come to nothing. However, the hint to Quine and to the analytic philosophy of the 1950s and -60s that (AJ) and (SJ) do not need any reformulation, but only have to be interpreted in the context of geometrical logic, also seems problematic for several reasons.

As Lu-Adler's study, in particular, has shown, many difficulties of interpretation remain when comparing Kant with certain geometrical logicians of the 18th century: The interpretations are based on comparisons of Kant with authors for whom it is not always comprehensible to what extent he exactly received or could have received them; or the interpretation is based on fragments of texts left behind, which often do not seem to be definitely interpretable; or the interpretation does not initially seem to

2.2 Analyticity – Analytic judgements, Containment Metaphors and Logic Diagrams

have been supported by any author of geometric logic in Kant's succession. In fact, the opposite seems to be the case: Maaß, for example, is counted among the geometrical logicians *and* he is a critic of the definitions of (AJ) and (SJ).

As I said, I am not concerned here with an improved exegesis of Kant, but with the question of the interpretability of (AJ) and (SJ) in the context of geometric logic. To this end, the latter interpretive difficulty seems to me to be of particular interest: Did any other geometrical logician at the time of Kant interpret (AJ) and (SJ) in terms of the thesis put forward by Lu-Adler et al.? Or do the post-Kantian logicians contradict the new approach of Kant research? I have already given a reason above towards the end of Chapter 2.2.3 of why Maaß, for example, could not contribute to the understanding of (AJ) and (SJ): He was not only a critic of the distinction between (AJ) and (SJ)[225] but also a critic of Euler's circle diagrams.

If one examines the not very luxuriant history of geometrical logic in the German-speaking world at the beginning of the 19th century – that is, especially the first generation after Kant's death – it soon becomes apparent that the combination of a Kantian and an Eulerian logic is exceedingly rare. The only thing that can be said about geometrical logic in the fifty years after the CpR was published is the following: Ulrich and Maaß use Lambertian rather than Eulerian diagrams; Krause and Fries, as far as I know, do not use the distinction between (AJ) and (SJ) in logic,[226] nor does the Kant critic Bachmann;[227] and the Kant followers Kiesewetter or Krug use logic diagrams almost exclusively to explain conversion rules.[228] To my knowledge, the first logician who both considered himself a Kantian – even if there are clear differences between the two – and used Eulerian diagrams extensively in all parts of his logic is Schopenhauer.[229] And to the best of my current knowledge, many of the subsequent geometrical logicians in the German-speaking world of the 19th century did not deal with (AJ) and (SJ) as intensively as Schopenhauer did.[230]

It is, however, astonishing that Schopenhauer was either almost completely ignored by later historians of logic in the 20th century or even portrayed, together with Hegel, as an opponent of geometric logic.[231] As could already be seen in Chapter 2.1.4, even philosophers of language, analytical philosophers and logicians have so far taken little

---

[225] Vide infra, Chapter 2.3.1.
[226] Cf. Karl Christian Friedrich Krause: Grundriss der historischen Logik für Vorlesungen. Jena et al. 1803; Jakob Friedrich Fries: System der Logik. Ein Handbuch für Lehrer und zum Selbstgebrauch. Heidelberg 1811, pp. 215ff.
[227] Cf. Carl Friedrich Bachmann: System der Logik.
[228] On Kiesewetter vide supra, Chapter 2.2.3; Wilhelm Traugott Krug: Denklehre oder Logik (System der theoretischen Philosophie. Teil 1). Königsberg 1806, p. 45, p. 311, pp. 397–407. The Kantian Mellin refers in his logic only to Kant's *False Subtlety*.
[229] Vide supra, Chapter 1.
[230] Between the 1820s and 1880s, there were still about twenty logics in the German-speaking world that used spatial logic diagrams (see the database mentioned in Chapter 2.2.1). Only one, as I will indicate at the end of Chapter 2.2.6, offers a short discussion of (AJ) and (SJ) in the form of diagrams supporting Schopenhauer's interpretation.
[231] Vide supra, Chapter 2.2.1 and esp. Theodor Ziehen: Lehrbuch der Logik, p. 229.

notice of Schopenhauer's logica major. In the 19th century, Ignaz Denzinger's *Institutiones Logicæ* alone contained a few noteworthy remarks on and adoptions of diagrams from "Schoppenhauer's" (sic!) minor logic.[232] At the turn of the 20th century, one finds isolated traces of Schopenhauer in the logics of Alexius Meinong and Alois Höfler.[233] In 1922, Alf Nymann in particular emphasised Schopenhauer's role in the development of logic diagrams after Kant.[234] However, what all the above-mentioned authors have in common is that they only reported on Schopenhauer's logica minor. Of all the logicians before the mid-2010s, Albert Menne was probably the only one who knew Schopenhauer's logica major, but he only published the judgement on Schopenhauer's logic quoted in Chapter 1.3.

But what about Schopenhauer's own knowledge of logic? Like Kant, Schopenhauer was, above all, familiar with the history of logic of the hundred years preceding him and had been logically socialised by it.[235] Thus he had made several efforts to present to his readers and students the prehistory of the diagrams he used. In this respect, the surviving notes on the history of Euler's diagrams in Schopenhauer's work diverge. In the logica minor of WWR I from 1819, Schopenhauer writes:

> The idea of presenting these spheres by means of spatial figures is very felicitous. It occurred first to *Gottfried Ploucquet* [1819 ed.: *Plouquet*], who used squares to do it; *Lambert,* who came after him, used plain lines positioned under each other; but it was *Euler* who perfected the idea by using circles [*Euler* führte es zuerst mit Kreisen vollständig aus.][236]

In the logica major, which Schopenhauer completed only about two years after the publication of his main work, one also finds the same chronology for the Euler diagrams, but without naming Lambert. This could have been because Schopenhauer did not want to confuse students by naming one-dimensional diagrams, as he clearly favoured two-dimensional diagrams at that time.[237] Interesting, however, is a later addition to the logica major, which must have been added after 1828[238] and in which Schopenhauer not only exemplifies line diagrams but also inverts the chronology of the history of development:

---

[232] Ignaz Denzinger: Institutiones Logicæ, Vol. II, p. 55, p. 245 and Tab. II.
[233] Cf. Alois Höfler, Alexius Meinong: Logik. Prague, Vienna, Leipzig 1890; Alois Höfler: Logik. 2nd rev. ed. Vienna, Leipzig 1922.
[234] Cf. Alf Nyman: Rumsanalogierna inom Logiken. En Undersökning av den Logiska Evidensens Natur och Hjälpkällor. Lund, Leipzig 1926, Chapter 5.
[235] Cf. Anna-Sophie Heinemann: Schopenhauer and the Equational Form of Predication. In: Language, Logic, and Mathematics in Schopenhauer. Ed. by Jens Lemanski. Basel 2020, pp. 165–181; Valentin Pluder: Schopenhauer's Logic in its Historical Context.
[236] WWR I, p. 65.
[237] Vide supra, Chapter 1.3.
[238] In the quote, Schopenhauer refers to Bachmann's *System of Logic*, published in 1828.

2.2 Analyticity – Analytic judgements, Containment Metaphors and Logic Diagrams

Lambert (Neues Organon [Leipzig 1764]) was the first to illustrate the conceptual relationships by means of lines:

Plouquet [sic] (Untersuchung und Abänderung der logikalischen Konstruktion[en] des Prof. Lambert, nebst Anmerkungen v. Plouquet [sic] 1765) introduced squares for drawing the conceptual spheres: – Euler used circles instead (Lettres à une princesse d'Allemagne 1770, Vol. 2. p 106) (according to Bachmann's Logik p 144).[239]

The quoted addition to the logica major is strange in several respects: First of all, it should be noted that Bachmann does not even establish a chronology of logic diagrams at the given reference, but rather evaluates different diagram systems. Even more strange, however, is that Bachmann, a Kant critic, even mentions Plouquet's writing from 1761, which the Kant follower Schopenhauer probably deliberately omits here. Since Schopenhauer advocated line diagrams rather than circle diagrams in later years,[240] this addition, which must have been inserted after 1828, can be seen either as the first document of his modified logical system[241] or as an acknowledgement of the history of the tradition of Kantian logic diagrams, in which Plouquet is usually not mentioned. It is also strange, however, that Schopenhauer refers to Bachmann here at all, since he knew the writings mentioned well.

Even before writing his dissertation (*On the Fourfold Root of the Principle of Sufficient Reason*), Schopenhauer had familiarised himself with many textbooks on logic, especially from the 18th century: around 1813, he was familiar with the great seminal logics (Aristotle, Ramus, Kant) as well as the Wolffian school of logic (Wolff, Reimarus, Platner), the post-Kantian logics (Jakob, Schulze) and algebraic logics (Leibniz, Maimon, Hoffbauer).[242] However, when Schopenhauer writes, following the above quotation from the logica minor of WWR I, that the "schematism of concepts [...] is already explained quite well in many textbooks", this is an exaggeration.[243] Since Lambert, Euler and Plouquet have already been mentioned up

---

[239] WWR2 I, p. 270.
[240] Vide supra, Chapter 1.3.3.
[241] Vide supra, Chapter 1.3.1.
[242] Cf. MR I, pp. 59–72. Cf. also Anna-Sophie Heinemann: Schopenhauer and the Equational Form of Predication.
[243] WWR I, p. 68. – For the context of the quote, vide supra, Chapter 1.3.3.

2 Logic and its Geometrical Interpretation

to the year 1819, only Maaß and Kiesewetter remain as geometric logicians whom the early Schopenhauer demonstrably knew.[244]

Since with Schopenhauer – as with Kant – there is no indication of whether or with which even earlier geometric logic textbooks with analytical diagrams he might have been familiar (Lange, Sturm, Weigel, Reimers, Keckermann, Vives etc.) and Leibniz's logic diagrams were only published in the early 20th century, his logical approach can initially be understood as a combination of Kant's systematics and Euler's diagrams.[245] (Only in Chapter 2.3 will it become apparent that the logica major is also strongly influenced by Aristotle and scholastic logic.) However, it is interesting to note first that Schopenhauer introduces (AJ) and (SJ) as logical forms of judgement in the sense of the CpR before he uses Euler's schemes. As discussed in Chapters 1.3.2 and 1.3.3, Schopenhauer introduces the distinction between (AJ) and (SJ) already in T1, Cap. 2 of his *Berlin Lectures*. In this way, Schopenhauer deliberately anticipates logic in order not to expect his audience to be confronted with completely undefined expressions until the lectures on the doctrine of judgement.

Schopenhauer defines (AJ) and (SJ) based on their common linguistic components (subject, predicate, copula) and on the basis of their etymological function (analysis/synthesis):

> In a judgement [Urtheil], i.e. in a statement [Aussage], a distinction is made between subject and predicate, i.e. between that which is stated and that which is stated by it. Both are concepts. Then the *copula*. Now the statement is either a mere division (analysis) or an addition (synthesis); which depends on whether what is said (predicate) was already thought of [mitgedacht] in the subject of the statement, or is only to be thought to [hinzugedacht] in consequence of the statement. In the first case, the judgement is analytical, in the second synthetic. All definitions are analytical judgements.
>
> E.g. Gold is yellow
>   "   " heavy   ⎫
>   "   " ductile   ⎬ analytical
>   ⎭
> Gold is a chemically simple substance: synthetic[246]

---

[244] Krug is mentioned a few times in Schopenhauer's complete works (mostly polemically), but there is no specific reference to logic. Due to some sentence cooccurrences, the assumption is obvious that Schopenhauer also knew Mellin. However, the thesis cannot be examined in more detail here.

[245] To my knowledge, logicians assumed between the 1760s and 1820s that the history of geometrical logic began with Euler, Lambert and Ploucquet, although Lambert himself referred to the *Nucleus Logicae Weisianae* in later years (Anlage zur Architectonic, vol. 1, p. 128, further: p. XIII, p. XXI). However, this reference became generally known only in 1836 by Drobisch (cf. Friedrich Ueberweg: System der Logik und Geschichte der logischen Lehren. [1st ed.] Bonn 1857, p. 225). Before Drobisch, however, there was the first reference to Vives and thus to pre-Euler diagrams already in Ignaz Denzinger: Institutiones Logicæ. Vol. 2, sect. 2, Leodii 1824, pp. 66f.

[246] WWR2 I, p. 123.

2.2 Analyticity – Analytic judgements, Containment Metaphors and Logic Diagrams

Here, Schopenhauer draws particularly on Kant's § 2 of the *Prolegomena*, in which both the supposedly mereological metaphors 'division' (Zergliederung) and 'addition' (Hinzusetzung) and the gold example are invoked. Schopenhauer also knows that the examples used can be problematic; I will take up this problem in Chapter 2.2.6 and ignore it here for the time being in order to examine solely the distinction between (AJ) and (SJ). This distinction is not determined by supposedly logical metaphors of containment or circumference, but by acts of thought that are convenient with mereological metaphors: With (SJ) something is thought of' (mitgedacht), with (AJ) something is 'thought to' (hinzugedacht).

Kant scholars distinguish between several criteria for (AJ):[247]

(1) Circumference: When the subject (1a) is contained in the predicate or (1b) is circumferentially identical with it;
(2) Explication: When a predicate implied in the subject is explicated;
(3) Truth: When the truth of the judgement can be determined with the help of the law of contradiction.

In T1, Cap. 2, Schopenhauer omits criterion (1) completely – probably because at this point in his lectures he had not yet introduced the circumferential schematic representations of Euler's circles. However, we find the criterion of explication (2) briefly following the above quote. There it says:

> In the meantime, this much is certain: in every judgement, knowledge of the concept of the subject is either merely specified, made *explicit* through a division [Auseinandersetzung] what is *implicitly* thought in it, or extended: accordingly, it is analytical or synthetic.[248]

I assign the quotation to criterion (2), although it expresses more than just a criterion of extension. The expressions 'explicit', 'implicit', 'specify' (verdeutlichen), 'extend' (erweitern) point to the distinction between amplicative and explicative judgements that is still common today; nevertheless, the quotation approaches criterion (1) through the likewise topological metaphors 'explicit', 'implicit' and – in this case even more clearly – through the local pronoun adverb "(thought) therein" (darin). However, the original quote with emphasis on the act of thinking ("durch Auseinandersetzung explicite des implicite darin *gedachten*") can also be associated

---

[247] Cf. Rico Hauswald: Umfangslogik und analytisches Urteil bei Kant, p. 291.
[248] WWR2 I, p. 124.

2 Logic and its Geometrical Interpretation

with the mereological metaphors mentioned above: Implied components of a judgement are 'thought in', extended components of a judgement are 'thought of'.[249]

Criterion (3) is also mentioned only very briefly by Schopenhauer in T1, Cap. 2. Only in the logica major does Schopenhauer take up the theorem of contradiction in connection with the four laws of thought[250] and treats it historically in connection with Aristotle and Wolff ("$A = -A = 0$").[251] In T1, Cap. 2, Schopenhauer only remarks in connection with (AJ):

> In such a judgment [sc. "A body occupies a space."] one need only develop the predicate from the subject according to the law of contradiction without resorting to experience.[252]

For the rest of T1, Cap. 2, Schopenhauer for the most part only addresses the validity of individual examples of (AJ) and (SJ) and discusses in detail their relation to logical (a priori) and experiential cognition (a posteriori). What is remarkable about T1, Cap. 2 is, above all, that Schopenhauer does not use circumference criterion (1) – he only adds it after the treatment of the schematically presented logic of concepts. Assuming that this is a deliberate strategy, one can cautiously interpret from this that Schopenhauer regarded supposedly metaphorical expressions of circumference and containment as incomprehensible without using the corresponding schemata, or that he even regarded the circumferential logical representation – in contrast to many analytical philosophers of the 1950s and -60s – as more superior than criteria (2) and (3).

The fact is, in any case, that Schopenhauer comes to speak of the distinction between (AJ) and (SJ) in the logica major and there explicitly links back to T1, Cap. 2:

> We have already discussed the difference between synthetic and analytical judgements [...]. Judgment consists in recognising the complete or partial identity of two or more concepts, or also their complete difference. Namely, the first thing we do is to compare concepts and find out that, in thinking one concept, we are also thinking the other, in whole or in part: "Iron is hard." Now, this is either so that with the first concept (subject) the other must necessarily be thought of [mitgedacht]; so the judgement is analytical: "a triangle has three sides; gold is yellow;" or it is so that the second concept can only be thought to the first, but the latter can also be

---

[249] Vide infra, Chapter 2.2.6.
[250] Vide supra, Chapter 1.3.3.
[251] WWR2 I, p. 262f.; Cf. also Anna-Sophie Heinemann: Schopenhauer and the Equational Form of Predication.
[252] WWR2 I, p. 124.

2.2 Analyticity – Analytic judgements, Containment Metaphors and Logic Diagrams

thought without it: Triangle is spherical: gold is fluid: then it is synthetic: and the connection requires another ground. But if both cannot be thought together at all, it is in opposition. "Gold is imponderable."[253]

In this resumption of the theme, Schopenhauer recapitulates all the criteria mentioned in Kant scholarship, but still avoids the supposed metaphors of circumference and containment. Criterion (2) and (3), in contrast to T1, Cap. 2 are summarised in a rather undifferentiated way: In (AJ) the predicate is "necessarily thought of" (nothwendig mitgedacht werden muß); in (SJ) it is the case that the predicate "can only be thought to" (nur mitgedacht werden kann). If one emphasises the form of thinking-of (mitdenken) in this distinction, one approaches criterion (2); if, on the other hand, one emphasises the modality (necessary, can), one approaches criterion (3). This criterion, which presupposes the proposition in opposition, is also taken up again in the last two sentences of the quotation. Criterion (1) continues to be used in the quote without much help from metaphors of circumference or containment and is defined via the concept of identity: "partial identity" or "thinking ... in part" takes place with (SJ); "complete identity" or "thinking ... in whole" takes place with (AJ).

Following the quote, which primarily serves the purpose of recalling the criteria for the distinction between (AJ) and (SJ) from T1, Cap. 2, Schopenhauer announces a determination of the possible relations between concepts in judgements. In this investigation, the four properties of judgements, namely 'quantity', 'quality', 'relation' and modality', are developed, and since quantity is always considered in judgements, it is "extraordinarily easy" to make use of "an intuitive representation", as found in Euler and Ploucquet.[254] Before Schopenhauer takes up his typology of conceptual relations in judgements from WWR I (Chapter 1.3.1), already outlined in Chapter 1.3.3, and supplements it with negative judgements,[255] he first discusses "what is actually expressed by this pictorial representation of the conceptual spheres and their relations".[256]

This section, which discusses the function of analytical diagrams or Euler diagrams in judgements, is interesting, in my opinion, for the discussions that started from the analytic philosophy of the 1950s and continue to this day, especially in Kant research. In this section, Schopenhauer exemplifies in detail the supposed metaphor of circumference and containment on the basis of the functioning of Euler diagrams, using the concrete example of "gold is yellow", which he uses as an example of (AJ), following Kant:

---

[253] WWR2 I, pp. 268f.
[254] Vide supra, Chapter 2.2.3.
[255] Vide supra, Chapter 1.3.
[256] WWR2 I, p. 270.

2 Logic and its Geometrical Interpretation

I said that in judgement we compare concepts in order to find whether in one the other is thought of wholly or in part or not. E. g. "Gold is yellow": i. e. in the concept of *gold* I always think of *yellow*; but not conversely in the *yellow* always the *gold*; but only sometimes: all gold is yellow; but only some yellow is gold. Hence we say: *yellow* is the wider [concept]: *gold* lies entirely in it, but does not fill it entirely: for there remains much yellow which is not gold; but no gold which is not yellow: therefore we represent the relation of these two concepts as follows:

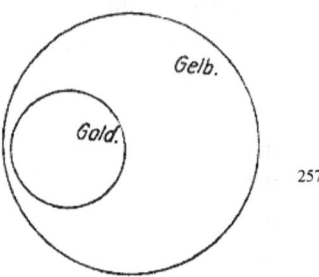

257

Here, Schopenhauer clearly uses logical expressions concerning circumference and containment such as 'wider concept', 'lie in something', 'fill in', etc., which are supposed to have a "continuous analogy" to the circle scheme cited.[258] Compared to Kant, the first advantage of this representation of an (AJ) in my opinion is that Schopenhauer uses the circumstantial logical expressions as far as possible only in connection with the corresponding diagrams and schemes, that he ties in more clearly with Eulerian diagrams, since he does not symbolise the copula with a circle (like Kant, for example)[259] and that the diagram can be interpreted quite clearly extensional due to the explanations that follow it.

But how can we tell that Schopenhauer is interpreting the diagram extensively? The question arises from the fact that the diagram is not clearly labelled: The adjective 'yellow' in the diagram can either stand for the set of properties `being yellow` (intensional) or for the set of objects that `are yellow` (extensional). In my opinion, the phrase used in the quote "for there remains much yellow which is not gold" already points to an extensional interpretation of the diagram, since here salva congruitate for 'there is still much yellow left' one can also substitute 'there are still many yellow objects left', but not 'there are still many yellow properties left'. Schopenhauer explicitly supports this reading twice following the quote:

---

[257] Ibid.
[258] WWR2 I, p. 269.
[259] Vide supra, Chapter 2.2.4.

2.2 Analyticity – Analytic judgements, Containment Metaphors and Logic Diagrams

> The relative size of the spheres, then, does not refer to the size of the content [Inhalt] of the concepts, but to the size of the circumference [Umfang]: not the concept in which we think the most (the most properties) has the wider sphere, that is, not the concept with the most thoughts; but the one through which we think the most things: that is, the one that is a property of very many things. [...] But the relative size of the spheres refers to the circumference, not to the content.[260]

That content is meant here in the sense of intension, circumference in the sense of extension, is explained by Schopenhauer in the context of this quotation. One could, he thinks, be confused by interpreting the sphere of the concepts intensionally:

I. 'Gold' is a wider concept than 'yellow' when determining the properties included:
   i. By 'gold' one thinks of "heaviness, fusibility, ductility, weldability, density, conventional value, indestructibility by rust, brilliance, solubility in nitric-salt-acid alone, etc."[261]
   ii. With 'yellow', on the other hand, one thinks only of the colour.

II. 'Yellow' is a wider concept than 'gold' when determining the objects included:
   i. By 'gold' one thinks of many things, but they must always be yellow.
   ii. By 'yellow' one thinks of all the things in (i), but in addition to "brass, tombac, ochre, yellow lead ore, gum-gutta, yellow flowers, yellow cloth, canaries, topaz, amber, etc.".[262]

Schopenhauer is thus aware that his analytical diagrams can be used for both intensional (I) and extensional representations (II), but that in each case – in accordance with the rule of reciprocity –[263] the conceptual assignment of the relative spheres would have to be reversed.[264] As the previously cited quotations show, the diagrams are to be understood in terms of circumference and not in terms of content. Schopenhauer thus argues for interpreting (AJ) and (SJ) purely extensional in the sense of (II). Whereas Kant scholarship – as can be seen particularly well in Rico Hauswald's paper – hardly advances to an answer to the criticisms of analytic philosophy, since there is

---

[260] WWR2 I, p. 271.
[261] Ibid.
[262] Ibid.
[263] Vide supra, Chapter 1.3.1, vide infra, Chapter 3.2.2.
[264] Cf. also the detailed commentary by Peter Bernhard: Euler-Diagramme, Chapter 4.1.

2 Logic and its Geometrical Interpretation

already disagreement on the question of whether Kant's logic must be interpreted extensionally, intensionally or otherwise, in Schopenhauer's case there is at least a recommendation for interpretation on the part of the author.[265]

The question of whether (AJ) must be interpreted in terms of objects or concepts, which is highly controversial in Kant research, has at least been clearly clarified for Schopenhauer's logica major. Schopenhauer answers this controversial question at a passage in the logic of judgement of his lectures, where he also rejects a theory of singular terms (proper names, labels), as introduced, for example, by the later Frege and then Russell, Whitehead and the early Wittgenstein as a starting point for a distinction between autocategoremata and syncategoremata.[266]

Schopenhauer clarifies the dispute of the status of logic between object- or concept-relatedness and auto- and syncategoremata with reference to the representationalist and nominalist tradition of his doctrine of concepts.[267] Even labels such as `this lectern` or proper names such as `Socrates` were only conceptions of a conception, i.e. abstractions from sensuality.[268] Therefore, it is wrong to assume a direct reference to an object in a concept and thus to want to introduce a uniqueness quantification or a definite description operator.[269] Judging, according to quantity, always consists only of quantities greater than one (general, particular), which are combined with quality (positive, negative) to form three quantifiers (positive: `Some`, `All`; negative: `No`).

Schopenhauer illustrates the use of the three quantifiers especially in the diagram form, which he also used for analytical judgements (see Fig. 1). Here it is particularly noteworthy that Schopenhauer, in agreement with Hamilton, Prantl or Venn[270] and many modern geometrical logicians,[271] shows that diagrams can have greater expressivity than verbal judgements: Whereas (AJ) such as `All `**`B`**`irds are `**`A`**`nimals` ("Alle Vögel sind Thiere") or `All horned animals are ruminants` ("Alle

---

[265] The fact that Schopenhauer unfortunately does not always succeed in an extensional interpretation of individual case studies (e.g. vide infra, Chapter 2.2.6) does not detract from the theory, however.
[266] Vide supra, Chapters 2.1.3–2.1.5.
[267] Vide infra, Chapters 1.2.3, 1.3.1 (Fischer's Nominalism-Thesis).
[268] Vide supra, Chapter 1.3.
[269] A diagrammatic syllogistic with cardinal (e.g. "Exactly $n\,x$...") or intersective quantifiers (e.g. "All x except..."), as first found in geometrical logic in the German-speaking world, as far as I know, in Carl Friedrich Bachmann: System der Logik, p. 175, would probably be rejected by Schopenhauer for reasons given in his philosophy of languages.
[270] Cf. William Hamilton: Lectures on Metaphysics and Logic. Vol. IV, pp. 255–317; Carl Prantl: Ueber die mathematisierende Logik. In: Sitzungsberichte der Bayerischen Akademie der Wissenschaften, Philosophisch-Philologische und Historische Classe 4 (1886), pp. 497–515, here: pp. 507f.; John Venn: Symbolic Logic, esp. Chapters 1 and 5.
[271] Cf. Jon Barwise, John Etchemendy: Visual Information and Valid Reasoning. In: Logical Reasoning with Diagrams. Ed. by Gerard Allwein, Jon Barwise. New York et al. 1996, pp. 3–27, here: 23f.; Jill H. Larkin, Herbert A. Simon: Why a Diagram is (Sometimes) Worth Ten Thousand Words. In: Cognitive Science 11:1 (1987), pp. 65–100; Atsushi Shimojima: On the Efficacy of Representation (PhD thesis). Indiana 1996.

2.2 Analyticity – Analytic judgements, Containment Metaphors and Logic Diagrams

gehörnten Tiere sind Wiederkäuer") can be correctly represented exclusively in a diagram like Fig. 1, a diagram such as Fig. 1 can do more, in that it can represent not only (AJ) with the universal quantifier ("All..."), but also judgements with existential ("Some...") and negative ("Nothing...") quantifiers.

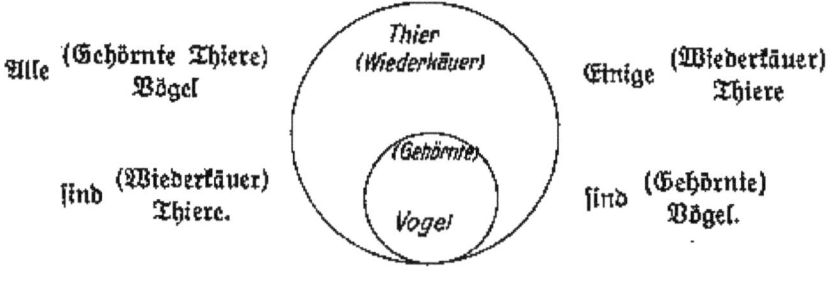

Fig. 1
WWR2 I, p. 273. *Left:* All B are A. *Right:* Some B are A. *Bottom:* Nothing that is non-A is B.

It should also be noted that Schopenhauer, by employing the negative quantifier in Fig. 1, sees even one more possibility of description than, for example, Hamilton or Venn. Unlike them, Schopenhauer does not explicitly quantify the predicate. However, since all three judgements together describe a diagram, the following contractions result:

```
Some B are A.  ⎫
               ⎬ All B are some A.
Some A are B.  ⎭
Nothing that is non-A is B.
All non-A are some non-B.
```

Why Schopenhauer uses a kind of negative quantifier but rejects the uniqueness quantifier becomes explicit in the following quote:

> But I maintain, on the other hand, that judgment is exclusively an operation of thought, not of intuition, and therefore remains exclusively in the domain of abstract concepts, not of individual things, and that, finally, a concept is always general, even if there is only a single thing that is thought by it, only *one single* intuition that gives it content or is a proof of it. My concept of this lectern is never this lectern itself: it remains an abstract, a *universal*. The concept never descends to the individual, to intuition, and in the verdict: "Socrates

is a philosopher," one could very well think of more people, different in shape, size and other properties, who would nevertheless correspond to the concept of Socrates.[272]

The quote explains questions that have so far had to remain unanswered in Kant research, since the textual basis there does not appear to be clear: Schopenhauer's extension logic refers – as Kant researchers following Robert Hanna would say – to a "notional comprehension" and not to an "objectual comprehension". In Kant's case, this classification has not been conclusively resolved. The sphere of a concept represented by geometric logic is not an object comprehension for Schopenhauer. He does point out that there is a "continuous analogy" between the conceptual spheres in the logic of judgement and the positions and surfaces in geometry,[273] and also that some concepts seem so concrete that one could almost take them for objects; nevertheless, according to Schopenhauer, a concept remains an abstraction:

> So, it is not because a concept is abstracted from several objects that it has generality; but conversely, because generality (i.e. absence of the determination into particulars that only intuition has) is essential to the concept as an abstract representation of reason, many different things can be thought through one concept.[274]

Schopenhauer's strict focus on the conceptual circumference (notional comprehension) thus excludes a two-part philosophy of language as found in Frege from the FoA onwards, in Russell and Whitehead in the *Principia Mathematica* and in Wittgenstein's Tlp.[275] Similar to the radical approach of New Aristotelianism emerging in the 1820s and especially -30s, Schopenhauer does not ascribe a referential role to uniqueness quantification in logic, which is set apart from contextual roles of quantifiers (Fischer's nominalism thesis).[276] Rather, the quantity remains limited to particular and general judgements. Thus, Schopenhauer erases the Kantian category of unity from quantity, just as he erases infinite judgements from the moment of quality as a "blind window" in Kant.[277] I will show in Chapter 2.3.4 that Schopenhauer follows Euler in particular with this approach, and I will show in Chapter 3.2 that rational representationalism in the field of social abstraction theory would also do well to jettison the conceptual doctrine of early analytic philosophy.

---

[272] WWR2 I, p. 276f.
[273] WWR2 I, p. 269.
[274] WWR2 I, p. 256.
[275] Vide supra, Chapter 2.1.
[276] Vide supra, Chapter 1.3.1.
[277] WWR2 I, p. 274.

2.2.6 The Principle of Extensionality and the Context Principle

I mentioned two of Quine's criticisms of (AJ) at the beginning of Chapter 2.2: The metaphor of containment and the subject-predicate relation. Five points of discussion emerged from this critique, especially in Kant research in the 1950s:

(AJ) and (SJ)
    (a) are defined differently,
    (b) apply only to judgements with subject-predicate form,
    (c) are not logically defined by the metaphorical expression 'containment',
    (d) are only psychologistic assumptions.

Among the points mentioned, especially (b) and (c) correspond to the critique of analytic philosophers in the wake of Quine and especially (b) to the discussion in cognitive linguistics and cultural studies. Based on a modal logic interpretation of Kant, I added further points of criticism at the beginning of Chapter 2.2:

(AJ) and (SJ)
    (e) cannot be unambiguously determined extensional or intensional,
    (f) cannot be determined unambiguously object- or concept-related.

I believe that I was able to refute several of these problems, not for Kant, but at least for Schopenhauer's logica major. The central point (c) could be rebutted by the historical context shown in Chapters 2.2.1, 2.2.2 and 2.2.3: 'containment' is not an empty metaphor in need of translation or reformulation, but a description of a scheme or, in more modern terms, the description of analytical diagrams in geometric logic.

I also believe that I was able to show, at least for Schopenhauer's logic, that although there are different definitions of (AJ) and (SJ) in the sense of (a), the geometric logic corresponding to (c) is the decisive definition.[278] The problems (e) and (f) currently discussed in Kant research were also explicitly answered by Schopenhauer for his logic: (e) the logic of containment is to be interpreted extensional and (f) the circumference does not refer to sets of objects, but only to sets of meanings or concepts that have an analogy to objects.

---

[278] This is also confirmed for Kant by Willem R. de Jong: Kant's Analytic Judgments and the Traditional Theory of Concepts. In: Journal of the History of Philosophy 33:4 (1995), pp. 613–641; Ian Proops: Kant's Conception of Analytic Judgment. In: Philosophy and Phenomenological Research 70:3 (2005), pp. 588–612; R. Lanier Anderson: The Poverty of Conceptual Truth, pp. 16f.

2 Logic and its Geometrical Interpretation

Schopenhauer's logic, however, still owes us answers to two points, namely to (b) the question of the restriction to the subject-predicate form and to (d) the question of whether (AJ) and (SJ) must or can be interpreted psychologistically.

Point (b) is usually interpreted to mean that (AJ) are problematic because they can only be related to simple judgements (atomic propositions) and not to complex judgements composed with connectives (molecular propositions).[279] Point (d) is usually understood to mean that (AJ) are problematic because they are reduced to psychologistic acts of thought that cannot be described logically.[280] Summarised in keywords, one could say that (AJ) refer directly or indirectly to the propositional-logical extensionality principle in (b), to Quine's critique of the dogma of reduction as well as to his theory of ontological relativity in (d). I will discuss (b) first and then (d) in the following.

(b) The thesis that (AJ) is restricted only to atomic propositions with subject-predicate form, but not to molecular judgments to which atomic sentences are connected, has been relativised or presented as a bogus problem by several authors. In recent years, to my knowledge, Ian Proops alone has attempted to restrict (AJ) to affirmative categorical judgments.[281] In contrast, Robert Hanna, for example, and Robert Larnier Anderson after him, have pointed out that categorical atomic propositions are the basis for the formation of molecular propositions in Kant and therefore (AJ) is not restricted to those.[282] If a categorial atomic proposition can now be evaluated as (AJ), there is no reason not to integrate it into molecular propositions or to formulate it as a molecular proposition. Hanna first takes the categorical judgement `Socrates is a human being` as an example of (AJ).

The example becomes explicit if it is presented as a conditional in which the antecedent and consequent are identical: `If Socrates is a man, then Socrates is a man`. This example is not a categorical but a hypothetical judgement, and although it is also governed by a subject-predicate form (in the antecedent and consequent), it extends the atomic antecedent by an equally atomic consequent. Pap also cites binary symmetrical relations as an example of an (AJ) conditional, such as: "If X is related to Y, then Y is related to X".[283] Moreover, negative statements such as `No triangle has four sides` are also (AJ) for him.

To Hanna's example, it can be added that conditionals can also be traced back to disjunctions because of the junction equivalence, which means that judgements such as "Socrates is a man or it is not the case that Socrates is not a man." could be candidates for (AJ). However, as I will discuss, in these examples much depends on the

---

[279] Cf. e.g. Cory Juhl, Eric Loomis: Analyticity, esp. Chapter 1.
[280] Cf. e.g. Robert Hanna: Kant and the Foundations of Analytic Philosophy, esp. Chapter 3.
[281] Cf. Ian Proops: Kant's Conception of Analytic Judgment. For the older research is particularly relevant Konrad Marc-Wogau: Kants Lehre vom analytischen Urteil. In: Theoria 17 (1951), pp. 140–157.
[282] Robert Hanna: Kant and the Foundation, pp. 61f., p. 140; Lanier Anderson: The Poverty of Conceptual Truth, pp. 20f.
[283] Arthur Pap: Semantics and Necessary Truth, p. 27.

2.2 Analyticity – Analytic judgements, Containment Metaphors and Logic Diagrams

particular conceptual scheme and also on the way the quantifiers, connectives, etc. are interpreted and used.

That Schopenhauer would not have limited (AJ) to affirmative categorical judgements can be attested to by several arguments. First of all, Schopenhauer agrees with Kant that he regards categorical judgements as the starting point of relations. After presenting the basic forms of judgement,[284] Schopenhauer writes:

> Strictly speaking, judgements have no other properties or determinations than the quality and quantity stated: for the judgement is the comparison of two concepts: and that, according to its form (apart from the content, the substance), would be exhausted here. Now, however, these simple, categorical judgements are separated and regarded as only one kind of judgement: Namely, the *hypothetical* and *disjunctive* are added to the *categorical* judgements, and this difference is understood under the title of *relation*. One can say, then, that just as a judgment can be different in quantity and quality, so it can also be determined in three different ways, categorically, hypothetically, and disjunctively. In fact, however, all judgements are categorical and there are no other simple judgements, for hypothetical and disjunctive judgements are already combinations of two or more judgements.[285]

Schopenhauer, like Kant, thus sees the categorical judgements as the basis of relational logic, which – as he writes following text passage of the quote – are not connected by the quantity and quality of the concepts, but by the connectives. In the sense of the Kantian table of judgements, he thus discusses the disjunctive and the hypothetical judgements as the sixth basic relational possibility of concepts, which are characterised by the fact that they have a sphere of a concept that is divided into two or more other spheres. It soon becomes apparent that both (true) hypothetical and disjunctive judgements are represented by Eulerian diagrams, which in turn represented (AJ). Schopenhauer is thus one of the first of many logicians in the 19th-century German-speaking world to use Eulerian diagrams not only for Aristotelian logic, which can be interpreted as a fragment of predicate logic but also for Stoic logic, which can be seen as a fragment of propositional logic.

---

[284] Vide supra, Chapter 1.3.3.
[285] WWR2 I, p. 274.

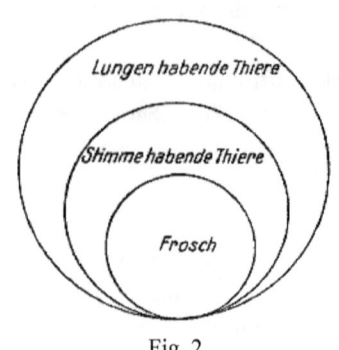

Fig. 2
WWR2 I, p. 281.
(Frosch = frog, Stimme habende Thiere = Voice-having animals, Lunge habende Thiere = lung-having animals)

In my opinion, the fact that molecular propositions can contain (AJ) is best seen in the diagram for hypothetical relations (Fig. 2). Schopenhauer's example judgement on this diagram is: "If all voiced animals have lungs, then frogs also have lungs" (Wenn alle stimmbegabt[en] Thiere Lungen haben; so haben auch die Frösche Lungen.). First of all, he hints several times in the context that one is standing here with one foot in the doctrine of judgement, but with the other already in the doctrine of inferences. This is not surprising insofar as Schopenhauer's example judgement can be interpreted as an enthymemic modus Barbara that dispenses with the propositio minor.

Schopenhauer does not say that this judgement or this diagram is an example of (AJ). But let us substitute in the diagram as follows: frog – gold, voice-having animals – yellow, lung-having animals – colour, then the kinship between the hypothetical conditional ("If yellow is a colour, then gold has also a colour") and the (AJ) becomes evident: the conditional consists of two explicit (AJ) namely the antecedent and the consequent, and one implicit (AJ), namely gold is yellow. The diagram shows the kinship of the hypothetical relations with the (AJ) better than the judgment itself, since it corresponds in form to two gold-yellow diagrams lying one inside the other or two Euler diagrams for *A*-judgments or also two combined diagrams of the second conceptual relationship of Schopenhauer's theory of judgement.[286]

If I have not overlooked more serious points of criticism, the problematic thing about my argumentation so far in (b) may have been that in Schopenhauer's approach I have justified (AJ) in hypothetical relations primarily through an enthymemic modus Barbara; critics might now object that the modus Barbara can only be justified through (AJ) located in it. Schopenhauer recognised this problem and offers a way out of it in his theory of proof, which will be discussed in more detail in Chapter 2.3.

(d) (AJ) appear problematic for many philosophers if they depend on acts of thought of the person making the judgement and not on the relationship of the associated concepts themselves. Maaß had already criticised that conceptual schemes were

---

[286] For the gold-yellow diagram, vide supra, Chapter 2.2.5. For the the second conceptual relationship of Schopenhauer's theory of judgement, vide supra, Chapter 1.3.1; For Euler diagrams for *A*-judgements, vide supra, Chapter 2.2.3. All these diagrams have the same form, in which one circle is completely contained in another.

2.2 Analyticity – Analytic judgements, Containment Metaphors and Logic Diagrams

subjective, person-related and therefore always distinct. To this day, this view is considered a point of criticism of the distinction between (AJ) and (SJ); for what is an (AJ) for one person may be an (SJ) for another judge. I had already given Bennett's tennis racket analogy as an example at the beginning of Chapter 2.2. According to Maaß, Bennett and many others, this relativity depends either on the metaphor of circumference or on the background knowledge of the person making the judgement: If a concept is broader for person A than for person B, it may be that A considers something to be an (AJ) what B considers to be an (SJ). But background knowledge can also play a role: If person A knows less about conceptual relations than person B, it is conceivable that A regards a judgement as ampliative, while B regards it only as explicative. With reference to Kant's relevant definition of (AJ) and (SJ) in the CpR, Maaß writes:

> But what does it mean to go beyond the concept of the subject? What does this figurative expression mean? If, therefore, transcendental idealism indicates an essential, not merely a relative, difference between analytic and synthetic judgments; if the same judgment is not analytic or synthetic [...]; then one must establish a universal rule according to which it can always be decided: whether the predicate B lies in the subject A or not? Whether it is thought by identity or not? Whether, then, the judgment is merely explicative or whether it is amplifying?[287]

Henry E. Allison popularised part of this quote, and in the following decades, various authors have brought Maaß' objection close to Quine's thesis of ontological relativity.[288] That Quine's ontological relativity is closely related to Maaß's thesis of psychologism can be shown in particular by some examples, the problems of which Hans Rott has presented in detail: Kant cites in Prol. § 2 the judgement `gold is a yellow metal` as an example of (AJ), since we should not connect the predicate `being a yellow metal` with the concept of gold, but since it is really present in it; in contrast, Locke, Leibniz, Hilary Putnam or Wolfgang Stegmüller would consider this judgement to be (SJ), because gold actually has no colour or because it is white.[289]

In Chapter 2.2.5, it was already pointed out that Schopenhauer was aware of this problem: I had cited a quote there in which Schopenhauer refers to acts of thinking

---

[287] J.G.E. Maaß: Ueber den höchsten Grundsatz der synthetischen Urtheile; in Beziehung auf die Theorie von der mathematischen Gewisheit. In: Philosophisches Magazin 2 (1789), pp. 186–231. (My transl. – J.L.)
[288] Cf. e.g. Cory Juhl, Eric Loomis: Analyticity, esp. Chapter 4. Allison translates only a part of the above given quote, cf. Henry E. Allison: The Kant-Eberhard-Controversy, p. 43.
[289] Cf. Hans Rott: Vom Fließen theoretischer Begriffe. Begriffliches Wissen und theoretischer Wandel. In: Kant-Studien 95:1 (2004), pp. 29–51.

such as thinking of (mitdenken) and thinking to (hinzudenken), which, however, depend on whether a concept is really contained in another or not. But how does one know whether a concept is *really* contained in another or not? Is there a rule for this, as Maaß demands, or does being contained or not being contained depend on the conceptual scheme, the knowledge and the psyche of the judge?

That Schopenhauer was concerned with these questions and was aware of the problem is shown by a quote he still gives in T1 C2, shortly after the example for (AJ) I cited in Chapter 2.2.5:

> Much of this [sc. in the examples for (AJ) and (SJ)] is obviously subjective-relative because it depends on how many predicates are already known to someone who hears a certain subject term, and what she accordingly thinks of the subject: hence to one the judgement
>
> "Gold is 19 times heavier than water"
>
> can be synthetic to the one, but analytical to the chemist, because this belongs to the characteristics which she thinks essential to gold.
>
> In the meantime, so much is certain that in every judgement the knowledge of the concept of the subject is either merely pointed out [verdeutlicht], through an explication [Auseinandersetzung] of what is implicitly conceived in it, or extended: accordingly, it is either analytical or synthetic [...].[290]

In this quote, Schopenhauer takes up Maaß's criticisms of the distinction between (AJ) and (SJ): What is an (AJ) for one may possibly be an (SJ) for others. As with Maaß, this depends on the level of knowledge of the subject and the conceptual scheme associated with it.

The fact that there is a semantic relativity and no fixed conceptual schemes does not pose a problem for Schopenhauer and does not force him to abandon the distinction between (AJ) and (SJ). Rather, both his nominalism and his conceptual theory come to his aid in this apparent problem. Schopenhauer says, for example: "Once the relation of two concepts is known, I can determine each of them more closely by means of the other [...]."[291] This determination is made with the help of the Euler diagrams and is itself the act of judging. According to Schopenhauer, to judge means to recognise and state the relationship of given concepts to one another.[292]

But how does one recognise or understand the relationship between two concepts? Schopenhauer seems to have two possibilities in mind here. On the one hand, a judgement can be an image of an intuitive cognition. For as soon as someone determines

---

[290] WWR2 I, p. 124. Due to the concept of property the quote suggests an intensional reading.
[291] Cf. WWR2 I, p. 273.
[292] Cf. WWR2 I, p. 260.

2.2 Analyticity – Analytic judgements, Containment Metaphors and Logic Diagrams

anew the relationship between two facts on the basis of an observation, this relationship can be expressed in a judgement or in the corresponding diagram. This is how one comes to new cognition in the first place or how one asks, on the basis of a concrete judgement about the empirical world, whether it discusses (AJ) or (SJ). On the other hand, basic conceptual schemes or the relationships between concepts are always learned during language acquisition. Ontological relativity can therefore also be a product of linguistic relativity: In the process of language acquisition, one has to "carve out entirely new spheres of concepts in one's mind", and it may even be that "conceptual spheres arise in us where there were none before".[293] And that also includes learning a technical language, such as that of the chemist.

Schopenhauer's theory seems to be limited to these two alternatives, at least I do not have more textual evidence at present that answers the problem of the subjective-relative assessment of (AJ) and (SJ). Nevertheless, one can combine both aspects into a circular theory: People learn conceptual spheres during language acquisition with the help of the use-theory and the context principle and thus understand the relationship in which the concepts can stand to each other.[294] These can be represented using Eulerian diagrams. Language is then used as the appropriate means to communicate with others about the intuitive given world. However, since people make discoveries in the intuitive world that do not correspond to their conceptual spheres or to the conceptual relations of others, they continually revise the conceptual schemata and discuss them with others. If the revision of the conceptual scheme is accepted, it is passed on, especially through language acquisition (use-theory of meaning, context principle). Schopenhauer's views on (AJ) and (SJ) thus also support the social aspect of the theory of language presented in Chapter 2.1.

Schopenhauer describes this process of discussion and persuasion in more detail in his treatises on dialectic.[295] However, since the question of (AJ) and (SJ) is not a dialectical question, it is only discussed based on – albeit subjectively relative – examples in logic. In order for this to happen, Schopenhauer specifically separates logical from empirical truth and assigns only one logical truth to (AJ) and (SJ).[296] The relation of a subject to the predicate expressed in (AJ) could also possess a "material truth", but this remains undecided. (AJ) and (SJ) thus do not depend directly on the intuitive, empirical or material world, but on the definition of the judge: If I define that the conceptual sphere `gold` is included in the conceptual circumference of `yellow`, I am obliged to recognise judgements such as `All gold is yellow` as analytical and not synthetic. If I define – despite the word difference – that the conceptual sphere

---

[293] WWR2 I, p. 246. Vide supra, Chapter 2.1.6.
[294] Vide supra, Chapters 2.1.5–2.1.6.
[295] Cf. Jens Lemanski: Logik und Eristische Dialektik. In: Schopenhauer-Handbuch. Leben – Werk – Wirkung. Ed. by Daniel Schubbe, Matthias Koßler. 2nd ed .Weimar 2018, pp. 160–169.
[296] Cf. WWR2 I, pp. 264ff.

of evening star is congruent with that of the concept of morning star,[297] I am likewise obliged to recognise the judgement The evening star is the morning star as analytical.

Whether and to what extent I am justified in asserting these definitions is not a question of logic for Schopenhauer, but of the empirical sciences, of the general use of language as well as of dialectics. Although the question of what (AJ) and (SJ) are can be explained through examples, the actual form of explanation is an intuitive presentation employing an analytical diagram in geometric logic. This explains why there is no need to reformulate and conceptualise the extensional vocabulary such as 'to be contained', 'to lie outside', further 'thinking to' etc. as metaphors in need of revision – as Quine had claimed. Anderson and Lu-Adler correctly noted that not reformulation or translation but schema completion is the correct way to understand (AJ) and (SJ). The vocabulary of containment is not deficient, but it needs the schemata to understand its potential. As far as I know – similar to Schopenhauer – (AJ) was also defined in the 1830s by Moritz-Wilhelm Drobisch and represented with the help of circle diagrams.[298] But Quine and his followers did not know these diagrams any more than they knew the analytical diagrams of Euler, Kant or Schopenhauer.

---

[297] As is well known, Schopenhauer does not give a diagram for synonymicity, although it is the first basic conceptual realationship in judgments, vide supra, Chapter 1.3.1 and also Chapter 1.3.3.
[298] Cf. Moritz Wilhelm Drobisch: Neue Darstellung der Logik nach ihren einfachsten Verhältnissen. Nebst einem logisch=mathematischen Anhange. Leipzig 1836, pp. 36ff.

2.3 Proof – Elementary Geometry, Syllogistic and Intuitive Proof Theory

Today, proof theory is considered the supreme discipline of mathematical logic. In it, "proofs as such are made the object of investigation" or – in other words – "it deals with operating with the proofs themselves".[1] The original aim was to carry out a "rigorous formalisation of the whole mathematical theory, including its proofs", following the example of the logical calculus.[2] Even if, on the one hand, the limits of this objective and, on the other, the unnaturalness of the procedure was quickly realised, the problem-solving of proof theory was recognised as groundbreaking and the study of "deduction used in practice in mathematical proofs" as profitable.[3]

The close relationship between mathematics and logic, which is particularly evident in proof theory, is usually regarded as a product of the modern era, although one can trace preliminary forms of proof theory as far back as Euclidean geometry and ancient logic.[4] Also, the logical calculus is not an invention of modern times but is often retraced to the Oxford Calculators.[5] Following the debate called *quaestio de certitudine mathematicarum*,[6] which continued the late scholastic concordance question concerning the doctrine of demonstration in Aristotelian logic and Euclidean geometry, the idea of calculability of thinking and reasoning by means of a calculus, which had become prominent through Hobbes' *Computatio sive logica* (*De Corpore* I, esp. 1.2), was consolidated.[7]

With the time of the approaching foundational crisis of mathematics, calculi with algebraic notation prevailed. The fact that calculi, decision procedures and formal proofs are not limited to the algebraic form, but can also be written arithmetically or

---

[1] David Hilbert: Neubegründung der Mathematik. Erste Mitteilung. In: Abhandlungen aus dem Mathematischen Seminar der Universität Hamburg 1 (1922), pp. 157–177, here: p. 169. (My transl. – J.L.)
[2] David Hilbert: Neubegründung der Mathematik, p. 165.
[3] Gerhard Gentzen: Investigation into Logical Deduction. In: The Collected Papers of Gerhard Gentzen. Ed. by M.E. Szabo. North-Holland, Amsterdam 1969, pp. 68–131, here: p. 68.
[4] Cf. e.g. Albert G. Dragalin: Proof Theory (Art.). In: Encyclopaedia of Mathematics. Ed. by. M. Hazewinkel. Intern. Ed. in 6 Vols. Dordrecht 1995, Vol. 4, pp. 596–599, here: p. 597: "The origin of proof theory can be traced to Antiquity (the deductive method of reasoning in elementary geometry, Aristotelian syllogistics, etc.)" Cf. also Dag Prawitz: The Philosophical Position of Proof Theory. In: Contemporary Philosophy Scandinavia. Ed. by R. E. Olson, A. M. Paul. Baltimore, London 1972, pp. 123–134, here: p. 124; Jan von Plato: The Development of Proof Theory. In: The Stanford Encyclopedia of Philosophy (Winter 2018 Edition). Ed. by Edward N. Zalta, URL = <https://plato.stanford.edu/archives/win2018/entries/proof-theory-development/>.
[5] Cf. e.g. Carl B. Boyer: The History of Calculus and Its Conceptual Development (The Concepts of the Calculus). New York 1949.
[6] Cf. Paolo Mancosu: Aristotelian Logic and Euclidean Mathematics. Seventeenth-Century Developments of the Quaestio de Certitudine Mathematicarum. In: Studies in History and Philosophy of Science Part A 23:2 (1992), pp. 241–265; Massimo Mugnai: Logic and Mathematics in the Seventeenth Century. In: History and Philosophy of Logic 31 (2010), pp. 297–314.
[7] Cf. Kuno Lorenz: Kalkül (art.). In: HWPh, Vol. 4, pp. 672–681; P. A. Verburg: Hobbes' Calculus of Words. In: Statistical Methods in Linguistics 6 (1970), pp. 60–65; Maarten Bullynck: Erhard Weigel's Contributions to the Formation of Symbolic Logic.

geometrically, has been pointed out by historically interested logicians in different epochs.[8] As shown in Chapters 2.2.2 and 2.2.3, Leibniz, Lange, Ploucquet, and Lambert had not only taken up the Hobbesian idea of a logical calculus, but had also sought to realize it through geometrical forms. Although Frege had used a two-dimensional notation reminiscent even of the diagrammatic logics of Krause or Schopenhauer,[9] with the onset of Russell-Whitehead-Hilbert logic, interest in geometric logics ended and algebraic notations prevailed. Soon the logician or formalist ideal of a calculus was no longer associated with geometric forms. Logic diagrams were considered at most a heuristic device, rather detrimental to formal systems of logic.

Logic diagrams played an inconspicuous but also not entirely unimportant role at the end of the 19th century and in the first half of the 20th century.[10] John Venn noticed that Euler's diagrams for the two particular judgments were based on one and the same geometric form, and so he developed a method of representing the universal judgments by this primary diagram as well: A primary diagram illustrating two concepts yields four regions, which could optionally be either negated by shading or affirmed by dotting. This made it possible to represent not only classical syllogistic but also the algebraic logic of George Boole and William Stanley Jevons.

The Peirce school succeeded in reducing the expressive power of Venn diagrams to an × notation, thereby transforming it into a complex tabular form. These ideas were taken up and developed further from the 1940s onwards in neurophysiology, in electrical engineering, and later in artificial intelligence.[11] Despite the logic diagrams used in these fields, algebraic notations continued to prevail in the following years and decades. As McCulloch and Randolph argued,[12] logic diagrams could replace and even improve upon truth tables, but the ideal of a logical calculus remained closely associated with linear algebraic notation in the minds of many logicians until the late 20th century.

It was Sun-Joo Shin's paradigmatic work *The Logical Status of Diagrams* in 1994 that put an end to diagrams being regarded as 'second-class citizens' in 20th-century logic. Her goal was to end the long distrust of diagrams in logic and mathematics and to show that they can form a formal system similar to linear algebraic notations. Methodologically, she showed that a syntactic and semantic dimension can be defined for

---

[8] Vide supra, Chapters 2.2.2 and 2.2.3. Cf. e.g. also Iohannes Christianus Langius: Nvclevs Logicae Weisianae, esp. p. 248, pp. 756–758; Mauritius Guilielmus Drobisch: De calculo logico. Lipsae, s.a [1827], pp. 3–6 (= Preamonenda); Volker Peckhaus: Logik, mathesis universalis und allgemeine Wissenschaft, esp. Chapter 3.
[9] Vide supra, Chapter 1.3.1 with the literature given there.
[10] The development outlined in the following sections is described in detail in Jens Lemanski: Logic Diagrams, Sacred Geometry and Neural Networks.
[11] Relevant to neurophysiology and AI is Walter H. Pitts, Warren S. McCulloch: A Logical Calculus of the Ideas Immanent in Nervous Activity. In: The Bulletin of Mathematical Biophysics 5:4 (1943), pp. 115–133. Fundamental to electrical engineering is Edward W. Veitch: A Chart Method for Simplifying Truth Function. In: Association for Computing Machinery, Pittsburgh, May 2, 3, 1952. Pittsburgh: ACM, 1952, pp. 127–133.
[12] Cf. Warren S. McCulloch: Embodiments of Mind. Cambridge/ Mass. 1965, p. 14ff., pp. 203ff.; John F. Randolph: Cross-Examining Propositional Calculus and Set Operations.

any representational system. Inspired by Peirce, she developed precise syntactic rules for manipulating diagrams, defined what a well-formed diagram was, and finally proved the soundness and completeness of two systems based on Venn diagrams.

Although the research of logic diagrams already had a long tradition in the Barwise school (Visual Inference Laboratory in Bloomington/Indiana), only Shin achieved great international attention with her Venn calculi. Already until the end of the 1990s, several graph systems were designed, which were not only based on Venn, but also on Euler diagrams or existential graphs (Peirce). From the turn of the millennium, numerous research viewpoints developed in which logic diagrams were studied in computer science, philosophy, linguistics, and psychology. To date, Brighton, Cambridge, Tallinn, Calcutta, Tokyo, and Mexico City stand out as strongholds of this research. And in artificial intelligence, in particular, it is hoped that great advances will be made by exploring diagrams since their very nature combines natural intuition and artificial computability.

When I speak of the fact that it is only with Shin that broad interest in geometric logic calculi begins again, this points to Volker Peckhaus' thesis that modern logic is a history of unconscious rediscoveries.[13] For just as the attempt to set up calculi with geometric figures, Peirce's, McCulloch's, and Randolph's reductionist notations have been repeatedly reinvented or derived from previous circle diagrams in the history of modern geometric logic.[14] The thesis of unconscious rediscoveries of geometric logic may explain, among other things, why Quine and many of his contemporaries did not associate the definition of analytical judgments discussed in Chapter 2.2 with analytical diagrams of geometric logic and Euler's principles: There always seem to have been periods in the history of logic and mathematics when diagrams and visual geometric figures and shapes were rejected and their customary applications were forgotten.

This periodization of geometric logic is not my invention: Hans Hahn and Klaus Volkert coined the phrase 'crisis in intuition' (Krise der Anschauung), which denotes a paradigm sceptical of intuitions that begins with the discovery and popularization of the so-called Weierstrass monsters around 1880, and continues in the natural sciences with counterintuitive particle physics and quantum mechanics.[15] Peter Bernhard and Catherine Legg have taken this scepticism on intuition to explain, among other things, why Euler diagrams, or analytical and logic diagrams in general, have only slowly returned to the focus of research since the late 1940s and increasingly since the

---

[13] Cf. Volker Peckhaus: Logik, mathesis universalis und allgemeine Wissenschaft, esp. p. 2, pp. 222ff.
[14] For example in Samuel Grosser (vide supra, Chapter 2.2.3) as well as in Karl Christian Friedrich Krause: Die Lehre vom Erkennen und von der Erkenntnis als erste Einleitung in die Wissenschaft, und bei Gustav Adolf Lindner: Lehrbuch der formalen Logik.
[15] Cf. Hans Hahn: The Crisis in Intuition. In: Empiricism, Logic and Mathematics. Vienna Circle Collection, Vol 13. Ed. by B. McGuinness. Dordrecht, 1980, pp. 73–102; Klaus Thomas Volkelt: Die Krise der Anschauung. Eine Studie zu formalen und heuristischen Verfahren in der Mathematik seit 1850. Göttingen 1986.

1990s.[16] This is consistent with the observation of Amirouche Moktefi, Sun-Joo Shin, and George Englebretsen that there was a golden age of logic diagrams between the 1760s and the 1880s.[17] And I believe I have shown in Chapters 2.2.2 and 2.2.3 that there were likewise golden ages of analytic logic diagrams roughly between the years 1530 and 1615 and between 1660 and 1715 – a thesis that is consistent with Kobzar's argument, among others, that Euler went to St. Petersburg in the 1730s partly because diagrams were taboo in Central Europe at that time.[18]

The history of logic diagrams in the last centuries seems consequently to be described by a 'business cycle': With the alleged 'invention' or popularization of analytical logic diagrams by Vives (c. 1530s), Weigel (c. 1660s), Euler (c. 1760s), and Shin (c. 1990s), the respective upswing phases set in, each of which then ends, for individual reasons, around 1615, 1715 and 1880, respectively. However, the evidentiary mortgage to be paid that accompanies the dogmatic assertion of such a business cycle is hardly manageable. Names like Reimarus, Peirce, Caroll, McCulloch, or Randolph can quickly be thrown in here, which cannot be well inserted into these periodizations. Nevertheless, the regular new or rediscovery of logic diagrams and the oblivion of their former use in the history of logic demands reasons. Why have geometrical forms been used again and again in the history of science, although algebraic and arithmetic notations and calculi were dominant for many decades?

Probably a historically and systematically approximately satisfying answer to this question is a lifetime project. However, if one examines only statements about diagram usage in the present paradigm – that is, after the rediscovery of logic diagrams in the 20th century – one finds several recurring arguments for diagram usage: McCulloch, for example, states that his reductionist Euler or Venn diagrams are an aid and a tool with which anyone can determine and verify truth values more quickly and easily than with Wittgensteinian tables.[19] The statement that diagrams are an aid and a tool with which one can operate faster, easier and safer is probably the main argument for the use of geometric logic in the 20th and 21st centuries.[20] Research since the 1990s argues for the didactic advantage that logic diagrams have in teaching philosophy, computer science, mathematics, and many other subjects.[21] This is one of the reasons

---

[16] Cf. Catherine Legg: What is a Logical Diagram?; Peter Bernhard: Euler-Diagramme, pp. 11–17.
[17] Cf. Amirouche Moktefi, Sun-Joo Shin: A History of Logic Diagrams; George Englebretsen: Figuring it Out. Logic Diagrams. In Cooperation with José Martin Castro-Manzano and José Roberto Pacheco-Montes. Berlin, Boston 2020, Chapter 2.2.
[18] Cf. esp. my papers *Logic Diagrams in the Weise and Weigel Circles* and *Periods in the Use of Euler-Type Diagrams*. For Kobzar's thesis vide supra, Chapter 2.2.3.
[19] Cf. Warren S. McCulloch: Embodiments of Mind, p. 15f.
[20] Cf. e.g. Jill H. Larkin, Herbert A. Simon: Why a Diagram is (Sometimes) Worth Ten Thousand Words; Atsushi Shimojima: On the Efficacy of Representation; Sun-Joo Shin: The Logical Status of Diagrams. Cambridge/Mass. 1994.
[21] Cf. R. Cox, K. Stenning, J. Oberlander: Graphical Effects in Learning Logic. Reasoning, Representation and Individual Differences. In: Proceedings of the 16th Annual Conference of the Cognitive Science Society, August 13–16, 1994, Cognitive Science Program, Georgia Institute of Technology. Ed. by A. Ram, K. Eiselt. Hillsdale/ N.J. 1994, pp. 188–198. Cf. also the numerous papers on logic diagrams in the journal *Teaching Philosophy*, e.g. Robert L. Armstrong, Lawrence W. Howe: A Euler Test for Syllogisms. In:

2.3 Proof – Elementary Geometry, Syllogistic and Intuitive Proof Theory

why visual learning is nowadays an independent research area in the debate between educational sciences and psychology.[22]

Representatives of the most influential analytical diagram types in logic today (Venn-I, Venn-II, existential graphs, concept diagrams, spider diagrams, GDS, etc.) emphasise that diagrams not only simplify and accelerate logical processes but also play a special role in proof theory. Whereas in the so-called 'pen and paper' age of logic diagrams already played a crucial role in proof theory, but for the most part appeared only as passive-factual illustrations,[23] current analytical diagram types are often completely autonomous in logic and proof theory:[24] Euler diagrams and other analytical diagrams can be used as a communication medium – even in human-machine communication –, as a posteriori verification of logical inferences as well as for decision procedures, and as an active-dynamic construction procedure of reasoning and proof.[25] In the metaphorical expression of Danielle Macbeth: Today we do not argue *with* diagrams alone, but also *in* diagrams.[26]

From the above, two main reasons for the use of analytical diagrams in today's sciences can be inferred:[27] (A) Analytical diagrams are an auxiliary didactic tool that can be used to perform logical operations more quickly and easily. (B) Analytical diagrams are used or seem to be (in the future) equally applicable to algebraic and arithmetic notations. Argument (A) is a rather weak didactic argument that indicates quantifiable advantages such as speed, ease, or simplicity over other notations; argument (B), on the other hand, is a strong argument that draws attention to qualitative equivalence to other notations in logic in different domains. However, reading the studies listed earlier as evidence, the overall impression is always that both arguments are made primarily as justifications for the use of analytical or general logic diagrams as opposed to arithmetic and algebraic notations.

The fact that geometric figures and diagrams appear to be in need of justification not only in logic but in general, maybe because historians of mathematics have in

---

Teaching Philosophy 13 (1990), pp. 39–46; Morgan Forbes: Peirce's Existential Graphs. A Practical Alternative to Truth Tables for Critical Thinkers. In: Teaching Philosophy 20 (1997), pp. 387–400; Marvin J. Croy: Problem Solving, Working Backwards, and Graphic Proof Representation. In: Teaching Philosophy 23 (2000), pp. 169–187.

[22] Information on this topic is provided, for example, by the Budapest Visual Learning Lab.

[23] Cf. Matej Urbas, Mateja Jamnik: Heterogeneous Proofs. Spider Diagrams Meet Higher-Order Provers. In: Interactive Theorem Proving 6898: Second International Conference, ITP 2011, Proceedings. Berlin et al. 2011, pp. 376–382.

[24] Cf. ibid.; Mateja Jamnik, Alan Bundy, Ian Green: On Automating Diagrammatic Proofs of Arithmetic Arguments. In: Journal of Logic, Language, and Information 8:3 (1999), pp. 297–321; Daniel Winterstein, Alan Bundy, Corin Gurr: Dr. Doodle. A Diagrammatic Theorem Prover. In: International Joint Conference on Automated Reasoning (2004), pp. 331–335; Judith Masthoff, Jean Flower, Andrew Fish, Jane Southern: Automated Theorem Proving in Euler Diagram Systems. In: Journal of Automated Reasoning 39:4 (2007), pp. 431–470; Ryo Takemura: Proof Theory for Reasoning with Euler Diagrams. A Logic Translation and Normalization. In: Studia Logica 101:1 (2013), pp. 157–191.

[25] Cf. also on the development Peter Bernhard: Euler-Diagramme, pp. 11ff.

[26] Cf. Danielle Macbeth: Realizing Reason. A Narrative of Truth and Knowing. Oxford 2014, Chapter 2.

[27] Cf. also Amirouche Moktefi: Diagrams as Scientific Instruments. In: Visual, Virtual, Veridical. Ed. by Andras Benedek, Agnes Veszelszki. Frankfurt/Main 2017, pp. 81–89.

many cases established a 'business cycle' for geometry, which was only later transferred to geometric logic. Hans Hahn and Klaus Volkert finally related their thesis of a crisis of intuition primarily to mathematics and further to physics: The period between the end of the 19th and the middle of the 20th century, called the 'crisis of intuition', started with the reception of Weierstrass functions, but was prepared by the shaking of elementary geometry by non-Euclidean geometries. Mark Greaves' thesis that the decision to use (logical or geometric) diagrams and figures depends on worldview or, somewhat more neutrally, metaphysical background conditions of the potential user, thus seems to contradict the pragmatic explanation of philosophers of science who have established pragmatic limits to diagram use in certain paradigms of the history of science.[28] However, as Ioannis Vandoulakis has pointed out to me, one can cite many other problems in mathematics around 1900 in which intuition and provability are disproportionate, for example, the four-colour theorem, Jordan curve theorem, or the Peano curve.

As another such paradigmatic example of a competition between intuitive geometric diagrams and the arithmetic-algebraic notations, Greek mathematics itself was consulted by many historians of mathematics in the 20th century. Following Hieronymus Georg Zeuthen, Oskar Becker had pointed out that Euclid and his successors would generally require a figurative construction to demonstrate mathematical entities.[29] Árpád Szabó, for example, had put forward the much-discussed thesis that Euclid in his *Elementa* continued the sensualism of Babylonian-Egyptian mathematics, but that the systematic-deductive construction showed only a time-related influence of rationalist and anti-empiricist Eleatism.[30] Wilbur Richard Knorr, despite all criticism of Szabó's thesis, had declared that diagrams were characteristic in Greek geometry, but that proof was in principle verbal and non-intuitive.[31]

Following the discussion of Szabó, Sabetei Unguru had criticized many previous historians of mathematics to the effect that if they read Euclid's diagrams only as algebraic notation, they would be interpreting him by modern means: "no diagrams, no geometrical way of thinking."[32] In philosophy, Pirmin Stekeler-Weithofer and Danielle Macbeth have recently also tried to prove – albeit with partly mutually exclusive lines of argument – that diagrams would play a crucial role in Euclidean elementary geometry and that the axiomatic method had been overrated in Greek

---

[28] Cf. Mark Greaves: The Philosophical Status of Diagrams. Stanford 2002.
[29] Cf. Oskar Becker: Grundlagen der Mathematik in geschichtlicher Entwicklung. 2. ed. Freiburg et al. 1964, pp. 90ff.
[30] Cf. Árpád Szabó: The Beginnings of Greek Mathematics. Transl. by A. M. Ungar. Dodrecht 1978, Chapter 3; ibid.: Die Philosophie der Eleaten und der Aufbau von Euklids Elementen. In: Philosophia 1 (1971), pp. 194–228.
[31] Cf. Wilbur Richard Knorr: On the Early History of Axiomatics. The Interaction of Mathematics and Philosophy in Greek Antiquity. In: Theory Change, Ancient Axiomatics, and Galileo's Methodology. Proceedings of the 1978 Pisa Conference on the History and Philosophy of Science. Vol. 1. Ed. by Jaakko Hintikka, D. Gruender, Evandro Agazzi. London et al. 1982, pp. 145–187.
[32] Sabetai Unguru: On the Need to Rewrite the History of Greek Mathematics. In: Archive for History of Exact Sciences 15 (1976), pp. 67–114, here: p. 76.

mathematics.³³ The more philologically oriented Euclid research, on the other hand, has taken a more meticulous path: in the 1990s, Ivor-Grattan Guinness rather discreetly pointed out that some diagrams would play a crucial role in the proof, but that so far little was known about diagrams in Euclid.³⁴ Research on Euclid's *Elementa* therefore currently concentrates mainly on the history of diagrams and states that many relevant Euclid editions (e.g. Heiberg) neither agree with each other nor with the surviving manuscripts.³⁵

As different as the diagrams in the Euclid editions are over the centuries, so different are the opinions about the figures and diagrams as well as about the connection between Aristotelian logic and Euclidean geometry, esp. concerning the term 'demonstratio' (proof). However, the debate in the history of philosophy and mathematics of the 20th century, sketched above on some of its main protagonists, already shows a similar justification situation of diagrams in geometry as it was previously recorded in contemporary geometrical logic: Not the verbal form of argumentation and proof, but the intuitive form is justified and defended within the debates. As far as I can judge from random samples, this compulsion to justify is found in all epochs.

If the justification of geometric diagrams is not accepted, as for example in the studies of the Aristotelian scholars Knorr, McKirahan and Golin, interpreters approach the thesis that Euclidean geometry has strong similarities with Aristotelian logic and philosophy of science:³⁶ McKirahan, for example, suggests an influence of pre-Euclidean geometry on Aristotelian logic, and an influence of Aristotelian philosophy of science on Euclidean geometry. The analogy between proof building blocks in Aristotle and Euclid (axioms/ principles, definitions, postulates) thus leads to an expendability of diagrams in logic and geometry.

The historian of mathematics Orna Harari has probably unconsciously pointed out an old empiricist-sceptical question, but one that arises in rationalist, anti-sensualist, or anti-representationalist positions: If the discursive proof theories of Euclid and Aristotle are now closely related, and if diagrams play no necessary role in geometry and logic, how can the deductive inferences needed for proof be justified?³⁷ That the question is apt and quite justified can be shown by the empiricist-representationalist tradition, which has raised this question as a sceptical argument in many different eras.

---

³³ Cf. Pirmin Stekeler-Weithofer: Formen der Anschauung. Eine Philosophie der Mathematik. Berlin et al. 2008; Danielle Macbeth: Realizing Reason.
³⁴ Cf. Ivor Grattan-Guinness: Numbers, Magnitudes, Ratios, and Proportions in Euclid's Elements. How Did He Handle Them?. In: Historia Matematica 23 (1996), pp. 355–375.
³⁵ Cf. e.g. the volume edited by Karine Chemla: The History of Mathematical Proof in Ancient Tradition. Cambridge 2012.
³⁶ Cf. Richard D. McKirahan Jr.: Principles and Proofs. Aristotle's Theory of Demonstrative Science. Princeton 1992; Owen Goldin: Explaining an Eclipse. Aristotle's Posterior Analytics 2.1–10. Ann Arbor 1996.
³⁷ Cf. Orna Harari: John Philoponus and the Conformity of Mathematical Proofs to Aristotelian Demonstrations. In: The History of Mathematical Proof in Ancient Tradition. Ed. by Karine Chemla. Cambridge 2012, pp. 206–228.

Empiricists and sceptics such as Sextus Empiricus, Francis Bacon, and John Locke had attacked the ultimate justification function of Aristotelian and Stoic logic, arguing that reductive proof procedures would either never end, or end arbitrarily, or in a circle. Empiricists and sceptics, therefore, doubted the validity of the proof theories of deductive logics, and thus the viability and persuasiveness of deductive arguments themselves. As a consequence, they demanded to give up the sovereign claim of purely deductive logic or to stabilize the faltering syllogistic by an inductive logic.[38]

Sextus Empiricus had argued especially against the Stoic propositional logic, that the five proof-requiring inferences of Chrysipp are to be proved by reduction to the *dictum de si et aut*,[39] but that the proof of the *dictum de si et aut* ultimately results in one of the tropes (infinite regress, dogma, circle, etc.).[40] Francis Bacon substituted his *inductio vera* for the Aristotelian *dictum de omni et nullo*, since the universal validity of the Aristotelian axiom of proof was only dogmatically asserted and axioms would always presuppose induction.[41] John Locke held that while the validity of the syllogism imposed itself factually on those who mastered it, syllogisms themselves could not explain why syllogisms were intelligible.[42]

In modern times, these arguments have been reintroduced especially by John Stuart Mill, Lewis Carroll, and Nelson Goodman and are sometimes treated under the title 'paradox of inference'.[43] John Stuart Mill drew from Bacon's critique the result that the *dictum de omni et nullo* was not an axiom but only a definition, and that the validity of all deductive inferences was therefore only borrowed.[44] Lewis Carroll, through a discussion of the Euclidean formulation of the Zenonian paradox, showed how, in a deductive argument, although its premises are accepted as true, one can nevertheless arrive at an infinite regress of reasoning, rather than a valid conclusion.[45] Following Mill, Goodman argued that there is no primacy of deduction over induction since deductive proof theory also ends in a circle or infinite regress.[46]

If empiricists and sceptics from Sextus to Goodman are right that the dicta (axioms of proof) of Aristotelian and Stoic logic are problematic – one can also simply apply

---

[38] As is well known, the last strategy can be explained with reference to Arist. Eth. Nic. IV.3 1139b. Cf. on the relation of induction and deduction also Jens Lemanski: Summa und System.

[39] Cf. also Jonathan Barnes: Truth, etc. Six Lectures on Ancient Logic. Oxford 2007, Chapter 5. Barnes argues that while there is no direct textual evidence to support a theory of proof by means of reduction of proof-requiring inferences to the *dictum de si et aut*, such a theory of proof is suggested in Stoic logic, especially on the basis of Sextus' descriptions.

[40] Cf. Sextus Empiricus: PH II 156ff.

[41] Cf. Francis Bacon: Distributio Operis; Novum Organum I 13ff., I 54, I 127; Advancement of Learning XIV (= Of Judgment); De dignitate et augmentis scientiarum II 3.

[42] Cf. John Locke: An Essay Concerning Human Understanding 4, XVII § 4.

[43] An overview of the dilemma or paradox of inference is given in Catarina Dutilh Novaes: Surprises in Logic. In: Logica Yearbook 2009. Ed. by Michal Peliš. London 2010, pp. 47–63.

[44] Cf. John Stuart Mill: A System of Logic, Ratiocinative and Inductive. New York 1848, esp. pp. 112–121 (= II 2).

[45] Cf. Lewis Carroll: What the Tortoise Said to Achilles. In: Mind 4:14 (1895), pp. 278–280.

[46] Cf. Nelson Goodman: The New Riddle of Induction. In: Ibid.: Facts, Fiction, and Forecast. 4th ed. Cambridge/ Mass. 1983, pp. 59–84, here: pp. 62–66 (= sect. 2).

2.3 Proof – Elementary Geometry, Syllogistic and Intuitive Proof Theory

the argument to axioms and rules of deduction of modern calculi – then Euclidean geometry is also problematic insofar as it is purely deductive-axiomatic. The problem of the justification of deductive arguments is still discussed today, especially in the circle of rationalism, and has sparked a multitude of other foci of discussion.

Michael Dummett has taken a position on this issue several times. Relevant, for example, is his paper *On the Justification of Deduction*, which can be understood as a way out of two semantic extreme positions:[47] Deductive proofs must neither be understood as a persuasion in which a semantic change follows from two accepted premises by the conclusion (as Wittgenstein would think) nor must the conclusion be only a rigid, non-knowledge-expanding explication of the semantics already present in the premises (as Mill and Goodman would have criticized). Nevertheless, there is an advantage of deduction over induction: Since we are always already convinced of the validity of deduction, we only need an explication for it; but since we always assume the invalidity of induction, its explication would first require persuasion to recognise it as valid.

Dummett's approach has been criticized in many ways: For example, his reduction of logic and argumentation theory to the 'induction/ deduction' distinction is questionable.[48] It is also debatable whether the rules of introduction and elimination of inferentialist calculi can be self-justifying or whether they are not ultimately based on semantically arbitrary definitions and rules that can be arbitrarily set and changed (Prior's tonk argument).[49] It is also questionable, for example, whether Dummett does not commit a fallacy when he counters Mill's and Goodman's scepticism about deductive arguments with the premise that we have always been convinced of deductive arguments.[50]

The introduction set up so far in Chapter 2.3 thus boils down to the very general question that concerns both geometry and logic: How can proof theories be justified? I will argue in Chapter 2.3 that, on the one hand, Schopenhauer can be placed in the history of the sceptical-empiricist tradition argument. On the other hand, however, he invokes transcendental arguments from Kant's theory of pure intuition to justify the equivalence – if not superiority – of the intuitive proof theory as opposed to the purely discursive one (Chapter 2.3.5). Thus, Schopenhauer does not fall into a seemingly

---

[47] Cf. Michael Dummett: The Justification of Deduction. In: ibid.: Truth and Other Enigmas. Duckworth 1978, pp. 290–318.

[48] Cf. e. g. George Bowles: The Deductive/Inductive Distinction. In: Informal Logic 16:3 (1994), pp. 159–184.

[49] Cf. e.g. Ebba Gullberg, Sten Lindström: Semantics and the Justification of Deductive Inference. In: Hommage à Wlodek. Philosophical Papers Dedicated to Wlodek Rabinowicz. Ed. by T. Rønnow-Rasmussen, B. Petersson, J. Josefsson, D. Egonsson. S.l. 2007. (www.fil.lu.se/hommageawlodek); Jaroslav Peregrin: Inferentialism, pp. 3–6. Vide infra, Chapter 3.

[50] Cf. e.g. Susan Haack: The Justification of Deduction. In: Mind 85:337 (1976), pp. 112–119. According to Sascha Bloch, Martin Pleitz, Markus Pohlmann, Jakob Wrobel: Deviant Rules. On Susan Haack's 'The Justification of Deduction'. In: Susan Haack. Reintegrating Philosophy. Ed. by Julia F. Göhner, Eva-Maria Jung. Cham et al. 2016, pp. 85–113, the debate has been continued in the 2000s by Paul Boghossian, Crispin Wright and Neil Tennant. The Kripke student Romina Padro is also currently trying to popularize the argument in a particular variant under the name 'adoption problem'.

2 Logic and its Geometrical Interpretation

hopeless alternative between inductive and deductive strategies of reasoning and proof as many other empiricists and sceptics do.

Rather, Schopenhauer attempts to establish a mediating position between intuition-based transcendental philosophy and the logicist rationalism of the late 18th and early 19th centuries (Chapter 2.3.1). He argues for the plausibility of the assumption that although proof problems in logic and geometry can be solved only by reference to figures and diagrams, these are not empirical but are themselves a priori forms of thought. Thus, intuition becomes the condition for the possibility of a geometry *more syllogismorum* in general (Chapter 2.3.6).

Schopenhauer's criticism of non-intuitive geometry and its deductive-systematic interpretation (Chapter 2.3.2) is still known to many mathematicians today. The evaluations in their reception correspond to the business cycle indicated above: While Schopenhauer's intuitive proof theory of geometry was positively evaluated in the golden age of logic diagrams, it is precisely followers and students of Weierstrass who brought his philosophy of mathematics into disrepute from the 1880s onwards, and it was not until the proof-without-words movement of the 1970s that it gained increasing interest again (Chapter 2.3.3).

Chapters 2.3.1–2.3.3 will present the historical context (2.3.1) of Schopenhauer's theory of proof in geometry (2.3.2) and its reception (2.3.3). In Chapters 2.3.4–2.3.6 I will confront these views on geometry with the evaluations of geometrical logic expressed by Schopenhauer's predecessors (2.3.4) and by himself (2.3.5). The last thing that emerges is that Schopenhauer takes a mediating position between representationalism and rationalism, in which he uses the empiricist-sceptical tradition argument to justify proofs and the theory based on them in geometry and logic (Chapter 2.3.6). (Readers who are more interested in the systematics than the historical details of these arguments should continue reading in Chapter 2.3.2 and refer to Chapter 2.3.1 for detailed questions.)

2.3.1 Geometria more syllogismorum? The Controversy of Leibnizians and Kantians

In the developments in the history of science up to 1800, there are already tendencies that can be interpreted as harbingers of the epoch that began in the late 19th century, which I called the crisis of intuition in the introduction to Chapter 2.3 with reference to Volkert and Hahn: The unfolding of pure analysis in the second half of the 18th century led to a steady rejection of visual methods of demonstration and towards a stronger formalization of geometry. Famous, for example, is Lagrange's declaration that in his *Mécanique analytique* of 1788 one would not find any intuitive figures, since the algebraic methods he used did not require constructions or geometrical or

2.3 Proof – Elementary Geometry, Syllogistic and Intuitive Proof Theory

mechanical considerations.[51] Gaspard Monge had also shown in his descriptive geometry of 1798 that although the most figurative representation of geometry was the simplest and most elegant; but as in calculus, further deductions could only be made with the help of the algebraic equations of geometry.[52]

Especially Hans Niels Jahnke has pointed out that there was nevertheless a clear figure-relatedness in the philosophy of mathematics of the early 19th century, which seems to contrast with a crisis of intuition in geometry and physics.[53] This figure-relatedness had been promoted, above, all by the Kantian concept of construction, which Jahnke presents in particular based on the reception history in Schelling, Fichte, and Herbart. This reception history of the Kantian concept of construction can be supplemented by many other studies.[54]

Beginning in 1787, a series of books and essays appeared that August Wilhelm Rehberg called the "Controversy over Leibnizian and Kantian Philosophy".[55] Although I will focus primarily on this dispute in this chapter, it will also become apparent in the following chapters that this discussion is only a fragment of a dispute that pervades almost the entire history of science in the modern era. Rehberg, however, has pointed out that the character of the dispute between Kantians and Leibnizians is based on the question of "the difference between synthetic and analytic judgments, and on the ground of mathematical evidence".[56]

Whereas I have already discussed the first point of criticism in Chapter 2.2 based on Eberhard and Maaß, I would like to discuss the second point of criticism in more detail here in Chapter 2.3.1, since the related discussion was a major motivation for Schopenhauer's elaboration of his own geometrical position, which I will present in Chapter 2.3.2. The most prominent Leibnizians who substantially criticized Kant's statements on geometry in the late 1780s and early -90s were first Tiedemann, Stattler, Bornträger, Feder, Weißhaupt, and Eberhard. Eberhard, in particular, not only gained more followers in the course of the dispute but also exerted a significant influence on the philosophy of geometry in the German-speaking world of those years.

To my knowledge, the whole dispute, especially the question of mathematical evidence, has not yet been fully reappraised.[57] But such a reappraisal would also be a

---

[51] Cf. M. de La Grange: Méchanique analytique. Paris 1788, p. vj.
[52] Cf. Gaspard Monge: Géométrie descriptive. Lecons données aux écoles normales, l'an 3 de la République. Paris 1798, pp. 15f. Relevant examples can be found in Michel Chasles: Geschichte der Geometrie. Hauptsächlich mit Bezug auf die neueren Methoden. Transl. by L. A. Sohncke. Halle 1839, pp. 192ff.
[53] Cf. Hans Niels Jahnke: Mathematik und Bildung in der Humboldtschen Reform. Göttingen 1990.
[54] Cf. e. g. Helga Ende: Der Konstruktionsbegriff im Umkreis des deutschen Idealismus. Meisenheim am Glan 1973 (also on Schopenhauer); Jürgen Weber: Begriff und Konstruktion. Rezeptionsanalytische Untersuchungen zu Kant und Schelling. Diss. Göttingen 1995.
[55] August Wilhelm Rehberg: Beantwortung von Herrn Eberhards Duplik, meine Rezension des philosophischen Magazins in der A.L.Z. 1789. No. 10 und 90 betreffend, im 2ten Bande 4tes Stück No. X seines philosophischen Magazins. In: Neues Deutsches Museum 4 (1791), pp. 299–305, here: p. 300.
[56] A. W. Rehberg: Beantwortung, p. 300.
[57] Some milestones of the question about the evidence of mathematics are provided by Darius Koriako: Kants Philosophie der Mathematik. Grundlagen – Voraussetzungen – Probleme. Hamburg 1999, § 24.

laborious and rather fruitless project since the literature is almost unmanageable and – as will be shown in this Chapter – the arguments become more and more entangled and confused by frequent reformulations. In what follows, therefore, I shall confine myself especially to a selected account of the attacks on and defences of Kant's claims that geometry is the only science capable of demonstrating its truths intuitively (CpR A 734) and that geometrical propositions are synthetic (Prol. § 2, c). All in all, the question of the similarity of logic and geometry is at the centre of the discussion to be presented, of which I will again present the essential main arguments systematically arranged in Chapter 2.3.2.

One of the first critics of the Kantian claim that geometric judgments are synthetic was Dietrich Tiedemann, who argued for semantic innatism in a paper published in 1784. Geometric axioms such as 'A straight line is the shortest distance between two points' are analytic in nature, because "[w]hen any relation is determined between two concepts, the reason for it (the *fundamentum relationis*, the scholastics said) already lies in the concepts themselves."[58] Thus, we know that the axioms of geometry are true only from the semantics of the concepts given to us, and if we were to fall into a dispute about the nature of the axioms, no reference to intuition would help, since this is, after all, subjective.[59]

In 1787, the Leibnizian Johann Georg Heinrich Feder adopted Tiedemann's criticism and radicalized its semantic innatism. He doubted Kant's thesis that philosophy could be guided by the proof theory of geometry since philosophy had to deal with "incompletely cognized real things".[60] Geometry, on the other hand, is "always a consequence of simplicity" and takes its complete distinctness and definiteness from its basic concepts. Even blind-born people, according to Feder, would be able to master geometry, since it is only a division of its basic concepts and a sensual examination of its results could not add anything to it in terms of certainty and evidence.[61]

In 1788 Johann Christian Friedrich Bornträger, following Jacobi and Mendelssohn, also attacked the distinction between analytic and synthetic judgments and declared that synthetic judgments are necessarily false since one cannot connect anything in a judgment using the copula that is not necessarily already contained in the concept.[62] In a test of the axiom that a straight line is the shortest distance between two points,

---

[58] Dietrich Tiedemann: Ueber die Natur der Metaphysik. Zur Prüfung von Hrn Professor Kants Grundsätzen. In: Hessische Beiträge zur Gelehrsamkeit und Kunst 1 (1785), pp. 113–130, pp. 233–248, pp. 464–474, here: p. 116.
[59] Dietrich Tiedemann: Ueber die Natur der Metaphysik, p. 116.
[60] Johann Georg Heinrich Feder: Ueber Raum und Caussalität, zur Prüfung der Kantischen Philosophie. Göttingen 1787, p. 44f.
[61] J. G. Feder: Ueber Raum und Caussalität, p. 58.
[62] Cf. J. C. F. Bornträger: Ueber das Daseyn Gottes in Beziehung auf Kantische und Mendelssohnsche Philosophie. Hannover 1788, pp. 25f. Such criticisms show, in my opinion, how useful it is to illustrate the different ways of representing the concept of 'containment', vide supra, Chapters 2.2ff.

2.3 Proof – Elementary Geometry, Syllogistic and Intuitive Proof Theory

the term 'between two points' must already *contain* the property that the shortest distance is the straight line.[63] Geometric judgments are therefore analytic in the sense of Leibniz's formula *praedicatum inest subjecto*.

Benedikt Stattler explained in his book entitled *Anti-Kant* that, according to Kant, proofs must always be both apodictic and intuitive, since, on the one hand, inferences from pure concepts are only factual and, on the other hand, non-conceptual inferences are never evident.[64] In this respect, he argued, mathematics, and geometry, in particular, was the model for Kant's theory of proof. Stattler, on the other hand, argued that geometry was only proof from pure concepts and that "mathematics, by constructing its concepts in an empirical intuition, does not accomplish a strand of hair more [kein Haar mehr leiste]" than philosophy.[65] Stattler repeated this argument in many variations and took as evidence the example that mathematics could not show infinite divisibility intuitively, but only discursively.[66] Therefore it was certain that both, geometry and philosophy, would prove purely discursively:

> Philosophy, then, has [...] just as securely and firmly founded demonstrations as mathematics; and only geometry, in fact, through the sensual designs or empirical constructions that are always at its service in most of its propositions (but also in many of them not), has no more completeness or certainty, but only a more vivid intuition of the sufficient reason of its demonstrated propositions.[67]

The intuitive demonstration in the form of diagrams in geometry is thus only an accessory and the demonstration takes place via the theorem of the sufficient reason: The proof takes place based on the innate concepts (ideae innatae) in such a way that one "clearly sees the containment of the predicate of the inference in the subject of the same in affirmative, or the contradiction in negative inferences".[68]

In 1788, in his book *Ueber die Kantischen Anschauungen und Erscheinungen*, Adam Weißhaupt had also questioned the special status of a visual proof theory in geometry and countered the Kantian theory, which wanted to bring discursive philosophy closer to visual geometry, with a *reductio ad absurdum*: If, according to transcendental aesthetics, space and time are only subjective properties, Weißhaupt argued, and if geometry requires an objective ground of proof, it can hardly be understood how spatial figures are supposed to assume an objective and universal function in geometry. In Kant's philosophy, geometry thus already presupposes an objective

---

[63] Cf. J. C. F. Bornträger: Ueber das Daseyn Gottes, pp. 30ff.
[64] Cf. Benedikt Stattler: Anti-Kant. Vol. 2. München 1788, pp. 289f.
[65] Cf. Benedikt Stattler: Anti-Kant. Vol. 2, p. 290.
[66] Cf. Benedikt Stattler: Anti-Kant. Vol. 2, p. 291.
[67] Benedikt Stattler: Anti-Kant. Vol. 2, p. 292.
[68] Benedikt Stattler: Anti-Kant. Vol. 2, p. 298.

2 Logic and its Geometrical Interpretation

space for its theory of proof, while philosophy still seeks to prove the existence of an external world.[69]

Eberhard continued these criticisms and founded the philosophical journal (*Philosophisches Magazin*) to emphasise the merits of Leibniz's critique of reason over Kant's. The main articles that appeared in the context of the Kant-Eberhard dispute have been compiled in five pages by Hans Vahinger, and Allison has provided a chronology of the main events.[70] Since the Eberhard controversy lasted the longest and Eberhard drew mainly notable mathematicians to his side, it probably had the most emphatic impact on the image of geometry during these years. Here, too, I concentrate only on central texts and arguments concerning the topics given above.

Eberhard explained the claim and goal of Leibniz' logicism in his paper *Ueber die logische Wahrheit*:

> Leibniz thought that nothing more was necessary for the perfection of metaphysics than to work on the fortification of the first principles of human knowledge, being completely calm about their transcendental validity or their logical truth. He [sc. Leibniz] concluded thus: the principles of contradiction and of sufficient reason have transcendental validity, consequently all truths that are built upon them must also have it, it merely depends on their being connected with each other and with their first reasons according to the rules of syllogistic.[71]

Kant, however, had wanted to abolish this logicist theory of proof with his demand for empirical verification (in the sense of an *adaequatio intellectus et rei*), although truths of reason would clearly lie outside the senses. Eberhard, however, holds to the primacy and autonomy of logical truth. With the help of pure logical truths, mathematicians would build up their whole science. This can be seen for example in the investigation of the conic sections by Apollonius and his interpreters.[72]

In 1789, in his paper *Ueber die apodiktische Gewissheit*, Eberhard claimed to "put the theology of Leibniz's critique of reason in its proper light".[73] Kant had declassified the metaphysical judgments about God, freedom, and immortality as empty analytic judgments since they supposedly could neither be demonstrated conceptually nor

---

[69] Cf. Adam Weißhaupt: Ueber die Kantischen Anschauungen und Erscheinungen. Nürnberg 1788, pp. 245ff.

[70] Cf. Hans Vaihinger: Kommentar zur Kritik der reinen Vernunft. Ed. by Raymund Schmidt. 2nd ed. Stuttgart 1922, Vol. 1, pp. 535–540; Henry E. Allison: The Kant-Eberhard Controversy, pp. 1–15.

[71] Cf. Johann August Eberhard: Ueber die logische Wahrheit oder die transscendentale Gültigkeit der menschlichen Erkenntniß. In: Philosophisches Magazin 1:2 (1788), pp. 150–175, here pp. 150f. (My transl. – J. L.)

[72] Cf. J. A. Eberhard: Ueber die logische Wahrheit, pp. 158f.

[73] Cf. J. A. Eberhard: Ueber die apodiktische Gewisheit. In: Philosophisches Magazin 2:2 (1789), pp. 129–186, here: p. 129.

2.3 Proof – Elementary Geometry, Syllogistic and Intuitive Proof Theory

proven intuitively – unlike the synthetic judgments a priori of mathematics.[74] Eberhard's use of Kantian terminology appears confused at many points in the paper, yet his strategy is discernible: he wants to show that mathematical propositions can never be verified and proved by intuition and that thus the difference between judgments of metaphysics and judgments of mathematics can be levelled. By levelling the difference between metaphysics and mathematics, Kant's thesis of the unverifiability of metaphysical judgments would ultimately also fall.

Eberhard argues with recourse to some of his previous works that the "pictorial in the concept of space cannot possibly be the sufficient reason for the absolute necessity of the truth of the geometrical axioms".[75] Geometry could be developed purely from concepts and definitions and geometric figures were only aids of the mind to clarify the concepts. For it is now the case,

> that we allow ourselves the crudest drawings from our own hand, without fearing that the certainty of a geometrical proposition, which we want to represent in it, is in the least lost. For these figures are only to serve as signs of certain concepts in which the mind recognises a certain property.
>
> No real line, we may draw it or imagine it by the faculty of imagination, is a perfect line, i.e. a mere length without width, just as no straight line is completely straight, at least we do not know it for sure.[76]

The quote proves that for Eberhard provability and certainty of judgments never depend on experience or on intuition. The inaccurate figures of geometry prove that these are only didactic aids, but contribute nothing to a theory of proof. To support his attack on Kant's intuitive proof theory and to find further arguments for a pure discursive proof, Eberhard had included in his *Philosophisches Magazin* several papers by Leibnizians and Wolffians.

In his paper published there in 1789, Maaß did not show a consensus with Eberhard, but in the broadest sense, he was able to contribute to the overall argumentation of the magazine. First, Maaß criticized a similar notion of circumference in geometry and in the logic of concepts; second, he raised doubts about the universality of a geometric calculus in the sense of Leibniz, Ploucquet and Lambert.[77] However, his conclusion

---

[74] Cf. J. A. Eberhard: Ueber die apodiktische Gewisheit, pp. 131ff.
[75] Cf. also J.A. Eberhard: Von den Begriffen des Raums und der Zeit in Beziehung auf die Gewißheit der menschlichen Erkenntniß. In: Philosophisches Magazin 2:1 (1789), pp. 53–92, here: pp. 82ff.
[76] J. A. Eberhard: Ueber die apodiktische Gewisheit, p. 161.
[77] Cf. Johann Gebhard Maaß: Ueber den Unterschied der Philosophie und der Mathematik, in Rücksicht auf ihre Gewisheit. In: Philosophisches Magazin 2:2 (1789), pp. 316–341. Vide supra, Chapter 2.2.

2 Logic and its Geometrical Interpretation

that philosophy and geometry each proceeded literally in proving by using the theorem of contradiction and identity appears to be following mainstream Leibnizianism; however, both differ in that logic justifies principles of proof by the *dictum de omni et nullo*, whereas geometry does so by a "dictum de partibus et toto."[78]

Rehberg and Reinhold then defended Kant's viewpoints against Eberhard in the *Allgemeine Literatur-Zeitung*. Kant himself supported these defences with further arguments in letters to his followers: To Reinhold, for example, he wrote on May 12, 1789, that Eberhard was wrong to invoke the *principium rationis sufficientis* in proving geometrical theorems. Kant had in mind Eberhard's paper *Ueber die Unterscheidung der Urtheile in analytische und synthetische*, although the argument could just as well have been read out of Stattler's *Anti-Kant* and other texts. Kant first criticized that the *principium rationis*, on the one hand, could not be a principle, since it was derivable from the *principium contradictionis*, and that, on the other hand, the *principium rationis* made no distinction between analytic and synthetic judgments.

Kant also took up another point that Christian August Crusius had already made many decades earlier in his critique of Leibniz and that Kant had taken up prominently in several places in his work, namely the difference between ideal and real ground:[79]

> In passing I remark (so that in the future people may more easily take notice of Eberhard's wrong track), that the real ground is again twofold: either the *formal* ground (of the *intuition* of the object) – as, for example, the sides of the triangle contain the ground of the angle – or the *material* ground (of the *existence* of the thing). The latter determines that whatever contains it will be called *cause*. It is quite customary that the conjurers of metaphysics [Taschenspieler der Metaphysik] make sleights of hand and, before one realizes it, leap from the logical principle of sufficient reason to the transcendental principle of causality, assuming the latter to be already contained in the former. The statement *nihil est sine ratione*, which in effect says "everything exists only as a consequence," is in itself absurd – either that, or these people give it some other meaning.[80]

Reinhold had this quote from Kant's letter published almost verbatim under his name in the *Allgemeine Literatur-Zeitung*, as was customary at the time.[81] What is amazing about this quote is that Kant or Reinhold uses the side-angle argument, which Crusius had also used to explain the difference between real and ideal grounds: "E.g. the three

---

[78] J. G. Maaß: Ueber den Unterschied der Philosophie und der Mathematik, pp. 337–339.
[79] Cf. e.g. Heinz Eidam: Dasein und Bestimmung. Kants Grund-Problem. Berlin 2000, esp. pp. 43ff., pp. 188ff.
[80] AA XI, p. 36 (Transl. taken from I. Kant: Correspondence. Transl. and ed. by Arnulf Zweig. Cambridge/Mass. 1999, p. 299). Crusius may also have influenced Kant's critique of the *principium rationis*.
[81] S.a. [Reinhold]: Philosophisches Magazin. Ed. by J. A. Eberhard. Drittes und Viertes Stück. Fortsetzung (Rez.). In: Allgemeinen Literatur-Zeitung 175, 12ten Junius (1789:2), Col. 585–592, here: Col. 588f.

2.3 Proof – Elementary Geometry, Syllogistic and Intuitive Proof Theory

sides in a triangle and their relation to each other make a real ground [sc. a ground that goes to the thing itself] from the size of its angles [...]".[82] With the help of this distinction of Crusius Kant resp. Reinhold argues against the conjurers of metaphysics, which unjustifiably oscillate between the ideal and the real ground.

Furthermore, in volume 1 of the *Prüfung der Kantischen Critik der reinen Vernunft*, Johann Schultz increasingly defended the view that geometry consists of synthetic judgments a priori. Schultz's writings are, in my judgment, the most interesting on the part of the Kantians, since Schultz was himself a mathematician and can thus come up with a variety of concrete arguments. For example, Schultz argues against Feder's thesis of analyticity of geometric concepts and judgments in a particularly striking way by seeing an illogical use of concepts in Euclid:

> The immortal Euclides tried to define them [sc. the basic concepts of geometry]. But a striking proof of how much this strict geometrician felt that these definitions cannot give us an idea of the explained things is already this, that he gave a double definition of them. First, he explains the point by that which has no parts, the line by a length without width, the plane by that which has only a length and width, and the solid figure by that which has a length, width and thickness. Afterwards, however, he again explains the plane by the boundary of the solid figure, the line by the boundary of the plane, and the point by the boundary of the line. But if we did not already have the idea of points, lines, surfaces, and physical space, we would never attain it through all those double definitions. The first class of them is even illogical [unlogisch].[83]

Schultz gives several examples of why the individual definitions in Euclid are illogical, i.e. circular in detail, contradictory or even incomplete. Interestingly, he assigns to Euclid's terms and definitions exactly the opposite semantics and a completely different status of reasoning to Eberhard's interpretation: Whereas Eberhard argued that the figures are inaccurate because they can never exactly fulfil Euclid's definitions, Schultz points out that the definitions are illogical because they would not correspond to the properties of intuitive figures. Here, the question about the priority in the relation between intuition and concept, world and logic, which was already mentioned in Chapter 2.2.4, comes up again.

In addition, Schultz discusses in volume 1 of his work over many pages the synthetic character of the Euclidean axioms, definitions and postulates. Especially with recourse to the axioms, he brings forward reasons against semantic innatism: "In all

---

[82] Cf. Christian August Crusius: Entwurf der nothwendigen Vernunft=Wahrheiten, wiefern sie den zufälligen entgegen gesetzt werden. 3rd ed. Leipzig 1766, p. 57 (§ 35). (my transl. – J. L.)
[83] Johann Schultz: Prüfung der Kantischen Critik der reinen Vernunft, Band 1. Königsberg 1789, p. 55.

2 Logic and its Geometrical Interpretation

these propositions, however, the predicate is not contained in the concept of the subject at all."[84] This can be seen in Bornträger's argument since the subject 'between two points' contains lines, but not only the straight line, but also a crooked one.[85] In this respect, Bornträger's argument is not compelling.

In the course of his analysis, Schultz takes on many of the Leibnizians' arguments already mentioned and discusses them using numerous examples that will not be treated further here. The individual examples are shrewd, if not always unproblematic. However, all these partial arguments ultimately only support Schultz's conclusion, namely that geometry is "not a product of any concept, but an immediate representation [unmittelbare Vorstellung]".[86]

That the dispute between Kantians and Leibnizians historically had a history going far beyond Kant is shown by Schultz's paper on the axiom of parallels, published already in 1784. In it, Schultz explained that historical approaches to a rationalistic geometry, i.e. one built purely from the semantics of concepts, in Ramus, Wolff, Segner, and others were inadequate, since the mathematician "does not demand discursive, but intuitive knowledge".[87] In Euclid (El. XI.1), for example, one finds – if one looks only at the purely discursive proof – a petitio principii, which Clavius and other commentators would have recognised. Clavius' reformulation *more syllogismi*, however, is itself only a petitio principii, which is why, in the final analysis, only intuition helps.[88] Schultz sees a confirmation of his argument that discursive proofs are inadequate in the eulogy on Georg Simon Klügel's famous dissertation written as early as 1763 by Abraham Gotthelf Kästner. In this, it is said that only the elaboration of the topology (geometria situs), which perished with Leibniz, will provide the proof of the parallel postulate.[89] As will be shown, many of the contemporary Leibnizians knew better than Schultz that Leibniz's geometria situs stood in a tradition of discursive proofs and was, therefore, an exceedingly problematic example.[90]

Finally, in 1790, Kant had also published a book against Eberhard, *On a Discovery whereby Any New Critique of Pure Reason Is to Be Made Superfluous by an Older One*. In it, Kant accused Eberhard of geometrical incompetence, which could be seen

---

[84] J. Schultz: Prüfung, p. 65.
[85] Cf. J. Schultz: Prüfung, p. 70.
[86] J. Schultz: Prüfung, p. 58.
[87] Johann Schultz: Entdeckte Theorie der Parallelen nebst einer Untersuchung über den Ursprung ihrer bisherigen Schwierigkeit. Königsberg 1784, p. 30. Cf. also AA XIV, p. 37.
[88] Cf. Johann Schultz: Entdeckte Theorie der Parallelen, p. 125; Johann Schultz: Prüfung der Kantischen Critik der reinen Vernunft. Vol. 1, p. 70.
[89] Cf. J. Schultz: Prüfung, p. 31. Schultz quotes "Habituros nos aliquando, veram eam cuius admoto geometriae lumine spectra dissipasti demonstrationem, vix speraverim nisi diligentius exculta doctrina situs, cuius analysis cum Leibnitio interiit." Kant seems to share this view, cf. AA XIV, pp. 33ff., esp. p. 37.
[90] Cf. Vincenzo de Risi: Leibniz on the Parallel Postulate and the Foundations of Geometry. The Unpublished Manuscripts. Cham et al. 2016, esp. Chapters 3.1, 5.2, 5.3. I recommend Risi's comments on the history of tradition up to Leibniz and on the history of reception following Schultz as supplementary reading to my account in the main text.

in his example of Apollonius. Eberhard was then forced to make some historical corrections but did not abandon the argument itself.[91] The Apollonius example was as poorly chosen by Eberhard as the Kästner quote by Schultz just cited. Following the debate about the Apollonius example, Eberhard did not cite any more examples from the history of geometry. From 1790 onwards, this was mainly done by Kästner, who, together with his student Klügel, was a respected mathematician and published several investigations in Eberhard's journal. Against Schultze's interpretation, Kästner thus clearly positioned himself on the side of the Leibnizians. Eberhard's and Schultz's examples are thus clear evidence of how overzealous and also unexpected the discussion was in the 1790s.

In the paper *Was heißt, in Euklids Geometrie möglich?*, Kästner argues that the postulates alone contain unprovable explanations of what is possible, while the rest of Euclidean geometry proves what is possible.[92] The whole paper, as well as all other essays published by Kästner in the *Philosophisches Magazin*, does not contain diagrams or geometric figures. Rather, Kästner explains, in the spirit of Eberhard, that diagrams in geometry have always been used only as didactic aids since they alone are prominent examples of ideally meant figures:

> Euclid's tasks do not really have the intention, because of which handicraft surveyors learn geometrical tasks, to draw, to make sensual pictures of geometrical concepts so exact that their strokes seem to the eye to be without width and thickness, their dots without extension. The sand, the old geometers' *pulvis eruditus*, did not permit such fine lines. But figures could be dug into it, which, rough as they were, helped the intellect to draw inferences. These figures always served the purpose of understanding the possibility. And this is the intention of the Euclidean tasks, for the mind, the so-called practical use notwithstanding.[93]

Kästner sees evidence for this, especially in Aristotle. However, he goes one step further and explains that also in geometry empirical or generally visual-figurative demonstrations can be problematic. To prove, so Kästner in agreement with Maaß, one must use the principle of contradiction. However, a contradiction does not always show up in the use of Euclidean tools, as one can see in the square construction with compasses or also in Klügel's famous critique of the proofs of the parallel postulate.

---

[91] Cf. AA VIII, pp. 191ff.; J. A. Eberhard: Berichtigungen einer Stelle in dem phil. Mag. B. I. St. 2. S. 159. mit Beziehung auf H. Prof. Kants Schrift über eine Entdeck. [...]. In: Philosophisches Magazin 3:2 (1790), pp. 205–211. Cf. also Gregor Büchel: Geometrie und Philosophie. Zum Verhältnis beider Vernunftwissenschaften im Fortgang von der Kritik der reinen Vernunft zum Opus postumum. Berlin et al. 1987, pp. 85ff.
[92] Cf. Abraham Gotthelf Kästner: Was heißt, in Euklids Geometrie möglich?. In: Philosophisches Magazin 2:4 (1790), pp. 391–402, here: pp. 391f.
[93] A. G. Kästner: Was heißt, in Euklids Geometrie möglich?, p. 393.

2 Logic and its Geometrical Interpretation

Thus, intuition did not show all contradictions, which was problematic for proof theory based on oppositions.

This leads Kästner to a radical rejection of the Kantian thesis that geometry can be proved intuitively – whether a priori or empirically. Proofs take place independently of the intuitive figure, which is only a didactic aid. This is even true for the Pythagorean theorem (Euclid, prop. I.47), whose general validity is justified by the preceding theorem (prop. I.46): "It [the Pythagorean theorem] would still be relevant if pencils, ink, and drawing pens were not in the world if no other squares were ever drawn than with a stick in the sand."[94]

Finally, in the paper *Ueber den mathematischen Begriff des Raums*, Kästner shows that his position is not only a critical examination of Kant's intuitive proof theory but connects to several debates in the philosophy of geometry of the 18th century, for example, the Molyneux problem or the debate between Hoheisel, Rüdiger and Körber, which is about whether geometric propositions can be proved *solo oculorum usu*.[95]

In the paper *Ueber die geometrischen Axiome*, Kästner brings to a head the debate about the analyticity of geometry on the basis of the question of the validity of Euclidean axioms: whereas empiricists like Locke would justify the Euclidean axioms by induction from intuition – a method that could be seen in Christian August Hausen and to which Jakob Bernoulli had taken up –, for him, following Leibniz and Wolff, axioms were self-evident propositions. Kästner supports this opinion by a radical variant of semantic innatism, which establishes analyticity in terms of a *pradicatum inest subjecto*: These axioms simply consist of 'clear' concepts such as line, points, etc., which would have an exact meaning similar to autocategoremata. The examples Kästner gives in the course of the paper for such 'clear' terms in judgments are all well-known analytic judgments.[96]

Also in his widely read *Geschichte der Mathematik*, Kästner shows that there was no compelling tradition in mathematics that committed geometry to drawings or relieved it of the logical justification of its propositions. Kästner found examples of this, on the one hand, in several historical geometries that have no diagrams, as in Boethius, and, on the other hand, in geometries from the time of the *quaestio de certitudine mathematicarum* that use "figures without letters", such as Scheubel's geometry.[97] In

---

[94] A. G. Kästner: Was heißt, in Euklids Geometrie möglich?, p. 398.
[95] Cf. Abraham Gotthelf Kästner: Ueber den mathematischen Begriff des Raums. In: Philosophisches Magazin 2:4 (1790), pp. 403–429, here: pp. 405ff. To my knowledge, there is no current reappraisal of this debate. Körber himself, however, discusses numerous arguments of his predecessors and opponents, cf. Christian Albrecht Körber: Archimedes defensus. Das ist Gründlicher Beweiß Daß das Theorema Archimedis Von der Verhältniß der Kugel zum Cylinder, So beyde einerley Höhe und Grund-Fläche haben, nicht solo oculorum usu, wie einige meynen, könne erfunden werden. [...]. Halle 1731. Cf. also J. A. Eberhard: Ueber die apodiktische Gewisheit, pp. 162ff., who also cites Moses Mendelssohn's *Abhandlung über die Evidenz in metaphysischen Wissenschaften* as a precursor in the controversy of the 1790s.
[96] Cf. A. G. Kästner: Ueber den mathematischen Begriff des Raums.
[97] Abraham Gotthelf Kaestner: Geschichte der Mathematik seit der Wiederherstellung der Wissenschaften bis an das Ende des achtzehnten Jahrhunderts. Vol. 1. Arithmetik, Algebra, Elementargeometrie, Trigonometrie, Praktische Geometrie bis zum Ende des sechzehnten Jahrhunderts. Göttingen 1796, pp. 266ff.,

2.3 Proof – Elementary Geometry, Syllogistic and Intuitive Proof Theory

both cases, the discursive proof is autonomous, the visual figure, on the other hand, is only a didactic aid in the last example.[98]

Kästner also gave an outline of the tradition representing geometry *more syllogismi* and reported that this tradition started with Petrus Ramus and Conrad Dasypodius and was continued via the Weigel circle (namely Sturm) with Leibniz and Wolff.[99] That there was a strong interest in this tradition in the circle around Kästner in the 1790s is evidenced, among other things, by the treatise on Dasypodius by Johann Georg Ludolph Blumhof, to which Kästner wrote a preface.[100] Both Kästner's and Blumhof's treatises on the history of Euclid interpretations, which were more logically structured than the original, prove that many of the mathematicians of the 'Leibniz-Wolff school' consciously wanted to bring about a crisis of intuition in geometry.

One will be allowed to believe Vaihinger's judgment that Kant must have been "very uncomfortable with the participation of the most respected mathematicians of that time in the anti-Kant journal [...]".[101] If one thinks of the quote from Kästner affirmatively reproduced by Schultz above, one can even assume that neither Kant nor Schultz reckoned with the fact that Kästner also regarded his advocacy of a *geometria situs* as a purely discursive endeavour. Kästner's Leibnizianism with respect to a non-apparent geometry must thus have been a disappointment for Kant and Schultz.

Kant's position against the Leibnizians, however, continued to be defended mainly by himself and by Schultz.[102] The main objection of all Kantians in the first half of the 1790s was that the Leibnizians had misinterpreted the CpR and, moreover, used vague terms such as that of the figurative.[103] Schultz, in his review in the *Allgemeine Literatur-Zeitung* in September 1790, which was largely dictated by Kant, wrote:

> Why does Mr. E. not show the correctness of this important assertion at least by a *single geometrical proposition*, since he is otherwise so liberal with examples from geometry? This was his task anyway, since the critique had explicitly declared that it wanted to consider itself refuted as soon as, for example, only a single proposition such as "in every triangle, two sides together are larger than

---

pp. 287f. Meant is Johann Scheybel: Das sibend/ acht vnd neunt buch/ des hochberümbten Mathematici Euclidis Megarensis [...]. S.l. [Augsburg] 1555.
[98] Cf. Abraham Gotthelf Kästner: Geschichte der Mathematik. Vol. 1, p. 647.
[99] Cf. A. G. Kästner: Geschichte der Mathematik. Vol. 1, pp. 332–345. On the tradition of a geometry more syllogismi cf. Maria Rosa Massa Esteve: The Symbolic Treatment of Euclid's Elements in Hérigone's Cursus Mathematicus (1634, 1637, 1642). In: Philosophical Aspects of Symbolic Reasoning in Early Modern Mathematics. Ed. by. Albrecht Heeffer, Maarten Van Dyck. London 2010, pp. 165–191.
[100] Cf. Johann Georg Ludolph Blumhof, Abraham Gotthelf Kästner: Vom alten Mathematiker Conrad Dasypodius: Ein literarischer Versuch [...]. Göttingen 1796.
[101] Cf. Hans Vaihinger: Kommentar zur Kritik der reinen Vernunft. Vol. 1, p. 538.
[102] Cf. Wilhelm Dilthey: Kants Aufsatz über Kästner und sein Antheil an einer Recension von Johann Schnitz in der Jenaer Literatur-Zeitung. In: Archiv für Geschichte der Philosophie 3:2 (1890), pp. 275–281.
[103] Cf. e. g. Kant's critique of Eberhard's concept of image in AA XX, p. 392, p. 416.

the third" was able to be demonstrated from the mere *definition* of the triangle, i.e. from the concepts of the plane figure, the side and the number three.[104]

Rehberg, who in my opinion is difficult to assign unambiguously to the Kantians or Leibnizians, also first made two criticisms against Eberhard: First, he believed that Eberhard could not present a purely discursive proof in geometry without falling into an infinite regress.[105] On the other hand, Eberhard's entire argumentation was based on the principle of sufficient reason, which could not be proved without falling into a *petitio principii*.[106]

As early as January 1791, a review in the *Oberdeutsche, allgemeine Litteraturzeitung* stated that the dispute between Eberhard and Kant had reached the point of declining interest, so that the reviewer, due to the "repetitions of what had already been said for a long time", could only repeat individual points, but not the content of every paper.[107] Although Kant and Eberhard had withdrawn from the geometrical disputes themselves, other opponents advanced. For example, the Wolffian Johann Christoph Schwab took up the challenge of Schultz and Rehberg to formulate an example of a geometrical proof only *more syllogismorum*.

In his paper *Ueber die geometrischen Beweise* of 1791, Schwab attempts such a purely discursive proof based on prop. I.20 of the *Elementa*.[108] However, as Rehberg criticizes in his responding paper *Ueber die Natur der geometrischen Evidenz*, it was not very clever of Schwab to start "also his explanation with the theorem explained by a figure".[109] In fact, in all the writings I have cited so far, I have found only one geometrical figure in the discussion around the year 1790, namely precisely in Schwab's paper, which was supposed to show the possibility of a purely conceptual proof.

Rehberg's own position in the paper *Ueber die Natur der geometrischen Evidenz*, however, tended in places more towards Leibnizianism than Kantianism. Rehberg first suggests that the concept of the triangle includes obtuse, acute and right-angled subtypes, but therefore one cannot conclude from the concept of the triangle to all

---

[104] S.a. [Johannes Schultz]: Philosophisches Magazin. Ed. by Johann August Eberhard [Rez.]. In: Literatur-Zeitung, Nr. 283 (26. Sept. 1790), pp. 801f. (My transl. – J. L.) The review is based in part on Kant's draft (AA XX, pp. 385–423).
[105] Cf. A. W. Rehberg: Beantwortung von Herrn Eberhards Duplik, pp. 302f.
[106] Cf. A. W. Rehberg: Beantwortung von Herrn Eberhards Duplik, pp. 304f.
[107] Vmg.: Philosophisches Magazin, [...] Dritten Bandes zweytes und drittes Stück [Rez.]. In: Oberdeutsche, allgemeine Litteraturzeitung IX, 21sten Jäner 1791, Col. 129–136, here: Col. 129.
[108] For the proof cf. Judson Webb: Immanuel Kant and the Greater Glory of Geometry. In: Naturalistic Epistemology. A Symposium of Two Decades. Ed. by D. Nails, A. Shimony. Dordrecht et al. 1987, pp. 17–70. On the course of discussion following the proof cf. Darius Koriako: Kants Philosophie der Mathematik: Grundlagen – Voraussetzungen – Probleme, pp. 321ff.
[109] August Wilhelm Rehberg: Ueber die Natur der geometrischen Beweise. In: Philosophisches Magazin 4:4 (1792), pp. 447–461, here: p. 449.

2.3 Proof – Elementary Geometry, Syllogistic and Intuitive Proof Theory

subtypes in the proof according to the logical law of reciprocity. The proof must be carried out for each figure separately:

> The figure can serve to make concepts distinct [deutlich] and to come to the aid of understanding by bringing the products of it before the senses in an example, but such a visualization can never be absolutely necessary if it is otherwise to remain a pure procedure of the understanding [Verstand]. But it is not enough that the senses themselves have no direct part in the demonstration; the imagination must not take their place either. The objects of the individual concepts that occur in the proposition, as line, angle, etc., must therefore be represented to the senses or the imagination, for they are concepts of sensuous objects, but the proof of the proposition that is to be led from concepts must not be of the kind that it can be led only from the composite (constructed) figure, but it must consist merely in development of the properties of those concepts. Therefore, the auxiliary means by which the Euclidean proof is carried out, the extension of lines, etc., are not at all admissible in a proof that is carried out solely from concepts, that is, with the understanding. I do not mean to say here that geometry should not be taught employing such proofs, (rather, I maintain that it is peculiar to it that it can be taught only by such proofs,) but in a proof by which it is to be shown that it is possible to demonstrate a geometrical proposition from concepts alone, nothing must be used but concepts.[110]

I have given the quote in full length because in my opinion it cannot be interpreted whether it reflects Rehberg's own opinion or whether it is only meant to explain to Schwab what one (esp. Leibnizians) understands by proof from concepts. In favour of the first interpretation are the assertoric propositions ("I do not mean to say here that..."; "rather, I maintain that..."); but in favour of the second interpretation is the context of the quote since Rehberg only pursues the negative goal of presenting Schwab's proof as insufficient in the sense of discursive proof. No matter to which position one ascribes the arguments now, nevertheless, above all their statement content remains of interest: Diagrams are a visual-didactic aid ("figure can serve to", "such a visualization can never be absolutely necessary" etc.) to make concepts distinct and to construct proofs; but neither the empirical intuition nor the imagination are necessary for a proof, but only the analysis of the already semantically prefigured concepts (the proof "must consist merely in a development of the properties of those concepts").

---

[110] A. W. Rehberg: Ueber die Natur der geometrischen Beweise, p. 450.

2 Logic and its Geometrical Interpretation

Schwab reacted to this criticism at first with the assertions of induction and the law of reciprocity: What is true in one kind of triangle is true for the triangle in general and what is true of the triangle in general is true for every kind of triangle. Thus Rehberg's whole criticism is already invalidated:

> Therefore, I could have done without the figure [sc. in the paper *Ueber die geometrischen Beweise*] and made the whole demonstration in my head, if it would not have been more convenient for me and the reader to have a sensual scheme in front of my eyes. But I repeat: it is not the triangle that the geometer has before his eyes, but the triangle in general, from which he proves his theorem: and he could also do this at best without drawing the figure, with mere words.[111]

As far as I know, Rehberg did not react to this defence of Schwab anymore. This can probably be called a wise decision, because just as Rehberg's anti-Eberhardian position sounded more Leibnizian than Kantian, so many statements in the paper of the Leibnizian Schwab can be understood as unintentional 'Kantianisms', e.g. he could have made the figure not on paper but also in his head.[112] The Rehberg-Schwab discussion might well be evidence for my above thesis that the dispute between Kantians and Leibnizians became more and more entangled and confused in the course of the discussion.

In 1792, in the second volume of his *Prüfung der Kantischen Critik der Reinen Vernunft*, Schultz continued to argue especially against Schwab, but also cautiously against Kästner. In this writing, two arguments, in particular, stand out, in my opinion. The first argument is directed primarily against Schwab: Schwab's proof may be comprehensible in its algebraized form, but as a consequence, the proof leads in the intuition to a geometrical absurdity ("geometrische Undinge"), namely to "circular lines of more than 360 degrees" and this, moreover, contradicts the Corollarium of Euclid's prop. I.15.[113] Schultz' first argument thus shows that there are purely discursive proofs that ultimately lead to results which no longer show or even can show any correspondence with reality. Therefore, Schwab's formalizations are rejected by Schultz because of their inapprehensibility; they become 'a *monster*, a pathological case, not a counterexample' to the visual proof theory of Kantianism.[114]

---

[111] Johann Christoph Schwab: Einige Bemerkungen über vorstehenden Aufsatz. In: Philosophisches Magazin 4:4 (1792), pp. 461–469, here: pp. 462f.
[112] Leibnizians would probably credit Schwab with having meant 'demonstration' in the first sentence of the given quotation not descriptively but purely discursively.
[113] Johann Schultz: Prüfung der Kantischen Critik der Reinen Vernunft. Vol. 2. Königsberg 1792, p. 123.
[114] Cf. Imre Lakatos: Proofs and Refutations. The Logic of Mathmatical Discovery. Ed. by John Worrall, Elie Zahar. Repr. Cambridge/Mass. 2015, p. 15.

2.3 Proof – Elementary Geometry, Syllogistic and Intuitive Proof Theory

The second argument that stands out is that a purely discursive proof, conducted independently of intuition, would undermine the semantics necessary for analytic judgments. Against Kästner and Schwab, Schultz makes a similar argument in each case:

> In such two individual intersecting geometric straight lines, the geometrician, however, sensually sees the general truth of the axiom that every pair of straight lines can intersect only in one point, and only from this intuition depends his immediate certainty that the predicate necessarily belongs to the subject [...].[115]
>
> If, therefore, geometrical demonstrations were possible without the construction of concepts, it would have to be possible to deliver a demonstrated system of geometry without knowing what the words in the definitions and propositions mean, indeed whether they indicate anything real at all or not, i.e. a geometry that taught only formal, but no real truth, but would be a mere logical play of ideas, approximately of the kind that Mr. Maimon imagines [...].[116]

Schultz argues that geometric definitions are only understandable by being able to visualize their basic concepts (line, solid figure etc.). Schultz's remarks already show very well how the topics of semantics and analyticity, which I treated in Chapters 2.1 and 2.2, culminate in the theory of proof:[117] Only intuition guarantees a semantics of geometrical concepts, and the semantics of concepts guarantees the *praedicatum inest subjecto* principle of the Leibnizians, which is so crucial for analytic judgments. But if one denies the correlation of concepts to intuition, as e.g. semantic innatism or Kästner's radicalized autocategoremata semantics do, then one also denies the relations of concepts themselves, which are decisive for the formation of analytic judgments, on which finally purely discursive proofs would be based.

Schultz explains that the principles of identity and contradiction cannot explain why and to what extent a predicate is necessarily contained in a subject without reference to the senses or to the imagination.[118] Thus he opposes the innatist semantics of the Leibnizians, in which the meanings are inherent in the concepts in a prefigured way, with a representationalist semantics, in which the meaning of the concepts is taken – in this case: directly and not indirectly – from intuition. And Schultz counters the possible counter-argument of sensory illusion regarding the imagination and to the non-empirical apriority of space, which legitimates the immediacy of intuitive forms.

---

[115] Johann Schultz: Prüfung der Kantischen Critik der Reinen Vernunft. Vol. 2, p. 48.
[116] Johann Schultz: Prüfung der Kantischen Critik der Reinen Vernunft. Vol. 2, p. 131.
[117] Vide infra, Chapter 2.3.6.
[118] Cf. Johann Schultz: Prüfung der Kantischen Critik der Reinen Vernunft. Vol. 2, p. 126, p. 128.

Schwab then defended his purely discursive proof against Schultz in the *Philosophisches Archiv* – Eberhard's successor journal to the *Philosophisches Magazin* – in late 1792. The paper, which was directed against Schultz, however, did not contribute anything more to the matter but pursued the sole strategy of showing that Schultz had misunderstood and misquoted him, which allegedly made Schwab even more convinced of his proof than he had already been before. Moreover, Schultz had not understood himself and actually admitted what he intended to criticize.[119]

Schwab's approach, however, did not seem to convince Leibnizians either. This can be seen, for example, in the fact that even sometime later Maaß made attempts to formulate a purely discursive proof. Maaß saw a successful example of such a geometric proof *more syllogismi* in the commentaries on Euclid and John of Sacrobosco, which were written by Clavius. In his article, Maaß does address arguments of Born and Wenceslaus Johann Gustav Karsten concerning the judgment under discussion;[120] however, Maaß does not discuss the fact that Schultz, in his writing on the parallel lines, took Clavius precisely as an example of purely discursive proofs falling into a *petitio principii*.[121]

I think one can already see from the excerpt from the discussion of the late 1780s and early 1790s, on the one hand, the multiplicity of sub-arguments, but, on the other hand, also the limitedness and problematic nature of the essential main arguments, which I will recapitulate in Chapter 2.3.2. My sampling of texts from the following years gave me the impression that a qualitatively new main argument has been put forward only in what Hans Vaihinger calls the second phase of Kant's arguments. Even if I have overlooked some – for other interpreters perhaps even important – sub-arguments from the debate, I believe that the now following chapters can remedy this deficiency. For especially in Chapter 2.3.3 it will become evident that the dispute between the Leibnizians and Kantians is again only a fragment of a much longer-lasting debate in which similar arguments regularly appear on other paradigmatic theories and problems in the philosophy of mathematics.

2.3.2 Conjuring Tricks, Mousetraps and Stilted Proofs

According to Hans Vaihinger's periodization, the first phase of the Kant disputes ends with the Eberhard-Kant controversy and the second phase begins with Herbart and Schopenhauer.[122] In my opinion, Schopenhauer actually made a new, albeit belated,

---

[119] Cf. Johann Christoph Schwab: Einige Bemerkungen über den zweyten Theil der Schulzischen Prüfung der Kantischen Vernunftkritik. – (Königsberg, 1792. bey Nicolovius.). In: Philosophisches Archiv 1:3 (1792), pp. 1–21.
[120] Cf. Johann Gebhard Maaß: Neue Bestätigung des Satzes: daß die Geometrie aus Begriffen beweise. In: Philosophisches Archiv 1:3 (1792), pp. 96–99.
[121] Vide supra.
[122] Cf. Hans Vaihinger: Kommentar zu Kants Kritik der reinen Vernunft. Vol. 1, p. 540.

2.3 Proof – Elementary Geometry, Syllogistic and Intuitive Proof Theory

contribution to the dispute between Leibnizians and Kantians by accommodating both sides to some extent and formulating a kind of harmony argument.

To be able to specify Schopenhauer's position, I will first recapitulate the main arguments of the two quarrelling groups presented in Chapter 2.3.1. Following this, in Chapter 2.3.2, I will present my interpretation of Schopenhauer's contribution to the dispute of the first phase of Kant's disputes, and in doing so I will try to clarify it by its similarity to the main arguments of the Leibnizians (L) and Kantians (K). In Chapter 2.3.3, my reading of Schopenhauer's philosophy of mathematics, which is derived primarily from its historical context, is supplemented with the more systematic views of its previous interpreters. Finally, in Chapters 2.3.4–2.3.6 it will be shown that many of the arguments elaborated in the present chapter can be transferred from the philosophy of geometry to geometrical logic.

I first summarize the main arguments of the Leibnizians according to Chapter 2.3.1 in bullet points and give their main representatives in brackets, if the argument is not a commonplace. In my opinion, the main arguments of the Leibnizians form the following structure: the overall argument is motivated by (L1) a positive logicism argument, which is based on a popular interpretation of Leibniz's theory of reason at that time, and by (L2) a negative didacticism argument, which is directed against Kant's theory of proof and is supported by (L3) another argument, which for the sake of simplicity I call by the catchword 'pulvis eruditus':

(L1) Logicism Argument:
    (L1.1) Logical Principles:
        (L1.1.1) Leibnizian principles of sufficient reason, contradiction, identity have transcendental validity and are therefore the basis for logical deduction and proof theory. (Eberhard, Maaß, Kästner)
        (L1.1.2) The principles of logic are more certain than the subjective transcendental aesthetics. (Tiedemann, Weißhaupt)
    (L1.2) Semantic Innatism:
        (L1.2.1) Geometric concepts form analytic judgments, since inherent in the concepts is the decision as to whether or not the predicate is necessarily contained in the subject. (Tiedemann, Feder, Bornträger)
        (L1.2.2) Geometric terms are autocategoremata, since their meaning and thus their relation in the judgment are clear. (Stattler, Kästner)

(L2) Didactic Argument:
    (L2.1) Geometric figures are only aids to understanding. (Eberhard, Kästner, Rehberg)
    (L2.2) Figures serve only for a more vivid visualization. (Bornträger)

(L.3) Pulvis eruditus backing:
- (L3.1) Figures are only imperfect signs of certain concepts, since, for example, the empirical line always has a width, whereas Euclid's defined line does not. (Eberhard)
- (L3.2) The way of representation by the old geometers in the *pulvis eruditus* did not allow such fine lines as Euclid's definitions demand. (Kästner)
- (L3.3) Empirical figures are only an image of the corresponding figure at all (Schwab)

Since the main positive arguments of the Kantians are based on Kant's own writings, I will not have to summarize them here in bullet points. Therefore, in the following, I only list the main arguments against the Leibnizians and name the main representatives in brackets as above. In my opinion, three negative main arguments can be put forward: The criticisms in (K1) can be summarized as an antilogicism argument since they are explicitly directed against (L1.1); the criticisms in (K2) can be named as scepticism-argument since they express the opinion that either purely discursive proofs or logicism itself lead into one of the Pyrrhonian tropes; (K3) I call monster-barring argument for the sake of simplicity since the listed criticisms express the opinion that equations or functions which do not correspond to any intuition (so-called monsters) are not provable objects and thus do not contradict a visual theory of proof.

(K1) Antilogicism argument:
- (K1.1) The principium rationis cannot be a principle, since it is derivable from the principium contradictionis. (Kant, Rehberg)
- (K1.2) Leibnizians arbitrarily switch between ideal and real ground. (Kant, Reinhold)
- (K1.3) Euclidean definitions are illogical because they do not correspond to the properties of intuitive figures. (Schultz)

(K2) Scepticism Argument:
- (K2.1) Purely discursive proofs end in a petitio principii. (Schultz)
- (K2.2) Purely discursive proofs end in an infinite regress. (Rehberg)
- (K2.3) The proof of the principle of sufficient reason ends in a petitio principii. (Rehberg)

(K3) Monster-barring argument:
- (K3.1) Concepts and judgments not corresponding to intuition are meaningless. (Kant, Schultz)
- (K3.2) Algebraic proofs without visual demonstration become geometrical absurdity, pathological cases. (Schultz)

2.3 Proof – Elementary Geometry, Syllogistic and Intuitive Proof Theory

I believe that these points represent the essential arguments that can be found in the first phase of Kant's disputes, divided according to Vaihinger, which ends with the Eberhard-Kant controversy. Certainly, one can argue about Vaihinger's periodization; however, the fact that Schopenhauer's early writings, which according to Vaihinger fall into the second phase of the Kant controversies together with Herbart's works, still represent a transitional form is given in the following: When Schopenhauer, in his 1813 dissertation *On the Fourfold Root of the Principle of Sufficient Reason*, interprets the principium rationis, so important for the Leibnizians, with Kant's doctrine of faculties, this approach can be understood as an attempt to harmonize the hardened positions of the dispute between the Leibnizians and the Kantians by means of a more precise differentiation. In contrast to the confused positions of Rehberg and Schwab presented in Chapter 2.3.1, where the school affiliation is blurred due to unfortunate formulations, Schopenhauer seems to know exactly what he is doing when he tries to use the Leibnizian principle to support the Kantian theory of proof in geometry. On the one hand, he tries to mediate in the dispute of the first phase, on the other hand, he tries to find his own point of view beyond the seemingly incommunicable positions.

In *On the Fourfold Root of the Principle of Sufficient Reason*, Schopenhauer initially adheres to the Wolffian thesis that chronologically a distinction between ratio cognoscendi (ground/ consequence) and causa efficiens (cause/ effect) was first made by Leibniz, which all preceding philosophers had overlooked.[123] Following this, however, Schopenhauer argues that both the post-Leibnizian (Wolff, Baumgarten, Lambert, and others) and post-Kantian philosophers (Hofbauer, Maaß, Kiesewetter, and others) did not sufficiently continue Leibniz's differentiation.[124]

> Suppose I ask, why are the three sides of this triangle equal? Then the answer is: because the angles are equal. Now is the equality of the angles the cause of the equality of the sides? No, since here the question is not of alteration and, thus, not of an effect that must have a cause. – Is it mere cognitive ground? No, since the equality of the angles is not mere proof of the equality of the sides, not mere ground of a judgement; indeed, from mere concepts it is never to be understood that because the angles are equal, the sides too must be equal, since the concept of the equality of the angles does not contain that of the equality of the sides. Thus here there is no connection between concepts or judgements, but between sides and angles.][125]

Schopenhauer denies that the question raised in the quotation must be understood as a question about a causa efficiens, since it does not imply a change. However, he

---
[123] Cf. FR, pp. 11–22 (§§ 6–9).
[124] Cf. FRpp. 23–27 (§§ 10–13).
[125] FR, p. 29 (§ 15).

2 Logic and its Geometrical Interpretation

immediately shows that the question cannot be understood as a question about the ratio cognoscendi either: For although the notion of angle may play a central role in answering the question about the notion of equality of sides in a triangle, there seems to be no correspondence decidable by means of containment expressions ("the concept ... does not contain...") or no significant intersections between the notions of the question and the answer. Briefly and strongly simplified: `angle` is not contained in `side` or v.v.

The problem raised by Schopenhauer to show the inadequacies of the previous debate at first only indicates that he does not share the semantic innatism of the Leibnizians (L1.2). After dealing with another problem that addresses the motive of actions, Schopenhauer sums up that "not all cases in which the principle of sufficient reason finds application can be reduced to logical ground and consequent and to cause and effect".[126] Motivated by the law of specification and homogeneity, Schopenhauer, therefore, endeavours, on the one hand, to further develop the Leibnizian differentiation, but, on the other hand, not to attribute an unmanageable number of individual cases to the principium rationis.

The result is well known: Schopenhauer extends the Leibnizian distinction into ratio cognoscendi and causa efficiens with two further "meanings" or "Form [Gestaltungen]" of the principle of sufficient reason, one of which – according to the two problematic examples indicated above – will refer to sensuousness, the other to the will. After inductive reasoning, Schopenhauer explains that the fourfold root finally corresponds to the four cognitive faculties that are based on Kant: "the principle of the reason of becoming, as the law of causality, lies in our understanding; the principle of sufficient reason of knowing, as the faculty for drawing inferences, lies in our reason; the principle of the reason of being lies in our pure sensibility; and finally, the law of motivation governs our will."[127]

I have claimed above that Schopenhauer's differentiation of the principle of reason into four meanings (according to the four types of faculties) could be understood as an attempt to harmonize the hardened positions within the dispute between Leibnizians and Kantians. This argument of harmony is historically problematic, since, as shown in Chapter 2.3.1, already the opponent of Leibniz, Crusius, inferred a differentiation of the principium rationis from the angle-side problem, in such a way that also geometrical propositions were explained by the existential ground (Existentialgrund). Kant and Reinhold had taken up Crusius' argument and made it public in 1789.[128] Schopenhauer himself had noticed these and other parallels between his and Crusius' works and documented them while still in his early creative period.[129] The parallels

---

[126] FR, p. 30 (§ 15).
[127] FR (only first ed., 1813), p. 193 (§ 51).
[128] Vide supra, Chapter 2.3.1.
[129] Cf. MR III, p. 327f. (152). Cf. also Katsutoshi Kawamura: Eine Wurzel der Vierfachen Wurzel des Satzes vom zureichenden Grund Schopenhauers. Schopenhauer und Crusius. In: Schopenhauers Wissenschaftstheorie. Der "Satz vom Grund". Ed. by Dieter Birnbacher. Würzburg 2015, pp. 59–74.

2.3 Proof – Elementary Geometry, Syllogistic and Intuitive Proof Theory

between Crusius and Schopenhauer are so astonishing that, especially in the late 19th century, it was often discussed whether they could really be based on coincidence.

Despite the parallels between Schopenhauer and the Leibnizian opponents Crusius, Kant and Reinhold, I believe that one can save the harmony argument if one realizes that Schopenhauer integrates logic into the realm of reason as the principle of the sufficient reason of knowing (or ground of knowing), but separates it from geometry, which falls into the realm of pure sensuality by the principle of the sufficient reason of being (or ground of being). Schopenhauer meets the arguments of the Leibnizians (L1.1.1) by recognizing the principium rationis as the "basis of all science",[130] which Kant and the Kantians would have overlooked; but he meets the arguments of the Kantians (K3.1) by wanting to explain and prove the meaning and validity of geometrical propositions not through logic, but through the a priori forms of intuition of the inner sense. In the end, one can also understand the harmony argument as if a Leibnizianism thought to its end must necessarily lead to an improved Kantianism.

Geometry plays a supporting role not only in the problematic example given above but also in the entire writing. Schopenhauer's main argument is that Leibnizians rightly analyse geometry with the help of the principium rationis, but they commit a kind of principle error (K1.2), if they analyse geometry only with the principle of the sufficient reason of knowing and not rather with the sufficient reason of being. The expression 'principle error' is debatable: On the one hand, it is true from Schopenhauer's point of view, since many geometers, when proving by means of principium rationis, mistakenly claim a logic for the intuition; but on the other hand, strictly speaking, it is not an error, but only a confusion, which then leads to unsatisfactory results. Thus, Schopenhauer is by no means fundamentally against the arguments of the Leibnizians (L1.1.1), but only proclaims an improved result of the arguments, if they are differentiated according to the Kantian doctrine of faculties and the differentiation is kept (K1.2).

Schopenhauer reproaches the geometricians, who rely on logic rather than on intuition for their proofs, with the fact that deductions in geometry can explain that something is so, but not why it is so. The argument Schopenhauer makes against a pure *geometria more syllogismorum* is strongly reminiscent of Locke's argument against syllogistic logic itself, presented in the introduction to Chapter 2.3: Pure logical proof, he argues, usually leads to insight into facticity, but rarely to insight into genesis. For Schopenhauer, the 'knowing that' of the logical proof could only be transformed into a 'knowledge how' by referring to the a priori intuition of the external sense.

Thus he does not completely exclude the possibility of proofs *more syllogismorum* in geometry – for it "goes without saying that insight into such a ground of being can

---

[130] FR, p. 9 (§ 4).

2 Logic and its Geometrical Interpretation

become a ground of knowledge" –,[131] but the discursive ground of knowing is criticized as deficient in comparison to the visual ground of being. This argument can also be interpreted as an attempt at mediation in the dispute of the first phase of the Kant controversy, since it does not completely reject Leibnizian logicism (L1.1.1) and even excludes the radical Kantian position against unapparent derivations (K2.1, K2.2, K3.2). Nevertheless, it becomes evident that Schopenhauer attaches much higher importance to intuition than Leibnizians do, who understand the geometric figures only as a didactic instrument (L2), which can never correspond to the logical proof (L3):

> Thus in geometry, only in the case of axioms is appeal actually made to intuition. All remaining theorems are demonstrated, i.e., a ground of knowledge of the theorem is specified, which compels one to accept them as true; thus a logical ground of the judgement, not a metaphysical ground, is provided. This, however, is the ground of being and not of knowing is never evident except by means of intuition. Therefore upon a geometric demonstration of this kind one indeed has the conviction that the demonstrated proposition is true, but in no way does one see why what the proposition asserts is as it is; i.e. one does not possess the ground of being, but usually by this point the demand for it has arisen. For proof through demonstration of the ground of knowledge produces mere conviction (convictio), not insight (cognitio): perhaps for this reason it would be more correct to call this *elenchus* rather than *demonstratio*.[132]

At the beginning of the quoted paragraph, Schopenhauer tends less to Eberhard's (L3.1) than to Schultz' (K1.3) interpretation of Euclid, when he speaks of the axioms having to be oriented to intuition and not vice versa. But he also approaches Leibnizian arguments (L1) as the derivations from the axioms are demonstrated deductively. Finally, Schopenhauer again moves close to the arguments of the Kantians (K3.1), but without completely abandoning the logicism of the Leibnizians: In addition to pure logic, with the help of which one knows *that* a proof is valid, one must, however, consult intuition in order to learn *why* the proof is valid.[133] Such proofs are thus rather a transfer of intuition into logic (elenchus in the sense of Soph. el. 168a17ff.) than a *demonstratio* understood in the etymological sense.

---

[131] FR, p. 124 (§ 36). Vide supra, Chapter 2.3.6.
[132] FR, p. 128 (§ 39). [I follow here the original version from 1813.]
[133] At this point, one is tempted to continue working with distinctions such as *demonstratio quod/demonstratio propter quid* or *knowing that/knowing how*, among others; but since these concepts of tradition do not have a binding historical point of reference (especially in Schopenhauer's case) and therefore carry some unintentional connotations, I will not give in to these temptations any further.

2.3 Proof – Elementary Geometry, Syllogistic and Intuitive Proof Theory

Schopenhauer picks out two propositions from Euclid's *Elements* as evidence for his thesis that visual demonstration is preferable to purely discursive proofs. Both in the problematic example already mentioned above, which motivated the doubling of the twofold Leibnizian root of the principium rationis, and in the resumption of the argument in § 37, Schopenhauer had argued that angle and side had no intuitively recognizable semantic overlap. In § 40, Schopenhauer now first tries to show that the apagogical proof of prop. I.6 (for △ABG holds, if ∠ABG = ∠AGB, then also AB = AG) is "a conviction merely grounded in induction";[134] but already with a roughly drawn figure (Fig. 1) the validity becomes evident.

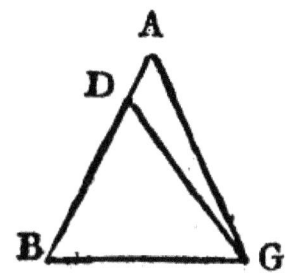

Fig. 1
FR, p. 129 (§ 39). [Figure taken from the 1813 edition.]

Furthermore, Schopenhauer discusses the vastness of the discursive proof of Elem. I.16 in contrast to intuition and sums up:

> Through all of this I have in no way proposed a new method of mathematical demonstration, no more than my proof will take the place of Euclid's; for by its whole nature [...] my proof is unsuitable as a new method; rather, I have only wanted to show what the ground of being is and how it is different from the ground of knowing, since the latter merely produces *convictio*, which is something completely different from insight into the ground of being. But the fact that in geometry one strives to produce only *convictio* (which, as was said, leaves a sense of dissatisfaction) but not insight into the ground of being (which like any insight, satisfies and delights) may be one reason, among others, why many otherwise brilliant minds have an aversion to mathematics.[135]

Even though these words conclude the paragraph on Euclidean geometry in Schopenhauer's dissertation, they are only the beginning of a comprehensive argument, which is continued and radicalized in § 15 of WWR I. There Schopenhauer demands a reduction of geometrical logic to intuition, whereas many Euclidean geometers would demand a reduction of geometrical intuition to logic. I in § 15 of WWR I, Schopenhauer takes up again prop. I.6 as an example and summarizes that Euclid did not give any insight into the essence of the triangle, but only a "laborious logical proof in accordance with the principle of non-contradiction [Satz des Widerspruchs]".[136] One

---

[134] FR, p. 129 (§ 39).
[135] FR, p. 131 (§ 39).
[136] WWR I, p. 95 (§ 15).

learns with the help of the principle of contradiction *that* the assertion made in prop. I.6 is the way it is, but not *why* it is the way it is.

> We have almost the same uncomfortable sensation people feel after a conjuring trick, and in fact most of Euclid's proofs are strikingly similar to tricks. The truth almost always emerges through a back door, the accidental result of some peripheral fact. An apagogic proof often closes every door in turn, leaving open only one, through which we are forced simply because it is the only way to go. As in Pythagoras' theorem, lines are often drawn without any indication of why: later they show themselves to be traps that spring unexpectedly to capture the assent of students, who must admit in astonishment what remains completely incomprehensible in its inner workings [...].[137]

Like Kant and Reinhold, Schopenhauer uses the conjurer analogy[138] to characterize the confusion between the logical and the demonstrative basis of proof (K1.2). Unlike Kant and Reinhold, however, Schopenhauer applies the conjurer analogy not to the metaphysicians (esp. Eberhard), but to Euclid himself: Mostly proofs are used in which Euclid infers from the impossibility of all alternatives to the facticity of the only remaining proposition, but without having tested its plausibility.[139] Thus one knows with the help of the *modus tollendo ponens* that the remaining proposition must be true because all alternatives are false; but one does not know why it is true, i.e. one knows nothing about the proposition itself.

The quote also shows that Schopenhauer, unlike many arguments of the Kantians, does not deny Euclid's logicism and relies on intuition; rather, Schopenhauer accepts the discursivity of Euclidean elementary geometry, but tries to point out its problems and limitations within proof theory. Above all, he attributes Euclid's logicism to a historical influence of Eleatic rationalism, which distinguished between what is observed (phainomena) and what is being thought (noumena) and disparaged the latter as deceptive and uncertain in contrast to the former.[140]

Only with Kant's transcendental aesthetics, it had been shown that the rationalists influencing geometry had unjustifiably used the argument of sensory illusion (L1.1.2) to justify rationalism, since the certainty of visual geometry was not learned from the empirical figure but was imagined a priori and produced by construction;[141] on the

---

[137] WWR I, p 95f. (§ 15).
[138] Vide supra, Chapter 2.3.1: "conjurers of metaphysics (Taschenspieler der Metaphysik)".
[139] In fact, this method of proof is found, for example, in the first book of the *Elements* in propositions 6, 7, 14, 19, 25, 27, 29, 39, 40. In my opinion, Schopenhauer seems to be thinking less of Arist. An. pr. I 6, 28b21f.; I 23, 41a23ff. but rather of an eliminative induction that is not verified by means of deduction (cf. on this Jens Lemanski: Summa und System).
[140] WWR I, p. 96.
[141] WWR I, p. 96ff.

2.3 Proof – Elementary Geometry, Syllogistic and Intuitive Proof Theory

other hand, "Euclid's logical way of treating mathematics is a useless precaution, a crutch for sound legs".[142]

As the last quote shows, in 1819, in WWR I, for the first time a unilateralisation of the original harmony argument of 1813 takes place since Schopenhauer's criticism of Eleatic-Euclidean rationalism also devalues Leibnizian logicism: Schopenhauer's criticism that logical reasoning is only a useless accessory seems to be the conversion of Eberhard's argument that intuitive figures are only imperfect auxiliary signs of unambiguous concepts and judgments (L3.1).

Against the criticism of the Leibnizians (L1.1.2, L2 and esp. L3) that the sensualism and representationalism presupposed by the geometrical figures are uncertain, Schopenhauer takes a direct stand. With recourse to Kant, he tries to show that his veto against purely discursive reasoning does not necessarily commit him to imperfect empirical intuitions, since geometric forms do not spring from experience or have to be mediated by empirical drawings. Rather, they are a priori forms of the mind itself. Therefore, while geometric figures are representationalist, they do not necessarily represent empiricism:

> Only now [sc. in post-Kantian philosophy] can we claim with certainty that what presents itself as necessary in the intuition of a figure does not come from the figure on paper (which could be very badly drawn), or from the abstract concept that we think as a result, but instead directly from the form of all cognition, something we are conscious of a priori. In every case, this form is the principle of sufficient reason; in this case, it is, as the form of intuition (i.e. space), the principle of the ground of being; but it is just as directly evident and just as directly valid as the principle of the cognitive ground, i.e. logical certainty.[143]

What is striking about this quote is that Schopenhauer agrees with the Leibnizians that drawn figures are defective (L3). However, in the sense of Kant's transcendental aesthetics, he invalidates the argument that geometric figures have to be drawn real at all. Rather, the geometric figure is an expression of the a priori conception of space and therefore initially only a purely internalized figure.

Similar to the dissertation, Schopenhauer also emphasises in the given quote that the ground of knowing in geometry only leads to the *knowing that*, not to the *knowing why*. But also here Schopenhauer tightens this aspect in contrast to his views around 1813: Knowledge of pure facticity is finally not scientific knowledge, he adds.[144] As an example of this difference, Schopenhauer points to the fact that knowledge of the

---

[142] WWR I, p. 97.
[143] WWR I, p. 97f.
[144] On Schopenhauer's definition of knowledge and scientific knowledge cf. Jens Lemanski: Wissen, Wissenschaft, Wissenschaftslehre. In: Philosophie als Wissenschaft. Ed. by Nora Schleich et al. Hildesheim 2021, pp. 113–133.

facticity of the column of mercury in the Torricellian tube is only an insufficient opinion if one cannot add the justification that the height of the column is determined by the air pressure and not by the horror vacui – an example that Kant used several times for a justified opinion verified by the illustrative experiment.[145]

Schopenhauer probably used the horror-vacui example to show what consequences a purely factual knowledge can have: Factual knowledge does not reduce the introduction of false explanatory hypotheses (e.g. in fluid mechanics), which in the worst case are then applied in other disciplines and hinder progress there (e.g. pump technology). And also with Euclid, one finds many times geometrical proofs, which aimed only at factual knowledge but are applied in many further disciplines (optics, astronomy, mechanics etc.). Schopenhauer sees one of several examples of purely factual knowledge of elementary geometry in Euclid's Elem. prop. I.47 – an example which was also discussed by Kästner before:[146]

> Similarly, Pythagoras' theorem tells us about a *qualitas occulta* of the right-angled triangle: Euclid's stilted, indeed underhand, proof leaves us without an explanation of why, while the following simple and well-known figure yields more insight into the matter in one glance than that proof, and also gives us a strong inner conviction of the necessity of this property and of its dependence on the right angle:
>
>
>
> even when the sides at the right angle are unequal we must still be able to achieve this intuitive conviction, as we can generally with every possible geometrical truth, because the discovery of a geometrical truth always starts out from an intuition of the necessity, and the proof is only thought up later. So all that is required to intuitively recognize the necessity of a geometrical truth is an analysis of the thought process that first led to its discovery.[147]

Schopenhauer does not argue here against Kästner, who built the justification of the Pythagorean theorem from the proof of Elem. prop. I.46. Rather, Schopenhauer states

---

[145] Cf. Jens Lemanski: Galilei, Torricelli, Stahl.
[146] Vide supra, Chapter 2.3.1.
[147] WWR I, p. 108.

2.3 Proof – Elementary Geometry, Syllogistic and Intuitive Proof Theory

that the proof of the validity of the theorem is provided by Euclid, but stops at facticity. Why what Euclid has produced is called "stilted [stelzbeinig]" can be interpreted by the parable I cited earlier as a one-sidedness of the argument from harmony: Logical proofs in geometry are an unnecessary crutch for Schopenhauer around 1819.

Thus, he argues against such theories of proof: Intuition without further explanation is sufficient to understand, why in the right triangle the square over the side opposite to the right angle is equal to the squares over the sides together, which enclose it. The figure or diagram is thus the justification itself. Schopenhauer understood his figure put forward in the quote above to be a purely intuitive proof that would allow "insight [Einblick]" into the nature of the triangle. The reason is thus given by the suchness of the intuition, and no purely logical deduction can correspond to this justification.

That Schopenhauer was not alone with a few Kantians like Reinhold or Schultz in his demand for clarity (K3), he tried to prove in 1819 by referring to the writings of Bernhard Friedrich Thibaut and Ferdinand Schweins.[148] Thibaut, a student of Kästner, had declared that the general postulate of geometry was the "original representation (intuition) of space" [ursprüngliche Vorstellung (Anschauung) des Raums].[149] From the points, the line develops, from the lines the plane and from the plane finally the physical space. This compositionality results in elementary geometry, which is the basis for all further geometric operations so that geometry "always creates its object itself using the faculty of imagination [Einbildungskraft], whereby it only permits a sensual intuition of it as an auxiliary means".[150]

While Thibaut discreetly composed the points to lines and the lines to planes and solids from intuition, the Heidelberg mathematician and philosopher Schweins proceeded polemically against Euclid and the rationalism of his successors. Schweins declared that he had found his "system of geometry" by only one 'axiom', namely by the power of sight (Sehkraft) or by the postulate that "objects should not be applied differently than they are presented".[151] Furthermore, Schweins replicated the reviewer of his previous works, especially in the preface of his *Mathematik für den ersten wissenschaftlichen Unterricht*. From these critical remarks, it can be seen that the dispute described above in Chapter 2.3.1 had continued into the early 19th century, albeit with diminished reference to Kant and Leibniz.

In the first edition of WWR I, Schopenhauer placed himself in the school tradition of Thibaut and Schweins, although he admitted that with Thibaut he still wanted "a

---

[148] In the second edition of the WWR, the paragraph on Thibaut and Schweins is deleted without replacement; in the third edition, there is a reference to Kosack (see below, Chapter 2.2.3). On Thibaut and Schweins cf. Moritz Cantor: Ferdinand Schweins und Otto Hesse. In: Heidelberger Professoren aus dem 19. Jahrhundert 2 (1903), pp. 221–242.
[149] Bernhard Friedrich Thibaut: Grundriß der reinen Mathematik zum Gebrauch bey academischen Vorlesungen. Göttingen 1809, p. 161. (My transl. – J. L.)
[150] B. F. Thibaut: Grundriß der reinen Mathematik, p. 164.
[151] Ferdinand Schweins: Mathematik für den ersten wissenschaftlichen Unterricht systematisch entworfen. 2 Vols. Darmstadt, Gießen 1810, Vol. 1, p. 11 (Vorrede).

much more decisive and thorough substitution of the evidentness of intuition in place of logical proof" and that Schweins' method was not consistent enough.[152] The classification in this school tradition is not surprising: In Schopenhauer's time, Schwein's *Mathematik* was a common textbook at high schools,[153] and Schopenhauer had studied mathematics with Thibaut in Göttingen in 1809/10.[154]

Although these remarks from the first edition of WWR I can be seen as an indication of a school tradition, Schopenhauer deleted them in the second edition and the third edition stated that he had been the founder of a school of geometry based on intuition.[155] How much Schopenhauer's philosophy of geometry changed in the course of the years, however, is not only shown by the question of school affiliation or in the comparison between the harmony argument of *On the Fourfold Root of the Principle of Sufficient Reason* (1813) and the rather anti-rationalistic tendency of WWR I (1819), but also by the changed tone that became noticeable in the later years – as was, unfortunately, the case everywhere in his works.

What in the first edition of WWR I was still called a "conjuring trick [Taschenspielerstreich]", receives in the later editions of *On the Fourfold Root of the Principle of Sufficient Reason* from 1847 a name relevant for the history of mathematics. In the revised dissertation thesis of 1847 Schopenhauer bundles all intuitive proofs from his works and explains based on the just mentioned diagram that this and its "the mere appearance of which without further discussion provides twenty times the conviction of the truth of the Pythagorean theorem than Euclid's mousetrap proof [der Euklidische Mausefallenbeweis]".[156] The original argument of 1813, which was supposed to harmonize the quarrelling Leibnizians and Kantians, turned into an anti-rationalistic one-sidedness in 1819, when Schopenhauer declassified Euclid's stilted conjuring trick as unnecessary assistance; he even called this conjuring trick 'mousetrap proof' (Mausefallenbeweis) in 1847, since one is beaten and trapped in facticity without having obtained the actually desired reason.

Chapter 13 of WWR II (1844) shows an even stricter tone. There, Schopenhauer dogmatically deals with a whole series of arguments and positions on geometry on a few pages, which are only summarized here in bullet points:[157] Euclidean geometry had become a parody of itself since every year mathematicians tried to prove the parallel postulate logically, although it was intuitively perfectly clear; amusing, therefore, was the useless logical proof in contrast to the intuitive one; one had rather to attack the eighth Euclidean axiom because coinciding ('Sichdecken') was either a mere tautology or had to be understood empirically-materially since it presupposed mobility;

---

[152] WWR I, p. 571.
[153] In addition, Schopenhauer had indirect contact with Schweins, who supported Schopenhauer's attempt to o apply for a position at the University of Heidelberg. (Cf. Ernst Anton Lewald an Schopenhauer, Nov. 4, 1819. In: SW, Vol. XIV, p. 263. (= L. 240 [142]))
[154] Cf. SW, Vol. VI, p. 631 (Biography, December 31, 1819).
[155] Vide infra, Chapter 2.3.3.
[156] FR, p. 131f. (§ 39).
[157] Cf. WWR II (1844), p. 139–142.

2.3 Proof – Elementary Geometry, Syllogistic and Intuitive Proof Theory

The Platonic doctrine of ideas is best shown in the relation between the a priori forms of geometry and their representation in empiricism; as William Hamilton also emphasised, mathematics has no further use except this purely indirect one and is therefore not a suitable means of education.[158]

2.3.3 Reception and Evaluation of Schopenhauer's Philosophy of Geometry

I analysed Schopenhauer's philosophy of geometry more closely in the previous chapter. Just as in logic, a revised doctrine emerged in Schopenhauer's writings in this area, which became ever sharper, especially in the way it was expressed, and ever more radical in its views. Whereas in the first edition of his dissertation Schopenhauer had still attempted a harmony argument in order to reconcile the scope of logically oriented Leibnizianism and intuition-oriented Kantianism, Schopenhauer radicalised further and further in the course of his creative period: if one reads individual passages of Schopenhauer's later published philosophy of mathematics independently of their historical and systematic context, they appear as a radicalised position of the Kantian philosophy of geometry focused on intuition.

To this day, Schopenhauer's proof of the Pythagorean theorem in particular, together with the expression 'mousetrap proof', appears at regular intervals in mathematics textbooks and in special treatises of the mathematical field. Nevertheless, a systematic assessment of Schopenhauer's philosophy of mathematics and, in general, his proof theory in geometry focused on intuition remains a delicate undertaking, not only for philosophers but also for mathematicians. One can see this especially in those special treatises which have dealt more intensively with Schopenhauer's philosophy of geometry in the last two hundred years.

In the following, I will present this reception history chronologically based on those writings from the German, English, French and Italian language areas which have either dealt with Schopenhauer's philosophy of geometry quantitatively over several pages or which have put forward such a qualitatively relevant opinion that it has played an important role in the reception history itself. At the end of this chapter, the papers and books discussed are systematically compiled.

This presentation will show that the evaluations of Schopenhauer's geometry in the history of reception follow the business cycle of visual geometry and geometric logic alluded to in the introduction to Chapter 2.3: While the assessments in the years between 1820 and 1880 are quite positive, from the 1880s onwards an almost abruptly negative and disparaging assessment sets in, driven especially by Weierstrass' students and followers. It was not until around 1950 that the taboo on Schopenhauer's

---

[158] Cf. Marco Segala: Schopenhauer and the Mathematical Intuition as the Foundation of Geometry. In: Language, Logic, and Mathematics in Schopenhauer. Ed. by Jens Lemanski. Cham 2020, pp. 261–285.

2 Logic and its Geometrical Interpretation

philosophy of geometry was slowly put into perspective, paving the way for a renewed, positive interest in Schopenhauer's philosophy of geometry from the 1990s onwards.

Probably the first serious reception of Schopenhauer's ideas on geometry appeared in 1822 in Adolph Diesterweg's mathematics education guide *Leitfaden für den ersten Unterricht in der Formen-Größen- und räumlichen Verbindungslehre*. Diesterweg saw geometry as a "more intensive means of education", since "it connects intuition and concept with each other".[159] On the one hand, the propositions of geometry received their "certainty, infallibility and evidence from the intuition", but on the other hand, they were not to be practised on the empirical figure but on the products of one's own imagination.[160] Many geometers had, therefore "regarded the addition of many figures as useless ballast", whereas Diesterweg wanted to encourage them to find themselves by intuition.[161] For these reasons, however, the original reading of Euclid is less suitable as an introduction to elementary geometry.

As a universal means of education, according to Diesterweg, geometry also imparts basic instruction in logic at schools and universities. Since Plato, there have been numerous authors who have taught logic employing concrete geometry or who have traced geometry back to logical rules. Diesterweg gives a not uninteresting list of texts by geometricians and philosophers, some of whom I have already mentioned in the introduction to Chapter 2.3, but many of whom published in the early 19th century: Christian Wolff, Iakob Harris, Maaß, Johann Andreas Christian Michelsen, Moritz Adolph von Winterfeld, Friedrich Johann Christian Schmeißer, Johann Gottlob Erdmann Föhlisch, Christian Heinrich Haenle and Johann Joseph Dilschneider.

According to Diesterweg, Schopenhauer is currently the only philosopher who does not share this logicism, since in Schopenhauer's opinion logic is understood intuitively and is therefore unnecessary – a typical misunderstanding, as shown in Chapter 1.3, which is based solely on the context-free interpretation of a few sentences of the WWR I. According to Diesterweg, Schweins and Johann Friedrich Schaffer also deviate from the mainstream of the logic-oriented geometers, since the former emphasises the genesis, the latter the product of the figurative intuition. Schopenhauer is also an exception:

> It would be good to settle the mathematical-philosophical dispute that Wagner and Schopenhauer have recently stirred up against the Euclidean mathematicians. In the theory of space, the aforementioned scholars demand intuitive perception, intuitive cognition, frowning upon discursive cognition based on the rules of logic. In

---

[159] F.A.W. Diesterweg: Leitfaden für den ersten Unterricht in der Formen-Größen- und räumlichen Verbindungslehre oder Vorübungen zur Geometrie für Schulen. Elberfeld 1822, p. 2. (My transl. – J. L.)
[160] F.A.W. Diesterweg: Leitfaden für den ersten Unterricht, pp. 4f.
[161] F.A.W. Diesterweg: Leitfaden für den ersten Unterricht, p. 6.

2.3 Proof – Elementary Geometry, Syllogistic and Intuitive Proof Theory

>the first part of this demand, Wagner and Schopenhauer are absolutely right, and those mathematicians who deny it do not understand what they are saying.[162]

Here, as in other places, Diesterweg refers to Schopenhauer's WWR I as well as to Johann Jakob Wagner's *Mathematische Philosophie*. It can be quickly demonstrated that Wagner and Schopenhauer set a high value on intuitive cognition. However, it is difficult to decide to what extent Diesterweg is right in his second partial assertion that Schopenhauer and Wagner also frown upon discursive knowledge. It is probably fair to say that he does not ascribe to Schopenhauer the mediating position of the early dissertation thesis but identifies him with the revised, rather anti-rationalist teaching of the WWR.[163] However, I do not wish to comment here on whether Diesterweg has correctly captured Wagner's complex oeuvre in these few words; one can at least raise doubts about Diesterweg's powers of judgement when, on the one hand, he praises Schopenhauer and Wagner but, on the other, strongly rebukes Friedrich Buchwald in a similar context because of his mathematical ignorance – 'Friedrich Buchwald' was, after all, one of Wagner's pseudonyms.[164]

Diesterweg himself finally tried to formulate an argument based on Schopenhauer, which was intended to mediate between the visual geometers (Schopenhauer, Wagner, Schweins, etc.), on the one hand, and the logical geometers (Wolff, Maaß, Dilschneider, etc.), on the other. "Mathematical knowledge is based on intuition etc. and concepts at the same time; mathematics as a science is created through the combined effect of the faculties of intuition and reflection, and one learns not only that something *is* the case, but also *why* it is the case."[165]

A good 20 years after Diesterweg's treatise, Schopenhauer's geometry experienced its most intensive epoch of positive reception and continuation. In 1847, the social pedagogue and linguist Karl Mager published *Die Encyklopädie, oder das System des Wissens*, in which Schopenhauer's views on Euclid were taken up and presented.[166] In April 1852, Karl Rudolf Kosack published the programmatic *Beiträge zu einer systematischen Entwickelung der Geometrie aus der Anschauung*, which explicitly contained a plane geometry according to Schopenhauerian principles and was submitted for public examination. Kosack explained that Euclid's geometry was not a natural but a highly artificial product which, given the arbitrariness of the proofs and the incoherence of the individual propositions with the axioms, only aimed at persuasion

---

[162] F.A.W. Diesterweg: Leitfaden für den ersten Unterricht, p. 8.
[163] Siehe oben, Kap. 2.3.1.
[164] Cf. F.A.W. Diesterweg: Leitfaden für den ersten Unterricht, p. 15, p. 51, Fn.
[165] F.A.W. Diesterweg: Leitfaden für den ersten Unterricht, p. 9.
[166] Cf. Karl Wilhelm E. Mager: Die Encyklopädie, oder das System des Wissens. Teil II. Zürich 1847, pp. 9–14. Cf. Ulrich von Beckerath: Eine Anerkennung der mathematischen Ansichten Schopenhauers aus dem Jahr 1847. In: Schopenhauer Jahrbuch 24 (1937), pp. 158–161.

rather than conviction.¹⁶⁷ For this reason, it was necessary to trace all geometric propositions back to the intuition of the productive faculty of imagination, as Schopenhauer had taught following Kant.¹⁶⁸

Kosack's programmatic writing is not pure Schopenhauerianism, but the absorption of Schopenhauerian proof theory motivated by a Kantianism in geometry that re-emerged in the 1850s.¹⁶⁹ For example, Kosack points to a programmatic paper by Friedrich Schmeißer in which the rationalist elements of Euclidean geometry were presented as a continuous corrupted text passage (Korruptele) and many logician approaches since Petrus Ramus were rejected.¹⁷⁰ Leopold Karl Schultz von Straßnitzki, Karl Snell, Karl Christian Friedrich Krause and others had also recognised this and had already designed good intuition-based geometries. Schopenhauer, however, according to Kosack, irrefutably eliminated logicism and completed the Kantian programme of geometry.¹⁷¹

Although Kosack's work contained only a first part on planimetry and thus remained unfinished,¹⁷² Schopenhauer and his circle of students or followers, who had only formed in these years, reacted almost exclusively emphatically to this programme paper.¹⁷³ Schopenhauer, who had already deleted the references to Thibaut and Schweins in the second edition of WWR I, included the reference to Kosack's elaboration of his approach in the third edition of the main work.¹⁷⁴ Julius Frauenstädt wrote in 1852 in the *Blätter für literarische Unterhaltung* that the pupils of Nordhausen were the first to learn geometry again without crutches and to be able to grasp the Pythagorean theorem as the ancient Greeks once did.¹⁷⁵ The later Frankfurt and Darmstadt mathematics professor Johann Carl Becker was similarly positive about Kosack's and Schopenhauer's approach in 1857.¹⁷⁶ Only Julius Bahnsen was not convinced by Kosack and argued against it that Schopenhauer actually disparaged mathematical knowledge as a whole since one finds in Chapter 13 of WWR II a critique of the usefulness of mathematical knowledge based primarily on William Hamilton's *On the*

---

[167] Cf. C. R. Kosack: Beiträge zu einer systematischen Entwickelung der Geometrie aus der Anschauung. In: Zu der öffentlichen Prüfung sämmtlicher Klassen des Gymnasiums zu Nordhausen [...]. Nordhausen 1852, pp. 1–31, here: p. 3.
[168] Cf. C. R. Kosack: Beiträge, pp. 5ff.
[169] Cf. C. R. Kosack: Beiträge, p. 5.
[170] Cf. Friedrich Schmeißer: Kritische Betrachtung einiger Grundlehren der Geometrie, wie sie meistens in Lehrbüchern vorkommen. Frankfurt/ Oder 1851.
[171] Cf. C. R. Kosack: Beiträge, p. 6.
[172] Cf. C. R. Kosack: Beiträge, pp. 11–31.
[173] On the Schopenhauer school cf. Fabio Ciraci, Domenico Fazio, Matthias Koßler (eds.): Schopenhauer und die Schopenhauer-Schule. Würzburg 2009.
[174] Cf. WWR I (1859), pp. 87 (= § 15).
[175] Cf. Julius Frauenstädt: Eine beachtenswerthe Erscheinung in der Mathematik. In: Blätter für literarische Unterhaltung 1852, No. 35 (28th August 1852), p. 836.
[176] Cf. J. C. Becker: Über Begründung und systematische Entwickelung der geometrischen Wahrheiten. In: Schulzeitung für die Herzogtümer Schleswig-Holstein und Lauenburg No. 14, 15, 2nd and 9th January 1853.

Study of Mathematics (german transl. *Ueber den Werth und Unwerth der Mathematik*) of 1836.[177]

The ensuing debate between Bahnsen, Kosack and many others was summarised years later in a chapter of Becker's *Abhandlungen aus dem Grenzgebiete der Mathematik und Philosophie*.[178] According to Becker, Schopenhauer and furthermore Herbart continued the Kantian ideas on geometry, which were also advocated by Schweins, Snell, Oskar Schlömilch and the Trendelenburgian Bernhard Becker. In contrast to Diesterweg, Becker believes that Schopenhauer accepted verbal proofs, but saw them as dependent on intuitive proofs. Schopenhauer demonstrated this with little skill using concrete examples, but it was Kosack who first successfully elaborated this programme.[179] He further developed it, published it and has been teaching it ever since.[180]

After 1877, Benno Erdmann's ambivalent judgement of Schopenhauer in his work *Die Axiome der Geometrie* received much attention: Erdmann, who apparently only knew Chapter 13 of WWR II, criticised Schopenhauer for having enjoyed a "bizarre" and "artificially one-sided education of Kantian theory".[181] At that time, Schopenhauer had stood in an intuition-oriented tradition of geometry with Carl Friedrich Gauss, Herbart, etc.; but even his later fame had not been able to prevent his addiction to wanting to justify everything by intuition. His objections to the eleventh and eighth axioms of Euclidean geometry, however, were evidence of "astuteness" and are still relevant today.[182]

Interest in the topicality of Schopenhauer's philosophy of geometry was particularly evident in the mid-1880s. At the beginning of October 1884, the mathematical section of the German philologists and schoolmen met in Dessau. In his lecture on the pedagogical school reforms in the subject of geometry, the Eisleben grammar school director and Leibniz researcher Carl Immanuel Gerhardt used the "not very elegant name mousetrap proofs'", but did not know from whom this name originated.[183] Rudolf von Fischer-Benzon, schoolmaster in Kiel, subsequently reported that the expression came from Schopenhauer and was related to the interscton method (Deckungsmethode). In a note to this conference report in issue 1 of the *Zeitschrift für*

---

[177] Cf. Julius Bahnsen: Der Bildungswerth der Mathematik. In: Schulzeitung für die Herzogtümer Schleswig-Holstein und Lauenburg, No. 21, 25, 26, 21th Feb., 21th and 28th March 1857. On Schopenhauer and Hamilton cf. Marco Segala: Schopenhauer and the Mathematical Intuition as the Foundation of Geometry.
[178] Cf. J.C. Becker: Abhandlungen aus dem Grenzgebiete der Mathematik und Philosophie. Zürich 1870, Chapter IV.
[179] J.C. Becker: Abhandlungen, p. 49.
[180] Cf. J.C. Becker: Abhandlungen, p. 51.
[181] Cf. Benno Erdmann: Die Axiome der Geometrie. Eine philosophische Untersuchung der Riemann-Helmholtz'schen Raumtheorie. Leipzig 1877, p. 29, p. 172.
[182] B. Erdmann: Die Axiome der Geometrie, pp. 65f.
[183] Friedrich Buchbinder: Verhandlung der Sektionen für mathematischen und naturwissenschaftlichen Unterricht auf der diesjährigen Versammlung deutscher Philologen und Schulmänner. Vom 1.–4. Oktober 1884 in Dessau. In: Zeitschrift für mathematischen und naturwissenschaftlichen Unterricht 16:1 (1885), pp. 66–76, here: p. 69.

2 Logic and its Geometrical Interpretation

mathematischen und naturwissenschaftlichen Unterricht, the editor, Volkmar Hoffmann, commented that Johann Carl Becker, who was very well-known at the time, had also used the "mocking name 'mousetrap proof'", that this went back to Schopenhauer and was not merely applied to the 'intersection [Deckung]' of geometric shapes. In the future, the topic would be dealt with in more detail.[184]

The report on the Dessau conference is remarkable in two respects: on the one hand, it shows that the Schopenhauerian expression 'mousetrap proof' was a well-known catchword in the 1880s, often used without reference to its originator; on the other hand, Gerhardt's remark, which was probably rather incidental, ignited an intensive treatment of the topic with several participants in issues 3 and 4 of the journal. Hoffmann first explains that he did not find the expression in question during his perusal of the WWR, but that he had discovered several passages (on the Pythagorean theorem, on the stilted proof, etc.) that would correspond factually to the "notorious bon mot".[185] Hoffmann also emphasises that Schopenhauer's reading would have been an excellent support for him if he had known it beforehand when he was writing his own papers.

Hoffmann adds that after his present article went to press, he received three more submissions in which the expression 'mousetrap proof' has been proven. Particularly noteworthy are the remarks of Carl Gusserow, who explains that Schopenhauer, with regard to the Pythagorean theorem, gives a "generally valid kind of proof from intuition, of the kind that the hypotenuse square is broken down into parts from which the cathetus squares can be composed";[186] and that he also demonstrated the general validity of this proof in his stereometry.[187] Hoffmann notes that this proof is also found in Carl Ludwig Albrecht Kunze, in Schlömilch "and in many other textbooks" – but all the proofs mentioned by Hoffmann were published at a later time than those of Schopenhauer.[188]

It was not until the paper by Hermann Märtens, a secondary school teacher in Naumburg, that the positive reception of Schopenhauer's geometry can be considered to have come to an end. According to Märtens, Schopenhauer's proof of the Pythagorean theorem was limited to the isosceles right triangle:

> For this case, however, the figure is quite simple, but without all the decomposition one does not arrive at the truth of the theorem even

---

[184] F. Buchbinder: Verhandlung der Sektionen, p. 69.
[185] Volkmar Hoffmann: Schopenhauer, der Philosoph, über die Euklidische Methode und die 'Mausefallenbeweise'. In: Zeitschrift für mathematischen und naturwissenschaftlichen Unterricht 16:3 (1885), pp. 105–107, here: p. 105.
[186] V. Hoffmann: Schopenhauer, p. 107 (Quote from the letter to the editor).
[187] Cf. Volkmar Hoffmann: Hebung eines Missverständnisses. In: Zeitschrift für mathematischen und naturwissenschaftlichen Unterricht 16:4 (1885), pp. 263–264 (Quote from the letter to the editor). Cf. Carl Gusserow: Leitfaden für den Unterricht in der Stereometrie mit den Elementen der Projektionslehre. Berlin 1885, pp. 94–96.
[188] Volkmar Hoffmann: Schopenhauer, der Philosoph, über die Euklidische Methode und die 'Mausefallenbeweise', p. 107.

2.3 Proof – Elementary Geometry, Syllogistic and Intuitive Proof Theory

in this case. One has to divide the square on the hypotenuse [Hypotenusenquadrate] into 4 triangles, the squares of the two other sides [Cathetenquadrate] into 2 triangles each and prove that the division of the square on the hypotenuse is equal to those of the squares of the two other sides.[189]

Schopenhauer had thus only proved a special case of the theorem, but not the entire proof. Moreover, Cantor et al. had shown, Hoffmann adds, that a similar but complete proof had already existed in the 12th century by the Indian mathematician Bhaskara II.

In addition to Märtens' treatise, the issue also contains several discussions of Schopenhauer, which, however, do not offer any real progress in knowledge. All papers, however, express either euphoric support, sceptical rejection or caution towards Schopenhauer's theses: Whereas Hoffmann, for example, gives a brief account of Schopenhauer's thesis, already advocated by Bahnsen, that a calculus does not imply understanding,[190] Friedrich Pietzker provides a detailed review of Kosack's book with special attention to the Schopenhauer allusions it contains.[191] Märtens provides the most negative reception of Schopenhauer of all the contributions.

In 1891, Heinrich Leonhard, a student of Carl Weierstrass, submitted a dissertation that dealt explicitly, albeit strongly negatively, with Schopenhauer and the Euclidean method of proof. Leonhard opened his dissertation with the thesis that Schopenhauer had made an attack on Euclidean elementary geometry of "such serious significance that, if its justification had to be conceded, it would have to be called a destructive one".[192] Schopenhauer's attack is not the same as the objections of analytical or projective geometry, since these could also be based on elementary geometry, among other things.

Leonhard reports that Kosack and Becker in particular accepted and elaborated Schopenhauer's geometrical approach and that these theories became known in the German-speaking world through Hoffmann's journal. Leonhard's central theme is the principle of reason: after explaining that Schopenhauer was an independent thinker who went far beyond Kant, he reviews Schopenhauer's relevant passages on the principle of reason from the second edition of the dissertation without regard to the historical problem context of Schopenhauer's early writings. He concludes that Schopenhauer has unjustifiably differentiated the principle of reason and that there is in

---

[189] Hermann Märtens: Schopenhauer über den 'Mausefallenbeweis'. In: Zeitschrift für mathematischen und naturwissenschaftlichen Unterricht 16:4 (1885), p. 181–186, here: p. 183.
[190] Cf. Volkmar Hoffmann: Schopenhauer, der Philosoph, über den Wert des Calcüls. In: Zeitschrift für mathematischen und naturwissenschaftlichen Unterricht 16:4 (1885), p. 186.
[191] Cf. Friedrich Pietzker: Ein Jünger Schopenhauers in der Geometrie. In: Zeitschrift für mathematischen und naturwissenschaftlichen Unterricht 16:4 (1885), pp. 187–190.
[192] Heinrich Leonhard: Beitrag zur Kritik der Schopenhauer'schen Erkenntnistheorie, insbesondere in ihrer Anwendung auf das Euklidsche Beweisverfahren. Bonn 1891, p. 1.

2 Logic and its Geometrical Interpretation

itself only the logical ground of knowing.[193] Schopenhauer is thus subject to a fundamental error with regard to his geometrical explanations.

Leonhard puts forward numerous theses against Schopenhauer and tries to expose contradictions and inconsistencies in Schopenhauer's philosophy. The central arguments concern, on the one hand, the proof theory itself and, on the other, the separation of logic and intuition. According to Leonhard, 'to prove' ultimately means no more than saying that a proposition is universally valid, and this must, above all, be verified logically-deductively.[194] Proofs are usually made by means of the theorem of contradiction. Schopenhauer, on the other hand, argues almost exclusively with psychologistic arguments that suggest subjective 'feelings' of truth and general validity.[195]

The separation of logic and intuition is also misleading, as can be seen in the geometrical examples of Schopenhauer, Kosack and Becker: all visual proofs are "as a result of frequent practice and because of the simplicity and transparency present in this case, an almost unconscious reduction of the theorem to the (conceptual) definition".[196] The intuitive method thus always presupposes the logical-discursive one; Schopenhauer and his followers, however, try to mask this by "obscurity of expression and incompleteness of elaboration".[197]

Leonhard's attack on Schopenhauer's visual reasoning was long considered convincing. Schlüter accepted Leonhard's arguments in *Schopenhauers Philosophie in Briefen* in 1900 and furthermore quoted a pessimistic letter from the late Schopenhauer to Becker in which the latter had classified his own and Becker's geometrical proofs as problematic and by no means universally valid.[198]

In 1904, the Weierstrass supporter Alfred Pringsheim repeated the essential arguments of Märtens and Leonhard in a paper, but without mentioning either by name and reported based on relevant text passages that Schopenhauer, like William Hamilton, was an enemy of mathematics.[199] Pringsheim's tone towards Schopenhauer fluctuates between amused, enraged and generally disdainful.

Nevertheless – or perhaps precisely because of this – Pringsheim's paper can be regarded as a milestone in the reception history of Schopenhauerian geometry around 1900. In 1907, Kewe did not raise any objection to Leonhard or Pringsheim, but only

---

[193] Cf. H. Leonhard: Beitrag, p. 45, p. 50.
[194] Cf. H. Leonhard: Beitrag, pp. 51ff.
[195] On the role of feeling cf. Laura Follesa: From Necessary Truths to Feelings: The Foundations of Mathematics in Leibniz and Schopenhauer. In: Language, Logic, and Mathematics in Schopenhauer. Ed. by Jens Lemanski. Cham 2020, pp. 315–326.
[196] H. Leonhard: Beitrag, p. 64.
[197] H. Leonhard: Beitrag, p. 65.
[198] Cf. Schlüter: Schopenhauers Philosophie in seinen Briefen, pp. 13ff. Cf. also Jason M. Costanzo: Schopenhauer on Intuition and Proof in Mathematics. In: Language, Logic, and Mathematics in Schopenhauer. Ed. by Jens Lemanski. Cham 2020, pp. 287–305, esp. p. 299.
[199] Cf. Alfred Pringsheim: Über Wert und angeblichen Unwert der Mathematik. In: Jahresbericht der Deutschen Mathematiker-Vereinigung 13 (1904), pp. 357–382. The title of Pringsheim's paper is an allusion to a translation of a Hamilton text that Schopenhauer quotes in WWR II (1844), p. 141.

2.3 Proof – Elementary Geometry, Syllogistic and Intuitive Proof Theory

pleaded for Schopenhauer's theses to be restricted to the didactics of elementary geometry.[200] In the transcription of Felix Klein's geometry lectures, one finds, with reference to Pringsheim's paper, a characterisation of Schopenhauer as an artistic speculator who wanted to invent his own intuitive theory of proof because he did not understand mathematics.[201] Although Klein refers to Schopenhauer's examples on several pages, he does not come to a positive judgement, for several reasons already mentioned by Märtens and Leonhard. Because of his reference to Hamilton, Pringsheim's paper and Klein's presentation of Schopenhauerian geometry were strongly received, especially in the English-speaking world.[202]

It was not until 1909 that Oscar Janzen presented a more moderate paper on Schopenhauer's logic and mathematics in the *Archiv für Geschichte der Philosophie*. It is difficult to characterise or summarise Janzen's writing: on the one hand, it features a large number of original ideas and theses, but on the other hand, they lack a systematic core thesis. Moreover, it is often not clear in the paper whether Janzen is referring to his own opinion, an opinion of another, or a critique of Schopenhauer. However, it can be stated that Janzen vacillates between admiration and rejection of Schopenhauer's theses. In doing so, he repeatedly makes suggestions for improvement but is just as eager to criticise.

It should be emphasised that Janzen, in contrast to Leonhard and Pringsheim, meets the demand I made with Lovejoy in Chapter 1.1.4 to contrast Schopenhauer's writings separately and in their development: Janzen comes to the conclusion that Schopenhauer's early writings have the intention of improving Euclid's theory of proof, while the later writings are directed against Euclid and Schopenhauer has the intention of founding his own proof theory.[203]

Two central themes of Janzen's paper concern the semantics and the intuition of geometric propositions: He first argues that Schopenhauer, in his dissertation thesis, must – without explicitly saying so – take the basic elementary geometrical concepts (point, straight line, plane, space) from intuition if he acknowledges Euclid's axioms and postulates.[204] Janzen then tries to improve Schopenhauer's – as he thinks – purely representationalist semantics in order to reject the ground of being,[205] since intuition only provides meaning, but not justification.[206]

In the first half of the 20th century, a positive, albeit largely rather tacit, reception of Schopenhauer can be found among the intuitionists, especially Luitzen Egbertus Jan

---

[200] Cf. Adolf Kewe: Schopenhauer als Logiker, pp. 73ff.
[201] Cf. Felix Klein: Elementarmathematik vom höheren Standpunkte aus. Vol. II: Geometrie. Ausgearbeitet v. E. Hellinger. Berlin 1909, pp. 257.
[202] Cf. Florian Cajori: A Review of Three Famous Attacks upon the Study of Mathematics as a Training of the Mind. In: Popular Science 80:22 (1912), pp. 360–372. The writing does not bring anything essentially new compared to Pringsheim and Klein.
[203] Cf. Oscar Janzen: Schopenhauers Auffassung des Verhältnisses der mathematischen Begründung zur logischen. In: Archiv für Geschichte der Philosophie 22 (1909), pp. 342–364, here: p. 348.
[204] Cf. Oscar Janzen: Schopenhauers Auffassung, p. 351.
[205] Cf. Oscar Janzen: Schopenhauers Auffassung, p. 355.
[206] Cf. Oscar Janzen: Schopenhauers Auffassung, p. 363.

2 Logic and its Geometrical Interpretation

Brouwer. The intuitionist reception history of Schopenhauer can be described as 'tacit' in that Schopenhauer is not publicly mentioned by Brouwer as a precursor of intuitionism until the late 1920s.[207] Nevertheless, already in Brouwer's early work, which forms the basis for intuitionism in mathematics and logic, theses can be found – such as the criticism of the universal validity of logical principles (especially the tertium non datur) as well as the demand for an intuition-relatedness in proof theory – which have often been presented as the influence of a Schopenhauerianism.[208] However, intensive research into Schopenhauer's influence on Brouwer and on intuitionists such as Hermann Weyl, Arend Heyting, Oscar Becker and others is currently in its infancy.[209]

In 1947, about sixty years after the devastating judgements of Weierstrass' student Leonhard and of Märtens, Pringsheim, Klein and others, the topologist Kurt Reidemeister published an article on *Anschauung als Erkenntnisquelle* in the *Zeitschrift für philosophische Forschung*: It is astonishing, he writes, that geometrical research, although an intuitive interpretation is repeatedly asserted, nevertheless does not possess a secure state of knowledge about the cognitive power of intuition.[210] On the contrary, there are so-called 'rigorous mathematicians' such as David Hilbert or Louis Hjelmslev, who want to base everything on logical proofs, although metamathematics supposedly regards intuition as the most reliable source of knowledge.[211]

Since there is consequently a lack of material and assured results in mathematics, Reidemeister falls back on Zeno, Plato, Dürer, Kant and Schopenhauer. As an example, Reidemeister takes Schopenhauer's intuitive proof of the Pythagorean theorem and explains:

> What is it that makes Schopenhauer's proof intuitive convincing, which can hardly be denied? It seems to me that one can very quickly read the proof from this figure: The figure is an excellent "characteristic" of the proof [...], i.e. a symbol with which the structure of the proof is precisely represented. In this way, it becomes understandable how the visualisation makes complicated contexts clear to us. Schopenhauer's visualization of the Pythagorean theorem and its proof is accessible to an exact clarification; it is a plane

---

[207] Cf. e.g. L. E. J. Brouwer: Die Struktur des Kontinuums. In: Ibid: Collected Works. Vol. 1. Philosophy and Foundations of Mathematics. Ed. by A. Heyting. Amsterdam et al. 1975, pp. 429–440.
[208] Cf. e.g. Mark Van Atten, Göran Sundholm: L.E.J. Brouwer's 'Unreliability of the Logical Principles'. A New Translation, with an Introduction. In: History and Philosophy of Logic 38:1 (2017), pp. 24–47.
[209] Cf. the largely systematic-religious-philosophical pioneering work of Teun Koetsier: Arthur Schopenhauer and L.E.J. Brouwer. A Comparison. In: Mathematics and the Divine. A Historical Study. Ed. by L. Bergmans, T. Koetsier. Amsterdam et. al 2005, pp. 571–595.
[210] Cf. Kurt Reidemeister: Anschauung als Erkenntnisquelle. In: Zeitschrift für philosophische Forschung 1 (1946), pp. 197–210, here: p. 197.
[211] Cf. K. Reidemeister: Anschauung als Erkenntnisquelle, p. 198.

2.3 Proof – Elementary Geometry, Syllogistic and Intuitive Proof Theory

broken down in an intuitive way, between whose parts intuitive relations exist.[212]

Reidemeister connects the isomorphism between intuition and logic, which he sees in Schopenhauer's proofs, with Lull's, Leibniz' and, above all, Lambert's and Ploucquet's characteristically intuitive logical calculi: the modern formalism of allegedly rigorous proof depends, developmentally speaking, on the intuition given.[213]

A few years after Reidemeister's paper, one also finds discreet confrontations with parts of Schopenhauerian geometry in French philosophy of science and American analytic philosophy. In 1953, François Rostand published a sympathetic interpretation of Schopenhauer's mathematical method of demonstration. Rostand first compares Schopenhauer's approach with Descartes, Malebranche, the *Logic of Port-Royal*, Pascal and others and comes to the conclusion that in the French tradition of early modern philosophy and mathematics one also finds a distinction between the ground of being (la raison intuitive) and the ground of knowing (la raison deductive), but that this was never as sharp as in Schopenhauer.[214] Only Locke, Hume and Euler would show a similar appreciation of intuitive knowledge as one finds later in Schopenhauer.[215]

Furthermore, Rostand discusses the question of whether intuition is a suitable method of communication or a psychological misstep in the history of science based on modern authors (e.g. Gaston Bachelard, Georges Bouligand, André Ombredane). The fact that Schopenhauer remained attached to the Kantian and not the Leibnizian paradigm of mathematics is said to be because in later years he was no longer able to sufficiently receive the development towards non-Euclidean geometry by Bernhard Riemann and Nikolai Ivanovich Lobachevsky in 1854 and 1855, as this was only popularised by Helmholtz – a thesis that is strongly reminiscent of Benno Erdmann's *Axiome der Geometrie*, although no reference is made to it.[216]

Around 1955, Max Black took Schopenhauer's expression of the mousetrap proof as an ideal example of what he called a "comparison view of metaphor". Black's aim was to use Schopenhauer's supposedly geometrical metaphor to make it evident that comparison theory was not a special case of the "substitution view of metaphor" – a theory I have already brought into play using Quine's critique of the containment expression at the beginning of Chapter 2.2. "When Schopenhauer called a geometrical proof a mousetrap, he was, according to such a view, saying (though not explicitly): 'A geometrical proof is like a mousetrap, since both offer a delusive reward, entice their victims by degrees, lead to disagreeable surprise, etc.'"[217]

---

[212] K. Reidemeister: Anschauung als Erkenntnisquelle, p. 206.
[213] Cf. K. Reidemeister: Anschauung als Erkenntnisquelle, p. 208.
[214] Cf. François Rostand: Schopenhauer et les démonstrations mathématiques. In: Revue d'histoire des sciences et de leurs applications 6:3 (1953), pp. 202–230, here: pp. 204f., p. 228.
[215] Cf. François Rostand: Schopenhauer, pp. 207ff.
[216] Cf. François Rostand: Schopenhauer, p. 229, also p. 216.
[217] Cf. Max Black: Metaphor, p. 283.

2 Logic and its Geometrical Interpretation

Donald Davidson took exception to Black's interpretation of Schopenhauer in 1978. He argued against Black (and also against Goodman) that his three paraphrases given in the quasi-quote (1. offer..., 2. entice..., 3. lead to...) are neither given in the metaphor nor in the elliptical comparison. In sensu stricto, Schopenhauer only tells us that the proof resembles a mousetrap; but if we take the expression sensu allegorico – as Black does – we do not learn what the difference is between a metaphor and a simile.[218] Due to the discussion between Black and Davidson, the alleged metaphor 'mousetrap proof' appears again and again in metaphor theories of analytical provenance. It is questionable, however, what the progress in knowledge of this discussion consists of for the individual alleged metaphors.

Rudolf Carnap also referred to Schopenhauer's mousetrap proof in several books and papers. Already in his early writings, including *The Logical Structure of the World*, Carnap had advocated several positions that have been identified in research as Kantian or neo-Kantian, but which not infrequently even seem Schopenhauerian in essence. The first explicit mention of Schopenhauer can be found in a paper in the *Kant Studien* of 1925, in which Carnap writes about the comprehension of the external world, more precisely *Über die Abhängigkeit der Eigenschaften des Raumes von denen der Zeit* (*On the Dependence of the Properties of Space on Those of Time*). To support this thesis stated in the title, Carnap uses symbolic-algebraic logic and at the end raises the question of whether this approach is not similar to Schopenhauer's critique on the "Euclidean mousetrap proof", i.e. substituting "formalistic analysis" for "immediate apprehension".[219] Quite in the Schopenhauerian sense, however, Carnap argues that the thesis is naturally grounded in the intuitive world, but must be translated into abstract logic for scientific reformulation.

In 1966, Carnap took up Schopenhauer's mousetrap proof again in an aside to emphasise the priority of intuition over logical proof, especially in didactics and heuristics: Schopenhauer's example of the Euclidean mousetrap proof shows someone who has been led into a maze by a mathematician, finds the exit at some point and then does not know how he got there. One could therefore learn from Schopenhauer that it is not the validity of the proof that is important, but the step-by-step visual reasoning of its validity.[220]

As far as I know, Reidemeister's cautious attempt to make Schopenhauerian geometry usable again for current questions in the German-speaking world was only taken up and continued in 1988 by Knut Radbruch, Professor of Mathematics and its Didactics at the Technical University of Kaiserslautern.[221] Radbruch emphasises that

---

[218] Cf. Donald Davidson: What Metaphors Mean. In: Critical Inquiry 5:1 (1978), pp. 31–47, here: pp. 39f.
[219] Rudolf Carnap: Über die Abhängigkeit der Eigenschaften des Raumes von denen der Zeit. In: Kant-Studien 30 (1925), pp. 331–345, here: p. 343.
[220] Cf. Rudolf Carnap: Philosophical Foundations of Physics. An Introduction to the Philosophy of Science. Ed. by Martin Gardner. New York, London, 1966, p. 112.
[221] Radbruch wrote further papers on Schopenhauer in subsequent years, but these are less informative than the essay discussed below.

2.3 Proof – Elementary Geometry, Syllogistic and Intuitive Proof Theory

Schopenhauer had already attributed great importance to mathematics in his dissertation and his main work, but that especially his unknown *Berlin Lectures* contain detailed remarks on geometry.[222] In the reception, however, one should not forget that the sixteen decades separating the recipient from the *Berlin Lectures* had brought about multiple paradigm shifts in mathematics; nevertheless, the interpretation shows "that a number of Schopenhauer's questions, insights and perspectives on mathematics are of astonishing topicality".[223]

Radbruch sees the optimism announced in the title of his paper primarily in Schopenhauer's belief that one must be able to trace all elementary geometric proofs back to a simple intuition. He, on the other hand, maintains that Schopenhauer's and also Kant's ideal of clarity could only be realised in the paradigm of classical Greek geometry, whereas Newtonian and Leibnizian mathematics, which were only just emerging at Schopenhauer's time, would later have clearly shown the "limits of intuition".[224]

Nevertheless, Radbruch, along with several cited authors, argues that mathematicians have been living in two different worlds since the foundational crisis of mathematics: On the one hand, nowadays one would have to believe that mathematical theorems are intuitionistic and can be made visual, but on the other hand, in cases of doubt, one immediately falls back on a formalism. Therefore, although Schopenhauer's affinity for intuition is certainly too radical – just like that of his contemporary Gauß – but it still basically meets today's longing for correlation and isomorphism between intuition and logic in mathematics.[225]

Finally, Radbruch notes that Schopenhauer's optimism also extends to proof theory. If Schopenhauer inspires the hope that it is not logical proof but intuition that can provide direct evidence in mathematics, then mathematics has a special position in the field of science since it is the only one that can dispense with logic because of its relation to intuition:

> With this direct access to the truths of mathematics, deductive proofs would then not be necessary to secure the truth. Schopenhauer did not explicitly state this, but it can be inferred from his explanations without a doubt: this possibility of being able to dispense with proofs in principle and to retrieve the truth entirely with intuitive contemplation is given only in mathematics. [...] There is no hint in Schopenhauer's texts from where he drew his optimism about a visual access to the whole of mathematics.[226]

---

[222] In Chapter 2.3.6, I will discuss some theses of Schopenhaue's philosophy of geometry that can only be found in the Berlin Lectures. However, most and most relevant remarks on geometry in the Berlin Lectures are already known from the dissertation (1813) and WWR I (1819), vide supra, Chapter 2.3.2.
[223] Cf. Knut Radbruch: Anschauung und Beweis in der Mathematik. Skeptische Anmerkungen zum Optimisten Schopenhauer. In: Schopenhauer-Jahrbuch 69 (1988), pp. 199–226, here: p. 199. (My transl. – J. L.)
[224] K. Radbruch: Anschauung und Beweis, p. 121.
[225] Cf. K. Radbruch: Anschauung und Beweis, p. 123.
[226] K. Radbruch: Anschauung und Beweis, p. 125. (My transl. – J. L.)

As insightful as the papers by Reidemeister and Radbruch are, they show that much of the 19th-century debate on Schopenhauer's geometry is no longer known and its results and lines of argument have been forgotten in the research enterprise. Radbruch, for example, does not discuss Schopenhauer's self-criticism from the letter I described above as 'pessimistic', which Schlüter had popularised; nor does Reidemeister address the fact that Schopenhauer's geometrical figure applies only to a special case of the Pythagorean theorem.

In 1996, Peter Baptist, Professor for Didactics of Mathematics at Bayreuth, wrote a paper on the question of whether the Pythagorean theorem actually has a qualitas occulta. He argues that Schopenhauer really hit a crucial point, because "unlike, for example, the intersection property of the median perpendicular of a triangle, in this case [sc. the verbalization of the Pythagorean theorem] the statement remains invisible at first, it is really a 'qualitas occulta'".[227] In his paper, Baptist focuses primarily on Schopenhauer's criticism of the auxiliary line construction: although it is clear in the case of axioms to what extent they must be understood by intuition, traditional elementary geometry dispenses with the normativity of intuition in the case of theorems. These are proved arbitrarily since it cannot be explained why – as in the case of the Pythagorean theorem – it is precisely these and not other auxiliary lines that are used.[228] In this respect, Schopenhauer is not wrong when he describes the lines as traps (Schlingen) and when he demands that the ground of knowing should be traced back to the ground of being. Moreover, Schopenhauer was not alone in his criticism of theories of proof with auxiliary line constructions, since Albert Einstein also exemplified this problematic nature with the standard proof of Menelaus' theorem. Schopenhauer, like Einstein, thus criticises the lack of connection between the proof strategy and the corresponding theorem in the case of auxiliary line constructions.[229] Like Radbruch, Baptist sees Schopenhauer's statement that his figure of proof only refers to the special case of isosceles right triangles as expressing the optimism that a figure can be found for all cases. Baptist goes on to say that Schopenhauer could have had a solid reason for his optimism if only he had looked more carefully at the literature. In 1741, Alexis-Claude Clairaut had advocated a heuristic-genetic approach that infers the general case from the intuitive special case. However, Clairaut's approach had remained unknown due to Voltaire's unfounded criticism, so that it was not until 1873 that Henry Perigal provided a complete visual proof without words for Pythagoras' theorem. Clairaut, Schopenhauer and Perigal were thus in the tradition of Thabit ibn Qurra, who had presented the first of the almost 400 purely intuitive proofs of the Pythagoras' theorem known until 1940.[230]

---

[227] Peter Baptist: Der Satz des Pythagoras – eine qualitas occulta? In: Der Mathematikunterricht 42:3 (1996), pp. 22–30, here: p. 22. (My transl. – J. L.)
[228] P. Baptist: Der Satz des Pythagoras, pp. 23f.
[229] P. Baptist: Der Satz des Pythagoras, p. 25.
[230] P. Baptist: Der Satz des Pythagoras, p. 29.

2.3 Proof – Elementary Geometry, Syllogistic and Intuitive Proof Theory

In the 1990s, Jean-Yves Béziau published several studies on logic and mathematics in Schopenhauer, which were not only motivated by his personal interest in Schopenhauer's philosophy but also favoured by his school affiliation: Heinrich Scholz re-examined the quarrel between Wolff and Crusius of whether the principle of sufficient reason was at all formalisable, derivable and thus a law of thought.[231] Newton Da Costa had refuted the thesis of the non-formalisability and non-derivability of the principle of sufficient reason by reference to his modal calculus C_n with formalisations such as $\forall p\ \exists q\ \neg(q \to p) \wedge \Box(p \to q)$ or $\forall p\ (p \to \exists q\ (q \to \neg(p \to q) \wedge (q \to p)))$.[232] As a follower of Da Coasta, Béziau valued his teacher's formalisation in a similar way to Schopenhauer's mediation and rehabilitation of the principle of sufficient reason in the 19th century.[233]

However, Béziau went further and interpreted Schopenhauer as having anticipated Árpad Szabó's thesis mentioned at the beginning of Chapter 2.3. The interpretation often repeated in the literature that Schopenhauer anticipated Brouwer's ideas should not be overestimated either, since Schopenhauer only described one of the four roots of the principle of sufficient reason as intuitionistic and did not completely reject the ground of knowing in geometry.[234] There is, however, a connection between Schopenhauer and Wittgenstein with regard to diagrams, which I have already pointed out above in Chapter 2.1.4. In the 20th century, the development of the Zermelo-Fraenkel set theory showed that intuition and not just logic proved to be the basis of mathematics.[235] In addition, Schopenhauer could be seen as a forerunner of 'universal logic', since both shared the interest in logical-philosophical basic research, intuition and differentiation of logical principles.[236]

The papers and the resulting reactions published in the *Mitteilungen der Deutschen Mathematiker-Vereinigung* in 2003 can also be seen as a repetition of a long-standing discussion. Here, too, the term 'mousetrap proof' used by Alfred Schreiber was the reason for a discussion that extended over several issues, and in a similarly incidental way as Gerhardt's remark in the *Zeitschrift für mathematischen und naturwissenschaftlichen Unterricht* of the 1880s.

What is particularly noteworthy about this discussion is that Schreiber, after many decades, has again drawn attention to the fact that Schopenhauer's proof only refers

---

[231] Cf. Heinrich Scholz: Geschichte der Logik. Berlin 1931, p. 59.
[232] Cf. Newton da Costa: Logiques classiques et non classiques. Essai sur les fondements de la logique. Paris 1997, p. 107.
[233] Cf. Jean-Yves Béziau: O princípio de razão suficiente e a lógica segundo Arthur Schopenhauer. In: Século XIX. O Nascimento da Ciência Contemporânea. Ed. by F.R.R. Évora. Campinas 1992, pp. 35–39; ibid: On the Formalization of the Principium Rationis Sufficientis. In: Bulletin of the Section of Logic 22:1 (1993), pp. 2–3.
[234] Cf. Jean-Yves Béziau: La Critique Schopenhaurienne de l'Usage de la Logique en Mathématiques. In: O Que Nos Faz Pensar 7 (1993), pp. 81–88, here: p. 85.
[235] Cf. J.-Y. Béziau: La Critique, pp. 87f.
[236] Cf. J.-Y. Béziau: Metalogic, Schopenhauer and Universal Logic.

to isosceles triangles.[237] Also profitable, or at least worthy of discussion, is Schreiber's observation that Schopenhauer must be placed in the context of the so-called "proofs-without-words movement"[238] that has formed since the 1970s from the Proofs-without-Words column of the *Mathematics Magazine* of the Mathematical Association of America.[239] Overall, the whole debate, which continued mainly in letters to the editor, shows no knowledge of previous research.

In 2008, Jason M. Costanzo published an account of Schopenhauer's philosophy of geometry, which implicitly alludes to Szabó's rationalism thesis, since in his opinion Greek geometry only took a turn towards synthetic mathematics with Euclid. Costanzo is also convinced that Schopenhauer's demand for analytical geometry and his rejection of Euclid's synthetic geometry can be traced back to Pappus.[240]

In 2012, Dale Jacquette wrote an article on Schopenhauer's logic and mathematics in which he states that Schopenhauer is not a comparable logician or mathematician to Plato, Descartes, Leibniz, etc. Nevertheless, Jacquette claims that Schopenhauer's logic and mathematics are "intrinsically interesting"; unfortunately, there is no justification for this judgement.[241] Jacquette speculates a lot about what mathematical knowledge Schopenhauer could have acquired in his studies; however, he forgets to state that Schopenhauer studied with Thibaut, for example, what exactly he read in terms of geometrical writings and what Schopenhauer demonstrably knew about elementary geometry.

In his chapter on logic, Jacquette relies solely on what I called 'appeasement arguments' in Chapter 1.3.3 above, which, if interpreted context-free, look as if Schopenhauer had no deeper understanding of and interest in logic. The fact that one can discover Euler diagrams, a distinction between logic and dialectics and a rudimentary independent theory of proof in Schopenhauer is not mentioned in Jacquette's overview.[242] Rather, in his chapter on geometry, he tries to establish a link between the theory of ideas in Book III of WWR I and the geometric figures in Book I, but he himself admits that this comparison is somehow skewed.[243]

In 2014, Francesco Saverio Tortoriello wrote a paper on Schopenhauer's didactics of geometry, recording his many years of teaching experience at a secondary school in the province of Avellino and trying to clarify it with Schopenhauer's philosophy.

---

[237] Cf. Alfred Schreiber: Vorsicht, Mausefalle!. In: Mitteilungen der DMV 11:1 (2003), pp. 58–59, here: p. 58 (cf. the letters to the editors by Roger Böttcher and Martin Lowsky, 2003). A reference to this aspect of Schopenhauer's proof can, however, also be found in Martin Gardner: Sixth Book of Mathematical Games from Scientific American. New York 1975, pp. 153f.
[238] Alfred Schreiber: Vorsicht, Mausefalle!, p. 58.
[239] Cf. Tim Doyle, Lauren Kutler, Robin Miller, Albert Schueller: Proofs Without Words and Beyond – A Brief History of Proofs Without Words. In: Convergence 11 (August 2014).
[240] Cf. Jason M. Costanzo: The Euclidean Mousetrap. Schopenhauer's Criticism of the Synthetic Method in Geometry. In: Journal of Idealistic Studies 38:3 (2008), pp. 209–220.
[241] Cf. Dale Jacquette: Schopenhauer's Philosophy of Logic and Mathematics. In: A Companion to Schopenhauer. Ed. by Bart Vandenabeele. Hoboken 2012, pp. 41–59, here: p. 43.
[242] Cf. D. Jacquette: Schopenhauer's Philosophy of Logic and Mathematics, pp. 46ff.
[243] Cf. D. Jacquette: Schopenhauer's Philosophy of Logic and Mathematics, pp. 52f.

2.3 Proof – Elementary Geometry, Syllogistic and Intuitive Proof Theory

He agreed with Schopenhauer that the didactic basis of elementary geometry is intuition since logical abstraction could only be learned gradually.[244] Schopenhauer also shared this view of his time with the pedagogical approaches of Herbart and Trendelenburg,[245] and in modern times it is still represented in Piaget's geometric development theory and in Van Hiele's theory of the five levels of thought.[246] In this respect, Schopenhauer remains a thoroughly current thinker from a pedagogical point of view.[247]

The volume entitled *Language, Logic, and Mathematics*, published in 2020, takes up many of the themes discussed here in isolation and could be seen as an initial offer for research to discuss Schopenhauer's theses not only in isolation and again and again anew, but to work through the arguments systematically, in context and in a targeted manner.[248]

A summary of this approximately two-hundred-year history of the reception of Schopenhauer's philosophy of mathematics seems just as problematic as a rudimentary 'objective' evaluation of it. Rather, one can see from the history of reception itself how dependent the respective assessments and evaluations are on (1) the selection of Schopenhauer's texts and the weighting of the statements they contain, (2) the disciplinary and school affiliation of the researcher and (3) the scientific paradigm in which the interpretation is undertaken. I would like to demonstrate only this by employing a brief summary focused on the key statements. Rather positive evaluations of Schopenhauer's philosophy of geometry are classified as (a), balanced and neutral evaluations as (b) and negative judgements as (c):

(1) (a) Above all, Kosack, Becker, Radbruch, Baptist and Béziau refer primarily to positive passages on mathematics from Schopenhauer's early work, in which Schopenhauer provides an improved or his own elementary geometry. (b) Klein, Erdmann, Hoffmann and Leonhard also refer to aspects of the early work, but also discuss negative evaluations of mathematics from WWR II (1844). (c) Bahnsen and Pringsheim focus mainly on the few statements critical of mathematics from WWR II (1844), giving the impression that Schopenhauer made no positive contribution to the philosophy of geometry.

(2) (a) Diesterweg, Kosack, Becker, Hoffmann, Brouwer, Carnap, Reidemeister, Baptist, Béziau and Tortoriello claim to have been more or less positively influenced by Schopenhauer or at least that he is useful for dealing with current research questions: Kosack, Becker, Brouwer, Reidemeister mainly

---

[244] Francesco Saverio Tortoriello: Schopenhauer e la didattica della matematica. In: Archimede: Rivista per gli insegnanti e i cultori di matematiche pure e applicate 2 (2014), pp. 86–91, here: p. 86, pp. 90f.
[245] Cf. F. S. Tortoriello: Schopenhauer, p. 90
[246] Cf. F. S. Tortoriello: Schopenhauer, p. 89.
[247] Cf. F. S. Tortoriello: Schopenhauer, p. 86.
[248] A summary of the results can be found in Dieter Birnbacher: Language, Logic, and Mathematics (Review). In: Schopenhauer-Jahrbuch 101 (2020), pp. 249–257.

because of their proximity to intuitionistic mathematics; Béziau mainly because of his proximity to universal logic; Diesterweg, Baptist and Tortoriello especially for reform pedagogical reasons; Carnap refers to Schopenhauer as an encouragement that logic without empiricism is meaningless. (b) A weighing of the advantages and disadvantages in Schopenhauer's philosophy of geometry can be found in Erdmann, Janzen, Radbruch and Schreiber as well as in the accounts of authors not mentioned by name here who have not made an explicit evaluation. (c) Schopenhauer's geometry is rejected, above all, by Weierstrass followers such as Leonhard, Pringsheim and Klein, but also by Leibnizians such as Gebhardt.

(3) (a) Whereas the positive reception of Schopenhauer's geometry peaked in the 1850s with Kosack and Becker, (c) a negative wave of interpretation set in with the crisis in intuition from around 1880 with Märtens and Leonhard; this increased to a devastating critique by Pringsheim and Klein around 1900. (a) The only exception to this is the intuitionist movement that began with Brouwer, but whose specific reference to Schopenhauer has not yet been researched in sufficient detail. (b) Around 1950, the first cautious approaches to Schopenhauer's philosophy of geometry were found with Reidemeister and Rostand; (a) this tendency intensified from the 1990s onwards with Radbruch, Baptist and, above all, Béziau, without one being able to justifiably speak of a renaissance of Schopenhauer's philosophy of mathematics or the like.

I believe that the scope for interpretation would have to be overstretched if one were to try to map a continuation of the arguments of (L) and (K) from Chapter 2.3.2 onto the current authors mentioned especially in (2). Nevertheless, one can probably take the view that one can see, both from the continuity of individual strands of argumentation and from the regular rediscovery of the individual topics discussed in Schopenhauer, that his philosophy of geometry can still be of interest today, not only in terms of the history of science but also systematically.

This interest is probably fostered, among other things, by the fact that not even a rudimentary consensus on the assessment of Schopenhauer's mathematical teachings has emerged. It is understandable that a literal succession of Schopenhauer, as with Kosack or Becker, no longer seems possible in the current paradigm. But that this can be seen as an argument in favour of the discursive one in the dispute between a visual and a discursive proof theory, which has lasted for many centuries, is not yet decided. It remains unclear or disputed to what extent Schopenhauer's demand for a 'picture proof' is at least tending in the right direction. That the question of the limits and possibilities of a theory of proof based on observation is not only relevant in the history of science, but also in (scientific) philosophy and mathematics, has been illustrated by several current positions outlined at the beginning of Chapter 2.3 (e.g. Stapelton, Macbeth, Stekeler-Weithofer).

I will claim in Chapter 2.3.4–2.3.6 that Schopenhauer made an argument related to his philosophy of geometry in the *Berlin Lectures* that can be seen as a strong attack on rationalism in logic, which is as strong as the arguments of rationalism against a non-logicist proof theory in geometry were. There may have been repeated discussions in recent years about how Schopenhauer defended an intuitive proof theory in geometry; but the question of why he considered the counter-argument, namely a purely discursive proof theory in logic and geometry, to be problematic has, to my knowledge, remained unconsidered until now.

2.3.4 Assessments of Geometric Logic from Reimers to Maaß

Schopenhauer's interest in logic, geometry and the connection between the two fields is not a special case in the history of science in modern times. If one examines the list of authors on geometrical logic that I presented in Chapters 2.2.1 and 2.2.3, one finds that most of those named there were concerned with both logic and geometry or sciences with an affinity to geometry (astronomy, mechanics, architecture, optics, etc.).

As my discussion in Chapters 2.3.2 and 2.3.3 has shown, attitudes to the question of the relationship between logic and geometry vary widely. If one disregards both extreme and moderate positions, the history presented so far can be summarised in two groups: Whereas one group has tried to emphasise Euclid's more rationalist and logicist tendencies and to trace geometry back to logic, the other group has rather emphasised the visual aspect of geometry and tried to strengthen the intuition of geometric propositions. The attitude towards geometry is also reflected in the evaluation of geometric figures and logic diagrams. Simplified and exaggerated on the basis of more extreme positions: For rationalists, geometric diagrams are at most didactic aids, whereas members of the other group accept a proof of geometric propositions solely based on an empirically existing or imagined diagram.

If there have been many authors in the history of science who have dealt with both geometry and geometric logic, and if geometers usually have to decide between a rationalistic and an intuition-based interpretation of their discipline, then the question arises as to what attitude geometric logicians usually have towards this choice of approach, above all, what value they assign to geometric diagrams in logic.

How difficult a purely interpretative answer to this question is can be illustrated, for example, by looking at parts of Leibniz research: As shown in Chapter 2.3.1, in the 18th century especially Leibnizian geometers advocate the rationalist theses (L1), logical principles form the basis for logical deduction and proof theory, and (L2), geometric diagrams are only a didactic crutch to learn a purely logical geometry. (L1) and (L2) are finally justified by the support (L3), which states that diagrams and figures cannot do justice to the semantics of geometric concepts and judgements. With

2 Logic and its Geometrical Interpretation

reference to the Leibniz-Wolff school, one can probably claim justification by asserting that Leibniz himself must also have held intuition-based proof procedures in low esteem.[249] In contradiction to this are interpretative approaches that attempt to transfer Leibniz's autarchy criterion from arithmetic and number theory to geometric logic to provide an explanation as to why Leibniz used logic diagrams and that he must have valued line diagrams better than logical circle diagrams.[250]

To prevent such problems of interpretation and my associated concerns, I have decided to examine the texts of the geometrical logicians presented in Chapter 2.2.3 to see whether and to what extent their authors themselves reflect on the use of geometrical diagrams. Since, of course, not all geometrical logicians reflect on and evaluate their actions, the following investigation is limited to the statements of Reimers, Weigel, Ploucquet, Lambert, Euler and Maaß and their analysis.

In Chapters 2.2.1–2.2.3 I put forward several reasons for the thesis that, on the one hand, it makes sense to be aware of the prehistory of analytical diagrams in antiquity and the Middle Ages, but that, on the other hand, one would do well to let the actual history of analytical diagrams begin with Vives and Reimers. Two central reasons can be briefly summarized here: The first diagrams demonstrably created and authorised by a writer to be used for the representation of logic are found in Vives; but since he refers to the triangular diagrams discussed in Chapter 2.2.3 in a self-evident manner and without much explanation, one can conjecture that he himself did not perceive his outstanding historical position for us because analytical diagrams were already established in his time.

Because of this self-evidence (in Vives), it is also only in Reimers that we find a first, albeit still very restrained, reflection that allows a cautious conclusion to be drawn about his evaluation of logic diagrams. After Reimers had given several examples of syllogisms in his *Metamorphosis Logicae*, he comes to the question of what the cause of syllogisms is ("De caussa [!] syllogismi").[251] In his answer, he emphasises that the derivation depends on the being-in or containedness of a whole in another whole, which is also called 'dictum de omni et nullo' ("id quod Philosophi τὸ ὅλον ἐν τῷ ἑτέρῳ ὅλῳ, id est, Totum in Toto (vulgo inesse, item Dici de omni et Dici de nullo)").[252]

As indicated in Chapter 2.3, Reimers orientates himself in the following on the logical vocabulary of being contained and not being contained ("Vocabula Logica, Inesse, &, non inesse"), which is the cause or reason ("caussa ac ratio") of the necessity of the syllogisms. He first demonstrates this with affirmative syllogisms, which

---

[249] An often cited proof that (L2) and (L3) are represented in Leibniz's writings is found in *Nouveaux Essais* IV 1, § 9.
[250] The latter theses are represented, for example, by Francesco Bellucci, Amirouche Moktefi, Ahti-Veikko Pietarinen: Diagrammatic Autarchy. Linear Diagrams in the 17th and 18th Centuries. In: Diagrams, Logic and Cognition. Ed. by J. Burton, L. Choudhury. CEUR Workshop Proceedings 1132 (2013), pp. 23–30.
[251] Nicolaus Raymarvs Vrsvs Dithmarsivs: Metamorphosis Logicae, p. 29.
[252] N. R. Ursus: Metamorphosis Logicae, p. 29.

2.3 Proof – Elementary Geometry, Syllogistic and Intuitive Proof Theory

build on the expression of being contained as well as on the dictum de omni; then he examines negative syllogisms, which make use of not being contained as well as the dictum de nullo. Both types of syllogisms are treated independently in one chapter each. These chapters are structured similarly: they each have a subchapter on the principle of inner insight ("Principium per intellectum internum"), in which the mode of expression is illustrated by one or more diagrams, and then a subchapter dealing with experience through the external sense ("Experimentum per sensum externum"). Both subchapters offer an idealistic and a realist criterion for syllogistic proofs.[253]

The second subchapter in each case indicates why Reimers speaks not only of reason ("ratio") but also of cause ("caussa") of the syllogisms: although the reasoning may be of linguistic nature, the proof of the same is nevertheless the evidenced experience from intuition ("dilucidum atque perspicuum Experimentum [...] ex inspectione").[254] For example, the affirmative syllogism is based on the same structure of being contained as is found in nature: the egg yolk is contained in the egg white and the egg white in turn in the eggshell, so that the egg yolk is also contained in the egg shell ("in eo namque intimus vi tell us inest intermedio albumini: ipsumque albumen inest extrema putamini: Ergo et ipse vitellus necessario inerit putamini, intimum puta extrema").[255] For the negative syllogism, Reimers constructs a similar example with the pupil of the right eye, which is not contained in the left eye.[256]

Following these subchapters, Reimers reflects on the significance of containing and not containing for the proof of a syllogism, and at the end of the reflection on the negative syllogisms, Reimers gives an overall conclusion. In this, he emphasises that all proofs are based on the dictum de omni et nullo or on the meaning of containing and not containing. All evidence is constructed by the evidence just given by means of containing and not containing. This had rightly already been indicated by Aristotle, who spoke in this context of "the Why of mathematicians" ("Ideoque recte a summo Philosopho dictum est τὸ δί ὅτι τῶν μαθηματικῶν").[257] Reimers thus builds a bridge between Euler's logical principles of being contained and the Aristotelian dictum de omni et nullo.[258]

As I have already pointed out, the reflection on the value of the diagrams is still very restrained here. Reimers does not explain why he uses diagrams in the subchapters in which he does so. Nevertheless, one can already read from what has been presented so far that in his opinion diagrams illustrate the natural principle of (non-)containment grounded in the senses (just like the egg and pupil examples), to which the syllogistic

---

[253] Cf. N. Reimarus Ursus: Metamorphosis Logicae, p. 2.
[254] N. Reimarus Ursus: Metamorphosis Logicae, p. 32.
[255] N. Reimarus Ursus: Metamorphosis Logicae, p. 32.
[256] Cf. N. Reimarus Ursus: Metamorphosis Logicae, p. 36.
[257] N. Reimarus Ursus: Metamorphosis Logicae, p. 37.
[258] N. Reimarus Ursus: Metamorphosis Logicae, p. 37.

2 Logic and its Geometrical Interpretation

proof must ultimately refer if it does not want to end up in an infinite regress of principles that need to be justified again and again. Reimers at least says this explicitly after using the first diagram to prove positive syllogisms:

Siquidem ipsa principia [sc. inesse et non-inesse], cum omnibus sana ratione praeditis hominibus per se aeque nota sint, nunquam Demonstrantur per alia principia (sic enim ipsa principia non essent principia, et Demonstratio in infinitum vagaretur) sed per sensus [...].[259]	Since these are the principles themselves [sc. containing and not containing], which are equally known by themselves to all humans of sound mind, they are never proved by other principles (for then the principles would not themselves be principles and the proof would progress to infinity), but by the senses [...].

Reimers already hints here at the sceptical-empiricist argument that I clarified at the beginning of Chapter 2.3 based on Sextus, Bacon, Locke, Mill and others: If one were to trace fundamental logical principles such as that of (non-)containment back to other logical principles, one would end up in an endless regress. But the principles mentioned are equally known by themselves (per se aeque nota sint), so that it is not necessary to justify them further logically. On the contrary, they become directly understandable through the senses (demonstrantur .... per sensus), as Reimers again illustrates from a realistic point of view with the egg and pupil examples.

Reimers only explicitly reflects this intuitive method in the title of his book: his *Metamorphosis of Logic* had set itself the task of omitting everything unnecessary of its predecessors and instead establishing a demonstration of the necessary syllogisms that is "solid, most evident and obvious" ("Cum solida, evidentissima, atque oculari demonstratione Syllogismorum necessario concludentium").[260] This demonstration, as the motto of the book suggests, is "the Why of mathematicians" ("Aristotle post. Anal. 1. cap. 7. τὸ δί ὅτι τῶν μαθηματικῶν"), which he later also cites in the context of the given diagrams.[261] This is not only implied by the context in which the diagrams are used, but also explicitly indicated by the author that the method used is a safer, clearer and thus improved form of logic compared to the conventional proof theory because of its reference to intuition.

Similar to Reimers, Weigel also oriented himself to the logical vocabulary when reflecting on his diagrams. Although he did not publish analytical diagrams, which he called 'Logometrum', until 1693, reports from Sturm and Leibniz indicate that he was already teaching them to his students in the early 1660s, either in his classes or in

---

[259] N. Reimarus Ursus: Metamorphosis Logicae, p. 32.
[260] N. Reimarus Ursus: Metamorphosis Logicae, Cover.
[261] N. Reimarus Ursus: Metamorphosis Logicae, Cover. Reimers is probably referring here to Anal. post. 79a3.

2.3 Proof – Elementary Geometry, Syllogistic and Intuitive Proof Theory

personal conversations.[262] In 1669, Weigel reported on his invention of the logometrum in his book *Idea Matheseos universae*, explaining the purpose and value of this invention in much greater detail than Reimers:

§ 17. Factum hinc est, ut veteres Mathematici quantorum abstractam rationem non abstracte, nec, ut directa methodus exigit, catholicis propositionibus, & quae sunt κατ' αυτό; sed quasi concrete tantum, indirecta methodo, per lineas & figuras, *tanquam per clariorem speciem doctrinae gratia tradiderint*, quod ex Euclidis libro tum secundo, tum quinto, nemo non agnoscit.

§ 18. Data mihi hinc est occasio cogitandi, *annon ad alia quaedam generaliora facilius tradenda* similiter adhiberi possint lineae vel figurae: Et illico vim earum in ipsis logicis Syllagisationibus alioquin abstractissmis expertus sum.[263]

§ 17. And from here it is explained that the old mathematicians did not teach the abstract relation of quantitative magnitudes in an abstract way, nor, as the direct method requires, in generally valid and by itself certain propositions, but merely in a way that is as it were concrete, according to an indirect method, namely, by lines and figures, as it were *for the sake of a clearer form of doctrine*; this method is recognised by everyone, in accordance with the second and also the fifth book of Euclid's *Elements*.

§ 18. From this, I have had the favourable opportunity to consider whether, *for the purpose of an easier exposition* of certain other more general relations, lines or figures can be similarly employed; and I have immediately examined their force precisely in those syllogisms which are otherwise very abstract.

In the transition from § 17 to § 18 of the quote, Weigel describes the application of intuitive geometry to logic. It is noteworthy that in § 17 Weigel takes what appears to be a clearly intuition-based position, which grew out of his interpretation of the anti-rationalist geometers of antiquity. After all, the ancient mathematicians would not have used an abstract and direct method that infers from axioms but would have used a concrete and indirect method that makes use of visual reasoning by lines and figures. As with Reimers, Weigel also interprets logic from the spirit of mathematics.

The allusion to the ancient mathematicians as well as to Euclid's second and fifth books of the *Elements* remains puzzling to me, and the context does not provide more

---

[262] Cf. e. g. G.W. Leibniz: Essais de théodicée sur la bonté de Dieu, la liberté de l'homme et l'origine du mal. Amsterdam 1714, Tom. I, p. 390f. (= § 212).
[263] Erhardus VVeigelus: Idea Matheseos universæ, p. 46. [My emphasis – J.L.]

2 Logic and its Geometrical Interpretation

information either, in my opinion: The reference to the fifth book points to Eudoxus; Weigel's oft-repeated confession to Pythagoras, on the other hand, points to even older geometers. Why the second and fifth book of the *Elements*, of all things, should be in line with the aforementioned intuition-based position remains a mystery to me.

Whoever the ancient mathematicians were that Weigel had in mind, they nevertheless used a concrete and indirect geometry, since they aimed at a clearer form of doctrine (tanquam per clariorem speciem doctrinae gratia tradiderint). As described in § 18, Weigel attempted – inspired by the ancient geometers – to apply the concrete and indirect form of the doctrine utilizing lines and figures to other subfields outside geometry, in order to achieve the purpose of easier transmission there as well (annon ad alia quaedam generaliora facilius tradenda). This intention had paid off especially in the teaching of syllogisms, which in itself was very abstract. And Weigel also gives a lengthy explanation (following on from the quotation given earlier) as to why the concrete-visual method could be applied to logic:

Cum enim coincidentiam & distantiam linearum figurarumve cum Identitate & diversitate Metaphysica similitudinem arctissimam habere deprehenderim, adeó quidem ut Identitas speciem Coincidentae (nempe praedicativam) & diversitas distantiae speciem simile, prae se ferre videatur, in quo utroque vis ac potestas universiae Syllogisationis juxta *dictum de omni & nullo* sita est; agnovi tandem, non gratis Aristotelem in Syllogismis tradendis usum esse vocibus Geomtrarum, (πέρας, σύνδεσμος, σχῆμα) sed omnes Syllogismorum modos per schemata figurasque geometricas *multo facilius discerni*, quam per *Barbara, Celarent*, multoque *succinctius demonstrari* (vulgo reduci) posse, quam per τὸ *Phoebifer axis obit terras athramque quotannis*:

adeò quidem ut, vera sit an falsa syllogisandi forma per nudam coincidentiam vel discoincidentiam sive distantiam figuralem ipsarum saltem literarum initialium cujusque termini (non enim

For I have found that the coincidence & distance of lines or figures bears the strongest resemblance to metaphysical identity & diversity, so much so that identity seems to drive before it the figure of coincidence (namely, of subject and predicate) and diversity, correspondingly, the figure of distance; in the two attributions consists, closely together with the *dictum de omni et nullo*, the force and capacity of the whole inference.

I have finally realised that Aristotle did not use the technical terms of the geometers in the surviving syllogisms by accident (boundary, connection, scheme), but also that all the modes of the syllogisms can be learned much *more easily* than by *Barbara, Celarent* and demonstrated (or reduced) *much more briefly* than by the *Phoebifer axis obit terras aethramque quotannis* through the geometrical schemes and figures:

and so much so – may it be a correct or incorrect form of inference – that by

2.3 Proof – Elementary Geometry, Syllogistic and Intuitive Proof Theory

opus est ut sint circuli vel Triangula) *veluti palpando statim deprehendere liceat,* [...].[264]

this it becomes possible to find out the modes of syllogisms mentioned by the mere coincidence or non-coincidence, i.e. by the figurative relation at least of the initial letters themselves and their boundary (for it is not necessary that they should be circles or triangles), *as if in the twinkling of an eye.*

At the beginning of the quotation, Weigel speaks of an analogy (similitudinem arctissimam) between geometry and logical metaphysics: just as diagrams, for example, can intersect or not intersect, so too can subject and predicate in judgement exhibit identity or diversity. Because of this analogy, it is possible to represent judgements by diagrams in such a way that logical identity is represented by the diagrammatic intersection, and logical diversity by the difference in diagrams. The analogy between geometric diagrams and the metaphysical logic of concepts is guaranteed by the dictum de omni et nullo, which connects logic and geometry.

An explanation of why the dictum de omni et nullo combines geometric and logical thinking can be found in the following paragraph of the quote: Aristotle, to whom our occidental history of philosophy goes back, had adopted the technical vocabulary from the geometricians and used it centrally to explain his deductive-logical reasoning. If one does not accept Aristotelian expressions such as boundary, connection, scheme, figures, etc. as background metaphors, but translates them back into visualisation, one has an advantage over the scholastic mnemonics: Diagrams can be learned more easily and much more simply in this way than through the scholastic words of art (Barbara, Celarent,...), and the validity of inferences can be proved more quickly than with scholastic apagogic proofs (Phoebifer axis...). Weigel summarises both aspects, simplicity and speed, in the last sentence of the quotation with the metaphorical phrase 'in the twinkling of an eye'.

The last paragraph of the quote already points to Weigel's unique invention, the so-called *logometrum*: proof does not even require geometric figures that are provided with variables or constants, but the variables or constants themselves can be written in such a way that they intersect or do not intersect. Similar to planimetric diagrams, it depends, as Weigel himself indicates, on the drawn boundary of the figures.

Weigel's logical approach thus shows itself to be strongly inspired by the intuition-based position of geometry presented in § 17. The intuition in geometry brings about the invention of a logical form that can be used to tie in with Aristotle and is intended to overcome scholasticism. The form represented by the logometrum itself eventually leads to a blurring between visual and discursive thought, since the variables or constants can themselves function as images and geometric forms.

---

[264] Erhard VVeigelus: Idea Matheseos universæ, II p. 46.

2 Logic and its Geometrical Interpretation

Another example of the evaluation of geometric diagrams in logic can be found in the so-called golden age of logic diagrams. In the dispute between Lambert and Ploucquet over the invention of a calculus more geometrico, already discussed in Chapter 2.2.3, Georg Jonathan von Holland took a position in favour of Ploucquet in his treatise *Abhandlung über die Mathematik, die allgemeine Zeichenkunst und die Verschiedenheit der Rechnungsarten*: Ploucquet had not only been the first to set up a calculus but had also used an improved algebraic variant since Lambert had confused subordination with extension in his line diagrams – as is well known, Maaß had taken up the argument in a similar way.[265]

Lambert, however, did not miss the opportunity to turn Holland's criticism of his calculus into a positive one and to present several aspects from the work as merits of his diagrammatic method:

> I then read Mr. Georg Jonathan von Holland's treatise [*On Mathematics*, 1764] with great pleasure. [...] I was particularly amused by the short note on the 28th page, which roughly says that one can only attain complete certainty in geometry, but that, according to general myth, this is too difficult for most people, and the most difficult of the sciences; and from this one can conclude how much one is content with the appearance of truth and empty words in the other sciences. In fact, this can be seen especially in those metaphysicians who want to construct geometry according to the concepts of their metaphysics. One still has the *means to discover the inconsistencies, because geometry soon conceals the fallacies.*[266]

In this quotation, Lambert alludes to several aspects of Holland's work, which he positively presents as his own merits. On pp. 27f. Holland had indeed declared that the geometry (Messkunst) was the most certain because its forms had been produced from the idealistic world itself, but that it was also considered the most difficult of all sciences. This, he says, is an indication that elsewhere one is only "content with the appearance of truth and empty words".[267] Unlike Lambert, however, Holland uses these remarks rather to show the merits of a characteristica universalis, as the historical end of which he sees Ploucquet's algebraic calculus.

Lambert, however, takes the advantages that Holland sees in Ploucquet's algebraic calculus and applies them to his geometric method, putting into Holland's mouth the

---

[265] Cf. Georg Jonathan Holland: *Abhandlung über die Mathematik, die allgemeine Zeichenkunst und die Verschiedenheit der Rechnungsarten. Nebst einem Anhang, worinnen die von Hrn. Prof. Ploucquet erfundene logikalische Rechnung gegen die Leipziger neue Zeitungen erläutert und mit Hrn. Prof. Lamberts Methode verglichen wird*. Tübingen 1764, p. 67. On Maaß vide supra, Chapter 2.2.3 and also Chapter 2.2.5f.
[266] J. H. Lambert: Neue Zeitung von gelehrten Sachen 1765:1 (3rd January). In: August Friedrich Bŏk (Ed.): Sammlung der Schriften, pp. 149–156, here: p. 150. (My emphasis – J.L.)
[267] G. J. Holland: *Abhandlung über die Mathematik*, p. 28.

2.3 Proof – Elementary Geometry, Syllogistic and Intuitive Proof Theory

quasi-quotation about the metaphysicians, which in the *Abhandlung über die Mathematik* referred in general to all sciences that cannot claim mathematical certainty. According to Holland, Ploucquet had succeeded in "guiding the intellect by an easy calculation in which it can overlook its errors with a single glance [mit einem einzigen Blick]".[268] Finally, the concluding sentence of the Lambert quote above reads similarly: Geometry is a means in logic to make fallacies visible.

Ploucquet responded to Lambert by dating his invention more precisely and applying the merits elaborated by Holland to his calculus:

> In 1758 I had the idea of drawing inferences and presenting them in figures, in order to bring them to visual knowledge in such a way that the *whole inference would be overlooked at a glance*, without thinking of consequences, and thus *all doubt about the infallibility of inferences would be completely removed*. If, for example, all M is P and all S is M, then, if the predicate in an affirmative proposition is regarded as a part of the concept of the subject, P is contained in M and M in S. Consequently, the construction may be this:

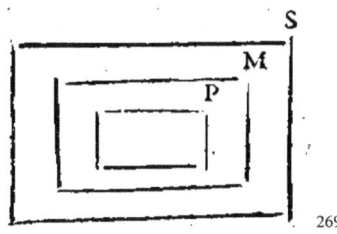

[269]

All three quotations (Holland, Lambert, Ploucquet) are similar in that they see analytical diagrams as a means to present inferences more simply, to avoid fallacies more easily or to recognise them more quickly. Although Lambert's critique of the metaphysicians' fallacies can be read as the critique of a sensualist, empiricist or representationalist, all three of these authors seem to have been motivated less than, for example, Weigel or Schopenhauer by a particular attitude towards geometry. I have not been able to find any indication that geometric logic would displace a certain scholastic or purely discursive technique in the writings of Holland, Lambert and Ploucquet.

Euler, on the other hand, was clearly a sensualist and anti-rationalist. The treatises against idealism, against the metaphysical monadic system and also on spatial expansion in the second volume of his *Lettres à une princesse d'Allemagne* show this in so

---

[268] G. J. Holland: Abhandlung über die Mathematik, p. 65.
[269] Gottfried Ploucquet: Untersuchung und Abänderung der logikalischen Constructionen des Hrn. Prof. Lambert. In: Bŏk (Ed.): *Sammlung der Schriften*, pp. 157–202, here: p. 157. (My emphasis – J.L.)

2 Logic and its Geometrical Interpretation

many places that I do not wish to single out individual examples here.[270] In his philosophy of language and logic, too, Euler initially followed the empiricist conceptual tradition: Ideas and concepts are abstracted from sensual impressions and form the basis of all judgements and inferences.[271] Euler thus builds his logic sensually and compositionally: Every person perceives objects through the senses, abstracts properties from them and forms independent ideas and concepts from them.

In the theory of abstraction, only a few differences to Schopenhauer's theory of language emerge: in Euler's case, it seems in places that every human individual goes through this process of abstraction from the senses to concepts, although Euler explicitly makes several arguments against a private language.[272] At least in Schopenhauer's logica major, abstraction tends to occur in the genus process, so that each individual of a new generation learns and understands the language created by the genus process contextually and from use.[273] Especially in comparison to naïve representationalism, it becomes clear how Schopenhauer's and Euler's semantic representationalism differ: While naïve abstraction theory presupposes an intimacy and privacy between the intuitive given world and the abstracting individual, non-naïve representationalism relocates this private language to a prehistoric time that can no longer be reflected upon, which marks the beginning of a phylogenetic natural, and thus also a social-cultural history of language development.[274] Euler also shares Schopenhauer's nominalism concerning the meaning of proper names: While the name 'Alexander the Great' may so far only belong to one individual, there are nevertheless innumerable 'Alexanders' who can also have the property 'to be great'.[275]

Euler's nominalism is the starting point for forming categorical judgements with 'quantifiers' such as `All`, `Some`, `No`, `Some ... not`. All basic types of judgements can thus be represented with circle diagrams and, according to Euler, have a strongly simplifying function:

Ces figures rondes, ou plûtôt ces espaces (car il n'importe quelle figure nous leur donnions), *sont très propres à nous faciliter nos réflexions sur cette matière*, & à nous découvrir tous les mysteres dont on se vante dans la Logique, & qu'on y démontre avec bien de la peine, pendant que par le moïen de	These circles, or rather these spaces, for it is of no importance of what figure they are of, are *extremely commodious for facilitating our reflections on this subject*, and for unfolding all the boasted mysteries of logic, which that art finds it so difficult to explain; whereas, by means of these signs, *the*

---

[270] Cf. Eberhard Knobloch: Leonhard Euler als Theoretiker. In: Mathesis & Graphe. Leonhard Euler und die Entfaltung derWissensysteme. Ed. by Wladimir Velminski, Horst Bredekamp. Berlin: Akademie, 2010, pp. 19–36.
[271] Cf. Leonhard Euler: Lettres à une princesse d'Allemagne, Vol. 2, pp. 86ff. (= L. C).
[272] Cf. Leonhard Euler: Lettres à une princesse d'Allemagne, Vol. 2, p. 90f. (= L. CI).
[273] Vide supra, Chapter 2.1.5.
[274] Vide infra, Chapter 3.
[275] Cf. Leonhard Euler: Lettres à une princesse d'Allemagne, Vol. 2, pp. 91f. (= L. CI).

ces signes *tout saute d'abord aux yeux*. On emploie donc des espaces formés à plaisir, pour représenter chaque notion générale, & on marque le sujet d'une proposition par un espace contenant A, & le prédicat par un autre espace qui contient B. La nature de la proposition même porte toujours, ou que l'espace A se trouve tout entier dans l'espace B, ou qu'il ne s'y trouve qu'en partie, ou qu'une partie au moins est hors de l'espace B, ou ensin que l'espace A tout entier est hors de B.[276]	*whole is rendered sensible to the eye.* We may employ, then, spaces formed at pleasure to represent every general notion, and mark the subject of a proposition, by a space containing A, and the attribute, by another which contains B. The nature of the proposition itself always imports either that the space of A is wholly contained in the space B, or that it is partly contained in that space; or that a part, at least, is out of the space B; or, finally, that the space A is wholly out of B.[277]

Although Euler is known for his circular logic diagrams and until today it is precisely the form of the diagram that is often used as a central criterion to decide whether a diagram is a so-called Euler diagram or not, Euler – according to the first sentence of the quote – does not place particular emphasis on the diagrammatic or geometric form. Rather, what is decisive is that diagrams are a facilitator for representing judgements. Whereas in ordinary logic one proves judgements with much effort (démontre avec bien de la peine), they are grasped intuitively by means of diagrams, so that – metaphorically speaking – their validity falls into one's eyes (saute aux yeux) – an expression reminiscent of Reimers. Euler's analytical diagrams thus form an aid to simplifying proof techniques in logic.

Euler explains how this simplification is realised in the second part of the quote: subject and predicate are each represented by a space or by a circle and its area. Unfortunately, Euler only explains why this is possible at all with a metaphor that is difficult to interpret: the nature of the proposition (la nature de la proposition) entails whether the relationship between subject and predicate can be represented by overlapping, partial overlapping, etc. This explanation – even if it is not very clear – can be used to explain the relationship between subject and predicate. Through this explanation – albeit more cautiously than, for example, in Weigel – one can also conclude in Euler an analogy argument that describes a similarity between the logic of judgement and the geometric area relation of diagrams.

Before I turn to Maaß, I would like to mention a puzzling passage from Euler's doctrine of inferences, in which the logical proof theory is contrasted with geometry. In Letter 105 (before the puzzling passage), Euler had stated that there were only nineteen valid types of inferences to which all valid syllogisms must be traceable. A

---

[276] Leonhard Euler: Lettres à une princesse d'Allemagne. Vol. 2, pp. 96–101 (= L. CIIf.). (My emphasis – J.L.)
[277] Translation taken from Letters of Euler, 1833, Vol. 1, pp. 397f. (My emphasis – J.L.)

valid inference can be recognised by the fact that a true conclusion necessarily follows from two true premises. First, construct all possible analytical diagrams for the two true premises and then see whether the conclusion is always represented by all diagrams that are possible to construct – usually, there is only one diagram. If this is the case, the inference is valid; if it is not, the inference is invalid.

Euler gives the following example of an invalid type of inference including three judgments (J):

(J1) Some (A) learned men are (B) misers.
(J2) But no (B) miser is (C) virtuous.
(J3) Therefore some (C) virtuous men are not (A) learned.

Premise (J1) is represented by one of the following diagrams ($\mathcal{D}1$) or ($\mathcal{D}1^*$) according to the four judgement types of Fig. 10 in Chapter 2.2.3:

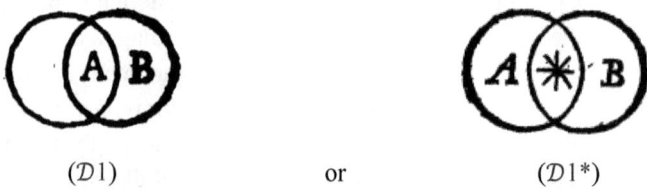

($\mathcal{D}1$)   or   ($\mathcal{D}1^*$)

Instead of the transposed letter ($\mathcal{D}1$), Euler also uses an asterisk ($*$) for particular judgements in order to present complex diagrams more clearly and unambiguously. We call the diagram in which an asterisk is used instead of the transposed letter ($\mathcal{D}1^*$).

Premise (J2) is now represented by the following diagram ($\mathcal{D}2$), also according to the four types of judgement given Fig. 10 in Chapter 2.2.3:

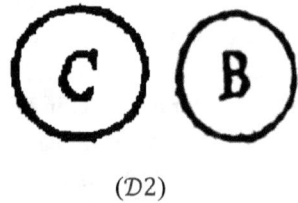

($\mathcal{D}2$)

Now ($\mathcal{D}1^*$) and ($\mathcal{D}2$) are combined with each other to form complex diagrams ($\mathcal{D}3$). Since the relative position of C and A is not clear from ($\mathcal{D}1^*$) and ($\mathcal{D}2$), all possible combinations that fulfil the circular relations of ($\mathcal{D}1^*$) and ($\mathcal{D}2$) must be drawn. These are:

2.3 Proof – Elementary Geometry, Syllogistic and Intuitive Proof Theory

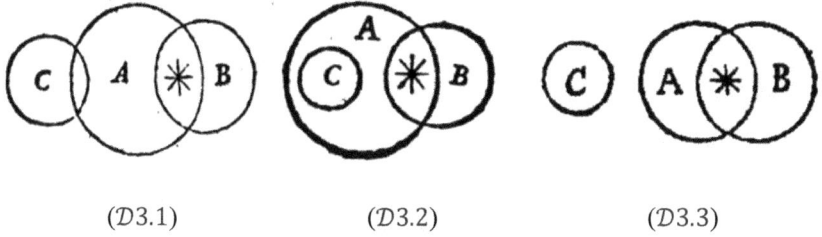

(D3.1)　　　　　　　(D3.2)　　　　　　　(D3.3)

In the last step, one has to check whether the *conclusion (J3)* can be read from all complex diagrams, i.e. (D3.1), (D3.2) und (D3.3). So now one only looks at the relation of the circles C and A in (D3.1), (D3.2) und (D3.3). If only one diagram for the relation of C and A shows a contrary or contradictory judgement to (J3) according to Fig. 4(a) in Chapter 2.2.4, the conclusion does not necessarily follow from the premises and the inference is invalid.

In the example given by Euler, the relation of C and A in (D3.2) is contradictory to the expected diagram for (J3), as Fig. 1 indicates. In other words, (D3.2) shows the judgment All C are A, although according to (J3) it should be Some C are not A.

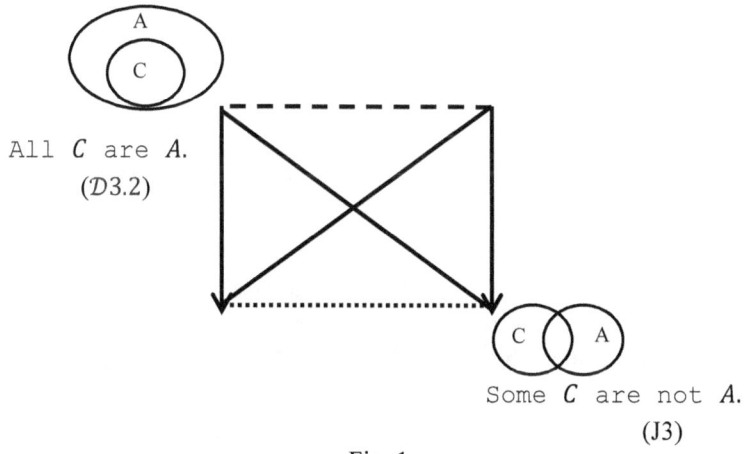

Fig. 1

All in all, Euler believes that with this method he has introduced a direct, i.e. non-apagogical, method of proof that decides on the validity and invalidity of syllogisms.[278]

---

[278] A detailed explanation of a proof procedure using Euler diagrams is given by Peter Bernhard: Euler-Diagramme, pp. 45–53 (= Chapter 3.2.2).

2 Logic and its Geometrical Interpretation

Although (J1) and (J2) are true, the falsity of the conclusion (J3) shines through and falls directly into the eye ("la fausetté de la conclusion saute aux yieux").[279] Euler further clarifies that the nineteen valid inferences guarantee that the conclusion is true if the premises are also true. It is to this statement that Euler now links what is, in my opinion, a puzzling quote:

D'ou V.A. comprend, *comment de quelques verités connües on arrive à des verités nouvelles*, & que tous les raisonnemens, par lesquels on démontre tant de verités dans la Geómetrie se laissent réduire des syllogismes formels. [280]	Hence you perceive, *how, from certain known truths, you attain others before unknown*; and that all the reasonings, by which we demonstrate so many truths in geometry, may be reduced to formal syllogisms.[281]

The quote is puzzling in that one can interpret the second clause in two quite different directions, depending on how one interprets 'syllogismes formels': If by formal syllogisms only the nineteen syllogisms written in the valid inferential forms of two true premises and one true conclusion are meant, then Euler's reference can be interpreted as a turn to logicism and rationalism: Geometry would then be, as the Ramists and Leibnizians think, traceable to unintuitive syllogisms. If, however, one interprets the partial theorem in such a way that 'syllogismes formels' refers to the purely diagrammatic form, then one can understand Euler in such a way that the logical propositions of geometry are also to be understood as intuitive as the figures and diagrams they represent: Geometry would then be, like Schopenhauer, Weigel and also some Kantians think, a purely visual science. Numerous good reasons can be given for both interpretations, but I would like to dispense with them here. It is already a mystery to me why Euler takes up the subject of geometry here at all since it otherwise plays no role in his *Letters*.

I now come to Maaß, whose attitude to geometry has already been determined in Chapter 2.3.1. Unlike Euler, for Maaß the form of analytical diagrams is not arbitrary. He favours triangular diagrams, which he uses in his *Grundriß der Logik* – a work that appeared in four editions between 1793 and 1823. The *Grundriß der Logik* is divided into three parts: (1) pure logic, (2) applied logic and (3) practical logic. In the introduction to the *Grundriß*, Maaß explains that pure and practical logic are systematically written in a complementary way. Thus, both are in contrast to applied logic, which is based on empirical principles taken from psychology. In a section that fills large parts of the introduction, Maaß compares his analytical triangular diagrams with those of his predecessors, criticising Euler and Lambert in particular:

---

[279] Leonhard Euler: Lettres à une princesse d'Allemagne, Vol. 2, p. 125 (= L. CV).
[280] Leonhard Euler: Lettres à une princesse d'Allemagne, Vol. 2, pp. 125f. (= L. CV).
[281] Translation taken from Letters of Euler, 1833, Vol. 1, p. 410.

2.3 Proof – Elementary Geometry, Syllogistic and Intuitive Proof Theory

> In applied logic, I await the verdict of experts on the new way of visualising the relations of concepts, judgments and inferences by means of drawings (§§ 365–381). As is well known, Euler and Lambert attempted the same thing. Euler's invention is not useful; Lambert's is much more perfect, but the signs that Lambert used still lack a complete analogy with the signified.[282]

Maaß's criticism of Lambert is similar to Holland's, which I have already mentioned above. According to Maaß, Lambert is trying to illustrate two metaphors in one diagram, but both metaphors are mutually exclusive: (1) one concept 'falls under' another and (2) concepts have an 'extension'. Only if both metaphors could be represented in a diagram would there be a perfect analogy between the sign and the signified. However, whereas Holland rejects the subsumption metaphor in favour of the extension metaphor and therefore favours Lambert, Maaß tries to improve Lambert's diagrams by seeking to unite both in one diagrammatic form – a similar project can also be found in Lange's *Inventum*.[283]

Although Lambert's diagrams come much closer to this analogy than Euler's, they are also deficient in that they do not illustrate both metaphors equally. The main argument that Maaß puts forward against Lambert concerns the fact that Lambert's lines are supposed to indicate the extension in the diagram, namely, for example, that *A* and *B* are identical; however, the lines are written one below the other, so that they de facto indicate that *A* and *B* are not identical but separate from each other. Only triangular diagrams can show both, since they can represent the subordination through the lines and the extension through the area that the lines delimit. The area thus also connects the lines that otherwise appear spatially separated from each other.

In his applied logic, Maaß returns to the assessment and evaluation of analytical diagrams. In § 364, he explains that a complete analogy between the sign and the signified is achieved precisely when all that is applied to a sign also applies to the signified. Following this definition, he reflects particularly on the advantages and value of signs and gives an example of a complete analogy between the sign and the signified by means of a description of analytical diagrams:

> Such a sign places the signified before our eyes, as it were, and thus immensely *promotes the distinctness and evidence* [*Deutlichkeit und Evidenz*] of the knowledge of the latter: [...] In the meantime, the sign, in addition to the aforementioned benefit, also *facilitates the invention of new truths* by allowing us *to overlook* the signed in all its relations, as it were, *at a glance* [daß es die *Erfindung neuer Wahrheiten erleichtert*, indem es uns das Bezeichnete in allen seinen Verhåltnissen gleichsam *mit einem Blicke übersehen* låßt.][284]

---

[282] J. G. E. Maaß: Grundriß der Logik, p. IXf.
[283] Vide supra, Chapter 2.2.3.
[284] J. G. E. Maaß: Grundriß der Logik, p. 245. (My emphasis– J.L.)

This quote also shows that for Maaß analytical diagrams are an aid to clarity and evidence. In particular, the fact that by means of diagrams the signified can be overlooked in a single glance is reminiscent of the metaphors of Weigel, Ploucquet and Euler. Maaß' insight that analytical diagrams facilitate the invention of new truths also recalls the first sub-sentence of the (enigmatic) Euler quote above. This statement according to which a new ampliative conclusion is generated from known true premises is not self-evident, however, if one thinks of Mill's critique of the syllogism, according to which the deduced conclusions of valid syllogisms can never contain more than what was already laid out in the true premises.[285]

2.3.5 Logica More Geometrico versus Geometria More Syllogismorum

In Chapter 2.3.1, I used the dispute between Leibnizians and Kantians to show the different valuation of figures and diagrams in geometry in the late 18th century: While Leibnizians at this time disdained the value of geometrical diagrams and tried to advocate a purely logical geometria more syllogismorum, Kantians saw geometrical diagrams as a veritable part of the mathematical discipline. In the course of Chapters 2.3.2 and 2.3.3, it has also become apparent that this dispute was not a historical special case that alone revived the discussion of the *quaestio de certitudine mathematicarum*, but that the discussion about the value of intuition in proofs has run through the history of – at least modern – mathematics and philosophy of mathematics and is still relevant today. In retrospect, the dispute between the Leibnizians and Kantians thus seems to be merely an episode in the modern dispute between logicists and representationalists, which, moreover, only related to the question of intuition in geometry.

Schopenhauer's opinion within this dispute also seems worthy of discussion. Although Schopenhauer primarily takes up anti-rationalist arguments and positions that were, starting from Crusius, latently represented in Kantianism, he nevertheless attempts to harmonise them with the rationalist principle of sufficient reason. Even though Schopenhauer never abandoned this argument for harmony, in the course of his creative period – I am convinced here by Oscar Janzen's thesis of a revised doctrine – the attacks against a non-intuitive and, above all, against a logically founded geometry become stronger.

In history, one may find geometries without diagrams, but I am not aware that there were also geometries only with diagrams and without algebraic or generally linguistic

---

[285] Cf. John Stuart Mill: A System of Logic, Ratiocinative and Inductive, pp. 122ff. Vide supra, Chapter 2.3.

2.3 Proof – Elementary Geometry, Syllogistic and Intuitive Proof Theory

signs.[286] Whereas rationalists quickly tend to the extreme position of eliminating intuitions from logic and geometry altogether in order to supposedly proceed in a strictly verbal manner, I have not found any author who attempted to design a non-verbal geometry or logic until the 20th century. To my knowledge, even Schopenhauer never argued for a purely intuitive geometry or even logic. Diagrams prove geometrical or logical judgements, but they do not thereby replace the judgements they are supposed to prove. It was only in the late 20th and especially 21st century that people became interested in purely non-verbal geometries or logics and began to seriously explore their possibilities.

But what role do visualisations play in geometry and logic? If even hardliners like Schopenhauer assume that visualisations cannot replace verbalisations, then what argument really speaks for saving and preserving intuitions in geometry and logic? In Chapter 2.3.4, I listed several arguments up to the early 19th century that philosophers and mathematicians have put forward in favour of approaches to logic based on visualisation. Almost all arguments can first be divided into two classes, (1) weaker didactic arguments and (2) stronger proof-theoretical moments.

(1) The weaker didactic arguments can be summarised as follows: Analytical diagrams

- *are solid, most evident and obvious* (Reimers),
- *are for the purpose of an easier exposition* (Weigel),
- *are much more easier to learn than by Barbara, Celarent* (Weigel),
- *makes it possible to find out the modes of syllogisms as if in the twinkling of an eye* (Weigel),
- *help us so that the whole inference would be overlooked at a glance* (Ploucquet),
- *extremely commodious for facilitating our reflections* (Euler),
- *are sensible to the eye* (Euler),
- *promote the distinctness and evidence* (Maaß),
- *allow us to overlook at a glance* (Maaß).

I have classified these quotes as didactic arguments because they agree in pointing to the ease, speed and simplicity of diagrammatic reasoning: It is easier, faster or simpler to teach or learn with diagrams than with other means. The traditional Aristotelian syllogistic, the scholastic mnemonic or the respective contemporary school of logic is usually used as an object of comparison.

I have called the quotes of (1) weaker arguments because I think they illustrate a problem: they roughly correspond to the formulations I summarised under the point (L2) in Chapter 2.3.2, which were used by rationalists and logicists to thereby weaken

---

[286] The Euclid edition by Oliver Byrne (London 1847) is at best close to a visual approach.

intuition in geometry. Diagrammatic thinking thus seems to have been valued differently by authors around the 18th century: In logic, ease, speed and simplicity are exceedingly beneficial tools, whereas in geometry they are perceived as an inessential tool that tends to obscure the rational core of geometry (L3).

(2) Among the stronger proof-theoretic arguments, the following statements can be summarised: Analytical diagrams

- *can demonstrate (or reduce) syllogisms much more briefly* (Weigel),
- *are means to discover the inconsistencies, because geometry soon conceals the fallacies* (Lambert),
- *completely remove all doubt about the infallibility of inferences* (Ploucquet),
- *show how, from certain known truths, one attains others before unknown* (Euler),
- *facilitate the invention of new truths* (Maaß).

The quotes from (2) all illustrate a problem: all the strong arguments given by the renowned logicians for the use of diagrams are (a) essentially either only reinforcements of the weak didactic arguments or (b) do not name any advantages that only diagrams possess. I will discuss the quotes from (2) in more detail in order to illustrate this problem.

The Weigel and Lambert quotes are (a) more strongly expressed didactic arguments applied to proof theory: The Weigel quote is a didactic argument as it emphasises the simplicity aspect (*more briefly*) in a diagram-oriented proof theory; the Lambert quote is a didactic argument as it emphasises the rapidity (*soon*) of an intuitive proof theory.

The quotes by Ploucquet, Euler and Maaß are (b) arguments that do not apply to diagrams alone: Even purely verbal syllogisms or formalised inferences can remove the doubt against the infallibility of inferences (Ploucquet), can show how to get from some known truths to new truths (Euler) and can facilitate the invention of new truths (Maaß) – think here, for example, of the discussion about Dummett that I outlined in the introduction to Chapter 2.3.

Of course, I cannot guarantee to have found all arguments for the use of diagrams in logic up to the 19th century. I have also left out Reimer's stronger arguments since he does not explicitly speak of diagrams in his doctrine of principles (*Inesse, &, non-inesse*). But even if the quotes listed in Chapter 2.3.4 are only a representative selection of reflections on analytical diagrams, the result of this selection is sobering, since there seem to be no really strong arguments for the use of diagrams in logic: Even in the so-called 'golden age of logic diagrams', strong arguments for an intuitive theory of proof can be traced back to didactic arguments, which have to be classified as weak insofar as they have been used in geometry by logicians precisely as a counter-argument for an intuitive proof theory.

2.3 Proof – Elementary Geometry, Syllogistic and Intuitive Proof Theory

Moreover, I have not been able to find many statements about geometry in the geometrical logicians presented: only Euler's enigmatic quote and Weigel's remark that diagrams were taught in geometry *for the sake of a clearer form of doctrine*, and lastly Reimer's Aristotelian quote alluding to diagrams being the mathematicians' 'Why'. Thus, there seems to be no strong case from geometrical logicians for the use of diagrams in geometry. Moreover, if diagrams did not have an independent logical function that could not also be fulfiled by purely verbal syllogisms or a formalisation, then this should also have weakened their position in geometry. To put it briefly and simply: if a logica more geometrico (or *solo oculorum usu*) fails, then this strengthens a geometria more syllogismorum. A rationalist line of argument can therefore be described roughly as follows: If geometry can never be communicated purely intuitively, then it requires a minimum of independent logical functions. Diagrams have no independent logical function. Therefore, diagrams are not suitable for conveying geometry.

In the following, I would like to argue that the first strong argument of a geometrical logician is found in Schopenhauer's *Berlin lectures*. His argument seems to turn the rationalist mode of reasoning on its head: A purely discursive geometria more syllogismorum is problematic as long as the logic is not proven intuitively more geometrico. If syllogisms can be proven intuitively employing a proof theory more geometrico, then geometry must also be provable intuitively.[287]

For Schopenhauer, the discipline in question is thus not geometry, but logic: if a logicist wants to prove geometry from logic, she must, in Schopenhauer's opinion, first explain how logic can be proved. However, Schopenhauer first builds up his representationalist argument against the logicist proof theory in an alternative way. In answering the question of why he favours an intuitive procedure in his logic at all, he first resumes some weak didactic arguments that Weigel, Ploucquet, Euler and others had already used before him. Moreover, he combines the weak didactic argument with a proof-theoretical argument that can indeed be called strong insofar as it explicitly speaks out against deductive proof procedures in logic and argues strongly for an analogy between discursive reasoning and visual-spatial perception. The relevant quote reads:

> [9]In particular, these intuitive schemata [sc. the analytical diagrams] [10]make it very easy for us to recognise the rules of syllogistic, [11]and relieve us of the proofs of the rules: for Aristotle [12]always gave a proof for every syllogistic rule, which [13]is actually superfluous, even impossible in terms of rigour; [14]for the proof itself is an inference and consequently presupposes the rules: [15]one can actually only make these rules distinct [16]and then reason immediately sees

---

[287] One may have found my use of the word 'theory' inappropriate (vide supra, Chapter 2.1), but it should become clear here at the latest that the term 'theory' in the context of the theses presented is precisely regaining its original meaning, which it increasingly lost in the 20th century.

2 Logic and its Geometrical Interpretation

their necessity, [17]because they are themselves the expression of the form of reason, i.e. of [18]thinking.

What Aristotle accomplished by his proofs, [19]the intuitive schemata will accomplish much better and much [20]easier for us: for, since they have a quite exact analogy to the [21]circumference of concepts, they let us see the relations of [22]concepts to one another in the easiest way, namely [23]intuitively, and we will thus bring the necessities, which arise from [24]these relations, to the easiest comprehensibility.

[25]The Aristotelian proofs have long since been omitted from [26]logic; but the clarification [27]by intuitive schemata has not yet been substituted for them as consistently [28]as I intend do.[288]

As in Chapter 2.1.5, I have also taken the line numbers (= L.) of the quoted Mockrauer/Deussen edition of 1913 and prefixed them to the respective line in the quote in superscript numbers for a more detailed discussion. To my knowledge, the quote represents a novelty in the history of analytical diagrams. That Schopenhauer rudimentarily noticed the significance of his undertaking[289] is evident not only in the last paragraph of the quotation (L.25–28) but also in the fact that he repeated several arguments from it in other variations several times in his logica major.[290]

Although the quote, on the one hand, (A) also takes up weak arguments, as they can be found similarly in the history of analytical diagrams; it also on the other hand, (B) names the first argument that can actually be classified as strong, since this (a) is neither only a reinforcement of the weak didactic arguments (b) nor names an aspect that could also be fulfiled by a purely discursive logic or a formalism. I first address (A) the weak arguments before examining (B) the strong ones.

(A) Schopenhauer first resorts to a seemingly weak didactic argument, which, similar to those of Weigel, Ploucquet, Euler and others, emphasises that diagrams reduce the complexity of logic with ease: (A1) Analytical diagrams will greatly facilitate the knowledge of the rules of syllogistic (L.9f.). In the second paragraph of the quote, Schopenhauer seems to repeat this weak didactic argument several times: (A2) The intuitive schemata will accomplish much more easily what Aristotle had carried out by his proofs (L.19f.). (A3) One could see the conceptual relations in the easiest way (L.22f.), (A4) and one could thus bring the necessities arising from the conceptual relations to the easiest comprehensibility (L.23–25).

That (A1) to (A4) are weak arguments is indicated by the concept of ease (Leichtigkeit) used in each case, which was also found in many of Schopenhauer's predecessors.[291] It is questionable, however, what exactly is meant to be facilitated.

---

[288] WWR2 I, p. 272. [Paragraphs were inserted by me – J.L.]
[289] I will discuss Schopenhauer's achievement for geometrical logic in more detail in Chapter 2.3.6.
[290] Cf. esp. WWR2 I, p. 357.
[291] Vide supra.

2.3 Proof – Elementary Geometry, Syllogistic and Intuitive Proof Theory

Here, seemingly different areas are initially defined: Rules (A1), proofs (A2), conceptual relations (A3) and their necessities (A4). In the following, I will focus particularly on (A1) and (A2), since they mainly concern proof theory; I will not take up (A3) and (A4) again until Chapter 2.3.6, where I argue that they address semantics.

In (A1), the notion of rule stands out, which is only used in the first paragraph of the quote – inflationary in (L.10, 11, 12, 14, 15) – and which perhaps does not necessarily have to be interpreted only as a didactic competence, since it is a central notion of Aristotelian syllogistic. It is problematic, however, that Schopenhauer, following the quote, interprets Aristotelian logic almost throughout with the scholastic mnemonic technique, which also shapes the interpretation of Aristotelian syllogistic to this day. What Schopenhauer has in mind with the concept of rules, which he applies to Aristotle, will become clear in the following, especially from two subchapters of the doctrine of inference of the *Berlin Lectures*, which bear the titles: *General Rules for the Inferences of All Figures and Special Rules (Allgemeine Regeln für die Schlüsse aller Figuren und Besondre Regeln).*[292]

Schopenhauer takes the *General Rules for the Inferences of all Figures* from Arist. An. pr. I.7, 29a19–29b28 as well as An. pr. I.23ff., 40b17ff. For example, the first general rule Schopenhauer states is: "1) The inference must have three termini or concepts; neither more nor less."[293] The general rule 1) refers to the proof given by Aristotle in An. pr. I.25, 41b36ff. *The Special Rules*, which refer to the three figures, Schopenhauer takes for the most part from Arist. An. pr. I.4–7, 25b26–29b28 or the scholastic interpretation. For example, the special rule for the first figure is: "a) Let the propositio major be universal. b) Let the propositio minor be affirmative. Sit minor affirmans nec sit maior specialis."[294] The special rule for the first figure refers to the proofs Aristotle gives in An. pr. I.4, 25b26ff.

Schopenhauer gives a total of nine rules, most of which can be taken from the Aristotelian text and which follow the usual eight from the scholastic tradition. Schopenhauer excludes only one separate general rule of modality (according to An. pr. I.12 32a8–12) and divides instead the one usually taken from An. pr. I.24, esp. 41b27–31 into two general rules no. 4 and 5.[295] Schopenhauer's omission of a separate general rule of modality (e.g. 'With respect to modality, a premise must be similar to the conclusion') is due to the fact – as discussed in Chapter 1.3.3 – that Schopenhauer deals with the problematic judgements of modal syllogistic separately and initially refers only to assertoric syllogistic.

Before Schopenhauer draws up the lists of rules based on Aristotle, however, he develops and explains them individually in a lengthy treatise with the help of the scholastic assumption that there are four figures and not just three – as is then asserted in

---

[292] WWR2 I, pp. 324f.
[293] WWR2 I, p. 324.
[294] WWR2 I, p. 325.
[295] As a comparison, consider, for example, Leonhard Rabus: Lehrbuch der Logik in neuer Darstellung, pp. 83ff. or Friedrich Ueberweg: System of Logic, p. 379ff.

his list of special rules.[296] From the combination of the four basic possibilities of judgement (An. pr. I.2, 25a1ff.: universal/ particular; affirmative/ negative) and the associated general rules for four figures arise a total of sixty-four modes with three judgements each ($4^3 = 64$), of which, however, only nineteen are valid.[297] Only at the end of this lengthy treatise does Schopenhauer conclude that the fourth figure is only "the straight inversion [gerade Umkehrung] of the first" and has only one rule, which, moreover, is in semantic conflict with the rule of the first figure;[298] therefore, Schopenhauer abandons the fourth figure and then draws up the Aristotle-oriented list with the special rules for only three figures.

In the lengthy treatise, Schopenhauer uses the scholastic mnemonic technique for the designation and explanation of the nineteen valid modes in four figures, which assigns individual vowels[299] to the basic possibilities of judgement and consonants to the rules of conversion;[300] this gives each syllogism its own name, namely B*a*rb*a*r*a*, C*e*l*a*r*e*nt, etc. Based on this approach, we can see that for Schopenhauer, Aristotelian and scholastic logic must be congruent in this lengthy section, albeit with the difference that that theory of proof focuses more on the rules, this one on the individual modes arising from them.

The scholastic mnemotechnical procedure has been sufficiently described by Schopenhauer himself, but also in many current works[301] so that I will leave it here with the explanations given. Schopenhauer evaluated the scholastic mnemonic technique differently: On the one hand, he has called it an essential improvement in comparison with Aristotelian rule logic, but on the other hand, he has also admonished its cumbersomeness and has regarded it as an interpretation of the original proof theory.[302] Probably for the latter reasons and because he rejected the fourth figure, which was only introduced in the time after Aristotle, he returned to the Aristotelian proofs by rules several times.[303]

Nevertheless, Schopenhauer uses the scholastic mnemonic technique following the quotation to explain the individual modes, which he additionally presents with his analytical diagrams. The difficulty of the above quotation and the expression 'rules'

---

[296] Cf. WWR2 I, pp. 299–324.

[297] Cf. WWR2 I, pp. 305f.

[298] WWR2 I, p. 321. Cf also p. 330.

[299] Schopenhauer cites, for example, Gottsched's mnemonic: "Das *A* bejahet allgemein: / Das *E*, das sagt zu Allem Nein. / Das *I* sagt Ja, doch nicht zu allen: / So läßt auch *O* das Nein erschallen." (WWR2 I, p. 287, verbatim: "The *A* affirms in general: / The *E* says No to all who are. / The *I* says yes, but not to all:/ and also *O* in No's recall")

[300] Cf. WWR2 I, pp. 284–293. (*S* = conversio *s*implex; *P* = conversio *p*er accidens; *M* = *m*utare; *C* = per contradictionem)

[301] Cf. e.g. Peter Bernhard: Euler-Diagramme, Chapter 2; Neil Tennant: Aristotle's Syllogistic and Core Logic. In: History and Philosophy of Logic 35:2 (2014), pp. 120–147; John Corcoran: Aristotle's Natural Deduction System. In: Ancient Logic and Its Modern Interpretations. Proceedings of the Buffalo Symposium on Modernist Interpretations of Ancient Logic. 21. and 22. April, 1972. Ed. by J. Corcoran. Dordrecht 1974, pp. 85–131.

[302] Cf. WWR2 I, pp. 305f.

[303] Cf. WWR2 I, pp. 319–324, pp. 327–330.

2.3 Proof – Elementary Geometry, Syllogistic and Intuitive Proof Theory

thus results primarily from the fact that in the quotation he refers solely to the Aristotelian procedure, but in the 'lengthy treatise' I have mentioned he always already explains it with the scholastic mnemonic technique. According to scholastic proof theory, he could thus also have formulated (A1) and (A2) as follows:

> In particular, these intuitive schemata (or analytical diagrams) make it very easy for us to recognise the syllogistic modes and relieve us of the proofs of these modes: for one always gave a proof for every syllogistic mode, which is actually superfluous, even impossible in terms of rigour; for the proof itself is an inference and consequently presupposes the validity of the mode used [....
> What scholasticism achieved with its artificial words, the intuitive schemata will accomplish much better and much easier for us ....

Schopenhauer, however, seems to have had a good reason for dispensing with such scholastic terminology in the original quotation and referring instead to Aristotle's rules. Schopenhauer will have initially found the scholastic mnemonic in the lengthy treatise to be a didactic advantage, in order to be able to introduce and name the actual proofs of rules in Aristotle more easily. But just as the scholastic mnemonic is only a didactic improvement on Aristotle's proofs of rules, so too are the analytical diagrams a didactic improvement on the scholastic mnemonic. In other words, the fact that Schopenhauer refers directly to Aristotle in the relevant quote is probably due to his belief that he no longer needs the diversions via the scholastic terms of art. Schopenhauer's weak arguments (A1–A4) thus state: Analytical diagrams become the actual interpretative instrument of the Aristotelian text for didactic reasons (Leichtigkeit).

(B) The real novelty in the given quote, however, concerns the strong arguments it contains for analytical diagrams in the logic of inference. Schopenhauer explicitly emphasises that the intuitive schemata or analytical diagrams (B1) "relieve us of the proofs of the rules" (L.11) and (B2) are "much better" (L.19) than the Aristotelian theory of proof. Schopenhauer thus goes considerably beyond his weak arguments (A1–A4), since he not only regards analytical diagrams as the actual interpretative instrument of the Aristotelian text, but also argues that intuitive schemata are better than the Aristotelian proof theory.

Schopenhauer, however, does not claim that analytical diagrams completely replace the Aristotelian doctrine of inference, but only that they make the purely rule-guided or mnemonic theory of proof dispensable. This is a crucial difference to today's approaches of purely intuitive diagram logics in the vein of Shin:[304] according to Schopenhauer, syllogisms are still communicated discursively and verbal, but are supposed to be proved intuitively.

---

[304] Vide supra, Chapter 2.3 (Introduction).

His integration of a visual proof theory into a verbal logic of inference thus does not have to dispense with the presentation of Aristotelian rules or the reference to scholastic terms of art for syllogisms, since the discursivity of language and the inferences it contains is not in question. Only the proof theory that justifies the validity of these inferences is to be written intuitively. In this way, however, individual components of the Aristotelian and scholastic logic of inference lose their significance. I will first discuss the loss of significance of the scholastic mnemonic and then the loss of significance in the Aristotelian doctrine of inference.

For scholastic logic, the loss of significance is most evident in the naming of the respective modes, which were composed of meaning-bearing vowels and consonants: Whether a judgement of an inference is universal or particular, affirmative or negative (vowel meaning), becomes apparent in a diagram by the fact that the areas representing the concepts overlap completely or partially or do not overlap. Without transforming the judgements of an imperfect inference in such a way that they are proven to be valid by a perfect inference (consonant meaning), it now becomes directly apparent whether all the conceptual relations of a syllogism expressed by the three judgements can be depicted in a diagram – or not.

Furthermore, as described in Chapter 1.3.3, the diagrams also show the differences between the respective figures and thus between the special rules. In the lengthy treatise, in which Schopenhauer still uses the scholastic mnemonic to designate the nineteen valid modes, he tries to show with the help of the analytical diagrams that, on the one hand, Kant is wrong when he wants to reduce the four figures to one, but on the other hand, scholastic logic is also wrong when it wants to extend the three Aristotelian figures by a fourth: for the inferences of the three figures, shown with the help of the analytical diagrams, each indicates their own appearance of the medius[305] and thus their own special rule and function: 1st Figure: ground of decision, 2nd Figure: ground of distinction, 3rd Figure: ground of elimination.[306] The fourth figure shows no unique visual function of the medius, which also supports Schopenhauer's thesis that its rule is only a reversal of the first figure and thus not an independent one. As a consequence, Schopenhauer rejects the fourth figure, as described above, and thus draws up a list of only three particular rules. Following this list, he no longer makes use of the scholastic mnemonic in his doctrine of inference.[307]

---

[305] Cf. WWR2 I, p. 329.
[306] Vide supra, Chapter 1.3.3. One can probably see a certain similarity between Schopenhauer's metaphoric of the three functions, i.e. 1. manipulator (Handhabe), 2. septum (Scheidewand) and 3. indicator (Anzeiger), and Zekl's vivid description of the running-over-the-path (Über-den-Weg-Laufens) of the premises, 1. relay (Stafette), 2. swarming out (Ausschwärmen), 3. star migration (Sternwanderung), cf. Hans Günter Zekl: Einleitung. In: Aristoteles: Erste Analytik. Zweite Analytik. (Organon Vol. 3/4). Hamburg 1998, pp. IX–CXXI, here: XXII.
[307] It is noticeable, however, that Schopenhauer does not revise his judgement from the longer treatise that nineteen valid cases result from the combination of the four basic possibilities of judgement and the associated general rules for four figures.

2.3 Proof – Elementary Geometry, Syllogistic and Intuitive Proof Theory

Furthermore, a loss of meaning also becomes apparent when one draws a comparison between Schopenhauer's visual and Aristotle' rule-guided proof technique: A large part of the general rules no longer needs to be considered since in the diagrammatic representation of an inference, compliance with or violation of the rules becomes directly visible. I will take two rules already mentioned above as an example: Whether, for example, a conclusion has exactly three termini (General Rule 1) becomes clear from the number of circles depicting the terms in a diagram; that, for example, a violation of the special rule for the first figure, namely that the propositio maior is universal, enables a valid representation of the premises, but then does not represent the conclusion, is shown by Schopenhauer in the diagram (see Fig. 1) for the syllogism: Some fish fly. All trout are fish. All trout fly.

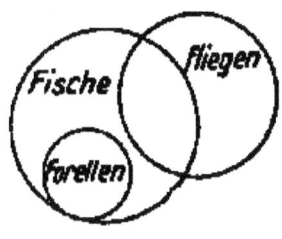

Fig. 1
WWR2 I, p. 298.

Fische = fish, fliegen
= fly, Forellen = trout

I have claimed above that the arguments classifying with (B) become strong arguments because they are neither (a) only essentially a reinforcement of the weak didactic arguments nor (b) designate an aspect that could also be fulfiled by a purely discursive logic. But it is precisely the argument of a loss of meaning that has just been put forward that indicates that (a) Schopenhauer's analytical diagrams may be facilitation or simplification, but their function (b) could basically also be fulfiled by the Aristotelian-Scholastic proof theory. What, then, makes Schopenhauer's (B) arguments strong arguments, apart from the explicit fact that they make classical proof theories dispensable (B1) and better (B2)? In short, why are Schopenhauer's analytical diagrams better than discursive proof techniques?

In the above quote, Schopenhauer gives two arguments for the preference of analytical diagrams over discursive proof techniques. The first argument is a negative one, the second a positive one. I will deal with the negative strong argument (NSA) first, and build on it to deal with the positive strong argument (PSA).

(NSA) At the beginning of Chapter 2.3.5, I had claimed that Schopenhauer argues, above all, for the fact that a purely discursive geometria more syllogismorum is problematic until logic is proven intuitively more geometrico. Schopenhauer's starting point for this argument is a problematisation of purely discursive logic. In this problematisation, he takes up the empiricist and sceptical tradition argument of Sextus, Bacon, Reimers and Locke, which Mill, Carroll, Goodman and others also discussed after him. These have already been mentioned at the beginning of Chapter 2.3 (and further in Chapter 2.3.4): Syllogistic cannot provide a proof theory in which syllogisms are proved by syllogisms since such a proof theory always leads into one of the classical tropes. The axiomatisation of deductive logics was seen as particularly problematic since reductive proof procedures of the individual syllogistic and

2 Logic and its Geometrical Interpretation

propositional modes always led to one or more principles such as the *dictum de omni et nullo* or the *dictum de si et aut*. Sceptics and empiricists therefore either rejected deductive logic altogether (Sextus, Locke) or opposed it wholly or in part with inductive logic (Bacon, Mill and further Arist. Eth. Nic. IV.3).

Schopenhauer takes up this criticism of reductive proof by carefully and radically strengthening a weak argument: Aristotle always gives a proof for every syllogistic rule, which is actually superfluous (L.11–13), according to the weak argument. He radicalises it, however, by denouncing the impossibility of such a proof procedure: according to the strictness of the argument, this reductive proof procedure is impossible, because the proof itself is an inference and therefore presupposes the rules (L.13–15). If we transfer this argument back into common philosophical language, it becomes even clearer: a reductive proof procedure is impossible because the proof consists of a mode and consequently presupposes a mode.

Schopenhauer does not reflect on this statement again in abstracto but I think one can regard this argument as an accusation of a classical error of proof: The entire theory of proof itself is based on a proof error, namely on a petitio principii. As Aristotle explains (An. pr. II 16, 64b34ff.), this error of proof occurs when something in need of proof is explained by something in need of proof and not by something that does not need proof (e.g. axioms). In my opinion, Schopenhauer's argument can be understood as follows: If inferences are proved in Aristotelian-Scholastic logic, they seem to be in need of proof. If this is true and inferences need proof, then something in need of proof is proven by something in need of proof, which should actually have been proven by something that does not need proof. Inferences must therefore be proven by something that is not itself an inference or implies an inference.

Schopenhauer seems to have two types of axioms in mind: 1) Either the axioms represent intuition, in which case they themselves do not require proof and everything that follows from them are – precisely speaking – not proofs but demonstrations;[308] 2) or the axioms do not correspond to intuition, in which case they must be justified. In any case, they should only be used in those sciences in which one has no access to that which does not require proof. For this reason, Schopenhauer also characterises axioms and principles from which proofs follow as a "makeshift" (Notbehelf) that is used "where direct knowledge through intuition is not accessible to us".[309] This can mean, for example, formal systems that can no longer be traced back to intuition. As was shown in Chapter 1.2, however, Schopenhauer's philosophy has access to intuition as an immediate source of knowledge and therefore does not need to resort to such a makeshift. Moreover, for Schopenhauer, any philosophy that denies this immediate cognition is the self-imposed limitation of a fictitious idealism.

Concerning philosophy, however, axioms and principles are also out of the question for a second reason: Philosophy, according to Schopenhauer, must be the science in

---

[308] WWR2 I, pp. 131f.
[309] WWR2 I, p. 549.

2.3 Proof – Elementary Geometry, Syllogistic and Intuitive Proof Theory

which every proposition may be a problem. This is precisely the differentia specifica of philosophy from all other sciences, especially those that are axiomatically structured.[310] This is important insofar as otherwise the boundaries between philosophy and those sciences with which it seems to share the same or at least a similar object of investigation, e.g. intuition, nature, the world or the like, would become blurred. For philosophy, Schopenhauer says, "from the outset everything is equally unknown and alien, it is grounded on no presuppositions at all".[311] To put it bluntly: Philosophy is the science in which everything is allowed to be a problem. Only through this characterisation can the questions that remain unasked in all other sciences and the problems that appear unproblematic in all other sciences become part of philosophical enquiry.

The axiomatic method of logicism may therefore be regarded as a strictly mathematical or perhaps also as a strictly scientific approach. However, it cannot satisfy philosophical demands, since they already follow presuppositions and methodological ideals. From a philosophical point of view, this is problematic: if the axioms do not represent intuition, they are either dogmatically set up, subject to the petitio principii described above, or they are supported by ever more axioms, principles or foundations, which, however, are themselves in need of proof.[312] This justification problem of logicism, which the Kantians had already demonstrated in elementary geometry, is transferred by Schopenhauer to Aristotelian syllogistic – a logic which, until its revival and further development in areas such as Generalised Quantifier Theory, Montague Grammar, Numerical Term Logic, Natural Logic, etc., was regarded, above all, by logicists only as a simplified fragment of predicate logic, but whose foundations cannot be justified so easily after all.

This sceptical-empiricist argument of Schopenhauer's can arguably be regarded as an (NSA) since it is (a) neither a reinforcement of a weak didactic argument (b) nor does it name an aspect that can also be solved by the discursive-deductive reasoning itself. Schopenhauer's solution to the problem, however, can be briefly summarised: Analytical diagrams accomplish something that discursive inference alone cannot. Schopenhauer thus does not merely criticise Aristotelian-Scholastic syllogistic like Sextus or Locke, nor does he oppose it with an inductive logic like Bacon or Mill, but he seeks a new basis of reasoning in the theory of proof employing analytical diagrams. The (NSA) of the sceptics and empiricists thus becomes the prelude to ushering in a (PSA) that builds on the tradition of Euler, Lambert and Ploucquet.

(PSA) In the above quote, Schopenhauer explains that there is only one way to prove the rules and modes of syllogistic, namely by 'making them distinct' (L.15: deutlich

---

[310] Cf. Jens Lemanski: Wissen, Wissenschaft, Wissenschaftslehre.
[311] WWR2 I, p. 549.
[312] Not uninteresting is the question, which would, however, require a separate investigation, to what extent Schopenhauer's critique of axiomatic systems corresponds to and differs from intuitionist approaches.

2 Logic and its Geometrical Interpretation

machen), whereby they 'can be seen' (L.16: eingesehen werden können). The composites of 'making distinct' and 'being able to see' remain undefined in the first paragraph of the quote, but the context of the quote already mentioned above – namely Schopenhauer's justification of the use of diagrams in the logic of inference – already indicates what he wants to be understood by the expressions: the intuitive schemata or analytical diagrams are the ones that must be put in place of traditional proof theory. The basic visual words of making distinct and seeing (Deuten und Sehen) in the composites 'to make distinct' and 'to be able to see' also point to the visualisation of inferences employing analytical diagrams.

Schopenhauer further explains that with the help of these analytical diagrams, reason immediately sees the necessity of rules and syllogisms because the rules themselves are the expression of the form of reason, i.e. of thinking (L.1–18). With the self-intuition argument thus put forward, he ties in with the discussion from geometry that I presented in Chapters 2.3.1–2.3.3: Analytical diagrams and geometrical figures are not empirical objects or abstractions of empirical objects, but forms of – Kantianly speaking – *äußerer Sinn*, i.e. outer or external sense. However, Schopenhauer had emphasised in numerous places in his dissertation thesis as well as in WWR I and WWR2 that the external sense is a property of the mind and that the spatial objects or appearances imagined through it and their definite or at any time determinable shape, size and ratio are a product of the human imagination.

Just as in his philosophy of geometry, Schopenhauer thus does not argue for empirical verification of discursive-deductive reasoning in logic, but he uses a transcendental argument: as purely a priori forms of our intuition, the analytical diagrams are the condition of the possibility of making distinct (deutlich) and seeing the rules and modes of logical thinking. Thus discursive thinking recognises the necessity of its rules and modes in the form of their own visual expression, and the proof of a deductive inference takes place in the insight of its possibility.

I believe that Schopenhauer's (PSA) in particular can put a stop to the thesis that diagrams have no independent logical function. Contrary to the rationalist figure of thought, in which evidence for a discursive geometria more syllogismorum is taken from the problems of ab intuitive logica more geometrico, Schopenhauer's visual proof theory lays the foundation for his intuitive geometry later explicated in WWR2. Schopenhauer's (NSA) becomes an argument for the fact that a purely discursive geometria more syllogismorum must be problematic since syllogistic itself requires an intuitive grounding. But if syllogisms can be justified intuitively by means of a visual proof theory, then there should also be nothing to prevent the syllogisms one uses in geometry from having an intuitive basis.

2.3 Proof – Elementary Geometry, Syllogistic and Intuitive Proof Theory

2.3.6 A Quite Exact Analogy to the Circumference of Concepts

In the previous chapter, I presented Schopenhauer's arguments that the logical ground of knowing (according to Schopenhauer's *Fourfold Root of the Principle of Sufficient Reason*) is ultimately based on the intuitive ground of being. A strong argument against logicism stated that deductive inferences must be based on something that itself no longer requires a proof and that axioms should only take on this role when they either represent the intuition or no other instance of science is available. A fundamental axiomatisation, therefore, seemed useful for several areas of the philosophy of mathematics, but useless as an application to philosophy as a whole. An extension of logicism to all the subject areas indicated in Chapter 1 would mean nothing less than limiting the space of reasons while constantly expanding the space of concepts. This cannot be a viable path for philosophy as a whole.

Schopenhauer had developed a positive argument based on this negative one and claimed that the ground of logic was an intuitive one. Analytical diagrams, as he used them in the tradition of Euler, Ploucquet and others, were an expression of thinking itself, since thinking produced logical inferences and verified them itself using the a priori forms of intuition. The self-intuition argument, according to which thought regards its own logic as a priori forms, does in fact point out that there is a procedure of proof in logic that is ultimately based on a ground that cannot itself be proven; but it is not an argument that explains why forms of intuition can be regarded as the foundation of logical inferences. Schopenhauer has established this problem in the proof theory of geometry and logic since 1813 and has explicitly admitted and formulated it as a problem several times.

The problem arises for the first time in the dissertation thesis at the point where Schopenhauer feels compelled to justify the quadruplicity of the root of the principle of sufficient reason. Schopenhauer explains that Kant solved such justifications by a deduction a priori, as he did with his categories. But: "But I admit that I do not see the possibility of a deduction a priori of the four classes of representations which are only given to us."[313] Although it may be easy to fake such deductions, he wants to base his division into the four classes of the principle of reason on induction, which is incapable of any other proof "than the challenge of finding some sort of object that does not belong under any one of the four classes I have advanced or presenting two of these classes as reducible to only one."[314] Following this reflection on methods, he also explains that "insight into such a ground of being can become a ground of knowledge", but he does not explain why this can be the case.[315]

In WWR I and WWR2 Schopenhauer discuss the problem between discursive and visual structures of reasoning in more detail. In the context of the two quotes from

---

[313] FR, p. 158 (§ 17 of the 1813 edition).
[314] FR, p. 158 (§ 17 of the 1813 edition).
[315] FR, p. 124 (§ 36)

2 Logic and its Geometrical Interpretation

WWR I and WWR2 already given in Chapter 2.2.5, in which Schopenhauer refers to his knowledge of the history of analytical diagrams (Euler, Ploucquet, Lambert), he makes the following confession, which varies slightly depending on the version:

WWR (1819), § 9	WWR2
I am unable to say what the ultimate basis is for this very exact analogy between the relations of concepts and those of spatial figures. But it is in any event a very fortunate circumstance for logic that the very possibility of all conceptual relationships can, [...] be presented intuitively and a priori by means of such figures.[316]	For between the possible relations that concepts can have to one another and the positions in which circles can be put together, there is a quite exact and absolutely consistent analogy. This is an exceedingly fortunate circumstance for the consideration we are now about to make; on what, however, it is ultimately resting, I do not know how to specify.[317]

To the best of my knowledge, this confession, published about two hundred years ago, has been taken up only once in the reception history of Schopenhauer's philosophy, in a paper written in 1949, which is, however, an example of the scepticism towards intuitions at that time, since it, unfortunately, testifies to a deficient knowledge of the history and systematic function of analytical diagrams.[318] Nevertheless, the author is justified in emphasising the peculiarity of the passage, which consists in the fact that the otherwise rarely modest Schopenhauer admits his ignorance when asked about the justification of the stated analogy. In what follows, the thesis will be defended that Schopenhauer has given a thoroughly satisfactory answer to the aforementioned problem, without, however, having recognised or at least emphasised the associated achievement of his answer.[319]

The aforementioned "exact analogy" (WWR) or "quite exact and absolutely consistent analogy" (WWR2) that Schopenhauer sees between the concepts and the "spatial figures" (WWR) or "positions in which circles can be put together" (WWR2) is discussed once again in the logica major: I had explained in Chapter 2.3.5 that I wanted to leave out the weak arguments (A3) and (A4) there; but now, in the present chapter, I want to come back to them, since they emphasise an aspect that is not, or not only, related to the proof theory discussed in Chapter 2.3.5: (A3) states that with

---

[316] WWR I, p. 65.
[317] WWR2 I, p. 69.
[318] Cf. Gerhard Klamp: Vom Symbolgebrauch geometrischer Figuren in der Logik. In: Schopenhauer-Jahrbuch 33 (1949/-50), pp. 39–65.
[319] Rather, as can be seen from the last paragraph of the quote given in Chapter 2.3.5, he saw his pioneering achievement in the consistent use of analytical diagrams in the logic of inference, although this must be put into perspective in comparison with the authors listed in Chapters 2.2.2 and 2.2.3 (esp. Weigel, Leibniz and Lange).

2.3 Proof – Elementary Geometry, Syllogistic and Intuitive Proof Theory

intuitive diagrams one can see the conceptual relations most easily, and (A4) applies to the fact that analytical diagrams bring the necessities arising from the conceptual relations to the easiest comprehensibility.

Both arguments are in a sentence within the second paragraph of the quote cited in Chapter 2.3.5 (L.21–25) and, simply because of their common position in the text, can hardly be meant as identical statements, although both arguments refer to both the ease and the conceptual relations. In (A3), however, only the conceptual relations are emphasised, whereas in (A4) the necessities arising from the conceptual relations are highlighted. The two weak arguments are also reinforced by a previous backing (BA): "since they have a quite exact analogy to the circumference of concepts" (L.20f., Sect. 2.3.5).

It may be a coincidence that Schopenhauer chose the sequence 1. (BA), 2. (A3), 3. (A4) in the second paragraph of the quote in question; however, all three points in their sequence correspond exactly to his compositionalist structure of logic, as I have presented it in Chapters 2.1, 2.2 and 2.3: (1) concept (quite exact analogy to the circumference of concepts), (2) judgement (relations of concepts to each other), (3) inference (necessities, which arise from these relations). If one takes the sequence seriously, one can take a hint from it as to what the very exact and absolutely continuous analogy can be based on.

Schopenhauer had explained in the doctrine of concepts that the sensualistic essence of the concept was to comprehend things: In the original formation of concepts, determinations or properties were separated from things in a process of abstraction (not to be understood individually), to which one then assigned one's own words (Chapters 2.1.5, 2.2.6). Individuals learn these terms by using them in the various contexts in which they have found the word. Only through this contextual use do they abstract the true meaning of the word and thus find the concept that the word denotes (Chapter 2.1.5).

Usage competence enables speakers to explicate relations of concepts in judgements, which can then in turn be revised by others or based on a newly appearing object reference (Chapters 2.1.6, 2.2.6). If speakers can give good reasons of why the sphere or scope of meaning of one term is necessarily or always contained in that of another, these are so-called analytic judgements; if they cannot, the judgements are synthetic (Chapter 2.2.5).

If concepts are put together several times in relation to judgements, inferences arise from them whose validity can be proved by means of the partially or completely overlapping or non-overlapping spheres (Chapter 2.3.5). The evidential power of analytical diagrams is based on the fact that the logical concepts can be presented a priori utilizing spheres and the geometric concepts by means of their specifically defined semantics and can thus obtain insight (Chapters 2.3.2, 2.3.5, 2.3.6).

Before I return to the last aspect, which suggests a relationship between the concept and the geometric diagram, I would like to give a possible answer to the question

2 Logic and its Geometrical Interpretation

raised in the dissertation, in WWR I and WWR2, with the help of the second paragraph of the quote given in Chapter 2.3.5, on what the analogy between discursive and visual thinking or the logical grounds of knowing and the intuitive grounds of being is based. The answer could be, for example, that concepts are originally representations of intuitions, which are learned through contextual use, form analytical or synthetic judgements in relation to each other, but can also produce new knowledge through inferences with multiple conceptual relations.

Finally, I would like to take up a thesis from Schopenhauer's *Berlin Lectures* in this chapter, which concerns the semantics of geometric terms alluded to above and which, in my opinion, makes a separation of terms into syncategoremata and autocategoremata. Although Schopenhauer had rejected this separation in the logica major (Chapters 2.1.3, 2.2.5) – especially with regard to proper names –, he takes it up again in his theory of space, since geometric terms have no analogy to "positions in which circles can be put together", or "spatial figures", but are themselves expressions of a priori forms of external sense. In this way, Schopenhauer takes up an argument that Crusius, Kant, Schultz and also many other geometers and philosophers after Schopenhauer have indicated (Chapters 2.3.1, 2.3.3).

As reported in Chapters 2.3.2 and 2.3.5, Schopenhauer had stated that in geometry only the axioms can appeal to intuition, but the rest is proved in this case. However, as far as I know, Schopenhauer only clarifies in such detail how the axioms are demonstrated in the *Berlin Lectures*. Space, Schopenhauer explains, is the only concept to which the category of unity or uniqueness quantification can be applied,[320] since it was not abstracted from experience but is based on a single intuition; and all so-called spaces are only parts of a single one, which is thus the condition for the possibility of experience in the first place.[321] Geometrical concepts, too, immediately refer to an intuition independent of experience, although the empirical intuition must be able to accompany it at all times:[322]

> Geometrical concepts are formed arbitrarily without experience, and then carried out in an intuition (which may or may not be supported at will by material means for the senses), but which now yields many more properties than the concept contained, which properties, however, are just as certain and independent of experience as the arbitrarily conceived concept. – The geometrical concept is the mere guide or rule to an intuition to be executed (in the faculty of imagination): when this is executed according to it, it stands there, as objectively as any object given in experience, with many essential properties which it did not specify and which can nevertheless no longer be diminished or increased, but merely discovered and found. Nevertheless, it is not a mere thing

---

[320] Vide supra, Chapters 2.1.3, 2.2.5.
[321] Cf. WWR2 I, pp. 128f.
[322] Cf. WWR2 I, pp. 131f.

2.3 Proof – Elementary Geometry, Syllogistic and Intuitive Proof Theory

of thought: for all real things that correspond to it in spatial relation also represent all the properties set with it.[323]

Schopenhauer uses numerous examples to explain the extent to which geometric concepts are mere instructions or rules. For example, he takes judgements such as A quadrilateral can have at most three obtuse or acute angles, but four right ones and explains that their truth and certainty do not result from the fact that the respective concepts of the angular properties are contained in the concept of the quadrilateral, but that one instantly follows the rules of the judgement in one's imagination and thereby sees whether or not one can demonstrate all conceptual rules with a diagram or figure. A figure can indeed have the property of being called a quadrilateral and possessing four right angles, but it is not possible to follow the rule of forming a figure in a conception that is called or can be called a quadrilateral and that at the same time also has four acute angles.

The suspicion now arises that geometric concepts possess properties that I claimed in Chapter 2.1 are not to be found in Schopenhauer's semantics: they resemble autocategoremata, as I found them in the labelling theories of Frege, Russell and Whitehead, and they compel the concept user to follow a kind of rule that is reminiscent of Wittgenstein's feature of the use theories of meaning as elaborated in Chapter 2.1.5. Nevertheless, one must be cautious about such a suspicion.

The proximity to autocategoremata can be explained by the fact that geometric concepts produce their semantics themselves and that this is not deduced from empirically taken intuitions or already existing meanings. As autocategoremata, geometric concepts force the user to convert the meaning of the geometric concept into intuition. However, both properties of geometric concepts show less a relationship than essential differences to the theories of post-Fregean philosophy of language: Schopenhauer's autocategoremata are not proper names, and Schopenhauer's rule-following is not a linguistic one.

The theory of proof or demonstration that Schopenhauer offers in geometry has the similarity with logic of referring to a single ground of being, but it differs from logic in that the concepts of geometry can be understood as rules and prompts, whereas logical concepts must first justify their metaphoricity and analogy to a priori intuitions. Of course, I can only leave it to the recipient to judge whether the explanation given above for the analogy between discursive and visual reasoning, or the logical ground of knowing and the intuitive ground of being, is a convincing interpretation from the second paragraph of the quote given in Chapter 2.3.5. However, the decisive result seems to me to be that Schopenhauer does not need such an explanation for geometry, since in his opinion it would be easier and better to trace the grounds of knowing back to the grounds of being in the logical theory of proof and semantics;

---

[323] WWR2 I, p. 132.

however, the geometrical concepts would first have to be transferred from the grounds of being to the grounds of knowing. Thus, for semantic reasons, visual geometry is ultimately the model for a logica more geometrico.

3 Logic and World

I have stated in the preface to this book that my aim is to argue for a variant of rational or non-naïve representationalism that can defend the thesis that the space of reasons must be larger than the space of concepts. This thesis provides the foundation for initiating a revision of the prevailing view, which is concerned with the relation of world and language, of metaphysics and logic. In Chapters 1 and 2, good reasons were put forward to show that it would be over-hasty to dismiss any representationalism as naïve merely because its object of research is not limited to the semantics of concepts, judgements and inferences. Nevertheless, I believe I have shown, especially in Chapter 2, that semantics plays a crucial role in judging whether a representationalism must be called naïve or not.

The criterion for deciding whether representationalism is naïve or not is usually determined by rationalist semantics.[1] The naivety of representationalism is characterised by the causal-theoretical assumptions that, on the one hand, the innocent eye is capable of capturing the world in a representative logic and, on the other hand, that logic represents the world when the causal transfer from the world to logic has produced a noticeable similarity between the two. From a logicist-inferentialist point of view, however, there is only one translation between the world and logic, and these two realms are only supposedly separate.

Since the world itself is logically constituted or can only be understood with the help of logic, naïve representationalism assumes a logically arrested eye and an optional equivalence between two languages, the first of which appears to be exogenous and the other endogenous or, to put it more in terms of intension, the first is said to have externalist and the second internalist attributes.[2] It is considered a naïve or unexplained attempt at explanation to simply assert that the expression `spider` is a representation of the object designated by the expression `spider` or an expression fully substitutable with it. From a rationalist point of view, the naivety of this assertion is based on the fact that with the assertion of a difference, neither good reasons for the difference between an externalist and an internalist vocabulary nor for an optional equivalence between the designation and the thing designated have been given or can be given.[3] Representationalism is consequently naïve when it attempts to justify a

---

[1] Cf. Jaroslav Peregrin: Inferentialism, Chapters 1.1, 2.4.
[2] Cf. A.J. Ayer: Foundations of Empirical Knowledge. London 1940.
[3] Cf. Wilfrid Sellars: Empiricism and the Philosophy of Mind. In: The Foundations of Science and the Concepts of Psychoanalysis (Minnesota Studies in the Philosophy of Science, Vol. I). Ed. by H. Feigl, M. Scriven. Minneapolis 1956), pp. 253–329, § 9 (*).

3 Logic and World

picture theory (or theory of representation) that already presupposes a semantic picture theory which, moreover, is also problematic. The logic of such representationalism thus pretends not to be primarily rational.

I have argued in Chapters 1 and 2 that at least one historical variant of representationalism can be presented that cannot be dismissed as naïve insofar as its picture theory is subject to a logic built on a non-representationalist semantics. There may thus be research programmes that aim to reflect the entire world in as few abstract terms as possible, without explaining the meaning of the terms only by a problematic reference to the objects they are supposed to denote. Such semantics has the advantage over naïve representationalisms of neither having to assert an optional equivalence between two languages nor having to give up the difference between externalist and internalist attributes.

I will show in Chapter 3 that such representationalism can not only defend itself against accusations of naivety but can also contribute to the justification of those programmes that limit their object of research to logic alone and conceive of the world at most as something that can only be conceived from logic. As discussed in Chapter 2.3, however, such logicist programmes tend to lapse either into an infinite regress of metatheories or into a circle of rules and axioms or into a dogmatism of ontological assertions. Chapter 3.1 will show that the internal critique of the inferentialist programmes of modern rationalism confirms this tendency and that the best alternative emerging from the critique seems to be the commitment to a grounding theorem of logic (Chapter 3.1.3). Chapter 3.2 will offer a way out of the justification problem of inferentialist programmes in the form of non-naïve representationalism based on this critique. This chapter will show that inferentialism and rational representationalism are built on the same grounding theorem; however, non-naïve representationalism emphasises that the grounding theorem has investments that extend beyond its rational commitment in the space of concepts.

Chapter 3.1 provides a comparison of those inferentialist programmes that purport to have good reasons for answering the question of the relationship between logic and the world in a one-sided way. The critique of inferentialist rationalism against more or less empiricist and causal-theoretical programmes will show that terms such as `logic` and `world` carry different spheres of meaning depending on the point of view: From the point of view of rationalism, naïve empiricist programmes fail because there is no complete overlap between logical space and worldly space, in such a way that the latter can also be used as a ground for assertions in the former. But rationalists themselves argue that the world is always already logically pervaded and that logical space is congruent with the space of reasons or – better said – exactly overlaps it.

Arguments are made in Chapter 3.2 *that the space of reasons extends further than the space of concepts* or, to remain in the picture of language, that the sphere of meaning of the concept `space of reasons` is wider than the sphere of meaning of the expression `space of concepts`. In contrast to inferentialists, I thus assume that the image of the overlapping of both spaces is just as inaccurate as the causal-theoretical notion of empiricists, which is dismissed as naïve by rationalism, since the latter claims that there is only a partial overlapping of both spaces. Following inferentialism, I do believe that our primary access to the space of reasons can be through the logical space of concepts; but unlike them, I point out that in this there are already indicators that point to a much wider space that goes by the name of `space of reasons`.

I will contend that these indicators can lead representationalists to take seriously the claim that the logical conceptual space already contains transfers from the space of reasons. Whereas these foreign objects in logical space can play a conceptual role in judgements, they are only superficially conceptual in nature; this becomes apparent when one looks, as it were, beyond their appearance, out of logical space into a much wider world of reasons.

In order to substantiate the seriousness of this claim, Chapter 3.2.1 supplements the use-theory of meaning, which is binding not only for inferentialism but also for non-naïve representationalism, with a theory of abstraction. Inferentialists, as Chapter 3.1 will show, reject the individual-subjective theory of abstraction together with causal and so-called transcendence-theoretical explanations and finally try to infer the thereby missing reference from logic. However, the theory of abstraction favoured in Chapter 3.2 is not individually conceived. Rather, I advocate a form of abstraction theory that can also be read out to some extent from Chapters 2.1 and 2.2 and that can be assigned to the so-called intersubjective, collective or "anthropological theory of the concept".[4]

By replacing the individual abstraction theory with an intersubjective and social variant, a way opens up to circumvent some of the problematic ontological claims of the inferentialists associated with a theory of concretion, proper names and definite descriptions. The intersubjective and collective semantics proposed here does not distinguish between singular and general terms, but it does distinguish the level of abstraction of concepts. How the criterion of a naturalness oriented towards reason, established in Chapter 3.1, can be fulfiled will be shown in Chapter 3.2.2.

Finally, in Chapter 3.2.3, with the help of the collective abstraction theory of meaning, I would like to make a distinction between translations and transfers, which in contextual terms concerns the distinction of the level of abstraction and expressiveness of concepts. In my opinion, one can justify the thesis that certain transfers from the space of reasons do take on a conceptual role, but that their level of abstraction is so concrete that they cannot be translated without their expressive power suffering. Finally, it is in line with my point of view that we find such transfers in the grounding theorem that underpins inferentialist programmes, as pointed out in Chapter 3.1.3. In contrast to inferentialism and logicism, however, I do not plead for translating or ignoring these transfers found in the grounding theorem, but for considering them as the

---

[4] Cf. Hans Blumenberg: Theory of Nonconceptuality. In: Ibid.: History, Metaphors, Fables. Ithaca 2020, pp. 259–297.

basis of the systematics and history of geometrical logic as presented in Chapter 2. Accepting these transfers in the grounding theorem not only expands the space of reasons beyond the limits of the space of concepts, but it also expands inferentialism by a ground that is the basis for geometrical logic, which in turn vouches for the rationality and non-naivety of representationalism.

3.1 Inferentialisms

Inferentialist programmes are defined by their common abjuration of the old superstition that there are immediate and non-inferentially constituted data that function as the ground of knowing.[5] In the context of this position, `ground` is understood either as the justification of individual assertions and opinions or as the final instance of references of knowing in general. In this context, it is initially irrelevant whether scientific or everyday knowing is meant by this justification or reference to a reason; rather, justifications and references are an activity that always involves the production or maintenance of knowing. Even if it were meaningful to speak of immediate data in a particular context, justifications would not be made with these immediate given data, but always with conceptually articulated data or facts. In this respect, data are always already conceptually strongly incorporated into inferences, since it would be incompatible with speech acts of reasoning if the immediate given could take on a role in the game of giving and demanding reasons.

It is essential to emphasise that for inferential programmes linguistic contexts are endowed and structured by *conceptual content*, by *propositional roles* and by *inferential rules*. Concepts, judgements and inferences form a space in which reasons and justifications can be publicly exchanged: they are accepted or rejected. This also makes it possible for contexts not only to have genealogies that can be traced back from an arbitrary recipient to an original speaker but also to give rise to traditions that make themselves felt as cultural phenomena in entire linguistic communities.[6] As a result, linguistic communities define themselves by a basic set of concepts, judgements and inferences that they either accept or reject.

The public game of reasoning and justifying constitutes a single coherent space that is filled by speakers first of all through that conceptual content that is put forward and accepted in judgements. The use of these concepts in judgements and the role of judgements in inferences are further determined by just such grounding rules, which in context decide how necessary reasons and justifications both the concepts used and the judgements made with them are. As a rule, judgements that are inferred coherently from other accepted judgements are in turn more likely to be accepted than those that

---

[5] Cf. Wilfrid Sellars: Empiricism and the Philosophy of Mind, §§ 3, 5.
[6] Cf. John McDowell: Mind and World, Kap. VI.7; VI.8.

3.1 Inferentialisms

have been put forward incoherently. The demand for coherence, gaplessness and indivisibility is thus an essential distinction of inferentialism and another grounding position. The abjuration of the old superstition that there are immediate and non-inferential data is not thereby taken as a gap since it is not immediate data but the conceptually divided datum in an inference that constitute the coherence in the public sphere where reasons and justifications are exchanged. The abjuration of the old superstition of immediate data is at the same time a commitment to a conceptual, propositional and inferential context in logical space. In the context of justifications, this is a commitment to a privileging of a space of concepts, and, furthermore, the abjuration of belief in non-conceptual given data is an essential feature of rationalism.

Inferentialism in its modern form is founded in three essential programmes, which are themselves divided into different standpoints and distinguish themselves from others. Its own justification takes place through the privileging of the respective contexts in the space of concepts, which is considered to be congruent with or to cover the space of reasons: The first inferentialism sees the ground of its programme in the concept, which is *formed* from the space of reasons and can be used in this way in judgement and inferences; the second programme emphasises that both the correctness of inferences and the stock of concepts referring to the space of reasons are *developed* from the materiality of judgements; the last inferentialism argues that the grounding of the concept and the materiality of the judgement are *derived* from the form of the inference so that a grounding rule is needed to determine the role of judgements and the use of concepts.

The critique of these forms of rationalism proceeds from within since inferentialisms examine their justifications according to the use, role and rule of their programmes just as they question the causal, transcendental and abstractionist standpoints from which they distinguish themselves. The history of this critique is the history of its own formation, development and derivation from a conceptual grounding, through propositional materiality and into a universally accepted inferential rule. The thesis that a performance of transfer is the grounding and material of this rule is argued for in Chapter 3.2.

3.1.1 Inferentialism of Grounding

Through the commitment to indivisibility, to coherence and to gaplessness, ruptures appear exceedingly problematic for the inferential programme. If the public space is expanded to include non-conceptual data, the practice of justification runs the risk of being broken through. Data require justifications through linguistic contexts, only through this do they enter the public space of reasons. In turn, given data, if they themselves are not supposed to have conceptual content and thus cannot be subject to inferential rules, are conceived as breaks or gaps in both logical and genealogical contexts: They can have the property of being neither unequivocally accepted nor rejected, since their conceptual content, which determines their public context, is precisely the property they lack. If the conceptual content is missing, this also leads to genealogical and ultimately traditional gaps, because publicly demanded reasons and justifications can no longer be traced back inferentially in linguistic communities.

An *example* of such a gap is the picture, in the form of a geometric figure, that represents such a non-conceptual given datum in a context of justification: A speaker demands proof of the truth of a geometric proposition in a public space, whereupon a geometric figure is painted in the sand for her. She accepted this as proof and is now convinced that the geometric proposition is demonstrably true. Many years later, she herself is asked to provide proof of her conviction, but she is unable to reproduce the figure. She can only refer to something that has been given to her without being able to justify that this given to her also has a conceptual content that enables her to develop its reason in a judgement and to coherently justify the inference in such a way that the proposition in question is considered proven.[7]

One can imagine similar situations in which gaps in justification do not arise from the inability of the one who is in need of proof, but are based on the change, expansion or disappearance of something that is thus given – perhaps also because the apparently given has never existed or only in imagination, or also because there are relations between the given and the conceptual that cannot always be maintained. What is decisive is that from the situation described or from similar ones, two points of view initially emerge, which can be called the *standpoint of excuse* and the *standpoint of conviction*.[8]

Both standpoints are based on a *confusion*: the *standpoint of excuse* assumes that one possesses a justified conviction of a given that is on a par with the content that plays a role in inferences. The standpoint of conviction assumes that the given only offers an excuse for what actually requires justification. For the first standpoint, a belief that cannot be justified publicly requires an apology to those who recognise gaps in justification in such data. The second point of view, however, then arrives at the conviction that a public apology points to data that cannot be justified.

The *standpoint of conviction* is supported by the proof that the conceptual would not be recognisable in experience if it were not also present in the conceptual. Excuses appear to exist for this point of view precisely when the conceptual is to be relieved either *by a given itself* or by the reference of *the conceptual to the given*. In both cases, the conceptual is relieved of responsibility by referring either completely or partially to something that it itself pretends not to comprehend or that it actually does not comprehend.

---

[7] Cf. Wilfrid Sellars: Empiricism and the Philosophy of Mind, § 37.
[8] Cf. John McDowell: Mind and World, Chapters I.3, II.2, III.4.

3.1 Inferentialisms

The conceptual is indeed relieved of the responsibility of justification in both cases, but only in the *first case* does it take away the complete freedom of assertion by burdening the given with justification. If it imposes the duty of justification on the given and takes all the freedoms of assertion from itself, then it can also turn its apology into an assertion, since this too, like every other publicly demanded reason, does not lie within itself but in the given. If the concept of this *theory of transcendence* lies in such a way that it makes the space of the concept appear as outside the space of reasons, inferentialism sees in it the position of a 'metaphysical realism'; if the conceptual is only a shadow of the reason, this position can even be called 'representational realism'.

The *second case*, however, in which the conceptual refers to the given and indexes it as its own instance of control, is an expression of the loss of control that accompanies complete freedom and thus an expression of concern. From the standpoint of concern, the freedom of realism is too great and the temptation to make assertions that do not need to be justified appears as an assertion of excuse itself. The conceptual, however, recognises that it rarely needs to assert such apologies, however, if it merely instantiates the given as a corrective and allows a public tribunal to decide the extent to which the assertion can claim validity according to which there has been a causal relationship between the given and the assertion that has led to a correspondence. Finally, the assertion of rare excuses can be understood as freedom of the assertion itself: If the causal relationship between an assertion and the given is not publicly endorsed, then the misconception in the form of the original assertion can only be excused by a new assertion. The conceptual consequently faces an active and a passive jurisdiction: that of the public and that of experience. If new assertions are introduced tacitly or if it is argued that it is not necessary to explain how assertions come about, this *theory of causality* is given the undeserved name of *'critical rationalism'*; if the assertions arise from the context of assertions that have been approved by the public jurisdiction, this standpoint is called *'semantic holism'* according to the default of inferentialist expression.

The two points of view that uses the given as a controlling instance *cannot be convincing*. However, *semantic holism* first offers an explanation of how the dualism between an assertion and a given can come about at all. It is the active jurisdiction that decides whether the relationship between the conceptual and its immediate controlling instance is legitimate. Since the immediate controlling instance is precisely passive, it is active public jurisdiction that imposes the new assertion on the conceptual, insofar as it invalidates the relation between the previous assertion and the controlling instance matched with it. This active public jurisdiction is thus sharply distinguished from the passive immediate controlling instance and called `language` or even `scientific language`. The context of scientific language consequently imposes new assertions on the conceptual for verification.

3 Logic and World

Semantic holism does not take the *standpoint of conviction*, since it is both in the service of apologising to the public and confronted with unnecessary dual jurisdiction. The second point, in particular, is raised to the real problem, since a controlling authority that is purely passive fails in its task if it has not fraternised with an active one. For it was actually the task of the given to control the activity of free assertion. Now, however, the fact itself shows that this passively given always needs an active controlling instance through scientific language (which, on the one hand, pretends assertions and, on the other, decides on the validity of the relationship between these and the given) so that it is just as relieved of the responsibility of justification as the standpoint of apology. Finally, the standpoint of conviction recognises that exonerations and controls, while offering advantages, remain excuses that must not be confused with reasons and justifications.[9]

A *third theory or fifth position*, which is close to the first or the first two, results from the inversion of the relationship between the conceptual and the given. The immediately non-conceptual given data are assumed to be inner private acts of thought and are conceptually occupied in such a way that a clear proximity and intimacy between the given and the concept is expressed. These acts of thought, in their double role as concept and as given, then form the grounding for an entire language, since from them ever further levels of abstraction from the intuitive world are carried out until one has reached the sphere of meaning of a few concepts, which also depict almost all other spheres of meaning of those concepts from which those few have been abstracted. Language then means having entered into an intimate relationship with the given, so that reflection on it remains intelligible only in the realm of the private, in the relationship itself.

But even this reference between the conceptual and the given is *not the standpoint of conviction*, because justification via abstraction from the world and thus via the meaning of the concepts themselves is again only a reference to the given, namely to the inner private acts of thought or to inner private grounding concepts or their abstracts. What intimacy with the given means cannot be publicly comprehended. As a result of this process of abstraction, the public loses weight, so that in the end, at most, the private could become the controlling authority. Provided that it is at all possible to speak meaningfully of a private instance of control, the proximity of the third theory or fifth conception to the second or third and fourth is evident – except that the roles and possibilities of the instances of control have shifted in the same way that the relationship between the conceptual and the given has shifted. Indeed, a public linguistic community whose tradition and genealogical context predetermines assertions and decides the relationship between assertions and givens is precluded by a theory of abstraction in which the meaning of words emerges from an intimacy whose memory

---

[9] Cf. John McDowell: Mind and World, Proscript I, 3f.

3.1 Inferentialisms

remains housed in a private treasure of language. For conviction, however, language must vouch not only for the absence of things but also for the absence of thoughts.[10]

The last point of view, *the theory of abstraction*, finally unites the disappointment and the one-sidedness of control. Disappointment at the lack of public justifications is based on the intimacy between the conceptual and the given in the act of thinking. Control, in turn, is based on this intimacy, which does not tolerate any further instance between the two partners. For the conceptual, the given alone is sufficient for control; but in this way, it itself loses the claim to public control and jurisdiction since it must disappoint all invitations to give or take assertions and reasons. This disappointment is the basis for the public conviction that the relationship between the given and the conceptual in the act of thinking from this standpoint was neither intimate nor private but reflected divisiveness that was brought out into the open and in which no agreement could be reached.

Through these disappointments, inferentialism has come to the conviction that the conceptual can only be recognised in experience if it has also put it there. The basic theories that inferentialism determines and from which it defines its own standpoint of conviction can be distinguished based on the spheres of the expressions `space of concept` and `space of reasons` in an intuitive schema of properties:

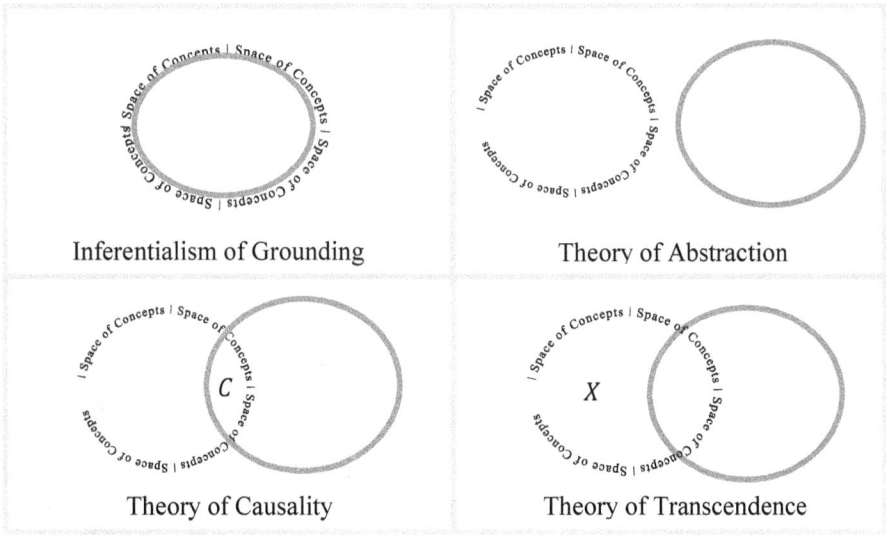

| Inferentialism of Grounding | Theory of Abstraction |
| Theory of Causality | Theory of Transcendence |

According to the definition of inferentialism, the *theory of transcendence* is characterised by the property that it regards the sphere of the space of concepts as partially separate from the space of reasons. Since some concepts, such as the object in itself

---

[10] Cf. Wilfrid Sellars: Empiricism and the Philosophy of Mind, § 38.

3 Logic and World

($= X$), appear only as a mode of representation, they are not grounded for inferentialism. This theory consequently symbolises the loss of all control and can never claim to be true together with inferentialism. Inferentialism's view of the *theory of causality* is that the sphere of the space of reasons is only partially identical with the sphere of the space of concepts, and only precisely when reasons that characterise this overlap are causal in nature. Inferentialism sees in the form of the theory of causality a part of its own truth, but it pushes causality ($= C$) completely towards rationality in order thereby to tame the uncontrolled influence of the world. However, the *theory of abstraction*, in which the sphere of the space of the concept is completely separate from the space of reasons, appears to him to be much more uncontrolled and ruleless. This abstractionism accuses the theory of causality of a false spatial conception, since the relation between the space of concepts and the space of reasons is only a remembered one, but cannot be a factually given one. Consequently, inferentialism does not see the slightest overlap between the spheres of the two expressions, the space of concepts and the space of reasons, in the theory of abstraction. Consequently, inferentialism takes the opposite standpoint to the theory of abstraction, since in it the spheres of the space of concepts and the space of reasons completely overlap.

In that inferentialism gains its conviction by delimiting those theories that have caused disappointment or suffered a loss of control, the theory outlined here is named '*inferentialism of grounding*'. The spaces of concepts and reasons that are congruent for it form the grounding and foundation for the *materiality of judgments* and for the *regularity of inferences*.

If one exchanges the intuitive schema of properties by which the inferentialism of grounding has determined itself for a diagram of semantic compartments, its conviction is that there is only a space of reasons and of concepts (3), but there is no space that has neither reasons nor concepts (1). Nor is there a space of concepts alone (2), nor a space in which only reasons apply (4).

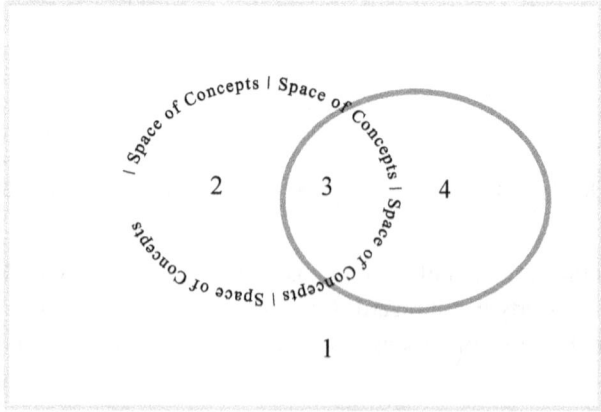

3.1 Inferentialisms

The inferentialism of grounding sees in the *theory of abstraction* a standpoint according to which the space of concepts (2) is constituted by the negation of the space of reasons and by the negation of the coexistence of both spaces. The *theory of causality*, like the inferentialism of grounding, focuses on the compartment (3) and declares that this compartment is a connection of the space of reasons with the space of concepts. Unlike the inferentialism of grounding, however, the theory of causality does not recognise that this connection is a logical conjunction but rather introduces the unjustified and problematic notion of causality as a connecting function. The *theory of transcendence*, on the other hand, disregards this conjunction and asserts either only the space of reasons (4) or neither the space of concepts nor of reasons (1), in each case reducing the common existence of both spaces and in each case one of the two spaces still in question to a mere appearance.

By interrogating the standpoints of these three theories, the inferentialism of grounding has come to the conclusion that with their reference to the given, they only offer excuses where justifications are actually required. For all other theories, the inferentialism of grounding is a disappointment because – as in the exemplary demand to reproduce a certain geometric figure – it calls into question their convictions of at least one immediate given datum. Both points of view are thus based on a confusion that explains why transcendence, abstraction and causality theories do not share the view of the intuitive schema of properties provided by the inferentialism of grounding, or why they would locate their position and that of their neighbours in a very different place of the overview.

Since the theory of abstraction sees in the space of concepts an exact copy of the space of reasons, only that the former contains a coarser structure than the latter, it would substitute itself for the inferentialism of grounding in the *intuitive schema of properties*. The theory of transcendence could be said to represent even the scheme that the inferentialism of grounding has assigned to the theory of abstraction since in this scheme the sphere of the space of reasons has the autonomy to regard the theories of transcendence as the actual meaning of the given. On the one hand, metaphysical realism emphasises that the sense of the spheres of both spaces is to indicate a correspondence, so that what takes place in one is also imagined in the other; on the other hand, however, it also emphasises that this correspondence is not based on an overlap of the conceptual spheres, so that the meaning of the two expressions remains completely different. Although representational realism also holds that the two spaces are each independent in their meaning, it usually refers to the explanation of abstraction theory in order to be able to justify this independence. For this reason, it is regarded by the inferentialism of grounding, on the one hand, as a scientific standpoint of the theory of transcendence, but on the other hand, as a subject of the theory of abstraction.

3 Logic and World

For the theory of causality alone, the inferentialism of grounding seems to have drawn an adequate picture insofar as it is not itself interested in the relationship between the two spaces, but only in the question of the extent to which a larger public totality considers its causal explanations of individual elements from both spaces to be adequate or not. The two forms of the theory of causality mentioned above, however, explain their dependence on the picture that the inferentialism of grounding draws of itself: Semantic holism gains meaning from causal explanations by embedding them in a web of meaningful propositions, which in turn has come to depend on a particular publicly endorsed overall explanation of nature. Critical rationalism, on the other hand, assumes that one can only speak of a network of meaningful assertions when individual causal explanations have at some point grown together into a publicly approved overall explanation of nature.

For itself, the inferentialism of grounding claims to be able to explain a complete overlap of meaning or congruence between the space of reasons and the space of concepts, which sets it apart from the three competing theories. Whereas causality, abstraction and transcendence theories accept a dualism between the space of concepts and the space of the given, the inferentialism of grounding believes it has demonstrated an undivided theory of understanding and reasoning. The inferentialism of grounding is itself aware that the theories presented, in their modern version, themselves often proclaim uniformity and undividedness. For it, however, this indivisibility is at most the result of the image that the inferentialism of grounding ascribes to these theories. If these theories want to lay claim to unity, they have no other option than to start from the dualistic picture, to regard this itself as a problem, and then either to extend the circumference of one of the two spheres to such an extent that the circumference of the other sphere is comprehended by the first or to synchronise and superimpose the spheres via their intensional content.[11] The conceptual content that is in the sphere of meaning of one of the two spaces, which are actually fundamentally different for each other, is thus interpreted according to the archetype of the exact opposite in such a way that either both spheres have a very similar conceptual content or that one sphere comprehends the other.

The inferentialism of grounding deals with the strategy of basic dualistic theories both *constructively and therapeutically*: Constructive for it are the reasons through which it gains its own standpoint, and that it defines itself through the grounding theories it intends to treat. Therapeutic is its approach in that it can make a construction between the space of concepts and the space of reasons, which does not have to start from a dualism of the spheres of meaning, but that results from the negative relations to the alternative theories of grounding. Consequently, the construction of this inferentialism is the result of the therapy of a syndrome of theories of grounding.

The only construction that remains after the exclusion of the other theoretical constructions, which can be described as grounding theories, offers this inferentialism the

---

[11] Cf. John McDowell: Mind and World, Chapter V.3.

possibility of referring to a dualism that is always already mediated because it is congruent or overlapping. When the inferentialism of grounding claims for itself to be able to explain the exact congruence between the spheres of meaning of the two spaces, it does so by invoking a 'collective' element of its theory: for both spaces are said to come together as natural givens in the human being. Here, inferentialism plays out the *ambiguity of grounding*: On the one hand, both the human and the animal being are grounded on the non-conceptual nature. On the other hand, however, this nature is grounded in the fact that the human is formed from the animal, so that the first appears as pure nature and the other as a quasi-nature, as an educated or cultivated nature. Naturalness thus becomes the decisive criterion of the inferentialism of grounding.[12]

The intensions of the two spatial expressions conveyed by the concepts of nature enable the inferentialism of grounding to assert a dualism according to the sense, but to refer to an undividedness according to the meaning: The two spheres of meaning, which are regarded as fundamentally divided in all other theories, are represented as undivided, congruent or overlapping in the inferentialism of grounding alone. For it, it may still seem reasonable to speak of two conceptual spheres, but ultimately the two overlap in such a way that they make possible the grounding form of an undivided theory. The purely natural space of reasons becomes comprehensible by being conceptually expressed and cultivated into a space of concepts.

The indivisibility of the spheres of meaning of the natural space of reasons and the cultivated space of concepts is described through a temporal process of transformation and synchronisation. Pure nature is recast by an acquired or cultivated nature in the course of time. This passage of time is the life course of that indivisibility developing into cultivated nature, which no longer organises its passage of time merely with the functions of its pure nature, but fills it with conceptual labour or work.[13] The space of reasons is thus filled with the space of concepts, or the space of concepts is formed in the space of reasons, precisely covering it. The work of this indivisibility is what the inferentialism of grounding calls 'education or formation' (Bildung), and it consists in reflection on the complete conceptual completion of the space of reasons in distinction from the theories of abstraction, causality and transcendence, in which this work is never completely done or should be done.

The last aspect is crucial to understand what the inferentialism of grounding means by formation. Formation is not the aspect of a view of the world that would take place independently of other competing views of the world. The world that the inferentialism of grounding offers as a view is at the same time a therapeutic refraining from theorising on the world itself. Its construction, after abjuration the old superstitions to which the other theory constructions had subscribed, is the only one left. As a result, the process of observing, refraining and excluding becomes the environment and the

---

[12] Cf. Allan S. Janik: Wie hat Schopenhauer Wittgenstein beeinflußt?.
[13] Cf. John McDowell: Mind and World, Chapter VI.4.

worldview of the inferentialism of grounding. On the one hand, this process is a temporal separation from pure nature, since the passage of time takes place in the confrontation with the existing theories of the world; on the other hand, it is also a spatial separation from it since inferentialism gains its undivided worldview in the conceptual process of observing, refraining and excluding of those divided spheres of meaning that it itself thereby loses.

3.1.2 Inferentialism of Matter

The inferentialism of matter sees in the *inferentialism of grounding only a further extended form of the theory of causality*.[14] Whereas the inferentialism of grounding had criticised the theory of causality for considering a partial connection of the space of reasons with the space of concepts and thus allowing an uncontrolled influence of the world in logical space, the inferentialism of matter now criticises the inferentialism of grounding for only extending the partial connection and thus completely substituting perceptual experience for rationality. Causality then gets into confusion with the concept of the rational when the space of reasons cannot be understood in any other way than in its cultivation, i.e. in an artificial transformation into a space of concepts.[15]

Naturalness in the inferentialism of grounding thus threatens to become a cult, for to the new form of inferentialism, education or formation is only the promotion of greater generality. In the worst case, formation congeals into a perceptual experience, an externalism, a schematism that is displayed as an altar in the holy of holies of the goddess of wisdom. To inferentialism of matter, therefore, formation is not a procession that strides from the space of reasons into the space of concepts, but only the recognition of what is already known to us and what passes our lips in every judgement we pronounce.

For the inferentialism of matter, the theory of causality first states that only that which is also domiciled in the space of reasons can have conceptual content. The assertion inherent in all inferentialisms that reflections presuppose the availability of conceptual content leads the inferentialism of grounding to a causal-theoretical extension of this assertion so that the reflection on the domicile in the space of reasons also leads to the availability of conceptual content. For the inferentialism of matter, this is the cultic grounding of the theory of causality.

But the fact that there is *no longer a division between the space of reasons and the space of concepts*, or that the givenness is nothing other than the concept, is interpreted

---

[14] Cf. Robert Brandom: Non-Inferential Knowledge, Perceptual Experience, and Secondary Qualities. Placing McDowell's Empiricism. In: Reading McDowell. On Mind and World. Ed. by Nicholas H. Smith. New York 2002, pp. 92–105; Richard Rorty: Truth and Progress. Philosophical Papers Vol. 3. Cambridge 1998, Chapters 6 and 7.
[15] Cf. Robert Brandom: Articulating Reasons, I.4.

3.1 Inferentialisms

by the inferentialism of matter as a loss of control: According to him, the inferentialism of grounding struggles with the problem that the space of reasons is completely overlaid by the space of concepts and that there is no longer any difference at all whether the role of conceptual content leads to what some call illusions and therefore reject, or to what others consider good or reliable and therefore generally approves of.

No matter what is the case, however, for the inferentialism of matter, the *judging* remains the decisive ground for conceptual content to be developed at all and then to play a role in inferences. Whereas the inferentialism of grounding wants to bring concepts passively into play, material inferentialism explains that conceptual capacities are dispositions that are actively brought forth, namely when a judgement plays a role in an inference. These judgements serve as grounding for the inferentialism of matter as concepts did for the inferentialism of grounding. An essential difference between the two inferentialisms, however, is that for the inferentialism of grounding conceptual content is formed in the space of reasons, whereas for the inferentialism of matter the space of reasons, together with its conceptual content, develops only from the act of thinking and speaking of judgement, and this judgement plays the decisive role in inferences.[16] The concept, which was once the ground of inferentialism, is now only developed from judgements whose correctness is not determined by a formal but always by a *principle of matter*.

This last point is another decisive distinguishing feature between the two forms of inferentialism mentioned so far. The inferentialism of matter sees the givenness not only in the immediacy of facts that are accessed by pointing gestures and references or that are equated with internal acts of thought to which one might have privileged access; but, unlike the inferentialism of grounding, it also pretends to be aware that the space of the concept can also take on a form of the givenness when it has not been constructed through the public process of judging. The space of the concept becomes a form of the givenness when its content stands in a purely signifying relation to its bearer. *Concepts* within such a space represent *givens*, *judgements* are compositions of concepts and form the *givenness*.

However, it would be a confused conception of the relation between the spaces of concept and of reason for the inferentialism of matter to admit a conception according to which the conceptual content has ascribed a role in *conditioned, trained or generally formed reactions*. If in response to the sound of a person sneezing, a machine reliably gives a kind of observational report such as `Bless you!` or `This person has a cold`, these statements are nevertheless to be understood only as words and sentences, but not as concepts and judgements. For the inferentialism of matter, this also applies to all simple reflex agents, i.e. also in a reversal of the stimulus-response scheme: If a handkerchief is reliably given in response to the statement `Cold!` or

---

[16] Cf. Michael Dummett: Frege, 1973, Chapter 10.

Bless you!, this reaction does not yet have to be understood as a function and product of the concept.

Only when reactions are integrated into such contexts in such a way that a context arises in which a speaker can either be held responsible for her or his judgement or in which it is possible for others to take responsibility for this judgement, can words be classified as concepts and sentences as judgements. Taking responsibility means being able to infer from the judgement to the intention a speaker intended to develop by using this judgement. Automated and unaccountable expressions remain words and sentences that do not reveal intentions. Although these expressions are based on programming, training, conditioning and formation, they do not reveal any intricate intentions that are answered for in a judgement. With the assumption and delivery of expressed responsibilities, material inferentialism ensures that conceptual content can be transmitted in a linguistic community. Conceptual content is predisposed in the linguistic community. However, *this is only the side of the speech act of judgements*.

For the inferentialism of matter, judgement is the basis of the concept because, on the one hand, it is the *form* in which conceptual content can play a role in premises or conclusions and, on the other hand, premises and conclusions express *content* that can be approved or rejected by speakers. The responsibility falls back on the speaker or is assumed by recipients. In the first case, the recipient does not approve of a speaker's judgement and demands that the conceptual content be marked in the space of reasons. The recipient demands an explanation of the definition of the conceptual content in the space of reasons. In the second case, the recipient approves of the speaker's judgement and includes the conceptual content in the space of reasons.

Judgements, therefore, have a progressive and a regressive function through the *speech act*. On the one hand, speakers with a judgement give recipients the possibility to be recognising or rejecting of this judgement in their own argumentation; and on the other hand, they can be obliged by the utterance itself to give reasons for it. In both cases, a judgement in the language game of giving and demanding reasons is a kind of move that can also cause recipients themselves to make a judgement about this move or about the speaker or the speaker's intentions.[17]

But to the inferentialism of matter, the conceptual content is also the entailment of a *thought act*, a kind of potential speech act. This is the *other side* in judging, and this standpoint ensures that the conceptual content or materiality is determined. Whereas speech acts are developed acts of thought, thought acts are said to be enveloped speech acts. Envelopments show up in two ways: first, as thought acts transposed into speech acts, and second, as thought acts predisposed or enveloped in speech acts and not developed. Whereas thought acts fill the entire space of the concept in an enveloped way, speech acts represent only a limited selection from it, but always refer to the

---

[17] Cf. Michael Dummett: What is a theory of meaning? (II). In: Truth and Meaning. Ed. by G. Evans, J. McDowell. Oxford 1976, pp. 34–93.

3.1 Inferentialisms

congruent space of reasons. Consequently, the limited number of speech acts is always contrasted with an apparently unlimited number of thought acts.

Since *speech acts* such as `Socrates is mortal` indicate a limitation in communicative situations because all thought acts can never be developed simultaneously or because the thought acts involved with them such as `Socrates is a human being, humans are rational living beings, rational living beings are living beings` etc. do not always play a compelling role in communicative situations, the thought acts associated with speech acts remain predisposed, involved or only potential. Speech acts such as `Socrates is mortal` can be regarded as limited and thought acts such as `Socrates is a human being` as non-compulsory if, for example, `Socrates is mortal` introduces an argument that is about Socrates' execution and that is not about whether there are also living beings that are immortal or the like. Thought acts, however, that concern Socrates' humanity or the relation of humans to other living beings may only be developed into speech acts when the speaker is required to give reasons and justifications for his speech acts: "Before you go on talking about Socrates' execution, why don't you explain: Why is Socrates mortal?"

For inferentialism of matter, the space of reasons is something that is constituted when the content of the space of the concept is questioned. It shares the opinion with the inferentialism of grounding that the space of reasons and the space of concepts are congruent or that the latter covers the former. But nevertheless, both inferentialisms are separated by the explanation of this congruence and the explanation of how both are overlapping: If the inferentialism of grounding explains that the space of concepts is formed from the space of reasons, the inferentialism of matter explains that the materiality and the content of the space of concepts are only developed congruently and overlappingly when the space of reasons is in question, i.e. when the conceptual is to become distinct. Just as the inferentialism of grounding relies on *formation*, the inferentialism of grounding relies on *development*.

Developments, however, are not only the consequence of the demand for reasons, but they are also, and precisely, evident in the giving of reasons. These are the two aforementioned standpoints of judgement. If a speaker commits herself to a certain judgement in a speech act, she develops it out of a seemingly unlimited abundance of thought acts. The particular development of the speech act is consequently an actualisation of the seemingly unlimited number of acts of thought. In this development, the speech act plays a double role. Not only is it, as a speech act, a public givenness in the game of giving and desiring reasons, but it is also a certain developed thought act that continues to be in a context to the seemingly unlimited number of acts of thought that have not been developed.

This context with all other enveloped acts of thought thereby structures the conceptual content of the developed speech act: the conceptual content of the judgement `Socrates is a human being` is determined by the fact that the judgement is

3 Logic and World

not only a developed speech act but that it also stands in a context with the enveloped thought acts `human beings are rational living beings, Socrates is a rational living being`, etc. or the judgement `Socrates is a human being` has been developed as a speech act from the context that it has structured together with the enveloped thought acts `human beings are rational living beings, Socrates is a rational living being`, etc.

The multiplicity of the enveloped thought acts determines the conceptual content of a developed speech act through the outline: what is the meaning of the judgement according to which Socrates is a human being is determined by the multiplicity of thought acts that determine Socrates as the *content of the conceptual sphere* of the human being and that determine the human being as the content of the concept of the rational living being. The conceptual content of the judgements that are developed in speech acts is consequently determined by the significant division of thought acts.

Even inferences that are not fully developed as speech acts can still be perceived as complete and convincing *in the spiritedness* as thought acts. A reason why Socrates is mortal already results from the development of the speech act `Socrates is a human being`: For with the two judgements `Socrates is mortal` and `Socrates is a human being` a context of the conceptual content of being mortal and being a human being has already been supplied; or both judgements subdivide the conceptual content of `Socrates` in such a way that in developing an inference with both judgements as a speech act, a judgement emerges as an intricate act of thought in the mind that determines being a human being on being mortal. A division of the conceptual content from being mortal, being human and being Socrates can thus take place from only two judgements, which can thus form a complete inference.[18]

On the ground of the structuring of the conceptual content, which takes place through developed or enveloped acts of thought, inferentialism declares the content and the *matter of the judgement* to be more decisive than the form.[19] According to this inferentialism, the conceptual content in a judgement is already determined by its context with another material judgement in an inference. If one wakes up one winter morning by the creaking of the carriages on the street and is thus prompted to judge that it has been a strong frost, then a material judgement was predisposed in the spiritedness as an enveloped thought act. The pure form of inference thus contributes little to conceptual determination, but at most shows what decisive role the conceptual content plays in enveloped thought acts when it is developed together with other judgements as a speech act.

The form of `If ..., then ...` is particularly revealing because this logical connection can be used to develop the enveloped thought acts in a reflective way, or to

---

[18] Cf. Wilfrid Sellars: Inference and Meaning, Chapter I; ibid.: Is there a Synthetic a Priori?. In: Philosophy of Science 20 (1953), pp. 121–138 (Repr. in: Science, Perception and Reality), § 8f.

[19] Cf. Robert Brandom: Inference, Expression, and Induction. In: Philosophical Studies 54 (1988), pp. 257–285.

3.1 Inferentialisms

develop speech acts that explain why an act of thought is correct: `If something is a human being, then it is mortal. Socrates is a human being. So Socrates is mortal.` Moreover, when reasons are required, they can inferentially contribute to clarification by giving alternatives: `If something is not a human being, then...` Finally, they give the opportunity to play out inferences in the case of approved inferences: `If Socrates is mortal, then....`

The possibility of being able to complete a judgement like `humans are mortal` in the spiritedness as a thought act in a speech act consisting of two judgements does not, however, justify the grounding-inferentialist assumption that *conceptual content can also be completed in the spiritedness*.[20] Speech acts such as `cold!` or `bless you!` are not incomplete judgements, but at most incomplete sentences – rather still autonomous discursive practices – since no conceptual content can be structured in them via the enveloped thought acts. Even if such speech acts are followed by actions that can be described with judgements such as `The machine gives the person a handkerchief`, the judgement together with the incomplete sentence does not form an inference from which an outline of `cold` or `bless you` would be recognisable as conceptual content. Fragmentary sentences do not form judgements, but only judgements in the context with others determine the conceptual content of the same. Inferences may only require two judgements, but they must develop two mutually distinguishable contexts between three concepts. Consequently, one can never have only one concept, but one must have at least three, if not an even greater number.[21]

According to the definition of inferentialism, incomplete sentences consist of one word or verbal phrase; and this becomes a concept when it is in the context to another word and thus develops a judgement, which in turn, in the context with another entangled or developed judgement, divides and structures the conceptual content of both judgements. In the ground form of the judgement, the conceptual content is assigned a specific role. The context that determines the conceptual content between two words separates the two contents from each other, both according to their form and their material. In the ground form of the judgement, one side decomposes into a given or a givenness that appears in conceptual form, while the other side is a determination of the given or the givenness.

The conceptually given and its determination in the context of a judgement can be distinguished with the help of stocks, according to the role that the conceptual content can take and play in the context of the judgement. Material inferentialism first divides these *stock roles* fundamentally according to the position of the conceptual content in

---

[20] Cf. Robert B. Brandom: Between Saying and Doing. Towards an Analytic Pragmatism. Oxford 2008, Chapter 2.3.
[21] Richard Rorty: Philosophy and the Mirror of Nature, Chapter IV.3.

the context of a judgement. Given can be translated reciprocally with given of the same conceptual content, whereas determinations can mostly only be substituted unilaterally: In `Socrates is a living being`, `Socrates` is the given that can be translated reciprocally with `the disciple of Diotima`, for example; the determination `is a human being` can be replaced with `is a living being`, but `is a living being` cannot necessarily be replaced with `is a human being` in all contexts.

Due to their mutual interchangeability, `Socrates` and `the disciple of Diotima` form a stock role for judgements. The individual parts or the specific conceptual content of the space of concepts can thus be occupied by all the holders of a stock role. This points in particular to the important function that the given occupies in composition: Although the determination explains the given in the judgement more closely, this given explains more closely that which it develops in the composition. Since givens or givenness in the judgement can be divided according to stock roles that can be occupied by different holders or in which different forms can be translated together and in combination, there is *not one form* or one holder of all stocks, but rather, according to the sum of the stocks, a *large number of givens* in the judgement.[22]

The division of stocks, however, does not only mean that there is more than one given, but also that it is less than a large number of givenness or givens that play a role in the judgement. Givens differs according to whether it is, for example, a proper name such as `Socrates` or a definite designation such as `the disciple of Diotima`; but in their role as separate givens in the context of a judgement, they form the *commonality of the stock*.

The inferentialism of matter does not understand the notion of givens or givenness in judgement as a transfer. *The given is not something that is supposed to come from outside the space of concepts*, but something that is related in the space of concepts in distinction to its determinations. It is itself a concept, plays a role in a stock role with the thing in itself or with the absolute, or the given and the givenness appear in conceptual form. Thus, for the inferentialism of matter, the indivisibility of the space of concepts and the space of reasons remains guaranteed, although the former is not formed from the latter, but the latter develops through the former. In this development, the given is always only a translation within the space of concepts, and where translations can take place, which assume certain roles in judgements, stocks can be described.

With regard to the space of reasons and thus to the question of where this connection comes from and how it can be asserted that, for example, being human can be substituted by being alive, but not vice versa, the inferentialism of matter drives a double strategy that is unsatisfactory for inferentialism of grounding: Genetically, it explains

---

[22] Cf. Robert Brandom: Making it Explicit, Chapter 6.IV; Gilbert Ryle: Categories. In: Ibid.: Collected Papers Vol II. Collected Essays 1929–1968. 2nd ed. London, New York 2009, pp. 178–194. Ibid.: Philosophical Arguments. In: ibid., pp. 203–222.

that the materiality on which it relies for all inferences is determined by a small number of judgements that the linguistic community has established as correct and that are determined as thought acts for every speaker of this community. From these determinations, it derives, according to the principle of materiality, all further judgements, which are then developed as speech acts. The inferentialism of grounding finds no answer to the question of descent, i.e. how the linguistic community justifies the correctness of the small number of determined judgements. Therefore, the inferentialism of matter, when compelled to answer such *ultimate why-questions*, falls back on a systematic explanation: why given things or givenness are there in the judgement at all and not rather not, is explained by the fact that logical contexts offer us the help to explain what we do and why we do it when and in what form we develop thought acts in speech acts.[23]

For inferentialism of grounding, this teleological explanation, which offers it a kind of minimal ontology, shows the justification problem of the entire inferentialism of matter. In its opinion, the inferentialism of matter falls back into the standpoint of a theory of causality when it either wants to explain the materiality of judgements by references with the help of a dogmatically derived minimal ontology or when it establishes the materiality of judgements by the materiality of other judgements. In the first case, material inferentialism vacillates between the standpoints of *critical rationalism* and *representational realism*; in the second case, it approaches the causal theory of *semantic holism*.

For the inferentialism of grounding, in the case of an asserted materiality of judgements, inferentialism *must explain* where the materiality of its judgements comes from, without referring to a simply given or pre-given and without falling into circularity or infinite regress, which only offers ever different or ever new translations in the space of concepts. From the perspective of the inferentialism of grounding, however, the inferentialism of matter does not manage to escape these two horns of the dilemma. It is threatened with the same fate as logicism.[24]

From the perspective of the inferentialism of grounding, the inferentialism of matter is a variant of the standpoint of *critical rationalism* when it cannot explain how the materiality of its judgements is grounded, but instead derives materiality from the process of judging. Whereas other theories of causality had seen judging as the expression of a relationship between a speaker and an object – a relation that can be said to be true or false – the critical rationalism of the inferentialism of matter replaces truth and falsity with the recognition and rejection of judgements. But how speakers come to make judgements in the first place, or why there can be judgements at all that are not rejected, remains a mystery to the inferentialism of grounding.

---

[23] Cf. Robert Brandom: Making it Explicit 2.I; 6.VII; John MacFarlane: Frege, Kant, and the Logic in Logicism, Chapter 3.
[24] Vide infra, Chapter 2.3.

The inferentialism of grounding sees an alternative to critical rationalism in the variety of *representational realism* to which material inferentialism refers several times. The notion of a given that is always in judgement or of a subject that is in each case different from a determination or a predicate is for it a shadow of a *theory of transcendence* that differentiates either between the absolute and its appearances or between substances and their accidents. For the inferentialism of grounding, it seems questionable why the conceptual content, which the inferentialism of matter confines to a stock of a subject, cannot also be translated into stock with which one assigns predicative roles. After all, the conceptual content of the term `living being` can only be structured in the inference discussed above by having a substantive role in one of the two speech acts and an accidental role in another.

For inferentialism of matter, a way out of the dogmatism of critical rationalism and that of representational realism initially seems to be offered by semantic holism. But here, too, the inferentialism of grounding sees further dangers: Because for the inferentialism of matter, a word or phrase becomes conceptual content when it is built into material judgements; and material judgements are not grounded on the formal validity of inferences, but on the conceptual content of their judgements. Material inferences are the condition of the concept, and the concept is the condition of material inference; or the conceptual content becomes the condition of the logic of inferences and the inferential reasoning the condition of the logic of concept. The logic of judgement, which was actually intended to play a privileged role in language, is left out of the relationship between the semantic and justificatory instances of the inferentialism of matter. Since there are now no real intermediate links between the logic of concepts and the logic of inferences, the inferentialism of matter remains trapped in this dilemma of mutual conditioning and presupposition.

3.1.3 Inferentialism of Form

The two preceding inferentialisms not only take a stand against each other, but both are also criticised by an inferentialism of form, since they allow themselves to make the validity and correctness of contextualized judgements dependent on the ground of the concepts or on the material of the judgements, without ascribing a decisive role to the form of reasoning. The inferentialism of matter, in particular, makes the mistake of privileging certain material inferences over formal inferences because of their already determined conceptual content. However, it would have to be considered that the structuring of the conceptual content of its material inferences already presupposes formal principles. Even material principles of reference or transformation rules always develop a technical vocabulary or show the use of the same in an enveloped way.

3.1 Inferentialisms

The conceptual content in an inference already presupposes an objectively given logical context, which in turn is determined by grounded compositions:[25] Whoever asserts the inference `Socrates is a human being, therefore he is mortal` is committing herself to the meaning of a logical contextual vocabulary which states that `Socrates is a human being` is *poorer in content* than `Socrates is mortal`. If one pronounces the inference "`Socrates is a man` is poorer in content than `Socrates is mortal`, therefore `Socrates is mortal` is *richer in content* than `Socrates is a man`", one commits oneself to a rule of conversion that determines the conceptual content of the binary relational terms `richer in content` and `poorer in content` via the *logical vocabularies* of symmetry and negation.

Whereas the two previous inferentialisms have particularly focused on the content of the space of concepts and connected it as congruent or overlapping with a space of reasons lying behind it by means of formation or development, the inferentialism of form explains form as the essential aspect for the space of reasons. Only the form of the space of reasons gives it a reason to speak of a content of the space of concepts.

For the inferentialism of form, the inferentialisms of grounding and matter are subject to the temptation to let *the space of concepts emerge from the space of reasons* using formation or development. For the inferentialism of grounding, the space of concepts forms like a second nature from the space of reasons, and for the inferentialism of matter, the space of reasons develops when certain contents of the space of concepts, which are already predisposed, are in question. In both cases, the space of reasons is only covered by the space of concepts, so that its role is only a one-sided one, which it has in the game of giving and demanding reasons. For formal inferentialism, however, the role of the space of reasons cannot always be given or demanded now and here; rather, it is only to be thought undivided from the space of concepts in such a way that the latter is covered by the form of the space of reasons and not vice versa. Whereas with the inferentialism of matter the content of the space of concepts determined its composition, the inferentialism of form explains the content of the space of concepts only through the reasoned form of the composition. Conceptual content is thus the result that can be read from the form of composition.

The question of whether this form of composition is transferred from the space of reasons into the space of concepts or whether both spaces must be thought to be congruent in such a way that the composition in one is also the composition in the other, is answered by formal inferentialism by referring to a stock of compositions that is specific to both spaces. According to its assertion, there is a special stock of contexts and compositions and these compositions are called *logical truths*. These logical compositions are to be thought of as grounded in both spaces so that they represent a grounding theorem for all other contents in the space of concepts.

---

[25] Rudolf Carnap: The Logical Syntax of Language, §§ 10 and 49.

3 Logic and World

A *grounding theorem* is a compound that must be declared valid or true independently of the compound of all other contexts. The context, that is, any other compound related to the grounding theorem, explains it more precisely, but the essence of the grounding theorem is to be justified and understood independently of the context. The explanation from the context is thereby made by recourse to the space of reasons supporting it. The reference of the grounding theorem in the space of concepts to the explanation in the space of reasons is called truth or validity. The explanation clarifies why the grounding theorem represents a grounded connection between the space of concepts and the space of reasons. The form of the grounding theorem in the space of concepts is the form of pure truth or validity in the space of reasons.

For inferentialism of form, the form exists independently of its explanation, since it too describes the relationship between the space of concepts and the space of reasons with the expressions of congruence and overlap. Only the explanation, however, shows that this congruence endures for the grounding theorem; and the form that it fundamentally takes in the space of concepts, it completely passes on to all compositions in the space of concepts. This means that *starting from the grounding theorem*, the space of concepts in its composition is, on the one hand, free of contradiction in itself, since no concept is not composed with other concepts via the grounding theorem; and, on the other hand, it is completely congruent with the space of reasons, since, through the contradiction-free composition, a true explanation of all compositions can also be supplied.

The inferentialism of form shares the idea of starting from grounding theorem with logicism. The predisposed and enveloped content on which material inferentialism has concentrated is only something secondary and ambiguous to logicism and inferentialism of form. For the conceptual content, in its various forms, may express something that does not influence the possible entailments, and it may be subject to deceptions to which its use gives rise. The uniformity on which the inferentialism of form insists is established by two values, namely by the form itself. It is to be used to be freed from its own constraint.

With unambiguous form, logicism and inferentialism of form hope for the grounded composition of all sciences up to perfect uniformity. The fundamental form becomes a means, for example, to analyse propositions including numbers.[26] From the propositions of motion and nature to the spoken word of everyday language, the fundamental form guarantees unambiguity and uniformity that the inferentialisms of grounding and matter lacked. Whereas inferentialism of grounding and matter still had the multiplicity of linguistic expressions in spiritedness, the inferentialism of form and logicism see in it only the relics of the old superstition in a world beyond provable quantities. Against these figures, which have sunk to the level of transcendence theory

---

[26] Cf. e.g. Peter Andrews: An Introduction to Mathematical Logic and Type Theory. To Truth through Proof. Dordrecht, Boston 2002, Chapter 6.

3.1 Inferentialisms

and are not permeated by pure inference, the inferentialism of form and logicism proceed with the rigour and uniformity of the grounding form.

How grounding compositions look like can be brought to a form if one takes out two words in the space of concepts ☐, Ⓐ and Ⓑ, which become concepts because of their relation, expressed in truth and falsity, to the congruent and overlapping space of reasons:

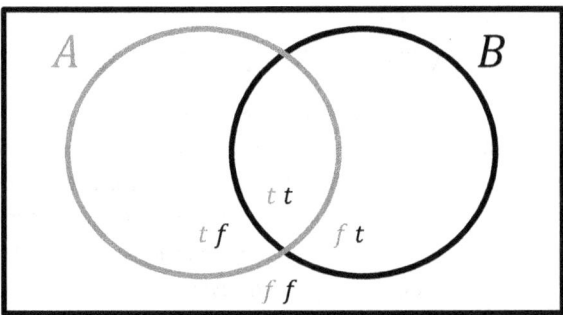

Since inferentialism does not want to give the impression through intuitive justifications that the space of reasons could be larger than the space of concepts after all, it expresses the grounding compositions through forms that reduce the intuition to the lower intersection of the common area of both concepts. Employing indicators • compositions can be represented from the reduced intuition, which represents both the unambiguous form (space of reasons) and the respective conceptual content (space of concepts). Such are, for example, A as ✘; B as ✘; A and B as ✘; A, but not B as ✘; neither A nor B as ✘ etc. A proposition that would always be true is denoted by ✖.[27]

On the one hand, grounded compositions are expressions of strong or weak elementary thought acts of a speaker, such as holding true, holding false, hesitating, doubting, deciding (wanting), etc.[28] On the other hand, however, they are completely independent of the speaker in their relation to the grounding theorem, and the speech act of true and false performed on them is only an empty game of combinations. In these grounded compositions, ordinary sentences are found that are related to the grounding theorem employing rules. For example, the rule of transformation can be brought into play, that if A, then B means the same as not-A or B. If → is the sign for if ... then, ¬ the sign for not, ∨ the sign for or and = the sign for identity, then

---

[27] This notation is inspired by the tradition beginning with Grosser, going through Krause and Lindner and elaborated in Peirce, McCulloch and Randolph, to which reference has been made several times in Chapters 2.2 and 2.3. The difference to McCulloch and Randolph can be seen in the more complex forms.
[28] Cf. Bertrand Russell: An Inquiry into Meaning and Truth. 5th ed. London 1956, Chapters IV and V.

with the help of the rule mentioned above, the following grounded theorem could be justified as a proposition that is always true:[29]

$$A \to (B \to A) = \neg A \to (\neg B \vee A)$$
$$= \neg ※ \vee (\neg ※ \vee ※) = ※ \vee (※ \vee ※) = ※ \vee ※ = ※.$$

But since the basic compositions can be replaced by the same forms, the liberation from constraint and the hoped-for unambiguity are in danger of being obscured by use and usage. This can be seen in another grounded composition, for example:

$$(\neg A \to \neg B) \to (B \to A) = \neg(A \vee \neg B) \vee (\neg B \vee A) =$$
$$\neg(※ \vee \neg※) \vee (※ \vee ※) = (\neg※ \vee ※) \vee ※ = (※ \vee ※) \vee ※ = ※ \vee ※$$
$$= ※ ※ ※ = ※$$

For an inferentialism that, on the one hand, wants to transfer the inferentialism of matter into a formal design, but, on the other hand, regards grounded compositions in each case as a different form of the given, which may only be proper to the standpoint of the theory of causality, these grounding theorems are artificial. Or, to put it another way, on the grounding theorems *the inferentialism of form splits* into an artificial one, insofar as it accepts the transfers between the space of reasons and the space of concepts as truth, and into a natural one, insofar as such transfers are regarded as an unnecessary relapse into the theory of causality.

For *natural inferentialism of form*, the grounding theorem represents only an artificial transfer from the space of reasons into the space of concepts and thus becomes for it a reversion to the theory of causality: just as the theory of causality derives its explanation of the space of reasons from a grounding theorem of the space of concepts, so too artificial inferentialism of form derives the entire composition of the content of the space of concepts from a grounding composition, which it regards as a context between the two spaces. Truth, after all, becomes a cypher for causality and dividedness to the natural inferentialism of form; and this cypher takes on a justificatory function where there should really only be rationality and undividedness.

Natural inferentialism of form also explains that in the game of giving and demanding reasons, it is actually never the case that speakers start from a grounded theorem given in the space of concepts. By limiting itself to grounded compositions, the inferentialism of form has distanced itself far from the natural use of conceptual content. It would be more natural to dispense with grounded compositions in the space of concepts and to regard all compositions as *rule-governed or self-positing*. The rule-governed compositions are already components that the artificial inferentialism of form cannot do without in order to be able to distinguish between propositions with richer and poorer content. If, for example, the rule applies that ※ can be inferred if ※

---

[29] Cf. Jan Łukasiewicz, Alfred Tarski: Investigations into the Sentential Calculus. In: Jan Łukasiewicz: Selected Works. Ed. by L. Borkowski. Amsterdam et al. 1970, pp. 131–152.

3.1 Inferentialisms

and $\neg \mathfrak{X} \vee \mathfrak{X}$ are set, then only from the grounding theorem $\neg \mathfrak{X} \vee (\neg \mathfrak{X} \vee \mathfrak{X})$ and the premise $\mathfrak{X}$ can $\neg \mathfrak{X} \vee \mathfrak{X}$ also be inferred.

Since the natural inferentialism of form has recognised that these rules are also inherent in artificial inferentialism of form, they must engage in a struggle to decide on grounding theorems and rules. While artificial inferentialism of form attacks the rule and forces its grounding theorem to unity, natural inferentialism of form disregards the principle and develops a multiplicity of rules. In this struggle for the number of grounded compositions, artificial inferentialism of form recognises its chance in bending the naturalness of language use to force unity of grounding theorems with the artificial repetition of the `and` `not` and to ward off the multiplicity of rules. For natural inferentialism of form, however, this is a short-sighted stratagem whose success is bound only to the narrow world of provable quantities and which would already be unthinkable in the broad field of naturalness.

Thus, the natural inferentialism of form shares the tendency towards naturalness with the inferentialism of grounding and matter. For natural inferentialism of form, the composition of individual contents in the space of concepts is a self-positing in the form of an assumption or a demand. If an assumption formula or demand appears in this self-positing that corresponds to the grounding theorem of the artificial inferentialism of form, this is not due to a grounded composition in the space of concepts, but solely to the feeling of naturalness when proving with certain quantities.[30] The demands and assumptions of natural inferentialism of form are, for example:

(1) If A and B are to be proved, then A must be proved and so B must be proved:

$$\frac{\mathfrak{X} \qquad\qquad \mathfrak{X}}{\mathfrak{X} \wedge \mathfrak{X}}$$

(2) If A or B is to be proved, then A must be proved or B must be proved:

$$\frac{\mathfrak{X}}{\mathfrak{X} \vee \mathfrak{X}} \quad \text{or} \quad \frac{\mathfrak{X}}{\mathfrak{X} \vee \mathfrak{X}}$$

(3) If If A, then B is to be proved, then B must be proved under the assumption of A:

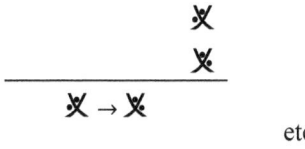

etc.

---

[30] Cf. Gerhard Gentzen: Investigation into Logical Deduction, II § 4.

3 Logic and World

The transition from the antecedent to the consequent or the fulfilment of the consequent is described by the inferentialism of form as self-positing. Everyone who knows that a proof of A and a proof of B is also a proof of A and B knows the meaning of composition with and by the use of (1). Anyone who knows that a proof of A is also a proof of A or B knows the meaning of composition with or by the use of (2). Anyone who knows that a proof of B assuming A is also a proof of If A, then B knows the meaning of the composition with If ..., then ... by the use of (3), etc. The meaning of and, or, if..., then etc. is thus the knowledge of their inferential use and not an expression of a thought act. In other words, the meaning of self-positings can be taken solely from the role they play in the context of inferences (1), (2), (3), etc.

Furthermore, self-positing opens up the possibility of the *opposition* of form to natural inferentialism of form. Introduced compositions can be dissolved back into their constituent parts in the opposite form. In this way, natural inferentialism of form asserts that the negation of that reality which it itself has composed in its demand for reasoning is legitimised by the principle of opposition. The space of concepts is thus composed into a reality that only exists as long as the negation does not sublate it.

(1*) If A and B are proved, then *A* can be proved and *B* can also be proved.

$$\frac{⊠ \wedge ⊠}{⊠ \qquad ⊠}$$

(2*) If A or B are proved, then *A* can be proved or *B* can be proved

$$\frac{⊠ \vee ⊠}{⊠} \quad \text{or} \quad \frac{⊠ \vee ⊠}{⊠}$$

etc.

As is peculiar to the first inferentialism of form, the latter sees artificial forms everywhere in the game of self-positing and opposing, which it assigns new forms to and assigns roles to them that the natural inferentialism of form does not recognise. Thus, *on the one hand*, the artificial inferentialism of form sees no difference between the intended proof and its asserted truth. For it, the natural inferentialism of form pretends that there is only one space of concepts *and* reasons, while at the distinctions between assumption and proof, proper and improper proofs, etc., the divergence of the two overlapping spaces is everywhere presupposed.

3.1 Inferentialisms

On the other hand, artificial inferentialism of form sees no reason why one should not assign a different role to composition than to opposition, or why compositions or oppositions should always play only one role.[31] If two components of the space of concepts are put together in a way utilizing a self-positing, then it is at least possible to assert a new kind of opposition with the composition. But if two components of the space of concepts are set together employing a self-positing, then it need not be ruled out that they will be *composed differently* at another time with the same self-positing. This artificiality of the natural inferentialism of form proceeds in such cases in an arbitrary game of asserting and contradicting, or in an arbitrary game of asserting in multiple ways.

Already in the use of the `or` in (2) an arbitrariness becomes visible that could be used as a gateway into the fortress of naturalness: If A `or` B is to be proved, the arbitrariness becomes apparent in the question of what happens if *A* and *B* either denote the same thing, or do not denote the same thing, or if no distinction was made between `or` and `either ... or`. The natural inferentialism of form objects to these four distinctions that artificial formal inferentialism has already applied its interpretation of judgements based on the problematic truth relation to (2), whereas it is more natural to interpret the meaning of the proof claim through the use of the logical composition that makes up the proof claim of A `or` B. This is not the case with the natural inferentialism of form. The use of the logical composition can therefore be distinguished from case to case. For this reason, the artificial inferentialism of form had not given four distinctions of one proof claim, but two distinctions each of two proof claims, `or` and `either...or`. Artificial inferentialism of form had interpreted the sign for the natural `or` in a divided way, whereas it had to be used undivided. But by doing so, the second argument of the natural inferentialism of form is a direct violation of the *principle of undividedness* inherent in inferentialism.[32]

Once this confusion had been unmasked by analysing the context of the proof, one could see from the use of `or` in the first two alternatives and `either... or` in the last two alternatives that both compounds could refer to either divided or undivided conceptual content. Thus, the use of compounds in the context of an inferential proof would allow conclusions to be drawn about conceptual content.

The artificial inferentialism of form rejects both arguments of the natural inferentialism of form and constructs a *third argument* from its rejection. Natural inferentialism of form, it argues, had only been able to produce its arguments against the arbitrariness of the proof by using the output of truth to distinguish the two times two alternatives of the composition `or`. Since it cannot distinguish even apparently natural compositions such as `or` by the word, it needs a semantics that is inauthentic

---

[31] Cf. Arthur Prior: The Runabout Inference-Ticket. In: Analysis 21:2 (1960), pp. 38–39.
[32] Cf. Nuel D. Belnap: Tonk, Plonk and Plink. In: Analysis 22:6 (1962), pp. 130–134.

3 Logic and World

to it, which clarifies the concept of the respective composition for it. The natural inferentialism of form had thus only been able to argue against the artificial inferentialism of form because, in the argument against arbitrariness, it presupposed a distinction in naturalness that it cannot itself explain. Together with the inferentialism of grounding and matter, the artificial inferentialism of form finally generalises that the inferences that provide the context of meaning can only be needed *because either the output of the inferences already has a ground or because this output can unconsciously fall back on a matter in judging.*

In the *first case*, there are several reasons why the meanings of the self-positings can already be taken as a given ground: The inferentialism of form, which claims to be natural, since it wants its proof procedure to be read off from the real or actual practice of reasoning, uses causal-theoretical argumentation from the point of view of the inferentialism of grounding. This is because the latter argues that there is an inferentialism of form that is artificial since it demands too strong a separation between the actually overlapping spaces of concepts and reasons utilizing the concept of truth. However, it itself also gains its natural property through an observation: it observes that speakers do not follow the artificial inferentialism of form, but *develop a technique of proof, which it declares to be a natural givenness.* The ground here is dialogical. The space of concepts thus becomes a representation of a traditional practice of reasoning, which cannot be ruled out as having established itself because its speakers have favoured it based on the artificial truth reference.[33] Should this tradition not be based on an artificial truth reference, it is ultimately not a false subtlety to consider the possibility that what formal inferentialism represents as a natural given in the space of concepts has always been an inauthentic justificatory performance that was once acquired and then always handed down as apparent authenticity.[34]

In the *second case*, the meaning cannot be abstracted from the context or from the use because the use is unreflective. In this case, too, two possibilities arise, which the artificial inferentialism of form takes as an argument that the natural inferentialism of form is never presuppositionless. First, a dualism between the context and the meaning of the conceptual content cannot be explained if the word is used regularly and therefore apparently reliably, but the use is only acquired and not reflected upon. A judgement about the reliability with which a word is used in certain contexts can only be ensured by explaining and justifying regularities. Second, a dualism between the context and the meaning of the conceptual content can only be explained if an explanation and justification of regularities is also provided, albeit based on the pure form of a grounding theorem.

The artificial inferentialism of form draws the conclusion from both problem cases that the verbal composition of conceptual content is itself part of the content of the

---

[33] Cf. J. T. Stevenson: Roundabout the Runabout Inference-Ticket. In: Analysis 21:6 (1961), pp. 124–128.
[34] Cf. Jaroslav Peregrin: Inferentialism. Why Rules Matter, Chapter 10.

3.1 Inferentialisms

space of concepts and thereby, according to artificial inferentialism's own determination, coincides with the explanation in the space of reasons.[35] Since, for the artificial inferentialism of form, this explanation bears the expression 'truth', compositions are not to run in arbitrariness, but in truth explanations of the compound conceptual content. If then, by means of an *output of truth*, it is explained in what relation compositions in the space of concepts stand to the space of reasons, then, in its opinion, not only the original content-related concepts can be defined, but also those concepts that determine compositions and oppositions.

For natural inferentialism of form, however, the recourse to the corresponding and compositional concept of truth reveals further leanings towards theories of causality that inferentialism had hoped to have eliminated long ago. Its ideal of covering the space of reasons by the space of concepts to such an extent that only the latter remains in the foreground prohibits it from accepting this explanation; the explanation is rejected as an artificial limitation. The suspicion of having fallen prey to an arbitrariness between the introduction and the elimination, or its reality and its negation, can be dispelled by introducing a *principle of harmony*, which itself, however, does not become part of the theory of proof regulated by it.[36]

This principle of harmony, like the principle of indivisibility, is a revision of the second argument that the artificial inferentialism of form had advanced against the natural inferentialism of form. The artificial inferentialism of form had not only seen an arbitrariness of compositions subject to the required proof of A or B, but also recognised an arbitrariness in the separation of self-positing and opposition. To this end, using the principle of indivisibility, he constructed a self-positing that would lead to the following opposition:

$$\text{If } \frac{\text{✘}}{\text{✘ ✘ ✘}} \text{ , then } \frac{\text{✘ ✘ ✘}}{\text{✘}}.$$

The natural inferentialism of form responds to the construction of such arbitrariness with a principle of harmony that is not itself a component of a proof regulated by it, but is close to the naturalness of reasoning. For it, inferential proofs are fixed on taking out of an opposition only that which was also previously put into the self-positing. This requirement of harmony, which limits the reality of the self-positings and the negation of the opposition, is, on the one hand, the advance over the disproportion of the seemingly endless number of enveloped thought acts in comparison to the few real developed speech acts in inferentialism of matter, but on the other hand, it is the step backwards in the argument against artificial inferentialism of form.

---

[35] Cf. J. T. Stevenson: Roundabout the Runabout Inference-Ticket.
[36] Cf. Michael Dummett: The Logical Basis of Metaphysics. Cambridge/Mass. 1993, Chapter 11.

3 Logic and World

The artificial inferentialism of form recognises in the demand for harmony a further inauthentic principle of the natural inferentialism of form since it is not itself part of the inferential proof procedure that it establishes. The naturalness, which was read off as a counter-programme to the artificial inferentialism of form from the givenness of natural reasoning, is grounded in the attempt to develop an actual inferentialism of form from the read-off forms only by referring to the read-off naturalness: "How is it grounded that the read-off or is used in this way? – It is justified by its naturalness. – How is it justified that the principle of indivisibility and harmony regulates the self-positing and opposition of the conceptual content? – Because it is natural that in oppositions no more is developed than what is present in their self-positing." For artificial formal inferentialism, even the talk of context and use is only an inauthentic and transferred way of talking about the ostensible naturalness of an inferentialism of form that always falls back on artificial justifications in its groundings.

In this constant confrontation, both forms of formal inferentialism win their victories in different fields: The artificial inferentialism of form insists on the applicability of its artificial systems just as the natural inferentialism of form praises its system of natural application. But the recognition in their own terrain is not enough for either of them. With an envious eye on the theory of transcendence, both share the intention of harmonising artificiality and naturalness. As the first inferentialism of form wants to transfer its technical victories to an artificial naturalness, the second inferentialism of form recognises in the spiritedness the idea of a naturally acting artificiality.

However, due to the many insoluble difficulties that appear in the part of the desired harmonisation, the challenged inferentialism finally returns to that form from which it had once started: it sees in its *organic origin* the conceptual ground, the materiality of the judgement and the form of the inference given. This origin is neither self-posited nor composed, but a natural given in which it recognises its undivided need for the form. This recognition evokes the conviction that the construction of unity is not the only, nor even the most important goal of inferentialism of form and logicism.[37]

Since inferentialism has learned in its development to extend the syllogistic form in many directions, it soon comes into competition with the forms favoured by the artificial inferentialism of form.[38] Inferentialism sees a stronger naturalness in syllogistic than in other forms of logic, because it follows the givens of linguistic traditions in a more natural way,[39] because this natural consequence is a better form of representing

---

[37] Andrzej Mostowski: On a Generalization of Quantifiers. In: Fundamenta Mathematicae 44:2 (1957), pp. 12–36.
[38] Cf. Robert van Rooij: Extending Syllogistic Reasoning. In: Logic, Language and Meaning. Ed. by M. Aloni, H. Bastiaanse, T. de Jager, K. Schulz. Berlin, Heidelberg 2010, pp. 124–132; Larry Moss: Completeness Theorems for Syllogistic Fragments. In: Logics for Linguistic Structures. Ed. by F. Hamm, S. Kepser. Berlin 2008, pp. 143–175.
[39] Cf. Michael Wolff: Abhandlung über die Prinzipien der Logik. Frankfurt/ Main 2004, Part II.

3.1 Inferentialisms

natural givens than that of artificial languages,[40] and because it provides a form of reasoning that underlies all artificial systems.[41]

The last point, however, opens up another battlefield between the artificial and the natural inferentialism of form. In the field of syllogistic, the artificial inferentialism of form tries to beat its opponent with its own weapons: it claims that the justification of syllogistic does not succeed by itself either, but that it needs a grounding in a form underlying it, which corresponds to the self-positings and and if..., then...: In if..., the two premises are connected by and, so that then... indicates the conclusion.[42] The proof of the validity of syllogistic, the grounding, must take place through the two self-positings (1) and (3), which the natural inferentialism of form has read out of the facts of natural reasoning. If these self-positings are the groundings for the validity of syllogisms, then they are not self-positings, but grounding theorems. However, since these grounding theorems do not originate from syllogistic, but from the self-positings of the natural inferentialism of form, they are transferred and inauthentic components in syllogistic.

The natural inferentialism of form, however, recognises a further artificiality in this strategy. For it, syllogistic is not a science built on grounding theorems, but the grounding of all sciences themselves, which are built with grounding theorems. For this reason, it does not need those principles that the natural inferentialism of form has previously read off in the natural facts of reasoning and that artificial inferentialism of form has tried to apply to it as a grounding. Rather, syllogistic is based on actual self-positings and not on grounding theorems. At the same time, the natural inferentialism of form certainly recognises the problem of the artificial inferentialism of form as a claim to the grounding of syllogistic: If syllogisms must be proved and should not be grounded on transferred and inauthentic grounding theorems, then the actual grounding must be laid in syllogistic.

To do justice to this claim, the natural inferentialism of form distinguishes between two forms of self-positings: *proper and improper*.[43] Improper self-positings of syllogisms are justified by proper ones. Proper self-positings of syllogisms *show* that the conclusion follows from the premises; they make this *distinct* without the need for anything further, such as an improper grounding theorem.[44] Improper self-positings are characterised by the fact that they cannot show that the conclusion follows from the premises. Rather, they must be translated into proper syllogisms and, in the translation, show a gaplessness and an all-connecting *thread* between their improper forms and the proper forms of self-positings in syllogisms.

---

[40] Cf. Jon Barwise, Robin Cooper: Generalized Quantifiers and Natural Language. In: Linguistics and Philosophy 4:2 (1981), pp. 159–219; Fred Sommers: The Logic of Natural Language. Oxford 1982.
[41] Cf. John Corcoran: Aristotle's Natural Deduction System.
[42] Cf. Jan Łukasiewicz: Aristotle's Syllogistic from the Standpoint of Modern Formal Logic. 2nd ed. Oxford 1957, Chapter 2.
[43] Cf. Michael Dummett: Logical Basis of Metaphysics.
[44] Cf. Timothy J. Smiley: What is a Syllogism?. In: Journal of Philosophical Logic 2 (1973), pp. 136–154.

3 Logic and World

The proper form of self-positings in syllogisms and their distinction from the improper forms finally establish the harmony between the two forms of formal inferentialism. For the natural inferentialism of form, the distinction it makes is a distinction between proper and improper self-positings, while for artificial inferentialism of form it is congruent with its distinction between the grounded compositions and the content compositions derived from them in the space of concepts. Proper self-positings and grounded compositions have in common that their validity shows itself. For the one, it shows itself in the natural, in the relation of the self-positing to its use in language; for the other, it shows itself in the artificial, in the relation of the composition to its justified truth.

The innocent request of how the validity of these relations is established brings forward a third form of form-based inferentialism, which leaves its standpoint open between naturalness and artificiality. This third form understands itself to be independent of the artificial and the natural inferentialism of form since it takes no stand on which of the two distinctions is binding for it, either that between the grounding and the content compositions derived from them, or that between the proper and the improper self-positings. However, it is important to it that, on the one hand, the innocent request of the inferentialism of form is answered and that the innocent request is addressed not only to the natural but also to the artificial interpretation of the syllogistic: Not only the proper self-positings that the natural inferentialism of form sees in the syllogistic need to be justified, but also the improper grounding theorems including and, if..., then etc. that the artificial inferentialism of form brings into syllogistic. The explanation for the validity of the proper self-positings which the formal inferentialism of form regards as proper in syllogistic is to it the *dictum de omni et nullo*, and the justification for the validity of the grounded compositions which artificial formal inferentialism regards as proper in syllogistic would be to it a *dictum de aut et si*. But whether this *dictum de aut et si* belongs to the origins of the inferentialism of form can only be proven by way of judgement. Since the inferentialism of form must therefore assume that it has been brought to the principles of the artificial inferentialism of form as something improper and that it was constructed solely on the model of the *dictum de omni et nullo*, its justification of the inferentialism of form stands or falls with the explanation of the *dictum de omni et nullo*.[45]

If the third form of the inferentialism of form takes seriously the innocent request for the justification of proper self-positing and grounding compositions, then the request of how the *dictum de omni et nullo* is justified cannot be dismissed as naïve. Inferentialism has not provided an answer to this non-naïve question, nor will it be able to. That it cannot provide this justification of its own activity is because it always proceeds from the premise, which is improper to it, according to which the sphere of

---

[45] Cf. Jonathan Barnes: Truth, etc. Six Lectures on Ancient Logic, Chapter 5.

3.1 Inferentialisms

the concept `space of reasons` may not go further than the sphere of the concept `space of concepts`.

3.2 Rational Representationalism

I do not want to use the dialectic given in Chapter 3.1 to show that either all inferentialist programmes together or all but one are wrong but to show that the picture that inferentialists present to us already indicates that it is only a section of a much larger one. This, at least, is indicated by the fact that all inferentialisms – or, more generally, rationalists programmes – always run into problems of justification at some crucial point that affects the basis of their own theories. Inferentialism certainly has its merits, and these should not be denied; but although its edifice already appears very magnificent, its picture does not show whether there is actually a solid foundation underneath. The disagreement that inferentialist programmes reveal about whether the foundation consists of concepts, judgements or inferences and how these relate to the ground is an indication that it is not the building elements of inferentialism that are in question, but its fundamental structure alone. Inferentialism stands on feet of clay.

Consequently, my aim is not to replace rationalism with representationalism, but with showing that inferentialism and logicism, on the one hand, *impose unnecessary limitations on themselves* and, on the other, *impose unattainable extensions on us*; these problems, however, can be circumvented by unburdening inferentialism with the help of a rational representationalism. The limitations of inferentialism concern the extent of the space of reasons, which is not supposed to extend beyond the space of the conceptual, and its extensions concern inferences that lie far outside the conceptual content of its premises and that result from the fact that the world is constructed not with but from logic.

It should be noted, however, that the limitations were self-imposed and not the result of external coercion. Of course, inferentialists have described the path to their theory as if it were the only way out of antinomies, aporias and dilemmas brought about by developments in modern thought about the relationship between logic and the world – a way out of the unexplained explainers of modern representationalism; but in the end, this apparent way out is grounded on heedlessness of the givens that have long been naturalised in the space that, according to the inferentialists, is supposed to be genuinely free of givens, as well as on the confusion of semantic theories of understanding and explaining.[1]

What I propose is *neither a variant of an exuberant metaphysics nor a down-to-earth form of physicalism*. On the one hand, I reject the positions that also the inferentialism of grounding and, building on it, the other inferentialisms rightly criticise since I also consider too strict a separation between the space of reasons and the space of concepts to be problematic. On the other hand, the rational representationalism

---

[1] Cf. Jens Lemanski: Concept Diagrams and the Context Principle. In: Language, Logic, and Language in Schopenhauer. Hrsg. v. Jens Lemanski. Cham 2020, pp. 47–73.

3.2 Rational Representationalism

proposed here also criticises the congruence and overlap that inferentialism tries to impose between the space of reasons and the space of concepts with all argumentative means.

The representationalism proposed here asserts that in the space of concepts we have always integrated essential components from the space of reasons so that this space of reasons does not coincide with the space of concepts. One could just as well affirm that there are contents in the space of concepts that do not actually belong to it, but which can nevertheless form a fundamental basis for all inferentialist programmes. That the representationalism proposed here is non-naïve and calling it 'rational' means that it *gives priority to inference over reference* in semantic questions. At the same time, however, it claims that philosophical or general scientific questions cannot be explained by semantics alone. On the contrary, reasons will be put forward to show that the grounding of inferences is based on representations and that inferentialisms have so far followed two unsuccessful strategies to deal with them: Either they have ignored them or they have tried to translate them into a pure conceptual content. This strategy has already been discussed in Chapter 2.2, the other in Chapter 2.3.

If I want to defend the assertion that the grounding of inferences is based on representations without falling back into an exuberant metaphysics or down-to-earth physicalism, I can only succeed in doing so by, on the one hand, sharing to a certain extent with the inferentialists the priority of inference over reference but, on the other hand, *not imposing the unnecessary limitation* according to which the world becomes intelligible only from logic. For within this limitedness it seems unattainable to extend the concept of logic so that it is applied to all that we usually call the world. By no longer sharing the *unnecessary restriction* of a priority of inference over reference from the degree at which inferentialists derive an ontology from logic and semantics, I also do not allow non-naïve representationalism the unattainable extension of evoking being where actually only language is present. Representationalism shares with the inferentialism of grounding, as was shown in Chapter 1, the attitude that language is a part of nature; but non-naïve representationalism has good reasons for considering nature not only as a part of our language.

As already emphasised at the beginning of Chapter 3, a defence of the claim that the grounding of inferences is grounded on representations succeeds through the *one single thought* that says that the conceptual constitution of our grounding of inferential proof theory is already based on inauthentic transfers. On the one hand, these transmissions pose a problem for rationalism, which it seeks to level or assimilate with translations; on the other hand, however, they form the basis for representationalisms, which have repeatedly become conspicuous in the history of logic by drawing on geometrical intuition. The proof of this inauthenticity in the logician vocabulary entitles me to assume that the space of reasons is larger than the space of concepts, without having to go beyond the space of concepts, which for inferentialists covers the space of reasons, in asserting this assumption. Because of this conceptual constitution of

3 Logic and World

non-naïve representationalism, I am spared from appealing to a merely down-to-earth and concept-distant physicalism or to an exuberant and purely conceptual metaphysics. Rather, transmissions that go beyond conceptual justification are already evident in the concept.

A demonstration of this expansion of the space of reasons beyond the space of the concept requires that we come to grips with the problems of rationalism concerning reasons by thinking the relationship between the world and logic differently than we have done so far. The unattainable extensions that rationalism has imposed on itself are based on the need to say something about the area that has so far been left to empiricists, natural scientists and metaphysicians – namely, the area that lies beyond the overlapping area of the space of concept and the space of reasons. However, if one sees the world as something that is not freely discharged from logic in the form of a derivation, but that is already connected to logic, then this view opens up access to a representationalism that need not be considered either exuberant or physicalist, that appears neither as a transcendental theory nor as a theory of causality.

However, my assertion of an expanded space of reasons and a related explanation of problems of reasoning in rationalism is accompanied by other theses that can be understood in different ways: On the one hand, they form a path towards the explanation of the one thought of transference developed in Chapter 3.2.3, but on the other hand, they are already the result owed to the habit of understanding these transferences themselves in a logical context. In order for this transference to come about at all and to have such a presence in our logic to this day, it required the historical development of an abstract language process. This abstract language process provides an explanation of why we find content in the space of concepts at all, which we can understand based on the materiality in our judgements and on the basis of the regularity in our inferences. Chapter 3.2.1 argues for the fact that we should not completely replace the abstract philosophy of language, dismissed by the inferentialism of grounding as uncontrolled, with a contextual theory, but should understand it as complementary to semantics of use theory. The inferentialism of grounding is certainly justified in criticising a theory of abstraction that is built purely privatively on internal thought acts and postulates an intimate relation between the concept and the given; but my proposal to establish a theory of abstraction as a complement to a contextual theory is at the same time accompanied by the demand to abandon this intimacy and privation and to socialise the theory of abstraction in its place. Such a collective theory of abstraction provides the basis for justifying, on the one hand, a developmental theoretical basis of judgements, as demanded by material inferentialism, and, on the other hand, for successfully establishing an individualised contextual principle for the kind of process of understanding demanded by almost all inferentialisms. Moreover, since I believe that non-naïve representationalism should not burden itself with unattainable extensions that arise from deriving givens from inference, I will argue against ontological commitments on the basis of semantics and instead argue for a principle of semantic commitment as a corrective that is developed based on geometrical logic. Being is nothing other than the logical representation of an intuitive relation.

Chapter 3.2.2 attempts to test the scope of this geometric logic in the area of conceptual representation using Matsuda matrices. The starting point for this is the thesis, already defended in Chapter 2, that the logical vocabulary has an affinity to fundamental transfers. The intuition thus transferred to logic is to be connected with abstraction theory, which is at the heart of neologicism. However, the abstraction theory of rational representationalism, on the one hand, proposes intuition as the starting point of conceptual models and, on the other hand, restricts it as the target point of these models not only to the concept of number but to all concepts.

Finally, in Chapter 3.2.3, the relationship between translations and transfers is to be set apart and the actuality of worldly transfers in the logical vocabulary is to be argued for. In my view, it is these transfers that entitle us to explain the model of geometrical logic, that will be presented in Chapters 3.2.1 and 3.2.2, in terms of abstraction theory and to understand it contextually. As was explained at the beginning of Chapter 3, the indication of fundamental logical transfers in the inferential vocabulary is to be understood as an analogy to the principle shown in Chapter 3.1.3: The basis by which the old and the new logics are repeatedly compelled to demand reasons can be understood by explaining their worldly parts – that is, the terms that index a much wider space of reasons – and recognising them both as the output of the problems of inferential logics and as the input of representationalist logics. Just as the world can only come to understanding on the grounding of logic, so the grounding of logic can only be explained by seeing its worldly transfers as actual constituents.

3.2.1 Abstraction and Being

As has been shown in Chapter 3.1.1, inferentialism gains its grounding theory by distinguishing itself from other standpoints, especially from the standpoint of abstraction theory, which is opposed to it. Advocacy and rejection of abstraction theory are, moreover, the essential distinguishing features of the two currently dominant rationalist positions, namely neologicism and inferentialism. From the point of view of inferentialism, there is not the slightest overlap in abstraction theory between the spheres of the two expressions 'space of concepts' and 'space of reasons'. Thus, for it, the theory of abstraction is a more uncontrolled and ruleless variant of the theory of causality, since in the latter the relationship between the space of concepts and the space of reasons is only one that is remembered in the individual but cannot be a factually given one.

It can be said that the *abjuration of classical abstraction theory* at the ground of inferentialism in the first place was the motivation for embellishing the inferentialist edifice with various forms and points of view. Since the time of the grounding, inferentialists have been confronted with an ultimate problem: Namely, answering the question of why there was meaning (in such plenitude) rather than no meaning. Even if inferentialists could give an acceptable answer to this, they ran into the next problem. They felt compelled to find an answer to the question of how inferentialism grounds itself, if meaning and being may only be inferred from the inferentialist grounds itself.

In this chapter – and in a much more concrete way following it in the next Chapter 3.2.2 – I will argue for a theory of abstraction that acts as a complement to the typical inferentialist semantics such as contextualism, the theory of use and the truth theory of meaning. In doing so, I advocate a theory of abstraction that points to a middle way, so that we neither fall back into a naïve representationalism nor get caught up in the justification problems of neologicism and inferentialism. This middle way is intended to prevent us from imposing *unnecessary restrictions on logic* on the one hand, on the basis of which the world then appears unattainable; on the other hand, it protects us from falling into an *exuberant metaphysics* in which propositions give us the impression that the world is always already something that must go beyond logic.

Before arguing for a more social approach to abstraction theory than has been the case so far, I will first recapitulate two problem areas that have already been discussed in more detail in Chapter 3.1. Both problem areas are the motivation for establishing *a collective abstractionism as a complement to an individualised contextualism*: I see one problem area of inferentialism as being the *commitment to meaning* to be gained without recourse to the world. The other problem area, in my opinion, is the *commitment to being*, which is to be gained only through recourse to logic.

The *first field* is to be illustrated by a myth that is to show the consequences of theories that have either deleted expressions such as 'abstraction from the world' from their vocabulary or that have put expressions such as 'concretion from logic' in their place. This gives rise to the second problem area, which makes commitments about being based on theories of proper names, definite descriptions or the copula 'is'. Contrary to these derivations, the thesis is advanced here that being is only the representationalist expression of an intuitive relation. In what follows, I will argue in favour of a theory of abstraction of meaning that comes alongside an individualised contextualism not as a competitor but as a complement. It is only through a collective, social and anthropological approach, in which language is understood not as a mere development and consequence of one's own concepts, but as a *form of life*, i.e. as an idea of the development of a living, actual essence, that abstraction theory assigns to contextualism its individual role of *understanding* and itself assumes the general role of *explanation*. Through this socio-historical conception, the theory of abstraction solves the *problem of descent* that has pushed inferentialism to form various theories that have accused themselves of not being able to give a complete explanation.

3.2 Rational Representationalism

Of course, it is possible, as many inferentialisms attempt, to remove abstraction theories from one's semantics entirely;[2] but then it is not surprising if an occult quality suddenly attaches to simple lexical definitions. The reference to such occult qualities is what at least the inferentialism of grounding and artificial inferentialism of form have criticised about material inferentialism in particular: How is it, then, that concepts such as `whale` and `mammal` can be meaningfully employed in a judgment such as `A whale is a mammal` or `All whales are mammals` without arousing suspicion of tautology or contradiction? Of course, it would be nonsensical if one wanted to impose on every human being that she must abstract the concept `whale` from the contemplation of a certain object by some inexplicable processes. It would be even worse if one wanted to burden oneself with an explanatory answer to the question of how every person in a cultural circle always manages to attach the same word to an object, which is also always subject to the same concept.

The explanation for such theories of abstraction put forward by rationalists as well as alleged empiricists points to a kind of inherence of the word and the concept in or on the objects. 'Abstraction' plays the role of a ground transfer for processes of dispossession and withdrawal in such linguistic attempts at explanation, which can be described even more intuitively in terms of *labelling allegories*: Things have slips of paper, name tags, badges, etc. attached to them, and each individual can take these off the things and file them away in a collective register, which is then given the name 'linguistic faculty', 'linguistic memory', etc.[3] Such explanations of such theories of abstraction are not unsatisfactory. They are explanatory moments in a theory whose relevance is shown by the fact that one can assign different acceptable roles to concepts in judgements or that one can sort them into certain compartments. After all, we generally disagree with judgements such as `All mammals are whales`. And this shows up even more drastically in judgements that speakers classify as analytical.

Whether we work with the technical term 'abstraction' or with intuitive labelling processes, it remains undisputed that such theories should provide an explanation of why concepts are used differently and how that we come to have different concepts that play a role in our judgements.

This *problem of the descent* of concepts goes hand in hand with the question of the *expressiveness of judgements*. Whereas inferentialism rightly depicts the understanding of concepts from usage, there seems to be a problem in explaining how the different usage of different concepts comes about. First of all, the term 'use' or 'usage' does not carry any active connotation in semantics, but denotes a kind of description of the state of language communities: The meaning of an expression is its established use.[4] In a second step, one restricts the speech of established usage to concepts that play a role in the judgement and leaves judgements themselves out of it. For whether

---
[2] Cf. Peter Geach: Mental Acts. Their Content and their Objects. London 1957, Chapters 6–10.
[3] Cf. Wilfrid Sellars: Empiricism and the Philosophy of Mind, § 26.
[4] Cf. Peter Geach: Mental Acts, Chapter 5.

a judgement is intelligible and meaningful is something that is not decided with the help of the statement that precisely this judgement is already established.

Even inferentialists cannot avoid explaining, in addition to possible truth attributions, the expressive strength of judgements *through the laws of regress of the rules and the composition of the conceptual content.*[5] Expressive power is characterised by the fact that, on the one hand, a construction rule can be formally applied recursively to individual components of the judgement and that, on the other hand, it is materially possible to compose an apparently infinite number of variants with a finite number of concepts. However, judgements are also said to be material if the concepts in them are used in their established way. But how is the established use achieved if it does not come from individual abstraction?

As shown in Chapter 3.1, the development of the various inferentialisms can be seen as attempts by rationalism to explain the removal of the expression 'abstraction from the world' from its vocabulary entirely or to replace the expression with a 'concretion from logic'. Whereas naïve representationalism formed ever wider spheres of meaning from individual givens utilizing individual abstraction, rationalism tends to concretise from a given and very wide sphere of meaning to such an extent that it eventually extends the concept even to the individual givens. This means that one can dispense with abstraction theories and still *make use of their achievements* by *reversing the direction of explanation*. One does not use logic as an underlying theory to explain the representation, but one develops the representation from an already fully present logic or a logical-linguistic faculty.

The differences in our conceptual scheme can then be explained in such a way that one does not start with the immediate givenness of facts, but develops the differences from language, which is mediated in such a way that it already appears to members within the language community as immediate naturalness and givenness. The inferentialism of grounding speaks of concepts that have always been meaningful, the inferentialism of matter of some material judgements determined by the language community, and the inferentialism of form of necessarily determined logical expressions. In order thus to be able to give a complete explanation of the differences in our conceptual scheme, one does not abstract, but one concretises; one does not take up what is given but takes it from logic. But, we have to ask ourselves, what has been gained now, apart from having replaced one problematic direction of explanation with another? I even suspect that the explanatory direction starting from logic can only with difficulty answer the question of how it was ever possible to come up with such a differentiated conceptual scheme in the first place that one was able to use it to discuss the consistency, coherence and constancy of conceptual schemes.

---

[5] Cf. Gottlob Frege: Gedankengefüge. In: Logische Untersuchungen. Ed. by Günther Patzig, 4th ed. Göttingen 1993, p. 72.

3.2 Rational Representationalism

To illustrate this problem, I would like to tell a story that, at best, can be considered a philosophical legend. The fact that I am downplaying the truth of this story in advance may be due less to the story than to the circumstances in which it came to my attention. At a time when I had not been enrolled at the university for long and had hardly made any acquaintance with the lecturers there, I attended a lecture during the first semester on the advice of a small number of senior students. Unfamiliar with the underlying writings and unread in the subject, I initially only took one episode from the lecture, which I already found extremely strange at the time, but which now seems so appropriate to the problem addressed that I feel I must reproduce it here to the best of my recollection. At the end of a lecture, a student took the liberty of asking whether a *holistic explanation*, which does not view the origin of language as fragmentary and atomic, but rather in all its parts as a whole and as having arisen in the same way, is still tenable today and what would speak for and against it. She formulated the question somewhat differently, namely why we can never have a single concept without already having a large number of concepts. Although these questions were not directly answered, the lecturer started the next week's session with a report that recognisably referred to the questions from the end of the last session.

The lecturer himself was in such good standing before his audience that he could allow himself, as he himself indicated, to confine to a merely probable narrative on the subject. It could already be characterised as probable by the fact that he had been told it many decades ago by an American ethnologist and that it was about a research trip in the Southwest Pacific undertaken by one of her old colleagues. This old colleague, a respected, if not necessarily famous, natural scientist with philosophical interests, had lived for several years with a tribe that lived isolated from civilisation on a narrow headland surrounded by rainforest and cultivated a peculiar language there.

The lecturer explained that although he had recalled the conversation with the American ethnologist several times in the following years and had gone through the details in a continuous and uninterrupted movement of thought, he could no longer remember details about people, places and times in particular, which is why he would add some details himself in the following account. He explained – since he himself was fascinated by the natural scientist's story – that he had written a letter to the American ethnologist a few years after the conversation asking for more details, but it remained unanswered. As he later learned, she had died shortly after their meeting. His research into who exactly the natural scientist was that she reported on, or the name of the tribe he had told her about, was also unsuccessful. As I have already said, I am only reproducing this legend so that it is not completely forgotten and because it illustrates the problems with which holistic theories of language development have to contend.

The protagonists of this saga have been called abstractionists by the lecturer and are people who speak a language which they share publicly with each other and which, in this respect, we would not call the result of an intimate relation of the concept with

3 Logic and World

the given. The characteristic of this language, however, is that it is not concrete and referential, that is, its constituents are not lexemes of which we could in any way imagine what object or set of objects and events they refer to. One could say that their 'gavagais' do not even have a recurring rabbit experience as a condition.

The language of the abstractionists is – I reproduce it here as the natural scientist is supposed to have once reported it – only a reflection on *abstract ideas in consciousness, on the idea of the idea*, on language itself and the essence of language, without anything like the world or something worldly being found in it. The expression 'something like the world' does not have to be specified here, since a 'more or less of the world' is also excluded with the expression. Rather, the language consists of generic expressions that have been characterised by the natural scientist as exceedingly abstract, as well as the basic connectives of the logic of judgement in their rhetorical variations. The processes in the life of the abstractionists are immediately coordinated and regulated, and precisely for this reason their language no longer requires reference to any corporeality or worldliness. Consequently, the world and its objects play no role in their lives.

The natural scientist came across this tribe one morning when he observed by chance a large group of indigenous people ascending a rough and steep footpath in the rainforest up to a windy hill, packed with clay jars full of purple garments, spring-fresh water and sweet food. Once at the top, they unloaded their luggage and used it to serve the resident abstractionists by clothing, washing and feeding them. As the shadows grew, as the sun began to set, they took the remains of the food they had consumed during the day and carried it back down the path they had come up in the morning. All this time no one spoke a word, neither the abstractionists nor their servants. Only when the servants had arrived back down in their caves at the foot of the hill did the members of the two tribes in the valley as well as on the heights each begin to talk among themselves: But while some of them were only concerned about how to organise the next day's duties, the others only reflected on the truth and reality of the ideas they had once and recently discussed.

Of course, the natural scientist initially knew nothing about the content of these conversations, nor about the strict division of roles between the two tribes. He observed the events described over a long period of time and gradually came into contact with both the abstractionists and the servants. Both tribes were well-disposed towards him, and since he was particularly interested in why the abstractionists were so willingly catered for by their servants and what role the dependent abstractionists played in this symbiosis, he decided at some point to live with them on the hill and learn their language.

As he reported, the so-called abstractionists possess two major social problems, the first of which conditions the second: First, many statements by alleged abstractionists are perceived as uninformative and uninteresting. This is because abstractionists are said to tend to fall into ways of speaking that the natural scientist is said to have described with the word *tautological*. The explanation he gave for this was considered by him to be in a sense trivial: if all concepts of the universalistic language have the same circumference of meaning because they are in the same way generic abstract, then the probability that these conceptual circumferences occupy the same place in the limited logical space is higher than in languages in which concepts differ not only by the place but also by the circumference occupied in the logical space. Second, because of this tendency to tautologise, there is a disreputable sect among abstractionists who have evidently recognised this problem and are pushing for a concretisation of the language or for a concretisation of a certain number of concepts in that language. They urge or demand – however, demands may sound in their language – to limit conceptual circumference by making objects of everyday life the object of language or by no longer allowing only language and its essence to be objects of language. They argue that this can be done without any problems, arguing that abstractionists would have a seemingly unlimited number of words anyway, with the only problem being that many of them mean the same thing. The disreputable sect calls itself concretionists and is responsible in the first place for calling all those who do not belong to their sect or who are servants 'abstractionists'.

The natural scientist reported that abstractionists and concretionists gave him the impression that their language had always existed in its present form, while the servants showed no understanding at all of the questions about the origin and development of their language. Neither tribe seemed to him to have developed any real historical awareness of their language. But perhaps, the researcher admitted, his skills in both languages were too limited to find the right words for such questions.

However, he said, in the course of the time he had lived mainly with the abstractionists, he had come up with several hypotheses to understand the strange symbiosis between the abstractionists and their servants. The most plausible one seemed to him to be that the abstractionists had arrived at their advantageous situation through their place of residence, since on their hill they escape with their lives during *floods and high tides*, while a large part of their servants are swept away by the sea or drown in their grottos in which perpetual waters glide. The tribe that dwelled on the hill and in high and dry places remained completely alive and was thus worshipped by those living in the valley as those favoured by fate. On the one hand, the floods meant that the servants living in the valley could never develop such an abstract and rich language as the abstractionists, since they were always occupied only with the concrete errands in everyday life and only with the life-and-death struggle in the exception; on the other hand, the abstractionists must have forgotten concrete language in the course of time, since everything was taken from them by the servants and they were able to live on the provisions offered to them even in times of flood. They seemed to try to compensate for the loss of conceptual concreteness by introducing a large number of word variations.

3 Logic and World

The natural scientist explained the fact that the two tribes never communicated with each other, and even when they did, for two obvious reasons: On the one hand, it seemed to him that neither the abstractionists wanted to move from their windy hill into the caves, nor the servants from their grottoes flowing with perpetual waters up to the craggy rocks of the emerging hill, so that they were both content with their respective places of residence and also with their respective roles, and did not want to endanger their harmonious coexistence by attempts at communication. On the other hand, such attempts at communication were probably doomed to failure anyway, since although the natural scientist noticed a grammatical and phonetic similarity between the two languages, understanding the abstract did not contribute much to understanding the servant language. For while every judgement of the abstractionists seemed familiar to the natural scientist after a while, every concept in the servants' language seemed like a new name he had never heard before.

How the need of the concretionists came about to reintroduce new or long-forgotten concepts into the language of the abstractionists to enliven the expressiveness of their language, the natural scientist found very difficult to explain. Of course, he felt it easiest to say that some abstractionists, moved by curiosity, had turned away from the heights of the inner contemplation of their language and descended in the evening to the lowlands, where they overheard their servants by the light of the fire in the caves and thus learned of the strangeness of their language. But how should abstractionists have conspired for such a coordinated enterprise? Or how, if only one of them might have heard the words of the servants by chance or even involuntarily, should he have told the others about it in an intelligible way, so that they later followed him and eventually even formed a sect? And how could an abstractionist or even a group of abstractionists have understood that the language of their servants was more concrete than their own language? Based on these questions, which were unanswerable to him, the natural scientist came to the conclusion that some of the abstractionists, with the help of reflection on language itself, had concluded that it lacked expressive power and that necessary extensions, therefore, had to be made.

Concretionists are disreputable among the so-called abstractionists for several reasons: Most abstractionists would not understand why their language should be given a higher expressive power since it had never occurred to them that their language lacked anything. Thus, while most so-called abstractionists are not at all interested in the proposals of the concretionists, there are a few abstractionists who assess the situation exactly the other way round: Actually, the so-called abstractionists are the true concretionists, and the sect of the so-called concretionists are in fact only abstractionists. The reason for the different assessment, they say, is that the terms that have apparently always been used have grown together in such a way that they cover the entire logical space, whereas the supposed concretionists want to subtract a large number of components of meaning from the few general concepts, thereby establishing a large number of special concepts with few components of meaning.

3.2 Rational Representationalism

Just imagine, the American ethnologist said to the lecturer at the time, how my old colleague might have met the folk just described. Of course, he first perceived a huge difference in language. The fact that abstractionists were not interested in his pointing gestures, which he had used in connection with his own language, soon forced him to learn their language without being able to use his own. As he reported, he had not been able to apply a picture theory of language in learning the abstractionist's language, but rather unconsciously employed language practices in learning that could be described as contextual, use-theoretical or perhaps truth-semantic. After all, he understood their language very well in the context of the topics they dealt with, but it seemed inexplicable to him how one could limit the expressive power of a language to this abstract form.

After a few years, the natural scientist had finally learned the language of the abstractionists and found out that there was precisely the said sect of concretionists who, in his opinion, were right to criticise the expressiveness of the abstractionist. After all, he too knew by comparison that his mother tongue was superior to the abstractionist in expressiveness, even though the two languages being compared referred in the same way to the principles of regressiveness of rules and composition of conceptual content. However, he could well understand the two social problems. More difficult to decide, however, he said, was the dispute over the rightful designation of individual groups and sects as abstractionists or concretionists. As he kept trying to placate curious members of the folk, from the point of view of his mother tongue, it was the case that members of the concretionist sect had a right to call everyone else an abstractionist. However, having learnt abstractionist's language, he could also understand the group that considered abstractionists to be very concrete. When asked how he, as a native speaker of a language that is supposed to be both concrete and abstract, understood the sect's efforts to make the abstractionist's language more concrete, he always replied with a smile, saying, "Everyone will meet halfway at some point, some going up the hill, others coming down" – an answer that probably remained unsatisfactory to abstractionists because of translation difficulties. Since neither the abstractionists, concretionists nor the servants could explain to him the extent to which his hypotheses were correct, which he could make about the origin and development of their strange coexistence, he left the narrow headland again and travelled back home via Friedrich-Wilhelmshafen, where he met the American ethnologist. At least that is what the lecturer told me and the other listeners at the time.

To what extent the story exceeds the truth of a legend or whether it was just an invented thought experiment, I may not judge. But even if I have only been taken in by a myth, I believe that the story can sharpen one's understanding of the problems that arise when one wants to substitute a holistic theory of the genesis of language for an atomistic one. At the very least, the explanation of the descent of meanings that language theories focus on need not be linked to the understanding of the languages in question. The saga, moreover, in my view, sharpens the critical question 'How can different elements of a judgment form a whole?' by extending questions such as 'How do we arrive at our order of the conceptual?' and 'How is it that the expressive power of the abstract seems limited even though the central principles of regress of rules and composition of conceptual content are fulfiled?' Even if, as this saga suggests, we turn our understanding of language upside down, we only reverse our problems, but we do not solve them.

Unlike the basic distinction between the language of the abstractionists and the language of the concretionists, however, I see an even more extreme distinction in rationalism. As was shown in Chapter 2.1, around the time that individual abstractionism was falling into disrepute, the dogma of a *distinction* between proper names and definite descriptions, on the one hand, and the more abstract terms, on the other, was established. With this distinction, inferentialists who do not want to make use of expressions such as 'abstraction from the world' are able to make a gradual distinction and conceptual division in their substitution and replacement tests that brings one of the two sides much closer to the individual and given than the other.

My plea is not to banish abstraction theory, but not to give disproportionate importance to the particular stock role called 'definite descriptions' and 'proper names' in our normal-language semantics and logic, since this subject holds psychological and ontological assumptions that are not or should not be contained in the concepts within it. In this way, I argue *against the second strategy of inferentialism*, which is to establish gradual differences of conceptual content without resorting to a theory of abstraction. Even though so-called proper names and definite labels may have a grammatical or formal-logical function of their own – I do not deny the possibility – I nevertheless believe that in solving normal language problems we are better off understanding the gradual difference of concepts via individual contextualism and explaining it using a new form of abstraction theory.

I do not want to claim with my plea that there are no individual objects or no sets of objects or that there are no living beings that speak of themselves in terms of qualities of unity and indivisibility. But individuality, personhood and persistence are, in my view, subjects of metaphysics and should remain so. We can talk about these things representationally, but we find them neither as conditions of possibility nor as presuppositions of meaningful semantic statements. I believe I have thus designated the neuralgic point that distinguishes non-naïve representationalism from many other forms of representationalism. Concerning the critique of the theory of descriptions, rational representationalism approaches the inferentialism of grounding, and with respect to the advocacy of purely contextual semantics, it approaches the inferentialism of matter, but without wanting to share its relapse into a theory of existential commitments.[6] In the end, however, it shares the preference for abstraction theory only with neologicism.

---

[6] Cf. John McDowell: Mind and World, V.6; Robert Brandom: Making it Explicit, Chapter 6.

3.2 Rational Representationalism

Even though I take up several critical figures of argumentation that are familiar in the debate about proper names, labels and their existential commitment, I nevertheless draw a different and much sharper conclusion: By employing a so-called 'anthropological theory of concept' (for the explanatory process) as a complement to individual contextualism (for the understanding process), I can dispense with common theoretical elements of philosophical semantics that I believe have so far only gotten us into trouble and that we would be better off avoiding for this reason. However, I will first answer the question of how we can talk about particulars without proper names, definite labels and existential determinations here by pointing out how we should not do it. In my opinion, the reference to singular objects results from the intersection of conceptual spheres in judgements, to which the geometrical logic of concepts entitles us, but which I will only outline in more detail in Chapter 3.2.2. As has already been noted, this benefits my view that we should initially understand some linguistic contractions, such as the expression 'being', as transfers of intuitive relations to which we have only assigned an independent conceptual role in the course of their history of use.

The speech about individuals or singularity does not arise – as the second strategy of inferentialism would have us believe – by conceiving the grammatical subject as something given or as a substance to which we ascribe determinations, properties and predicates. Existential assertions cannot be circumvented by predicates either, since they only occur when we commit ourselves, through the use of expressions such as Pegasus, Sherlock Holmes or even Juwiwallera, to the fact that they can play a semantically meaningful role in the judgement – and we do so by ascribing to them an extensional relation to other meaningful terms. Even logatoms have a conceptual sphere, although they certainly do not refer to any set of givens, such as the paradoxical `blithyri has no meaning` or the antinomy `blithyri has meaning if and only if blithyri has no meaning`.

I am not taking on an existential or ontological mortgage – in the sense that there is or ought to be a place for blithyris in "my ontology" – but I am taking on a *semantic commitment* that there are truncated judgements, such as `blithyri is a logatom` or `The sentence has a logatom as subject`, of which I can claim to be able to give at least one intuitive interpretation that clarifies the relation between the grammatical subject and the predicate. Whereas inferentialists of form, in particular, think that the language shortcut of such judgements is that the conditions for existence, characteristics and perhaps even uniqueness contained in it are only implicated, I think we are on the wrong track if we are looking for anything at all like ontological and substance metaphysical conditions of sentences. What we are committing ourselves to are not existences, properties and possible individuals, but relations that exist between the subject and the predicate of a judgement and that can be intuitively represented. Involved in the aforementioned judgements is only the fact

3 Logic and World

that, for example, `blithyri` is contained in `logatom`. The translation of such relations into ways of speaking, according to which there are some or a something or no something that is blithyri and that has the property of being a logatom, leads us in the worst case to the erroneous assumption that the world is something that can be found in a tangled way only in our logic – without, however, finding an explanation in this assumption as to how it got there in the first place.

I suspect that this becomes clearer if one chooses transfers such as '`blithyri`' is contained in '`logatom`' as an example – but I will come to that later. It seems more important to me at present that by translating the transferred speech from 'ontological commitment' to 'semantic commitment', even judgements that seem to consist only of logatoms can be understood as meaningful: `gostak is like blithyri`. I believe that my last example has clarified, from context alone, the roles that `gostak` can play in a judgement. In doing so, I have neither referred to an object I call `gostak`, nor given a lexical definition or a history of ideas of `gostak`, nor established, applied, etc. rules for `gostaks`. From the knowledge that `blithyri` is contained in the expression `logatom` and that `gostak` is like `blithyri`, the intuitive relation of the conceptual spheres of `gostak` and `logatom` should be understandable – and in the wider context, it should also be understandable what 'having meaning' and 'having no meaning' mean.

I also believe that it is precisely logatoms that make it clear to us how we have to deal with universals and with names – namely, no differently than with all other concepts. When I say: `Socrates was a doctor`, I do not commit to saying that there is only one that was Socrates and that was a doctor etc., but I take on the commitment that the judgment is meaningful because I have to be able to explain in what fundamental relation the concepts `Socrates` and `doctor` stand to each other. Here, at the latest, begins the social dimension that rational representationalism shares with many kinds of rationalism. It may be that I had a precise object of reference in mind when I pronounced the judgement and thus asserted its meaningfulness, but this intention is part of a representational enquiry that we would do better to leave to epistemology or philosophy of mind or perhaps even better to other disciplines altogether. The judgement `Socrates was a doctor` refers to a conceptual sphere of `Socrates` that is contained in the extensional meaning of the concept `doctor`. And if we are to believe the arguments of the rationalists given in Chapter 2.3, our geometrical mode of representation does not allow us to distinguish between points and spheres in such a way that we attribute general concepts to those and individuals to these, since both always have an extension.

But if one objects that it would have made more sense to say, for example, `Socrates was a philosopher` or `Socrates was in Athens`, i.e. if it is required to concretise the reference of the judgement, then I will have to *give further spheres of meaning* that explain the set of possible references of `Socrates` and `is-`

3.2 Rational Representationalism

`a-doctor` that come into question. I could concretise and say `the one who was a paediatrician` or `the one who was South American Footballer of the Year in 1983`. It may be that most people are satisfied with these explanations. But one will *never be able to give a complete explanation*, because ultimately concepts, as concrete as they may appear, are always abstractions from the world they describe and whose meaning we can only understand in context by clarifying the relations between the concepts used. This explanation, however, is ultimately one that introduces reasons in the space of concepts that are not themselves conceptual in the literal sense. This will be clarified later in Chapter 3.2.3.

We would be making it too easy for ourselves if we were to switch from the logic that compels philosophers to refer to substances in so-called proper names and definite descriptions to a strictly extensional logic. While sets of objects already seem easier to handle than sets of properties, I still think that we would thereby again contradict our non-naïve representationalist principle, which says that concepts, however concrete they may seem, never have a factual equivalence to a single concrete object. 'Sets of meanings' sounds rather vague at first, but becomes intuitive even when we imagine, for example, that `1983-South American-footballer-of-the-year` is contained in `1983-South American-footballer-of-the-year` and this, in turn, is contained in `footballer-of-the-year` etc., and of course `1983-South American-footballer-of-the-year` is contained in `1983-footballer-of-the-year` and probably in something like `footballer-of-the-year-1983` and so on. I am, of course, using vague expressions like 'probably contained in something like' here because I fully agree with inferentialists in that I also consider these relations to be unfixed; rather, our need to elicit fixations from other people and speech communities is a key reason why we communicate at all. And we do this until we have sets that are not confused with objects, but with which objects given in intuition can be described to such an extent that the linguistic representation *satisfies* us.

I would like to emphasise the last expression once again: Rational representationalism should share the opinion of the inferentialism of matter that judgements can be understood as a game of giving and demanding reasons. In this respect, there are judgements that we approve of if the reasons satisfy us. With the inferentialism of grounding, however, I not only share the scepticism towards classical theories of representation, but I increasingly claim that there are no definite descriptions. The famous problem of the King of France was not that he was bald yesterday and dead today, but that for the definite description of the man or for the description of the object, especially in a logic including uniqueness assertions, one has imposed on oneself the inductive burden of having to judge all relations of the referred concepts to all other possible terms. So-called definite description theories (in logics with uniqueness assertions) may well be useful for specification in human-machine or machine-machine communication – I have never questioned usefulness – but they do not seem to me to be suitable for explaining human logic or natural language in a sustainable way.

Moreover, definite descriptions are never as definite as they may seem to us. Examples of this abound in the literature, although one rarely finds the consistency to abandon this problematic category of terms or to treat them as one treats all other terms. By `the highest mountain on earth`, one may mean Mount Everest, if defined `in terms of sea level` or the like, but according to other definitions one may also mean Mauna Kea or Chimborazo, and perhaps Pythagoreans would have quite other candidates on offer. In traditional rationalism, one treats definite descriptions like analytical judgements, in which one specifies a predicate that can only belong to a single subject. Is `the discoverer of America` is the predicate necessarily contained in `Christopher Columbus` – unless perhaps one does mean Leif Eriksson or even Bjarni Herjúlfsson. And of course one can debate whether it is a true story or satire to claim that the first man on the moon lived in the 2nd century CE.

My examples may be taken as arguments that stand for theories of proper names and labels known as bundle theory. In contrast to such theories, however, I am not interested in describing a proper name with a bundle of descriptions, but in completely dispensing with so-called singular terms to establish a general theory of terms instead, which can then be represented by geometric logic or which is already given by relations that we find in our intuitions. My examples are thus intended to show, on the one hand, that definite descriptions are not as definite as they pretend to be and, on the other hand, that 'proper names' are not necessarily proper to a single object.

My strategy is neither to interpret proper names in the same way as descriptions, nor to replace proper names with descriptions, but to reject in principle that there should be or need be at all, a special subject called 'proper name' in normal language logic. I believe that abjuring the old superstition of proper names is the first step in escaping the unnecessary limitation according to which we must study not the world but logic alone, and also in avoiding the unattainable extension according to which the world is something that must be derived from logic. Inferentialists and modern causal theorists have defined in singular terms the intention to refer to exactly one object. Some among them, who, like me, are sceptical of talk of intension in semantics, have dismissed such talk as a picturesque notion, since it is, after all, only meant to remind us of grammatical peculiarities that singular terms exhibit in the context of judgements.

That so-called singular terms exhibit peculiarities on the grammatical level I cannot deny; however, I believe that many other words also behave in a grammatically peculiar manner and that singular terms also share many of their peculiarities with words that we do not mark with the strange property of occupying a singular role. I will go into this in more detail in a moment. Beyond that, however, I believe I have shown in Chapter 2.1 several reasons that legitimise a clearer separation of logic from grammar.

3.2 Rational Representationalism

While so-called singular terms may have a special form in grammar, this does not prove that we should also assign them a special function in logic.

However, I would like to discuss the semantic mode of operation in more detail using an example and, in doing so, already refer to the model that I see as the starting point of the transfer described in Chapter 3.2.3 and as the way to the examples of use described in Chapter 3.2.2. Semantic use theories were once known to have a methodological predilection for looking at processes of language acquisition in children. If one observes how children deal with names, the example just presented becomes even clearer. Let's imagine that Linn is two years old and has an uncle named Werner. Werner is almost forty years older and lives in Stuttgart. Linn does not live in Stuttgart and only knows one person named Werner. For Linn, Werner is what is usually called an individual with a proper name, and is the uncle living in Stuttgart is a kind of (definite) description. At the age of three, Linn goes to kindergarten and meets a boy of the same age named Werner. In this short thought experiment, we have now already experienced several satisfactory identifiers to be able to use them to distinguish between the individual persons. For example, I can label Linn (1) satisfactorily by placing her in relation to the set of meanings three years old (2), living in Stuttgart (3), knowing a three-year-old Werner (4), knowing Werner who is almost forty years older (5), being called Linn (6):

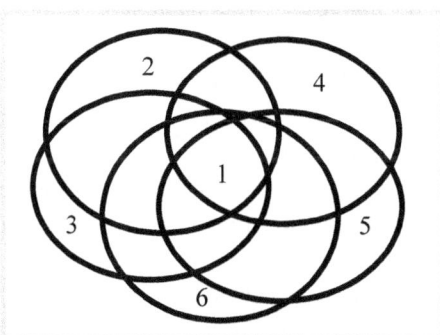

Of course, this diagram of an intuitive semantics is not entirely free of problems, since it has to justify not only the function of the diagram but also the interpretation of the thought experiment. However, let us assume here for the time being that we intuitively understand the diagram from the logic described in Chapter 2, without having given precise rules for the interpretation so far. Does the thought experiment given before really justify the uniformity of the intersection between the conceptual spheres (2), (3), (4), (5) and (6)? Why, for example, was no concept such as knowing only one Richard at the age of two or the like inserted into the picture?

3 Logic and World

One can certainly argue about how adequately this diagram represents the thought experiment. Nevertheless, I believe that these or similar intuitive semantics can save us from confusing (6) with (1), as I accuse much of today's prevailing labelling theories of doing. We save ourselves from confusion by not ascribing a proper term to (1), since I regard a satisfactory labelling of unities and undividednesses – that is, what previously proper names were supposed to do – as the result of a relation of terms in possible judgements: (1) labels Linn not as an ontological substance or as my psychological intention, but as a relation of meaningful concepts in possible judgements. And these possible judgements are those that were factually given in the thought experiment but can be recombined differently. (1) – i.e. the Linn, which I do not understand as a name, for that is (7) – is a satisfactory description, but not a representation of an undividedness, a unity, or even an exemplified assertion of uniqueness.

I call satisfactory descriptions those in which the relation of concepts gives rise to one or more representations that are interpreted as satisfactory in the context of the description. Satisfactory descriptions are thus never definite, since representatives, i.e. persons or objects, always have more properties than we are capable of giving for the definite description and than we have to give in order to arrive at a satisfactory description. The first point is that we know that there are more properties for describing the mental and physical world than we can give.[7] The second point is that human beings educate themselves to be pragmatic beings who are satisfied with a finite, i.e. very limited, number of labels. In principle, inferentialism of matter already points to these facts with its distinction between the seemingly infinite number of thought acts and the finite number of speech acts.

One can also imagine this distinction with simple examples: The descriptions (2)–(7) given above may not only apply to one person named `Linn`. Perhaps there is a Linn in Bremen and a Linn in Munich to whom (2)–(7) apply. And maybe there are even two Linns in Bremen to whom all the descriptions apply, and so on. One can quickly get lost in an infinity of modal arguments in such considerations. Nevertheless, in identity proofs, we are already satisfied with a finite set of descriptions. And we also know, for example, to what extent our diagram applies to exactly one person in the discourse universe of the thought experiment, or that it makes sense to first identify this representative by (7) since the alternative persons are already in relation to the representative by (5) and (6).

At this point, I would like to return to the content of the thought experiment, as I believe that the reflection on the story so far has already provided us with a key that explains why Linn, through her encounter with the three-year-old Werner, develops a semantics similar to the one I have just proposed to describe Linn. By learning at the age of three that there are two representatives for one name, Linn learns a strategy that enables her to compensate for the ambiguity of the name in her judgement: She excludes labels from the context by evaluating relations with probability.

---

[7] Cf. Michał Dobrzański: Begriff und Methode bei Arthur Schopenhauer, Chapter 7.2.

3.2 Rational Representationalism

This strategy is almost trivial. She attributes statements like `Werner drives home from work` to her uncle, provided she knows that all kindergarten children do not go to work. `Werner drives to you from Stuttgart` she attributes to her uncle, provided she knows that three-year-old Werner does not stay in Stuttgart. Names will be considered satisfactory descriptions provided that the context of relational concepts excludes ambiguities with a certain probability. Of course, it is possible that three-year-old Werner was in Stuttgart with his parents; but we also usually know how many additional identifiers we have to include in a statement in order to assume that our interlocutor can interpret this statement satisfactorily.

My thesis that names have no special status in the space of concepts is a hallmark of rational representationalism in that it states that concepts are context-bound and that this must apply to all concepts and thus also to names. The fact that such an exceptional status has been attributed to names, and that the thesis put forward here has not been known until now, is, in my opinion, due to the problematic mode of representation of formalist semantics and logics. As simple as our diagram was, as complex are the formalisations of the thought experiment described at the beginning. These not only force us into a confusing linear representation but also into problematic intensional representations as well as into an ontologically interpretable quantification or even a higher-level predication.

Of course, one can say that expressions like (2), i.e. `three years old`, should be taken as a property. But in an explanation of how the diagram works, the question of to whom we should attribute this property would prove tricky. The classical picture assumes a subject of substance to which one ascribes the property like `is three years old` or `being three years old`. According to the picture I favour here, however, it makes no sense to speak of properties, since what I have just called `the substance subject` does not occur in language: talk of substance subjects is a translation possibility from the transferred way of speaking about intuitive relations by means of the abbreviated expression of being. The conception of such an instance called substance subject only arises from the satisfactory description of a certain relation of properties.

In Chapter 3.2.2, I will strive for a more detailed examination of the logic diagrams used here only intuitively, in the hope that geometric logic will eventually save us from some of the linguistic problems that algebraic logic once imposed on us. In any case, the strategy indicated so far is to transform the radical difference between proper names and abstract concepts into a differentiated theory of levels of abstraction. In the process, it should also become clear that the relations that are represented in semantics between concepts using geometric logic have a strong correspondence to the relations that we find in the intuitive given world.

I have drawn on two essential elements of non-naïve representationalism in the example discussed above: First, to geometrical logic as the central organ of representationalism and to contextualism as the central method of a non-naïve theory

3 Logic and World

of understanding that does not want to fall back into an exuberant metaphysics or a down-to-earth physicalism. If one were to stop at this point, however, one would only have replaced the organ of inferentialism with a representationalist logic. Moreover, by criticising the theory of concretion, proper names and definite descriptions, one would have saddled oneself with additional problems, since one would have moved even further away from a solution to the problem of descent than inferentialism has done. How is it, then, that we can ascribe different content to concepts that inferentialism, with the help of contextualism, can classify into different role subjects and that non-naïve representationalism, with the help of contextualism and intuition, understand as spheres of meaning? The ultimate why-question of non-naïve representationalism, then, seems to be: Why do concepts exist at all and not just empty words?

Inferentialism has attempted to answer these questions in each case by recourse to intensional logic: it has ascribed to the concept the property of non-conceptual reason, to the subject the determination of the substance and to the connective the function of fundamental connection. The explicit attempt to answer the ultimate why question arose only in the inferentialism of matter, which was criticised by the inferentialism of grounding with the argument that it presupposed goal and intention in its logic and that it could not have taken a step towards explanation if it did not impute goal and intention to logic. Since the inferentialism of matter derived its ontology from the logical-grammatical distinction between givenness and determination, in which subjects and individuals – especially accompanied by the use of proper names and definite descriptions – played a decisive role, it restricted the ultimate why question to the so-called singular concept.

I believe that the embarrassment of inferentialism in the face of ultimate questions of justification is that, on the one hand, it rejects with good reasons the individual speech of an 'abstraction from the world', but on the other hand, it cannot provide sufficient justification for the speech of a 'concretion from logic'. I believe, however, that this problem can be solved by freeing the talk of an 'abstraction from the world' from individuation and thus embedding it in a social theory of the concept. The fact that this has so far been regarded as a rather inconceivable solution may be due to the fact that a collective theory of abstraction can refer to the result of a process, but cannot genetically accompany the decisive epoch of the development of language history. For us, the decisive epoch of linguistic development for abstraction processes is a non-conceptual time of stillness, silence and oblivion.

Because of this apparent non-conceptuality of the approach, semantic abstraction theories have been seen as describing a historical process of the individual that led to absurd explanatory approaches. For, as I mentioned above, one easily falls into diffuse strategies of justification if one wanted to explain how every individual can abstract the same concept from the intuition of certain objects or how it is possible that every

3.2 Rational Representationalism

person in a cultural circle always attaches the same word to an object to which they also always assign the same concept.

In my eyes, the inferentialism of grounding has come closest to a collective theory of abstraction, since it has understood language as part of our natural history. However, unlike the inferentialism of matter, it did not conceive of language intersubjectively, but individually, and history not as a process, but as a result. And unlike the inferentialism of form, it was not interested in the rules and laws, but in the product of natural history and in the ability to produce this product. As a result, the inferentialism of grounding conceived of the natural history of language not as the development and process of rules, but as the formation and product of a process. Thus, the development of language became a rather fleeting epoch of the individual, at the end of which the concept or conceptual ability in the mind and world of the speaker emerged as the meaningful product in each case.

When I share the claim of the inferentialism of grounding that we should understand commanding, questioning, narrating and philosophising as part of our own natural history, I do so from the perspective of the inferentialism of matter and form. This means that I am more interested in the phylo- than the ontogenetic facet of language, and that in this I consider the rules of generic development to be more decisive than the product of individual formation. Even if no stories from the conceptual dawn of the world resound, the regularity of this generic-intersubjective natural history offers an approach to explaining the descent of conceptual content.

The myth of the abstractionists then shows how crucial this problem of descent is, if one accepts the arguments that problematise an explanation of the problem of descent by means of substitution and replacement tests and the like, namely because of the associated concretion, proper name and description theories. Concepts, after all, must mean something, but the meaning was not developed from the concept. Rather, the myth of the abstractionists shows that the expressive power of a language and logic cannot be explained by the laws of regress of rules and composition of conceptual content alone. The determination of the conceptual content originally had to take place in relation to the world and the gradual strengthening of expression results when concepts can take on different relations or the conceptual content is not fixed by only one form of the sphere of meaning – after all, it is more appropriate to speak of a logical space of concepts than of a space of the concept.[8]

For a solution to the problem of descent, it is not enough to point to the factuality that a concept in its scope of meaning comprehends several other concepts within itself, or to determine which concept comprehends which concepts within itself; rather, a solution to the problem of descent requires a theory that explains how it comes about that all this is the case. A collective theory of abstraction assumes that certain laws can explain the descent of the content of concepts and that philosophy, unlike

---

[8] Cf. Martin A. Nowak, David C. Krakauer: The Evolution of Language. In: Proceedings of the National Academy of Sciences 96:14 (1999, Juli 6), pp. 8028–8033.

3 Logic and World

the many other theories and disciplines in the field of language formation and development, has privileged access to these certain laws.

Although theories of language formation and development draw from numerous disciplines such as behavioural research, evolutionary and developmental biology, archaeology, linguistics, psychology and, more generally, the neurosciences, evaluating diverse material such as modern pidgin and creole languages, emergent sign languages, animal behaviour, individual language acquisition, human artefacts, fossil finds or neuronal imaging techniques,[9] *philosophy nevertheless has privileged access* to certain questions of word use and concept development. This is because the historical documents of collective abstraction theory are the history of philosophy itself, in which the concrete concepts of everyday life, as well as myth, have been relieved of their intuition over generations and, moreover, can always be individually questioned in their semantic scope. Philosophy, if it does not want to be dogmatism, is finally the scientific discipline in which every concept, every judgement and every inference may be raised to a problem.

Of course, the material of collective abstraction theory *does not show continuous progress from the concrete to the abstract*, but is repeatedly corrected and thus regulated by paradigms of concretion, in favour of the preservation of its own expressive capacity. Nevertheless – and here collective abstraction theory bears witness to the prevailing opinion of the other disciplines of modern language formation and development theory –[10] this expressive capacity is not the concretised product of an originally abstract holism of meaning, but the result of a development that has lifted with concrete concepts and grounded content.

The fact that philosophy always pushes beyond this concrete beginning is due to its inherent exuberant realisation that the abstract can explain more about the concrete and intuitive than the latter itself gives to understand.[11] The very groundedness of philosophy, being able to apply the abstract to the concrete, becomes a regulative idea of itself. In this tension between the concrete beginning and the abstract end, forces such as exuberant metaphysics and the down-to-earth form of physicalism keep philosophy in a balance of content, but push it towards an increase in expressiveness.

The fact that concepts are derived from intuitions – even if even the most concrete concepts must never be confused with intuitions themselves – cannot be justified genetically. This is because the original process of abstraction or the representation of intuition through concepts falls into an epoch of human development in which and through which linguistic communities are first constituted and whereby this epoch can thus naturally not be reflected conceptually: The *initial baptism*, i.e. the original construction of the concept or the transfer of an intuition into a word that takes place in

---

[9] Cf. Rudolf Botha: Language Evolution. The Windows Approach. Cambridge 2016.
[10] Cf. M. A. Nowak, J. B. Plotkin, V. A. Jansen: The Evolution of Syntactic Communication. In: Nature 404:6777 (20. März 2000), pp. 495–498.
[11] Cf. Michał Dobrzański: Begriff und Methode bei Arthur Schopenhauer, Chapter 7.3.

this epoch, is something that the concept can only ever conceive reflexively and as a post-construction. A collective abstraction theory cannot provide a causal explanation by the very fact that in the initial baptism situation there was not the conceptual capacity to generate factual knowledge that forms the prerequisite for explanatory knowledge. But one can factually infer that there must have been an epoch in which the conceptual meaning was so intuitive that only through iterative application of abstraction steps did meaning distinctions become possible. These repeated abstractions, which were later recognised as *regularities and laws*, were the precondition for the expressive power of language that has been preserved until today and that continues to develop because we pass it on from one generation to the next in individual education and formation.

The transfer from worldly non-conceptuality to the logical concept is initially so intuitive that it cannot be reflexively accompanied. Finally, for this epoch, a psychological and perhaps even ontological talk of proper names and definite descriptions or generally singular terms may be applicable insofar as only an individual theory of abstraction can have prevailed in this primaeval time since a linguistic community only constituted itself subsequently through tradition. If someone wanted to explain this beginning and say how language came into being in it, she would have to go back to this beginning and, in accompanying explanation, begin again herself from the beginning before language came into being. But since this can only be a profitable explanation by using an already existing language, for philosophy, this past lies hidden in the dark.

However, the beginning of language development and original language formation, in which many of the above-mentioned disciplines are interested, is *not a decisive epoch for the collective theory of abstraction*. This is because usually abstraction theories only imagine the intimate relationship of the conceptual individual to her or his intuition, thereby excluding themselves from so-called 'collective' or 'anthropological' theories, or they speculate about a language community formed in the tradition of concrete meanings, thereby excluding them from abstraction theories. As has been argued at length in Chapter 3.1, the strategy of modern philosophies of language has been to transform the problem of semantic descent or the question of the explanation of meaning into a problem of linguistic understanding or a question of semantic comprehension. This, however, unnecessarily *limited* logic in such a way that its *extension* with secular components seemed almost *unattainable*. A way out of this fatal situation, however, is provided by the material of collective abstraction theory. In abstraction from individual intuition, the concept becomes an institution of collective memory over many epochs, but also an institution about which and with which individuals dispute.

As was shown in Chapter 1.1.3, it is not sufficient to trace the abstracting and regulative development of the concept on one document or on several documents of an individual. Changes that an individual makes to the sphere of the meaning of a concept in one or even several documents are not perceived by the linguistic community as a rule or regularity, but as an expression of inconsistency and contradiction, unless this change has itself become or can become a semantic monument in the tradition of the concept. Only in the succession of such monuments and milestones, which have been documented over several generations, do the regularities and laws of abstraction (and also of regulation) become distinct, and with them also the principle of descent of conceptual content.

The documents first show that the descent of conceptual content is based on a process of abstraction, which, however, in the further historical course of its literal use is repeatedly regulated by concretisations to maintain the expressive force of a conceptual scheme. Astonishingly, however, they also show that the regularities of abstraction are the same as the principles that are responsible for the expressive force of judgements, namely regressivity and combination or *recursivity and compositionality*.[12] The circumference of a concept results from the fact that abstractions, such as the original abstraction of the concept from the intuitive world, can be applied to it several times and that the meaning of a concept is the composition of all the processes of abstraction that have been recursively applied to it after the original abstraction of the concept from the intuitive world.

If, as inferentialism teaches us, our linguistic ability is part of our natural history, and if we assume that other species such as bats or cats have acquired similar outstanding abilities in the course of their *natural history*, which cannot simply be replicated by other species, then we must not treat our linguistic ability differently from echolocation or night vision, and this initially only means attributing *regularities* to it. Such regularities, however, are not only apparent in the period of development of these abilities, but continue, especially in the case of language and conceptual ability, in the course of their overall development. The concept has the property of recording traces of these regularities in words and signs and transmitting them in documents over generations.

No matter which documents are consulted, their monuments show the regularities of abstraction, as they must always have occurred in the history of the development of the concept. *Every conceptual sphere relates to its historically preceding sphere of the concept* in a way that can be described with the laws of abstraction or also of the concretion that regulates it. In this context, it is true that in regulation no concept can be more concretised than was the case in what was probably the decisive epoch of the emergence of language. Moreover, it is not possible to extend a concept in a context even further than in such a way that no traces of intuition are left in it and there is no longer any relation to concepts whose relation to intuition can be explained.

---

[12] Cf. Kenny Smith, Simon Kirby: Compositionality and Linguistic Evolution. In: Handbook of Compositionality. In: Oxford Handbook of Compositionality. Ed. by Wolfram Hinzen, Edouard Machery, Markus Werning. Oxford 2012, pp. 493–509.

In addition, however, there are two further regularities, both of which have already been hinted at and which will be discussed in more detail in Chapter 3.2.3, namely the *law of transfer and that of translation*. Put simply, the law of transfer means that there is a qualitative difference in the original abstraction of the concept from the intuitive world, whereas the law of translation only states that there is a quantitatively precisely determinable set of translations of a concept, corresponding to the compositional meaning of the abstraction. However, transfer and translation are already known insofar as inferentialists deny that there are transfers at all, while instead of translations they speak of substitutable concepts in stock roles.

Since, according to the principle of descent, the conceptual content is based on an original transfer from intuition, the content traces the form of intuition and is thus either wider according to the degree of recursion or it intersects with other concepts according to compositionality. How the conceptual content can be traced intuitively is given by the semantics of geometric logic, as demonstrated especially in Chapters 2.1.5f. and 2.2.5f. In these chapters, abstraction processes were presented in individual case studies, but not based on a material evaluation in the sense of the collective abstraction theory outlined here. A first step towards specifying the regularities mentioned here, modelling them, and linking them to the collective abstraction theory is taken in Chapter 3.2.2.

3.2.2 Intuition and Concept

In the previous chapter, I made a plea for a linguistic theory of abstraction which, based on historical material, should deal with abstracting and regulative conceptual processes not individually-subjectively but collectively-intersubjectively. With the help of this abstraction theory of meaning, I have tried to come a step closer to solving the problem of descent; this problem inferentialism had either tried to solve by resorting to theories of concretion and singular terms or had intended to circumvent by transferring it to a theory of contextual understanding. My aim of establishing a collective abstraction theory of meaning is based on the problem that proper names and definite descriptions are themselves based on presuppositions which, on the one hand, are not inherent in the concept itself and which, on the other hand, commit speakers to certain logical models that do not do justice to the transfer of the concept.

My perspective, which I have indicated in the previous chapter and already developed based on a representationalist theory in Chapter 2, is based on the argument that logic can be represented intuitively, namely in the form of geometric figures, i.e. analytical diagrams. That I favour such a representationalist view of logic is not because I am seeking to establish something like a counter-model to rationalism, but is rooted in the conviction that inferentialist logic and philosophy of language in practice obstruct our analysis of modern problems in that many linguistic problems arise from

3 Logic and World

the fact that our vocabulary has an affinity with fundamental transfers that accommodate geometric and representationalist logic.

At the end of Chapter 2, several arguments were also formulated that made further discussion of one of the two schools of modern rationalism seem unnecessary. It was objected to logicism and neologicism that deductive inferences are themselves in need of justification and that therefore the attribution of elementary mathematical statements to logic would not be a viable method if logic itself appeared to require justification. At the beginning of Chapter 3, I, therefore, announced a further critique of logicism as unnecessary. However, in Chapters 3.1 and 3.2.1, an element of the theory that is essential to neologicism has repeatedly imposed itself. This refers to the theory of abstraction, which is at the heart of neologicism and which even sees itself as abstractionism because of this theoretical element. However, as essential as abstraction theory is for neologicism, it is also problematic for it. Numerous principles of abstraction have been formulated in the past decades without any of them being considered as purposeful.[13]

How a geometric logic of rational representationalism could be constructed will be presented below on the basis of a model for the logic of concepts. This logic of concepts of non-naïve representationalism, given as an example, shares with neologicism the preference for a theory of abstraction and with inferentialism the goal of collectivity. This theory of abstraction takes over from neologicism the essential distinction into an objective and a conceptual abstraction.[14] However, as should quickly become clear in the following, the abstraction theory of rational representationalism also differs from that of neologicism in numerous points. This is because the abstraction theory of rational representationalism constructs a model of the intuitive representation as its starting point and is not limited to the concept of number as its target point, but focuses on the concept in general.

Matsuda diagrams or matrices, which are based on the so-called Rule 30, provide a useful model for intuitive representation.[15] Rule 30 was discovered by Stephen Wolfram in the 1980s and is characterised by generating chaotic structures with recurring patterns.[16] Therefore, these diagrams or matrices provide a suitable model for the seemingly infinite abundance of sense data, which is also structured by recurring patterns that we recognise as objects, for example. Katsunori Matsuda explored that there is an interpretation of rule 30 that conforms to the three principles of intuitive representation as presented in Chapter 1.2.3: Space, time and causality. The following matrices show the time states T vertically ($T = \{I, II, III, \dots\}$), the space states S horizontally ($S = \{a, b, c, \dots\}$) and the causality through rule 30 to be explained further

---

[13] Cf. Paolo Mancosu: Abstraction and Infinity. Oxford 2016, Chapter 4.
[14] Cf. Kit Fine: The Limits of Abstraction. Oxford 2002.
[15] Cf. Katsunori Matsuda: Spinoza's Redundancy and Schopenhauer's Concision. An Attempt to Compare Their Metaphysical Systems Using Diagrams. In: Schopenhauer-Jahrbuch 97 (2016), pp. 117–131.
[16] Cf. Stephen Wolfram: The Mathematica Book. 5th ed. Champaign, Ill., London 2003, Chapter 3.8.6.

3.2 Rational Representationalism

below. Here, however, we will first focus on the representation of time and space with the help of a $T \times S$ matrix that contains 48 elements:

$$T \times S \text{ matrix} = \begin{matrix} \ldots & Ia & Ib & Ic & Id & Ie & If & Ig & Ih & \ldots \\ \ldots & IIa & IIb & IIc & IId & IIe & IIf & IIg & IIh & \ldots \\ \ldots & IIIa & IIIb & IIIc & IIId & IIIe & IIIf & IIIg & IIIh & \ldots \\ \ldots & IVa & IVb & IVc & IVd & IVe & IVf & IVg & IVh & \ldots \\ \ldots & Va & Vb & Vc & Vd & Ve & Vf & Vg & Vh & \ldots \\ \ldots & VIa & VIb & VIc & VId & VIe & VIf & VIg & VIh & \ldots \end{matrix}$$

To fill the $T \times S$ matrix, which has so far only been designated with positional determinations, with content and 'matter', I use a bitstring semantics based on the binary code $\{1, 0\}$.[17] This initially gives each position in all columns S_{a-h} of the first row T_I a bit. The causality is expressed by a logical formula that explains how a spatial state is occupied in the next time state. To determine the bit of a particular position in a row, a triple of spatial states – I call them x, y, z – must be read in the row situated above the corresponding position, where y is vertically directly above the bit to be read. I call the element to be determined the Q-bit, so the following formula applies according to the standard rules of Boolean algebra:

$$\text{Q-Bit} \overset{\text{def}}{=} 1, \text{ iff } x \text{ XOR } (y \text{ OR } z) = 1$$

Let us assume, for example, an initial situation T_I with eight spatial states, S_{a-h}. Each of these eight elements of T_I is now occupied by the following bits: $Ia = 0, Ib = 0, Ic = 1, Id = 1, Ie = 0, If = 0, Ig = 0, Ih = 0$. We thus obtain the bitstring 00110000. To now causally determine Q-bit = IIb, we read $x = Ia, y = Ib$ and $z = Ic$. This means that $IIb = 1$ if either Ia or Ib or $Ic = 1$. But if neither Ia nor Ib or $Ic = 1$, then the Q-bit $IIb = 0$. According to the rules of Boolean algebra, we read for Ia, Ib, Ic (0 XOR (0 OR 1)) = 1, and therefore $IIb = 1$. With this method, we can now determine any position in the matrix. (If x or y cannot be read in the matrix, e.g. for the columns a or h, 0 is automatically set for x or y). Let us now extend the example so that we come back to an 8×6 matrix, which again is only a section of a much larger one. If we assume that $T_1 = \ldots 00110000 \ldots$, we get the following matrix according to rule 30:

---

[17] For bitstring semantics cf. Fabien Schang, Jens Lemanski: A Bitstring Semantics for Calculus CL. In: The Exoteric Square of Opposition. Ed. by Jean-Yves Beziau, Ioannis Vandoulakis. Basel (forthc.).

3 Logic and World

$$T \times S \text{ matrix} = \begin{matrix} \ldots & 0 & 0 & 1 & 1 & 0 & 0 & 0 & 0 & \ldots \\ \ldots & 0 & 1 & 1 & 0 & 1 & 0 & 0 & 0 & \ldots \\ \ldots & 1 & 1 & 0 & 0 & 1 & 1 & 0 & 0 & \ldots \\ \ldots & 1 & 0 & 1 & 1 & 1 & 0 & 1 & 0 & \ldots \\ \ldots & 1 & 0 & 1 & 0 & 0 & 1 & 1 & 1 & \ldots \\ \ldots & 1 & 0 & 1 & 1 & 1 & 1 & 0 & 0 & \ldots \end{matrix}$$

The bit matrix can be extended at will and thus provides a model that represents the complexity of intuitive representation, as the three capacities of the mind, i.e. space, time and causality are represented by S, T and rule 30. For Matsuda, one or more vertical rows of bitstrings represent objects perceived by a subject. However, since there is no compelling reason for Matsuda matrices to extend objects to entire vertical rows on the one hand, and to limit them to only one vertical row on the other, we go back here to Wolfram's insight according to which, despite the chaotic behaviour of rule 30, there are nevertheless recurring patterns (triangles, L-shapes, etc.) that extend over several rows and lines.

Let us now assume that objects in our Matsuda matrices can normally have an extension over several S-rows and T-columns. I now imagine that I have learned to recognise certain sense data given in intuitive representation as a certain object. We label this object O^I, and the following bit matrix is a model for O^I:

$$O^I = \begin{matrix} 1 & 0 & 1 \\ 0 & 0 & 1 \\ 1 & 1 & 1 \\ 1 & 0 & 0 \end{matrix}$$

Let us now assume that the matrix M^I is a model that describes all my actual sense data, which does not differ in size and bits from the $T \times S$ matrix given above. Since M^I represents this sense data in space, time and causality, and since I currently have no more data than M^I, I might be led to say that this is the whole world as my representation. But in what follows we assume that this sense data can also be grasped and processed by others and that we thus have a general model of sense data, which we will continue to call M^I. If we now compare our bit matrix O^I with the sense data from M^I, we quickly realise that O^I is actually also contained in M^I, namely in $IIc, IId, IIe, IIIc, IIId, IIIe, IVc, IVd, IVe, Vc, Vd, Ve$. To better identify this in our model, we treat the bit matrix M^I like a diagram and label the bit matrix O^I with a geometric shape. Due to the shape of the matrix, a polygon is appropriate, in this case a square. We provisionally assume that M^I also has a boundary and therefore also draw a square around all the updated sense data. The result is the following diagram $D1$, which describes the actual world of representation:

3.2 Rational Representationalism

$$M^I = \begin{pmatrix} \cdots & 0 & 0 & 1 & 1 & 0 & 0 & 0 & 0 & \cdots \\ \cdots & 0 & 1 & 1 & 0 & 1 & 0 & 0 & 0 & \cdots \\ \cdots & 1 & 1 & 0 & 0 & 1 & 1 & 0 & 0 & \cdots \\ \cdots & 1 & 0 & 1 & 1 & 1 & 0 & 1 & 0 & \cdots \\ \cdots & 1 & 0 & 1 & 0 & 0 & 1 & 1 & 1 & \cdots \\ \cdots & 1 & 0 & 1 & 1 & 1 & 1 & 0 & 0 & \cdots \end{pmatrix}$$

$\mathcal{D}1$

$\mathcal{D}1$ shows us a model of the sense data of our intuitive representation M^I in which an object O^I, represented by the inner square, is recognised. Two possibilities of determination arise: The object of our intuitive representation O^I can be determined by a bit matrix or O^I can be delimited from all other objects employing a polygon. We started from the first variant, but the theory of abstraction that follows should make it clear that the other starting point would also have been conceivable. The decisive point is that we can separate both, i.e. bit matrix and geometric figure, from each other mentally, but still perceive them as a unity in our intuitive representation. In the model, this means: The structure of the bit matrix already has the geometric figure, which is explicated by the square, and the square has a shape that can be symmetrically filled by a bit matrix. If we assume that $\mathcal{D}1$ is a model of our sense data, then we can understand the inner square of the bit matrix thereby conceived as the boundary of an object.[18]

I mentioned at the beginning of this chapter that rational representationalism shares with neologicism the division of abstraction theory into two types of abstraction: objective and conceptual abstraction. Since objects are part of the intuitive representation and this finds a model in the bit matrix M^I, the first abstraction step results from the abstraction of the bit matrix. The abstraction from the objects takes place in the model through the abstraction of M^I. After the objective abstraction, the diagram $\mathcal{D}2$ remains, in which no sense data and no objects are represented anymore, but only the boundaries that refer to these objects.

---

[18] It is probably not important for our model whether we conceive of the object and the boundary as natural or artificial. Cf. Barry Smith: Boundaries. An Essay in Mereotopology. In: The Philosophy of Roderick Chisholm. Ed. by Lewis H. Hahn. Chicago, Ill. 1997, pp. 534–561.

3 Logic and World

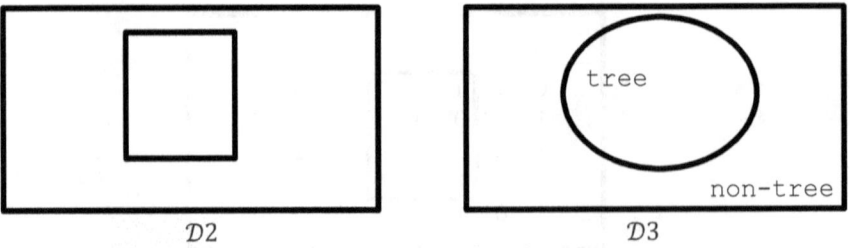

We assume that the object O^I is designated by reason or the conceptual representation with the word tree. The word tree is a concept if it has a boundary, as described in Chapter 2.1.6: reason apprehends the bit matrix of O^I, defined by the understanding (Verstand), and designates it with the word tree. $\mathcal{D}2$ is now no longer a model of the intuitive representation in which the limits of the object are designated, but after the abstraction from sense data, i.e. from space, time and causality, a model of the conceptual representation.

To make this recognisable in the diagram, we should define that an object is represented with polygons, but a concept with closed curves or spheres. The objective abstraction of $\mathcal{D}1$, therefore, leads to $\mathcal{D}3$. In $\mathcal{D}3$, the concepts are now marked by words. $\mathcal{D}2$ only serves us as an auxiliary diagram that makes the abstraction step between $\mathcal{D}1$ and $\mathcal{D}3$ recognisable. We can understand the entire abstraction step from $\mathcal{D}1$ to $\mathcal{D}3$ as a *transcoding* that creates a qualitative and quantitative difference between the two diagrams: In contrast to $\mathcal{D}3$, $\mathcal{D}1$ represents qualitatively different information (namely about objects and not about concepts) and quantitatively different information (namely a large number of bits, as opposed to a small number of concepts).[19]

What is probably striking about $\mathcal{D}3$ is the fact that the outer square of $\mathcal{D}2$ was not transferred into a sphere. This is based on the fact that the infinitely large amount of possible sense data in the model M^I (marked by the ellipses in $\mathcal{D}1$) can be conceptualised at all. However, since our preliminary assumption that M^I has a limit is also questionable, it remains undecided whether the outer square denotes an object or a concept. We, therefore, define that the outer square or frame of any diagram $\mathcal{D}$ initially only indicates the boundary of $\mathcal{D}$ designated by F.

The step of abstraction from $\mathcal{D}1$ to $\mathcal{D}3$ also indicates that the conceptual sphere designated by the word tree in $\mathcal{D}3$ is a concretum. According to Chapter 1.3, a *concretum* was a concept directly extracted from the intuitive representation. Since every limited and definite concept has a sphere, the concrete concept tree also has such a sphere, which we represent by a circle and which we designate by the word tree in $\mathcal{D}3$. Since we have found criteria through our conception represented in

---

[19] Cf. Michał Dobrzański: Begriff und Methode bei Arthur Schopenhauer, Chapter 6.

3.2 Rational Representationalism

$\mathcal{D}1$ that allow us to say what belongs to the object of the tree and what does not, we can refer to tree in $\mathcal{D}3$, on the one hand, and to non-tree, on the other. However, since the region in $\mathcal{D}1$ goes to infinity outside the bit matrix for tree (as indicated by the ellipses), the description non-tree also remains undetermined and is thus not delimited in $\mathcal{D}3$ with a sphere, but only lies outside the only conceptual sphere known so far that indicates the concept tree.[20]

Let's take another step back and assume that we identify another object O^{II} in the intuitive representation. We determine O^{II} with the help of the following bit matrix:

$$O^{II} = \begin{matrix} 0 & 0 \\ 0 & 1 \\ 1 & 1 \\ 1 & 0 \end{matrix}$$

If we compare this object with our actual sense data, we see that O^{II} is located outside of O^{I}, but is nevertheless a component of M^{I}. O^{II} is located in the matrix of M^{I} at the positions $Ia, Ib, IIa, IIb, IIIa, IIIb, IVa, IVb$. We now imagine a square around the designated positions of O^{II} that has no intersection with the square of O^{I}. After the objective abstraction step, we are left with a diagram that has a similar relation to geometric figures as in $\mathcal{D}3$. We, therefore, call this diagram $\mathcal{D}3^{*}$ and merge both diagrams, $\mathcal{D}3$ and $\mathcal{D}3^{*}$ in $\mathcal{D}4$. The object O^{I} identified by the understanding (Verstand) is now assigned the word table and classified by reason with as a clearly defined concept.

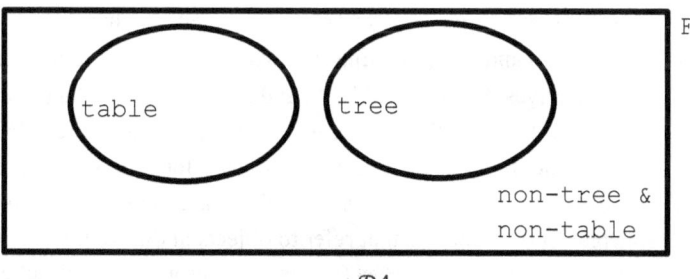

$\mathcal{D}4$

$\mathcal{D}4$ is a model of conceptual representation in which table falls in the region of non-tree and tree in the region of non-table. Outside both spheres for tree and table is the realm of non-tree & non-table.

---

[20] Diagrammatic conventions for infinite terms were introduced by Johann Christoph Sturm, see above, Chapter 2.2.3.

3 Logic and World

But let us go back to the intuitive representation one more time, to our model M^I, and let us assume one more object O^{III}, which is determined with the following bit matrix:

$$O^{III} = \begin{matrix} 0 & 0 & 1 & 1 \\ 0 & 1 & 1 & 0 \\ 1 & 1 & 0 & 0 \end{matrix}$$

O^{III} is also an object of our actual sense data and we recognise in M^I that O^{III} partly intersects with O^I, but also partly with O^{II}: O^{III} shares with O^I the positions $IIc, IId, IIIc, IIId$ and with O^{II} the positions $Ia, Ib, IIa, IIb, IIIa, IIIb$. If we again abstract from objectivity and first look only at the relationship of O^{III} to one of the other two objects, we get the following diagrams $D5$ and $D6$.

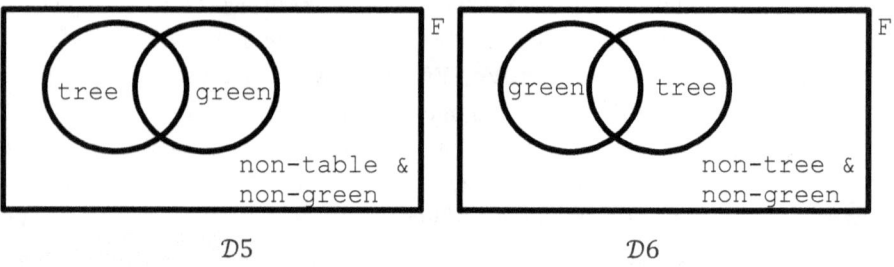

$D5$ \hspace{4cm} $D6$

As can be seen in $D5$ and $D6$, we assume that the object O^{III} was apprehended by reason with the word green. However, two putative problems go hand in hand with these diagrams: Firstly, $D5$ and $D6$ (including the word green) no longer indicate only nouns but also adjectives. Secondly, $D4$, $D5$ and $D6$ no longer seem to be easily transferable back to $D1$ in terms of their form, just as, for example, $D3$ can be transferred to $D1$ via $D2$. The first problem can be solved, for example, by clearly distinguishing between matrices or diagrams for objects and diagrams of concepts. Diagrams with spheres denote concepts that refer to objects in different ways, for example, to objects without explicit properties (nouns) or to properties without explicit objects (adjectives). Another solution to the problem would be a kind of nounification of the adjectives, for example, so that green is the abbreviated way of speaking of green object or stands for something green.

The second problem is that the position of the circles does not seem to correspond to the position of the squares. But this need not be the case, because what is important about sphere diagrams is that they make the relation between the spheres also correspond to the relation of the bit matrix squares. In $D5$ and $D6$ we see a partial overlap that corresponds to the partial overlap of the bitstring matrices of O^{III} and O^{II}, on the

3.2 Rational Representationalism

one hand, and of O^{III} and O^{I}, on the other.[21] In order to be able to analyse this more precisely, however, it is first important to separate the syntax and semantics of concept diagrams.[22]

To be able to represent the respective components of the diagram more precisely, the syntax of the diagrams is to be separated from the semantics and variables are to be attributed to the concepts in $D5$ and $D6$. The result is diagram $D7$, which has the same diagrammatic form as $D5$ and $D6$:

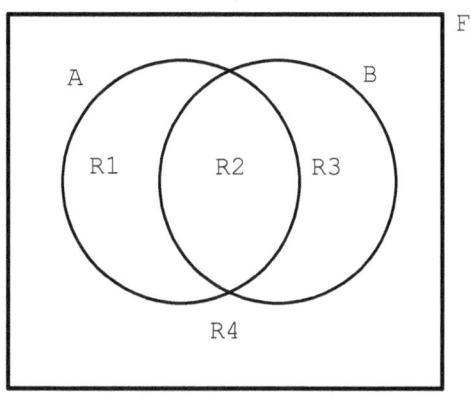

$D7$

$D7$ shows two circles (A, B) in F, which together give four areas that can be described as regions (R), i.e. {R1}, {R2}, {R3}, {R4}. This results in several possible descriptions for regions or associations of regions for $D7$:[23]

(Syn1) For the region {R1} it is true that it represents the conceptual abstraction B's of A, i.e. $A \setminus B$.

(Syn2) For the region {R2} it holds that it represents the intersection of A and B, i.e. $A \cap B$.

(Syn3) For the region {R3}, it holds that it represents the conceptual abstraction of A's from B, i.e. $B \setminus A$.

(Syn4) For the union of regions {R1, R2, R3} it holds that it represents the union of A and B, i.e. $A \cup B$.

(Syn5) For the region {R4} it holds that it represents the conceptual abstraction of L from A and B, i.e. $F \setminus (A \cup B)$.

---

[21] The relations of partial or complete overlap or difference of diagrammatic elements was first introduced by Weigel, vide supra, Chapters 2.2.3 and 2.3.5.
[22] Cf. Sun-Joo Shin: The Logical Status of Diagrams.
[23] I adopt here the set-theoretic notation of Lorenz Demey: From Euler Diagrams in Schopenhauer to Aristotelian Diagrams in Logical Geometry, Sect. 3.2, but also share his assessment that the diagram types discussed here are not set diagrams.

3 Logic and World

We see that Syn1-5 are also possible descriptions of $\mathcal{D}5$ and $\mathcal{D}6$, as these diagrams have the same regions or associations of regions:

Syntax	Region	Semantics of D5	Semantics of D6
Syn1	{R1}	$table \setminus green$	$green \setminus tree$
Syn2	{R2}	$table \cap green$	$green \cap tree$
Syn3	{R3}	$green \setminus table$	$tree \setminus green$
Syn4	{R1, R2, R3}	$table \cup green$	$green \cup tree$
Syn5	{R4}	$F \setminus (table \cup green)$	$F \setminus (green \cup tree)$

But let us go back to the matrices or diagrams for objects and assume that we would still conceptually call O^I a table and O^{II} a tree, even if the part of the bit matrix that O^I and O^{II} have in common with O^{III} would change. For example, we assume that in some cases O^{III} can be replaced with a particular bit matrix O^{IV}, which we call brown for conceptual distinction. If O^{IV} takes the place of O^{III}, we designate the object O^{Ia} but continue to call it a table.

If we are satisfied with this extension of our model, we can now introduce a distinct difference between the object and conceptual diagrams: If in intuitive representation we recognise an object in which certain elements can change, then in the conceptual representation the relation between object and element is to be drawn by a diagram according to the syntax of $\mathcal{D}7$. {R2} in $\mathcal{D}5$ indicates the concrete object O^I; {R1} in $\mathcal{D}5$ indicates that in $\mathcal{D}1$ there are still objects similar to O^I, but which do not coincide with O^{III}, e.g. O^{Ia}; and {R3} in $\mathcal{D}5$ indicates that in $\mathcal{D}1$ there are still objects like O^{III}, which are related to whole objects other than O^I, e.g. the object denoted by {R2} in $\mathcal{D}6$, which corresponds to O^{II} in $\mathcal{D}1$.

The difference between object matrices and conceptual diagrams should have become somewhat distinct as a result. According to the objective abstraction, diagrams still show the same relations, but they are more general, they can be applied to a variety of objects in $\mathcal{D}1$. However, as we saw in Chapter 2.2, other conceptual regularities should be expressed by geometric logic. The following regularities, which were already announced in Chapter 3.2.1, stand in a contrary relationship to each other:

(R-Cont) If in an object diagram an object O^1 *sometimes* occurs together with another object O^2, but *sometimes not*, then O^I corresponds to concept A, and O^{II} to concept B in concept diagram $\mathcal{D}7$.

(R-Ness) If in an object diagram an object O^1 *always* occurs together with another object O^2, then O^I corresponds to concept A, and O^{II} to concept B in the concept diagram $\mathcal{D}8$.[24]

---

[24] For the interpretation of $\mathcal{D}8$ vide supra, Chapters 2.2.4f.

3.2 Rational Representationalism

(R-Imp) If in an object diagram an object O^1 *never* occurs together with another object O^2, then O^I corresponds to concept A, and O^{II} to concept B in concept diagram $\mathcal{D}9$.

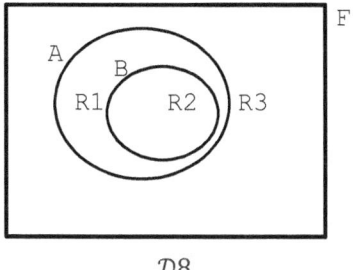
$\mathcal{D}8$

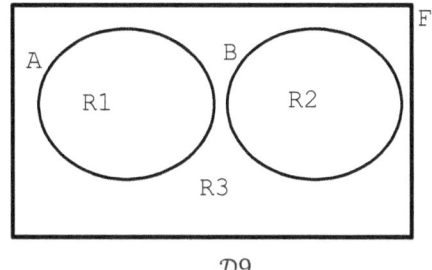
$\mathcal{D}9$

With these rules, it must, of course, be taken into account that the application of the abstract rules to the concrete individual case can always be a point of contention between concept users: $\mathcal{D}4$, for example, may be a faithful reflection of the concretely determined bit matrix of the conceptual representation of $\mathcal{D}1$. However, it may well be that there are concept users who know a different section of $\mathcal{D}1$ (e.g. areas that are only indicated by the ellipses) and have noticed an overlap of O^I and O^{II} there and therefore do not accept $\mathcal{D}4$ as a diagram for the general concepts of table and tree. It is also conceivable, as we have seen in Chapters 2.2.5f., that a concept user would not allow $\mathcal{D}8$ to be a diagram for the terms gold (= A) and yellow (= B), but would argue that the relationship between these two concepts corresponds to $\mathcal{D}7$. Be that as it may, the diagrams proposed here are nothing more, but probably also nothing less than a means of clarification: they can be seen as a means of clarifying conceptual relations, of clarifying approaches to the philosophy of language or also as a means of clarifying the relationship of world and language based on concrete models.

If the above rules are supplemented by two further ones that result from the subordination relationships of R-Cont, namely

(R-Poss) If in an object diagram an object O^1 sometimes occurs together with another object O^2, then a concept A may correspond to a concept B.

(R-NNess.) If in an object diagram an object O^1 sometimes does not occur together with another object O^2, then a concept A need not correspond to a concept B.

The corresponding relations of A and B can be seen in Fig. 1, in which the oppositional relations are also drawn, corresponding to a modal pentagon of opposition.

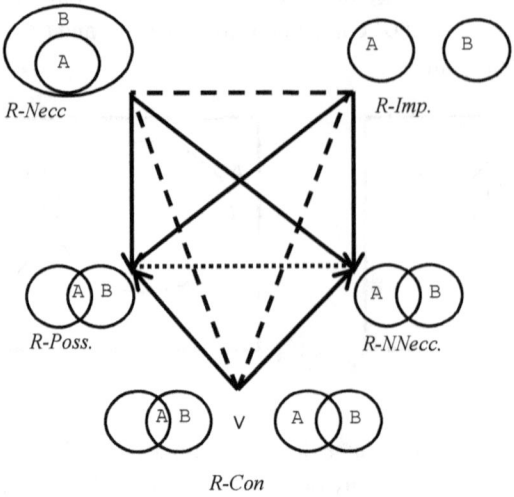

Fig. 1
Modal Pentagon of Opposition

However, before Chapter 3.2.3 reflects on how the relationship between the world and logic can be thought, it remains to consider the conceptual abstraction already mentioned, which the collective abstraction theory shares with that of neologicism. Conceptual abstraction has already been addressed in Syn1, Syn3 and Syn5. In the following, we will refer less to models in order *to understand* the relation between the intuitive and the conceptual representation, but only to certain relations within the conceptual representation. To understand these conceptual relations, contextual or use-theoretical aspects, which were thematised in Chapter 2.1, must be presupposed. We first see from Syn1, Syn3 and Syn5 that they no longer express a relation between $D1$ and $D5$ or $D6$, but that they alone clarify the relations within $D7$. More precisely, regions are determined by subtracting certain parts of the diagram. Concepts are thus not only designated by circles or spheres but certain parts within and outside certain circles, namely the regions and the associations of regions, are also given meaning.

In Chapter 1.3.1 it was worked out that those concepts can be called *concreta* that have originated directly from an objective abstraction, whereas *abstracta* are those concepts that are not directly deducted from intuition but from other concepts. Let us take $D7$ as an example and see how certain concepts can be determined as *concreta* or *abstracta*.

Let a *concreta* be a region or a union of regions in $D7$, $\{R1\} \vee \{R2\} \vee \{R1, R2\} \vee ...$, then it holds that since $\{R1\} = A \setminus B$, $\{R1\}$ is not an objective abstraction, but an abstraction of the concept of A from B and thus an *abstracta*. The same can be

3.2 Rational Representationalism

applied to {R3}. Since {R1, R2} = $(A \setminus B) \cup (A \cap B)$, the abstraction step of {R1} is already presupposed. The same can be applied to {R2, R3}, {R1, R2, R3}, {R1, R3}. If now all regions with {R1} or {R3} presuppose conceptual abstractions, then {R2} must be a *concreta*, i.e. $A \cap B$.

But as we have already seen, the division into *abstracta* and *concreta* is too imprecise. Finally, we see in {R4} that we have several conceptual abstraction steps here, namely the abstraction of A and B or the abstraction of the association of regions {R1, R2, R3} and which can be explicated further and further as follows: $F \setminus (A \cup B) = F \setminus ((A \setminus B) \cup (B \setminus A) \cup (A \cap B)) = (F \setminus (A \setminus B)) \cap (F \setminus (B \setminus A)) \cap (F \setminus (A \cap B))$. Thus, to determine the function term of *abstracta* more precisely, we add a *level* determined by the number of abstracta steps: L0-concept, L1-concept, L2-concept, ..., Ln-concept. For each conceptual abstraction step (\), the respective degree of abstraction now increases.

For example, in $\mathcal{D}7$ {R2} = $A \cap B$ has no abstraction step and is, therefore, a *concreta* or L0 concept. {R1} and {R3} each have an abstraction step, namely $A \setminus B$ and $B \setminus A$, and are therefore L1 concepts. As could be seen from the explicit expression for {R4} above, there are three abstraction steps here, so we speak here of a L3 concept. (Whether a still meaningful L2 concept such as $(A \setminus B) \cup (B \setminus A)$ is relevant in terms of philosophy of language will not be discussed in detail here).

In the following, I will show how one can determine the level of abstraction on more complex diagrams and unite several already known diagrams for this purpose. Let us first assume that in our model of intuitive representation M^I we find three objects that are marked by the three squares in $\mathcal{D}10$.

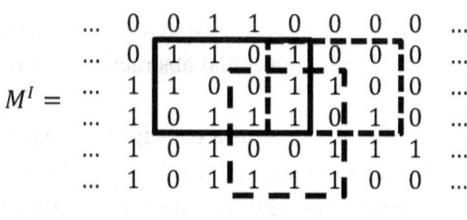

$\mathcal{D}10$

Through comparisons with other areas of $\mathcal{D}10$ that are not further identified here, we find out, for example, that none of the three objects always occurs in connection with one of the others. Thus (R-Necc) is already excluded and (R-Imp) is already refuted for the three objects by $\mathcal{D}10$. Thus (R-Cont) is the appropriate rule to translate the relation of the three objects of D10 into the conceptual diagram $\mathcal{D}11$ with three concepts A, B, C. Since in (R-Cont) the position of the words actually plays a relevant diagrammatic role, as can be seen in Fig. 1, I will put A, B, C outside the circles to simplify matters.

409

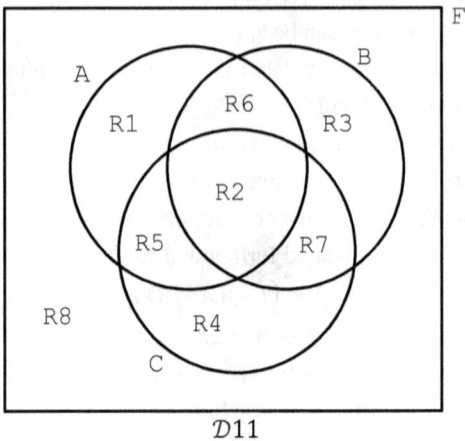

$\mathcal{D}11$

The three concepts of $\mathcal{D}11$ now form 8 regions: {R1}: $A \setminus (B \cup C)$; {R2}: $(A \cap B \cap C)$; {R3}: $B \setminus (A \cup C)$; {R4}: $C \setminus (A \cup B)$; {R5}: $(A \cap C) \setminus B$; {R6}: $(A \cap B) \setminus C$; {R7}: $(B \cap C) \setminus A$; {R8}: $F \setminus (A \cup B \cup C)$.

We now adopt the semantics of Fig. 2 from Chapter 2.1.6, so that A denotes the concept tree, B the concept green and C the concept flower-bearing. {R5}, for example, denotes all those objects to which the concepts green, flower-bearing, but not tree apply. {R6}, on the other hand, denotes those objects to which the terms tree and green but not flower-bearing apply. According to the assumption we have just made, *concreta* or L0-concepts are those that have no or as few abstractions as possible, but the most intersections. In the case of $\mathcal{D}11$, we can see from the 8 regions shown that {R2}, is a *concretum*, since the term has many intersections $(A \cap B \cap C)$ and no conceptual abstraction, but only the objective abstraction.

One can read the level of abstraction of a concept in a diagram like $\mathcal{D}11$ by how many convex and non-convex boundaries it has.[25] If one decomposes $\mathcal{D}11$ into individual regions, as in $\mathcal{D}12$, a total of four different diagrammatic forms of regions show abstraction levels. These levels can be determined by the number of non-convex boundaries. To indicate the number of convex or non-convex boundaries, line segments can be used to connect the outermost points of the diagrammatic form: If all points of the line segment lie within the diagrammatic shape (and do not intersect any other adjacent region), it is a convex boundary. If at least one point of the line segment lies outside the diagrammatic form (and thus intersects another neighbouring region), it is a non-convex boundary. The levels of abstraction are:

---

[25] I am deliberately avoiding the term convex set here, as these are conceptual diagrams and not set diagrams. The idea is unmistakably borrowed from Peter Gärdenfor's approach, but differs significantly from conceptual spaces in application and function.

3.2 Rational Representationalism

(**L0**) L0 does not denote a concept at all, but the intuitive representation, as it was represented as a model in $\mathcal{D}10$, for example.

(**L1**) {R2} is a *concretum* or L1-concept, since it has only convex boundaries, which are drawn in grey in $\mathcal{D}13$;

(**L2**) {R5}, {R6} and {R7} have two convex and one non-convex boundary, marked with two grey and one black line in $\mathcal{D}14$. These regions are thus 1st level *abstracta* or L2-concepts; i.e. they exhibit a conceptual abstraction from {R2}; but they are more concrete than regions of higher levels;

(**L3**) {R1}, {R3} and {R4} have one convex and two non-convex boundaries and are thus 2nd level abstracta or L3-concepts; i.e. they are conceptual abstractions of several lower-level concepts. The two non-convex boundaries are shown in black in $\mathcal{D}15$;

(**L4**) As can be seen in $\mathcal{D}16$, {R8} has a peculiar structure in relation to all other concept, which distinguishes it as the most abstract element of $\mathcal{D}12$: {R8} is not bounded externally by a sphere, but only internally by all the other conceptual spheres of $\mathcal{D}12$, i.e. {R1-7}. Already through this diagrammatic determination, {R8} is denied its conceptual character. By the absence of the limiting sphere, it is not clear where the outermost points of the diagrammatic form lie, except that they are outside {R1-7}. But if we assume that {R8} is a concept that stands in relation to all other concepts {R1-7}, then, as $\mathcal{D}16$ shows in two examples, each line segment would indicate a non-convex boundary of {R8}. In this case, we speak of L4-concepts.

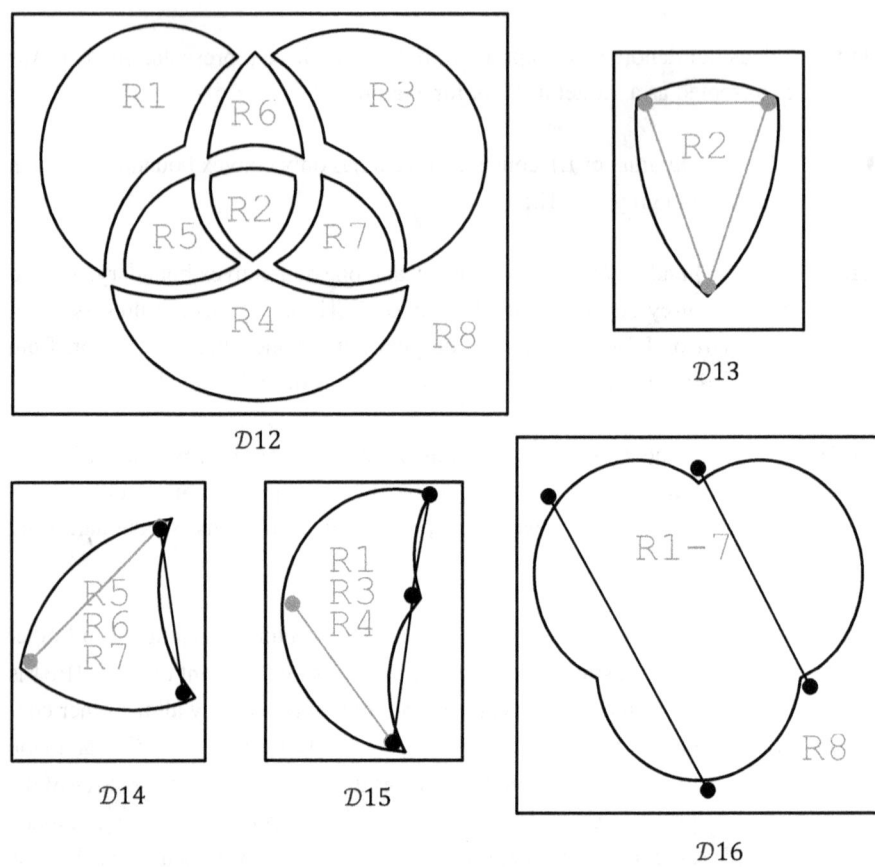

Analytical diagrams for concepts are not object diagrams, nor are they set diagrams, which denote the set of objects. Nevertheless, the level of abstraction also indicates a level of intension and extension, which is determined by the law of reciprocity (see Chapter 1.3.1). The law of reciprocity said: The higher the degree of extension C_{Ext}, the lower the degree of intension C_{Int} of a concept. Since we are only working here on a model that started from Matsuda matrices, the degree does not show the actual set of objects or properties, but only the relation between C_{Ext} and C_{Int}. This degree can be transferred in a *reciprocity function*:

If C_{Ext} can be described by a natural number x of a sequence from 0 to n ($[0, n]$:= $\{x \in \mathbb{N}_0 \mid 0 \leq x \leq n\}$), then $f(x) = n - x$ holds for C_{Int}. If the number of concepts levels L is known, then a suitable quantity can be given for n with the following formula: $n = L - 1$.

Let us take the syntax of diagram $\mathcal{D}11$ with the semantics of Fig. 2 from Chapter 2.1.6 again and try to connect the results mentioned so far with the collective abstraction theory that was argued for in Chapter 3.2.1. If we distinguish between objective

3.2 Rational Representationalism

and conceptual abstraction, we can say that the step from $\mathcal{D}10$ to $\mathcal{D}11$ is an objective abstraction. Let us now also assume that initially no differentiation between the particular degrees of concept can have taken place and that conceptual differentiations have only emerged in the course of a collective process of abstractions in order to increase the expressivity of language with linguistic refinements. Conceptual abstraction thus takes place on the concrete concept and assigns it certain properties that can also be applied to other concepts and, in the end, also to other objects. Only after this process of differentiation can an allocation into several *abstraction steps (AS)* be made, which we illustrate here again with our example:

(**AS1**) Objective abstraction: L0-concept to L1-concept;
(**AS2**) Conceptual abstraction: L1-concept to L2-concept;
(**AS3**) Conceptual abstraction: L2-concept to L3-concept;
(**AS4**) Conceptual abstraction: L3-concept to L4-concept.

For the case $\mathcal{D}11$, these abstraction steps can be represented in a graph. This graph would have as many nodes as it has regions. Such a graph is $\mathcal{D}17$, which has 12 edges because, due to AS4, there is an abstraction step to L3 from each L2 region, i.e. {R1}, {R3} and {R4}. Since one cannot now assume that all concepts of a concept level were abstracted simultaneously in a conceptual abstraction step, the diagrammatic explanation of $\mathcal{D}17$ must be transformed into a causal scenario. As explained in Chapter 3.2.1, however, this only transfers the non-causal explanation to a causal-hypothetical scenario. Conceivable, for example, is a descent of the terms {R7}, {R3} and {R8} starting from {R2} based on the sequence of abstraction steps AS2, AS3, AS4. This hypothetical-causal scenario is shown in $\mathcal{D}18$.

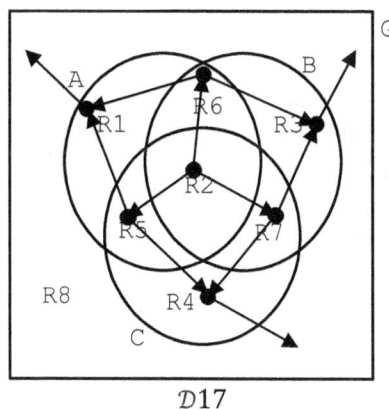
$\mathcal{D}17$

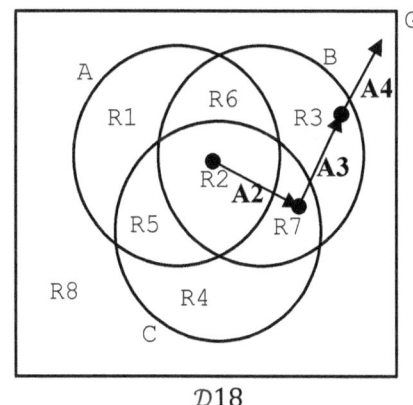
$\mathcal{D}18$

3 Logic and World

The hypothetical-causal scenario of $\mathcal{D}18$ is now to be linked to the collective abstraction theory, as announced in Chapter 3.2.1. For this purpose, the abstraction steps can now be transferred into a highly simplified dialogical scenario with several moments of interaction. This scenario is highly simplified insofar as the previous theory only operates with concepts and intuitions, not with judgements. Furthermore, the intergenerational collective dialogue is broken down into a simple dialogue between two parties. In this dialogue, a concept is to be compared with an intuition, so that one party makes a conceptual proposal through an abstraction step and another party compares the concept with the individual intuitive representation and reacts to it according to the regularities given above, i.e. (R-Necc.), (R-Poss.), (R-Cont.), etc.

P1 and P2 indicate the parties developing moments of interaction, which may or may not be verbal. Moments of interaction (I) include inferences, judgements, concepts, verbal signs, gestures, etc., whose meaning is contextually understood. Let us now imagine the simplified hypothetical interaction scenario for $\mathcal{D}18$ as follows:

P1: I_1 Do you call the object you are looking for green, tree and flower-bearing?

P2: I_2 I don't call it a tree.

P1: I_3 Then either green and flower-bearing or not green at all?

P2: I_4 No, not flower-bearing at all.

P1: I_5 Do you call it green then?

P2: I_6 No, not green either.

P1: I_7 Then you call it neither green nor tree nor flower-bearing.

We can imagine that the dialogue breaks off after I_7 or continues in another direction because at this point P1 has received what we called a satisfactory description according to Chapter 3.2.1. The satisfactory description results from P2's exhaustive responses to the concepts offered in I_1.

We can clarify this scenario again by using dialogue or interaction diagrams from Ludics. These are appropriate because they conform to the historical source, which I also took as a starting point in Chapters 1 and 2, and because they have an affinity to the terminology used so long in Chapter 3.[26] Since moments of interaction are usually speech acts, these interaction diagrams are represented with a circle, as in $\mathcal{D}18$. The reception of a speech act by another dialogue party is a thought act to which a responding speech act is expected. These thought acts are shown in the diagram as interaction moments without a circle. Arrows in the diagram indicate, on the one hand, the communication path between the speech and thought acts of the parties, and on

---

[26] Myriam Quatrini: Une relecture ludique des stratagèmes de Schopenhauer. In: Influxus (2013), No. 65; Christophe Fouqueré, Myriam Quatrini: Ludics and Natural Language. First Approaches. In: Logical Aspects of Computational Linguistics. LACL 2012. Ed. by D. Béchet, A. Dikovsky. Berlin, Heidelberg 2012, pp. 21–44.

3.2 Rational Representationalism

the other hand, also relations of reasoning and justification. Decision moments in which interaction moments are made in the form of several utterances (here in the form of the alternative question I_3) for thought acts of one party are represented by lines.

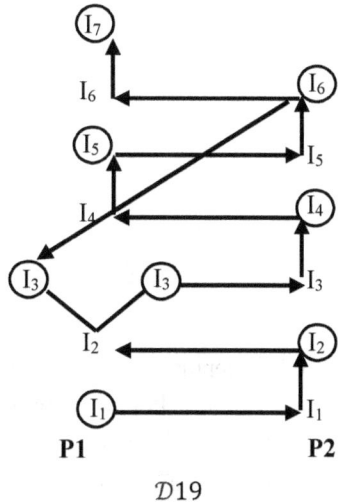

$\mathcal{D}19$

This hypothetical interaction scenario is highly simplified, but it represents a way of analysing the sources of collective abstraction theory mentioned in Chapter 3.2.1 with their interaction moments. The simplification, however, also consists in the fact that only the abstracting development of the concept has been represented and no regulative developments, for example, by P2 tracing an L2-concept back to an L1-concept.

However, the fact that regularities can be established here in the abstraction steps or in the abstracting development of the concept, as announced in Chapter 3.2.1, can be demonstrated, for example, by the reciprocity ratio for $\mathcal{D}17$: Since $L = 4, n = 3$. Therefore, according to the *reciprocity function* given above, for L1 concepts $C_{Int} = 3$ and $C_{Ext} = 0$, ..., for L4-concepts $B_{Int} = 0$ and $B_{Ext} = 3$. The Cartesian coordinate system below thus illustrates the relation between the degree of intension and the degree of extension for diagrams such as $\mathcal{D}11$. If it is known that regions like $\{R2\}$ are described by most of the possible properties, then the highest value for C_{Int} can be given, for example as $y = 3$, with which the value for C_{Ext} can then be calculated via the equation $x = y - 2$. That $\{R2\}$ has the highest C_{Int} in a diagram like $\mathcal{D}11$ can be seen, for example, in the high number of intersections of the properties ($A \cap B \cap C$) or in $\mathcal{D}13$. The fact that $\{R2\}$ now has a $C_{Ext} = 0$ does not mean that $\{R2\}$ does not designate a set of objects, but that fewer objects can be designated with a term like $\{R2\}$ than with terms of higher levels of abstraction, e.g. L2-, L3-, L4-concepts.

3 Logic and World

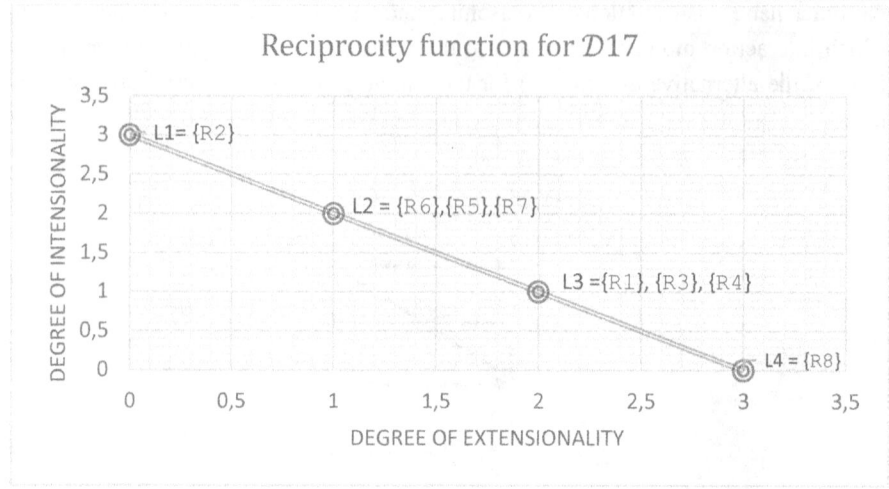

Concepts with a high degree of conceptual abstraction, such as {R8} or I$_7$, refer to many areas of the intuitive representation, but no longer have any limitations of the conceptual representation. {R8} in $\mathcal{D}11$ refers to everything that is not a tree, is green or bears flowers. Thus, this so-called infinite concept has the highest C_{Ext} in relation to the other concepts in $\mathcal{D}11$, but the lowest C_{Int}.

In the context of a large-scale conceptual system, which could be called a philosophy, concepts with a high degree of conceptual abstraction occur in an inflationary manner precisely when the latter is deflationary with regard to *concreta*. Large object areas can be *explained*, but conceptually they offer little to *understand*. We can say, for example, that I$_2$, I$_4$, I$_6$ of P2 are explanations of I$_1$, but we can also say that P1 has not understood exactly what P2 has in mind through I$_7$ or {R8}. Strictly speaking, even these most abstract concepts are not concepts at all, but words that only have a negative relationship to all concrete concepts. Since their abstraction step – in the case of $\mathcal{D}11$ it is AS4 – follows after the other abstraction steps, they have appeared very late in the collective process of language abstraction.

I believe that one can see from what has been said in this chapter how analytical diagrams or a geometric logic of rational representationalism can provide a non-causal explanation for the problem of descent and thereby complement the theory of understanding presented in Chapter 2.[27] Many areas – for example, the areas of judgement and inference – remain unexplored, and many of the explanations given here can be

---

[27] I would like to point out here that I have developed some of the theses presented in this chapter together with Michał Dobrzański. Some of the points mentioned here have already been discussed in more detail in the following papers: Jens Lemanski, Michał Dobrzański: Reism, Concretism and Schopenhauer; Michał Dobrzański, Jens Lemanski: Schopenhauer Diagrams for Conceptual Analysis. In: Diagrammatic Representation and Inference. Diagrams 2020. Lecture Notes in Computer Science, vol. 12169. ed. by Ahti Veikko Pietarinen, P. Chapman, Leonie Bosveld-de Smet, Valeria Giardino, James Corter & Sven Linker. Cham (2020), pp. 281-288.

3.2 Rational Representationalism

further refined. Moreover, the ideas presented here have only been developed on a model, namely R30, and not on concrete data taken from intuition. The diagram form in this chapter also only amounted to one of the forms given in Chapter 2.1.6. Yet Chapter 1.3 indicated several other shapes that can be included in a geometric logic. I would like to illustrate this with just one more example.

One could also imagine, for example, how someone has conceptually partitioned his sense data as we did above on the model with Matsuda matrices – in such a way that this person has reached ever-higher levels of abstraction in the process. If we disregard the many levels of concreta and lower abstracta and focus only on the most general concepts, then we might be left with a diagram that could look like $D20$. For such a diagram, different rules should apply than those mentioned above for $D11$. For example, one could understand its syntax in such a way that every separated region, let us call it Rx, in $D20$ excludes every other region Ry of equal or greater size, provided Rx is not contained in Ry. Regions could be formed by this exclusion. For example, $\{R1\}$ would then be the region for which it holds that $A \setminus B$ and $A \setminus (B \cup C)$ and $A \setminus (B \cup C \cup D)$ etc. But if regions of the same size could not only exclude but also unite, we could form a region like $\{R2\} = (A \cup B)$ for which $(A \cup B) \setminus C$. If $(A \cup B \cup C) = \{R3\}$, then $(A \cup B \cup C) \setminus D$ also holds. But since $\{R4\} = (A \cup B \cup C \cup D)$ does not find an equal or larger area in $D20$, only $\{R4\} = E$ holds.[28]

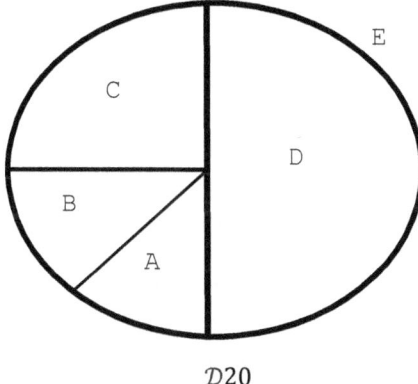

$D20$

But if we now imagine that someone were to develop a semantics for this syntax, so that $\{R1\}$ is called, for example, `causality` and $\{R2\}$ `understanding` and $\{R3\}$ `representation`, we might be inclined to concede that a suitable expression for $\{R4\}$ `World` might be, since this concept includes everything that exists, and there is nothing that this concept would exclude. But if one now assumes, as above in $D11$, that the circular area does not designate everything and that there is a region $\{R5\}$ that lies outside the diagram, then one would have to be able to designate this region by a

---

[28] Much more detailed research on these partition diagrams can be found in Lorenz Demey: From Euler Diagrams in Schopenhauer to Aristotelian Diagrams in Logical Geometry, Sekt. 3.2; Jens Lemanski, Lorenz Demey: Schopenhauer's Partition Diagrams and Logical Geometry.

term as well. This term {R5} would certainly be a relative one, for which only the following applies: no D and no $(A \cup B \cup C)$, i.e., no E. But for those to whom {R4} is being, to them, after the complete negation of $(A \cup B \cup C \cup D)$, this concept is – nothing.

3.2.3 Translation and Transfer

From Chapter 3.2.1 onwards and on the basis of the representationalism elaborated in Chapters 1 and 2, I have begun to argue for a theory of rationality that defines the space of reasons more broadly than the space of concepts. In doing so, I was not only concerned, as one might assume based on Chapter 2.2.2, with enhancing the value of intuition as opposed to the concept, but also with reconciling two essential aspects of rationalism with representationalism: On the one hand, the theory of abstraction and, on the other, collectivity.

I have argued for substituting an abstraction theory of meaning based on intuition for a theory of singular terms. To do justice to the arguments of the inferentialism of grounding, the critique of the private language problem was transferred from the individual abstraction theory to a collective one, with the aim of *anthropologising the abstraction theory*. The question of what the different meanings of concepts derive from, or why the conceptual spheres are so different, could finally be answered in favour of an original intuition with a recursive-compositional process of abstraction. This process of conceptual abstraction and differentiation is thus not reflected in the linguistic use of the individual, but only in the reenactment in a linguistic community cooperating verbally over many generations – in other words, the development of a living, actual essence. The *history of philosophy* was presented as relevant documentation material of this process since its essence consists in the process of conceptual and propositional adaptation and recombination, in which conceptual abstractions take place over many generations, but are also regulated by processes of concretion or reification.

In Chapter 3.2, the method of semantic analysis with the help of analytical diagrams was demonstrated on certain models. In contrast to inferentialism, which accepts a strong difference between the concrete and the abstract, I recommended the use of *geometric logic*, which only traces semantic, but no ontological commitments: Being is something that occupies a prominent role in language as an expression of intuitive relations. Concepts are *understood* as the extensional circumference of meanings in relations to other concepts, and these relations of meaning or the differences of the conceptual spheres represented can be *explained* as the result of an abstraction from an intuition that has always been continued and regulated in an intersubjective-generational process. The explanation of these collective steps of abstraction serves as the basis of contextual understanding.

3.2 Rational Representationalism

With these arguments, which taken together form the theoretical basis of a theory of rational representationalism, I have not intended to oppose inferentialism, but to expand its picture of the world and logic. I believe that this expansion is helpful and necessary in several respects: for by giving up the unnecessary limitation of the inferentialist picture of the world and logic, we can expand our understanding of the space of reasons in such a way that answers the *problem of justification* pointed out in Chapter 3.1, the *problem of translation* presented in Chapter 2.2, and the problem of original intuitive relatedness indicated in Chapters 2.3 and 3.2.1 become easier for us.

I would like to begin by addressing the last problem, which I have so far only hinted at; but it is a problem that I believe must not be painted over in a picture that philosophy presents to us if it is to harmonise with the paintings of other sciences in a single exhibition space. Indeed, the fact that the rationalist picture of a complete overlap and congruence of the space of concepts and space of reasons requires expansion is suggested not only by purely philosophical problems that are currently becoming louder in the philosophy of mind[29] but also by experimental studies from psychology, ethology and evolutionary anthropology. For several decades, numerous experiments have documented that primates, domesticated and also wild animal species are not only able to learn transitive inferences, but also possess an apparently natural predisposition for these inferences.[30] Since the species studied were not said to have the ability to form abstract conceptual content, but only what in the philosophy of mind is subsumed under the notion 'nonconceptual mental content' or what in zoo-semiotics is subsumed under 'unidirectional communication', there has been speculation as to whether there is something like 'inferential abilities without logic'.[31]

The psychological theory, which bears a family resemblance to the representationalism advocated here, states that the nonconceptual mental inferential faculty is not a mysterious apriority or something like logic without logic, but is based on the comparison of spatial intuition:[32] In language akin to collective abstraction theory, these studies describe how an object visually lights up to a particular experimental animal and an intelligent action can be performed with it under the condition that it places this object in relation to others. This theory is supported by several studies that show

---

[29] Cf. José Luis Bermúdez: Animal Reasoning and Proto-Logic. In: Rational Animals? Ed. by Susan Hurley, Matthew Nudds. Oxford: Univ. Press, 2006, pp. 127–137; Elisabeth Camp: A Language of Baboon Thought?. In: The Philosophy of Animal Minds. Ed. by Robert W. Lurz. Cambridge: Univ. Press, 2009, pp. 108–128.
[30] Cf. Brandan O. McGonigle, Margaret Chalmers: Are Monkeys Logical?. In: Nature 267 (1977), pp. 694–696; Douglas Gillan: Reasoning in the Chimpanzee: II. Transitive Inference. In: Journal of Experimental Psychology: Animal Behavior Processes 7:2 (1981), pp. 150–164; Hank Davies: Transitive Inference in Rats (Rattus norvegicus). In: Journal of Comparative Psychology 106:4 (1992), pp. 342–349.
[31] Cf. P. N. Johnson-Laird: Reasoning without Logic. In: Reasoning and Discourse Processes. Ed. by T. Myers, K. Brown, B. McGonigle. London 1986, pp. 13–50.
[32] Cf. C. B. De Soto, M. London, S. Handel: Social Reasoning and Spatial Paralogic. In: Journal of Personality and Social Psychology 2 (1965), pp. 513–521.

that the simplest and thus basic type of transitive inferences are spatial relations:[33] That is, inferences with relations such as 'higher/ lower' or 'containing/ being-contained' are easier and simpler than inferences with relations such as 'lighter/ darker' or 'faster/ slower'.

One can take these experimental results as empirical evidence that there is a closer relationship between representation and rationality than rationalists usually want to admit.[34] This does not mean, of course, that we have to drop logic and rationality as a specific difference that defines our species in contrast to other creatures; but it should sensitise us to be careful about what exactly we mean by terms like 'thinking', 'rationality' and 'logic'. It seems that we share the basics of propositional and quantifier-based reasoning, which we describe with intuitive relations such as 'containing' and 'excluding', not only with our kind but also with other living beings.[35]

Unlike other living beings that are capable of similar inferences, however, we have the possibility to justify the applied forms of rationality representationally: E.g. by rendering the space of reasons as a space of concepts, or by tracing the forms of the space of reasons. In both cases, we not only apply our inferential disposition, but add something that classical philosophers have described with the terms 'reflection' and 'consciousness', referring to an ability that distinguishes us from other rationally acting beings.

The fact that I am more interested in the rational relationship between us and other living beings than in what separates them is based on the fact that I see in this close connection the solutions to the translation and reasoning problem of inferentialism. In my view, the problem of grounding inferential programmes stems from the fact that they have always been limited to themselves and have therefore never gone beyond dogmatic, circular or regressive grounding in explanatory and evidentiary procedures. Rather, even translation strategies have been mustered to bring intuitive relations and expanding transfers into the limitation. In this concluding chapter, I will argue that transfers in the logical vocabulary are often not everyday or mythical remnants that need to be translated into a purer form of logic, but that they can also be *grounded elements of language* whose function becomes understandable by making use of their intuition, for example, employing geometric forms – in other words, by recalling what other living beings only claim in an unreflective and unconscious manner for grounded rational processes.

If intuition has a use for logic, it may be because logic has often made unnoticed use of this intuition. In what follows, then, the point is that intuitive-geometric logic need

---

[33] Cf. P. Shaver, L. Pierson, S. Lang: Converging Evidence for the Functional Significance of Imagery in Problem Solving. In: Cognition 3 (1975), pp. 359–375. I only indicate here the classic text of the *imagery debate* and refer to the current studies mentioned in Chapter 2.3.

[34] Doubts about several studies can, of course, be raised with regard to individual differentiations (e.g., which transitive inferences are meant? Should the conclusion only be validly inferred or also reflected? Does the conclusion have to be inferred verbally? etc.).

[35] Cf. Elisabeth Camp: A Language of Baboon Thought?.

3.2 Rational Representationalism

not be explained by a purely conceptual and completely abstract logic, but rather that the abstract forms of logic can be explained by a geometric logic. For the *one thought* that I add in this chapter to the transfers worked out in Chapter 2.3.4 is that already in the principle on which the inferentialist logics have agreed as a basis, *grounded transfers* can be found that can be assigned to the stock role of being contained. Therefore, on the one hand, these transfers cannot be meaningfully translated into a non-transferred language without the logical vocabulary losing its expressive power; on the other hand, the transfers explain the motivation of many interpreters, which was discussed in detail in Chapter 2 to place logic in the vicinity of the intuition of geometry; and finally, they show that our logical reflections do not accommodate the saltationism of the inferentialists, but only testify to a higher degree of intuitive thinking, which has asserted itself and developed to varying degrees in the natural history of some living beings.

The difference between translation and transfer can first be explained without looking at their *anthropological roles*. *Translation* is a relation that does not exist between the space of reasons and the space of concepts, but only within the space of concepts, in such a way that words that play the role of conceptual contents within this space are *restructured* by other words and thus take the place of the first. *Transfer* is generally a one-way relation that exists between the space of reasons and the space of concepts (without presupposing any partial or complete identity between the two) so that reasons are *transposed* in such a way that they can play the roles of conceptual content. Translations and transfers are in any case a change proposed by speakers and a change realised in linguistic communities, either of words or of grounds.

If conceptual roles are to be restructured, words redivide the content of the space of concepts. Restructuring thus defines or interprets the space of concepts in a different way than before: if the division is done by explanation or justification, the space of concepts can be considered redefined; if the division shows up with the help of interpretation in context, the space of concepts must be reinterpreted by recipients. If, as is often the case in technical languages, only a small part within the space of concepts is redefined or interpreted, then so-called stocks, i.e. conceptual subspaces, are created in which translated words are to play *standing roles* of conceptual content. If the entire space of concepts is redefined or interpreted, as in the case of text translations, for example, the division into stocks should remain as far as possible, in such a way that the translated words restruct the standing roles. If the translation is done in such a way that the conceptual content of standing roles can be completely restructured, one speaks of facultative equivalence. If the conceptual content cannot be translated at all, we speak of zero equivalence.

If grounds are to be transposed into conceptual content, they must take on roles that were not previously occupied. Definitions and interpretations of transpositions are problematic: for one thing, speakers cannot explain transfers without resorting to translations and without then pretending that transpositions are nothing more than restructions. For another, recipients can never fully interpret the interpretation in context, since transfers have always been transposed from their original context in the space of reasons. Unlike translations, transfers can only ever make up a small part within the space of concepts, and the fact that they can play roles of conceptual content does not mean that they are actually conceptual content. After all, 'playing a role' means making someone believe you are someone or something else without necessarily having to be so.

Ideally, when transfers play the role of conceptual content, they make us believe that they are stocks and that we are dealing with a technical language in which a small part is redefined or interpreted within the space of concepts. We then believe that the transfer is an integral part in the space of the concept and plays a standing role of conceptual content. In this belief, transfers can be handled like translations: If one is disturbed by a concrete word, it is restructured by one that is supposed to correspond more 'clearly' and distinctly to the conceptual content or fits better into the abstract context of the technical language. If restructuring succeeds, then the word, which has remained in the context of the increasingly abstract technical language for a while, can be counted among the *residual stock* of concrete concepts. If, however, this restruction does not succeed, because no equivalence emerges in which the translation corresponds more distinctly to the conceptual content than before, or no conceptual content at all emerges from the abstract context, or even one transfer is only ever translated by another, then the word has only played the role of conceptual content. In this case, the word can be interpreted as if it were a transfer of a grounded or *basic stock* that only refers to the space of concepts within and detached from the space of reasons.

I suggest, however, that the relationship between transfer and translation should also be understood in *anthropological* and intersubjective terms. Transfers are *not causal* but rational in structure. The intuition may have formally imposed itself, but in the end, it was the speaker's achievement that she put the intuition into words and placed it in the logical space of concept. In this respect, it is erroneous to claim that the transfer was causal in nature, for even though it may seem plausible that the concept is subordinate to intuition, there was neither an impulse nor a necessary connection that could have been observed between the intuition and the concept. This was already correctly recognised by the inferentialism of grounding, although it denied that there could also be a rational transboundary influence.

Originally, it must always have been *individual speakers* who transferred the intuition into the space of the conceptual. They opened up the space of the concept and formally invited the intuition to enter. In the course of time, the linguistic community has become habituated to these intuitions and has either *adapted* a large part of them by abstraction from the original intuition in the logical space or *selected* them at some point due to their resistance to abstraction processes. By custom and through the recursion of abstraction processes, the transfers soon took on a conceptual role in which

3.2 Rational Representationalism

no relation to the intuition was needed anymore. While transfers also stand in relation to intuition, the meaning of concepts is determined solely by the level of abstraction that expresses them in relation to other concepts in the judgement.

Since it is primarily through the recursion of abstraction processes that the sphere of the meaning of the concept is determined, it is no longer the original intuition that is decisive for the contextual determination of concepts in judgements, but rather the relationship that concepts assume in relation to other concepts. In our acts of thought and speech, we finally assert that the concept living being has undergone more levels of abstraction than the concept Socrates and that living being thus occupies a larger part of the space of concepts than the concept Socrates when we make judgements such as Socrates is contained in living beings or more indeterminately Socrates is a living being. Conversely, these judgements also do not require intuition of an object, since the level of abstraction of Socrates is determined by the fact that it is contained in the concept living being and thus has fewer steps of abstraction.

Inferentialists are right, of course, to suggest that we learn to *understand* conceptual relations through the use of language; but the process of understanding spheres of meaning is not the *explanation* of how the multiplicity of concepts that we can relate in a judgement by means of words comes about in the first place. Conceptual circumferences that we understand and that we ourselves influence in use have been handed down over many generations, have been abstracted further away from intuition or have been regulated towards intuition. Transfers, however, can play a different role in judgements than concepts such as Socrates or living being.

The fact that transfers can play a different role in the judgement than concepts, but do not have to, is based on the *distinction* that belongs to the transfers themselves. They can be residual or basic, depending on whether the conceptual content in a judgement has only been restructured by a transfer or whether the transfer plays the role of conceptual content but cannot be translated into other concepts. I am primarily interested here in *basic or grounded transfers*. I further argue that such grounded transfers exist in logic and that restructuring has failed because their expressive power lies in going beyond the space of concepts.

Transfers of any kind are conspicuous in the judgement either by having relations of degrees of abstraction that do not fit the division of concepts or by offering a conceptual division through the relationship in the judgement that does not correspond to the understanding of conceptual spheres. With recourse to the standard analysis of knowledge, one could perhaps even say that judgements with transfers are relations of which a speaker can be convinced or offer a justification or also know that they cannot be true in the sense of a conceptual outline. If one claims, for example, that the great revelations of quantum mechanics lay in the discovery of discontinuities in the book of nature, then the uneasy feeling is fostered by the fact that, on the one hand, the claim expresses a strange overlapping of the conceptual circumferences of 'nature'

3 Logic and World

and 'book' and that, on the other hand, this claim implies that the formulation 'revelations of quantum mechanics' is also circumferentially identical with the discoveries of discontinuities found in this book.

One can certainly understand the expression of transfer, if one understands it conceptually, more broadly than I do here. As I said, I am concentrating primarily on the double role of the grounded transfers in order to bring to bear the significance of a single thought for the development of intuition in logic. Crucially, however, these transfers, which not only play a conceptual role but also evoke non-conceptual reasons, cause difficulties for inferentialists of matter and form. Namely, certain assertions with transfers are perceived as an offence and as foreign bodies in the judgement that needs to be *replaced*, since they disorganise the conceptual structuring that is supposed to take place from the material judgements into inferences and from the rules of the logical vocabulary. Replacement and substitution are thereby a means of restoring the required 'clarity' of expression and the 'rigour' of thought.

The fact that the space of reasons, the space of the conceptual, the concept, the judgement and the inference, the implication, development, formation and the many other expressions of inferentialism have not yet fared as many other residual transfers have either distinguishes their still welcome expressive power or indicates that one has become habituated to them in their conceptual role. If one draws the attention of inferentialists to such foreign bodies in the judgement, they generally pursue a causal-theoretical strategy to justify the use of transfers: they allude to custom or even to a *customary law* that is supposed to allow them to treat transfers as pure concepts, although in doing so they are precisely limiting their expressive power. They thus indicate that it is enough to limit transfers to their conceptual role, to ascribe to them from the context a sphere of meaning that is sometimes uncertain, and to refrain from the intuition that they might evoke. It is precisely the emphasis on a double role of transfers in the form of verbal nouns with standing roles in technical vocabulary that is often dismissed as an etymological fallacy.

The *expressive power* we attribute to transfers is based on a function we attribute to all concepts and all signs: they make the absent present, and they make actions over distances possible.[36] Speech acts of reasoning are also made possible by signs. Since a reason that requires intuition cannot extend to something that is distant from this intuition, except insofar as intuition reaches what is distant through mediation, the absent reason also requires a *mediation through signs*. The expressive power of transfers is characterised by the fact that the absence of intuitive grounds is made possible in conceptual contexts in which intuition itself can play no role, or at most only a very limited one.

The argumentative strategy of emphasising the conceptuality of transfer by stressing habit and treating it like all translations must be seen as legitimate to a certain extent since otherwise, the very statement that one treats transfers as pure concepts seems

---

[36] Cf. Michał Dobrzański: Begriff und Methode bei Arthur Schopenhauer, Chapter 7.5.

tautological; however, this strategy is much more likely to tempt one to see the *space of concepts as a hermetically closed space* and to pretend that there has never been any recourse to intuition and that there has never been any need to enrich language with intuition conceived in words.

But those who look at transfers only from their conceptual side will never get out of the interior of logic and block their intuition to the much wider space of reasons. This is an unnecessary limitation that at the same time demands extensions that seem unattainable. One resembles someone who walks around enclosed in a room, searching in vain for an exit and describing the walls in the meantime, because one believes that these are already the limits of the world. But it is the metaphor of transfer that gives us the *key to intuition*. In the transfer, language and logic are given to us in two quite different ways: first, as a concept among concepts, so that the transfer can play the role of the conceptual content in the judgement and has an extension which, at most in its level of abstraction, provides a relation to an intuition but which has long since been forgotten and has no function for the meaning and understanding of this transfer; secondly, however, in a quite different way, namely as a ground located among the concepts, which has no extension and directly evokes the representation of an intuition.

I have shown in Chapter 2, with a particular interpretive direction of transcendental philosophy that replacement strategies fail because the untidy transfers keep imposing themselves, so that periodically in the history of logic there are rediscoveries of achievements in justification with the help of intuitive geometry. This is because transfers possess a strength of expression in justifications that familiar logical concepts lack when confined to the space in which they are supposed to play a restricted role. Finally, the intrusiveness and expressiveness of transfers point to the fundamental nature that made them uncomfortable transfers in the first place. Transferences impose themselves because they express reasons that the concept if it is not itself understood as a transfer, is not supposed to access. The concept in its untransferred form is supposed to experience precisely this limitation of not being intuitive and not being conceptual so that it can be defined out of itself. The fact that uncomfortable transfers now impose themselves is consequently due to the intended limitedness of the expression `concept`.

That the transfer does more than playing a conceptual role that can be translated in many ways, and that it must be understood as a grounding relic, an incursion from the space of reasons, is made clear by the expression of 'something necessarily being the case' or 'something not necessarily being the case' indicated in Chapter 2.2, which goes back to a semantics dealt with in Chapter 2.1 and extends to the grounding of proof theory indicated in Chapter 2.3. The history of this logical semantics is a sign that the dispute about the foundation of modern rationalism is not decided in the question of the priority of the concept, judgement or inference, and that the justification of rationalism cannot be provided in an unproblematic way from within it. The history of logic shows that inferentialism and logicism contain transfers that cause difficulties if they are met only with translations; for it is rather in their nature to refer beyond the space of the concept to realities in the space of reasons. Such fundamental transfers evoke intuitions, or intuitions are enabled by transfers to occupy a role in the space of concepts.

In *Summa und System*, I have tried to show how reflections on methods have developed from intuitions and have remained in all scientific reflections on methods up to the present day using bottom-up and top-down transfers. That there are also bottom-up transfers in logicism was shown most clearly in Chapter 3.1 in the explanation of that calculus of natural deduction which, for inferentialism, represents the actuality and grounding theorem of rational thought. This grounding theorem shows by its transfers of containing and comprehension that intuition and concept are mediated, that the world is already a part of logic, or even that the space of reasons to be referred to is larger than the space of concepts:

> τὸ δὲ ἐν ὅλῳ εἶναι ἕτερον ἑτέρῳ καὶ τὸ κατὰ παντὸς κατηγορεῖσθαι θατέρου θάτερον ταὐτόν ἐστιν. λέγομεν δὲ τὸ κατὰ παντὸς κατηγορεῖσθαι ὅταν μηδὲν ᾖ *λαβεῖν καθ'* οὗ θάτερον οὐ λεχθήσεται· καὶ τὸ κατὰ μηδενὸς ὡσαύτως.
> (An. pr. 24b25–30; my emphasis – J.L.)

Appendix

Bibliography

Most historical texts with logic diagrams are collected in the digital repository:
History of Euler-Venn-Diagrams
<https://www.zotero.org/groups/history_of_euler-venn-diagrams/>.

Allison, Henry E.: *The Kant-Eberhard-Controversy*. Baltimore et al. 1973.
Alstedio, Johanne-Henrico: *Logicæ Systema Harmonicum* […]. Herbornæ Nassoviorum 1614.
Anderson, R[obert] Lanier: Containment Analyticity and Kant's Problem of Synthetic Judgment. In: *Graduate Faculty Philosophy Journal* 25:2 (2004), pp. 161–204.
Anderson, R. Lanier: It Adds up After All. Kant's Philosophy of Arithmetic in Light of the Traditional Logic. In: *Philosophy and Phenomenological Research* 69:3 (2004), pp. 501–540.
Anderson, R[obert] Lanier: *The Poverty of Conceptual Truth. Kant's Analytic/Synthetic Distinction and the Limits of Metaphysics*. New York 2015.
Andrews, Peter: *An Introduction to Mathematical Logic and Type Theory. To Truth through Proof.* 2nd ed. Dordrecht; Boston: Kluwer 2002.
Angelelli, Ignacio: Critical Remarks on Michael Dummett's Frege and Other Philosophers. In: *Modern Logic* 3 (1993), pp. 387–400.
Aristoteles Stagaritae Peripateticorum: *Principis Organon*. Ed. by Iulius Pacius. Morgia 1584.
Armstrong, Robert L./ Howe, Lawrence W.: A Euler Test for Syllogisms. In: *Teaching Philosophy* 13 (1990), pp. 39–46.
Atten, Mark Van/ Sundholm, Göran: L.E.J. Brouwer's 'Unreliability of the Logical Principles'. A New Translation, with an Introduction. In: *History and Philosophy of Logic* 38:1 (2017), pp. 24–47.
Atwell, John: *Schopenhauer on the Character of the World. The Metaphysics of Will*. Berkeley 1995.
Ayer, A[lfred] J[ules]: *Foundations of Empirical Knowledge*. London 1940.
Ayer, Alfred [Jules]: *Language, Truth and Logic*. London 1936.
Bachmann, Carl F.: *Ueber Hegel's System und die Notwendigkeit einer nochmaligen Umgestaltung der Philosophie*. Leipzig 1833.
Bachmann, Carl Friedrich: *System der Logik. Ein Handbuch zum Selbststudium*. Leipzig 1928.
[Bacon, Francis:] *Francisci De Verulamio Summi Angliae Cancellarii Instauratio magna*. Londini 1620.

Bacon, Francis: *Of the Proficience and Aduancement of Learning Diuine and Humane*. London 1605.

Bahnsen, Julius: Der Bildungswerth der Mathematik. In: *Schulzeitung für die Herzogtümer Schleswig-Holstein und Lauenburg*, 21, 25, 26 (21 Feb., 21 and 28 Mar. 1857).

Baptist, Peter: Der Satz des Pythagoras – eine qualitas occulta? In: *Der Mathematikunterricht* 42:3 (1996), pp. 22–30.

Barnes, Jonathan: Commentary. In: *Porphyry's Introduction, Translated with a Commentary*. Oxford 2003, pp. 21–312.

Barnes, Jonathan: *Truth, etc. Six Lectures on Ancient Logic*. Oxford 2007.

Baron, Margaret E.: A Note on the Historical Development of Logic Diagrams. Leibniz, Euler and Venn. In: *The Mathematical Gazette* 53:384 (May 1969), pp. 113–125.

Barwise, Jon/Etchemendy, John: Heterogeneous Logic. In: *Diagrammatic Reasoning. Cognitive and Computational Perspectives*. Ed. by J. Glasgow, N. Hari Narayanan, B. Chandrasekaran. Cambridge/Mass. 1995, pp. 209–232.

Barwise, Jon/Etchemendy, John: Visual Information and Valid Reasoning. In: *Logical Reasoning with Diagrams*. Ed. by Gerard Allwein, Jon Barwise. New York et al. 1996, pp. 3–27.

Becker, J[ohann] C[arl]: *Abhandlungen aus dem Grenzgebiete der Mathematik und Philosophie*. Zürich 1870.

Becker, J[ohann] C[arl]: Über Begründung und systematische Entwickelung der geometrischen Wahrheiten. In: *Schulzeitung für die Herzogtümer Schleswig-Holstein und Lauenburg* Nr. 14, 15 vom 2. und 9. Januar 1853, Nr. 14, 15 vom 2. und 9. Januar 1853.

Becker, Oskar: *Grundlagen der Mathematik in geschichtlicher Entwicklung*. 2nd ed. Freiburg et al. 1964.

Becker, Oskar: Zur Logik der Modalitäten. In: *Jahrbuch für Philosophie und phänomenologische Forschung* XI (1930), pp. 497–548.

Beckerath, Ulrich von: Eine Anerkennung der mathematischen Ansichten Schopenhauers aus dem Jahr 1847. In: *Schopenhauer-Jahrbuch* 24 (1937), pp. 158–161.

Beiser, Frederick C.: *Weltschmerz: Pessimism in German Philosophy, 1860–1900*. Oxford 2016.

Bellucci, Francesco/ Moktefi, Amirouche/ Pietarinen, Ahti-Veikko: Diagrammatic Autarchy. Linear Diagrams in the 17th and 18th Centuries. In: *Diagrams, Logic and Cognition*. Ed. by J. Burton, L. Choudhury. CEUR Workshop Proceedings 1132 (2013), pp. 23–30.

Belnap, Nuel D.: Tonk, Plonk and Plink. In: *Analysis* 22:6 (1962), pp. 130–134.

Bennett, Jonathan: Analytic–Synthetic. In: *Proceedings of the Aristotelian Society* 59 (1958–9), pp. 163–88.

Bennett, Jonathan: *Kant's Analytic*. Cambridge 1966.

Bennett, Jonathan: On Being Forced to a Conclusion. In: *Aristotelian Society Supplementary Volume* 35 (1961), pp. 15–34.

Berg, Robert Jan: *Objektiver Idealismus und Voluntarismus in der Metaphysik Schellings und Schopenhauers*. Würzburg 2003.

Berger, Anna Maria Busse: *Medieval Music and the Art of Memory*. Berkeley 2005.

Bermúdez, José Luis: Animal Reasoning and Proto-Logic. In: *Rational Animals?* Ed. by Susan Hurley, Matthew Nudds. Oxford: Univ. Press, 2006, pp. 127–137.

Bernhard, Peter: *Euler-Diagramme. Zur Morphologie einer Repräsentationsform in der Logik*. Paderborn 2001.

Bernoulli, Jakob: *Parallelismus ratiocinii logici et algebraici*. Basileae 1685.

Béziau, Jean-Yves: La Critique Schopenhaurienne de l'Usage de la Logique en Mathématiques. In: *O Que Nos Faz Pensar* 7 (1993), pp. 81–88.

Béziau, Jean-Yves: Metalogic, Schopenhauer and Universal Logic. In: *Language, Logic, and Mathematics in Schopenhauer*. Ed. by Jens Lemanski. Cham 2020, pp. 207–257.

Béziau, Jean-Yves: O princípio de razão suficiente e a lógica segundo Arthur Schopenhauer. In: *Século XIX. O Nascimento da Ciência Contemporânea*. Ed. by F.R.R. Évora. Campinas 1992, pp. 35–39.

Béziau, Jean-Yves: O suicídio segundo Arthur Schopenhauer. In: *Discurso* 28 (1997), pp. 127–143.

Béziau, Jean-Yves: On the Formalization of the Principium Rationis Sufficientis. In: *Bulletin of the Section of Logic* 22:1 (1993), pp. 2–3.

Béziau, Jean-Yves: The Power of the Hexagon. In: *Logica Universalis* 6 (2012), 1–43.

Birnbacher, Dieter: Language, Logic, and Mathematics (review). In: *Schopenhauer-Jahrbuch* 101 (2020), pp. 249–257.

Birnbacher, Dieter: Schopenhauer und die Tradition der Sprachkritik. In: *Schopenhauer-Jahrbuch* 99 (2018), pp. 37–56.

Black, Max: Metaphor. In: *Proceedings of the Aristotelian Society, New Series* 55 (1954/55), pp. 273–294.

Bloch, Ernst: *Leipziger Vorlesungen zur Geschichte der Philosophie* (1950–1956). Vol. 4. Frankfurt/ Main 1985.

Bloch, Sascha/ Pleitz, Martin/ Pohlmann, Markus/ Wrobel, Jakob: Deviant Rules. On Susan Haack's 'The Justification of Deduction'. In: *Susan Haack. Reintegrating Philosophy*. Ed. by Julia F. Göhner, Eva-Maria Jung. Cham et al. 2016, pp. 85–113.

Blum, Paul Richard: *Studies on Early Modern Aristotelianism*. Leiden et al. 2012.

Blumenberg, Hans: *Die Lesbarkeit der Welt*. Frankfurt/ Main 1986.

Blumenberg, Hans: Theory of Nonconceptuality. In: Ibid.: *History, Metaphors, Fables*. Ithaca 2020, pp. 259–297.

Blumhof, Johann Georg Ludolph/ Kästner, Abraham Gotthelf: *Vom alten Mathematiker Conrad Dasypodius: Ein literarischer Versuch* [...]. Göttingen 1796.

Bôk, August Friedrich (Ed.): *Sammlung der Schriften, welche den logischen Calcul Herrn Prof. Ploucquets betreffen, mit neuen Zusâzen.* Frankfurt, Leipzig 1766.

Booms, Martin: *Aporie und Subjekt. Die erkenntnistheoretische Entfaltungslogik der Philosophie Schopenhauers.* Würzburg 2003.

Bornträger, J[ohann] C. F.: *Ueber das Daseyn Gottes in Beziehung auf Kantische und Mendelssohnsche Philosophie.* Hannover 1788.

Botha, Rudolf: *Language Evolution. The Windows Approach.* Cambridge 2016.

Bowles, George: The Deductive/Inductive Distinction. In: *Informal Logic* 16:3 (1994), pp. 159–184.

Boyer, Carl B.: *The History of Calculus and Its Conceptual Development (The Concepts of the Calculus).* New York 1949.

Bradwardine, Thomas: *Preclarissimum mathematicarum opus* [...]. S.l. [Valenica] 1503.

Brandis, Christian A.: Über die Reihenfolge der Bücher des Aristotelischen Organons und ihre Griechischen Ausleger, nebst Beiträgen zur Geschichte des Textes jener Bücher des Aristoteles und ihre Ausgaben. In: *Abhandlungen der Königlichen Akademie der Wissenschaften zu Berlin. Aus dem Jahre 1833.* Berlin 1835.

Brandom, Robert: *Articulating Reasons. An Introduction to Inferentialism.* 2nd ed. Cambridge/Mass. 2001.

Brandom, Robert: *Between Saying and Doing. Towards an Analytic Pragmatism.* Oxford 2008.

Brandom, Robert: Inference, Expression, and Induction. In: *Philosophical Studies* 54 (1988), pp. 257–285.

Brandom, Robert: *Making it Explicit. Reasoning, Representing and Discursive Commitment.* 4th ed. Cambridge/Mass. 2001.

Brandom, Robert: Non-Inferential Knowledge, Perceptual Experience, and Secondary Qualities. Placing McDowell's Empiricism. In: *Reading McDowell. On Mind and World.* Ed. by Nicholas H. Smith. New York 2002, pp. 92–105.

Brandom, Robert: *Tales of the Mighty Dead. Historical Essays in the Metaphysics of Intentionality.* Cambridge/Mass 2002.

Brandt, Reinhard: *Die Urteilstafel. Kritik der reinen Vernunft A 67–76; B 92–101.* Hamburg 1991.

Bronzo, Silver: Bentham's Contextualism and Its Relation to Analytic Philosophy. In: *Journal for the History of Analytic Philosophy* 2:8 (2014), pp. 1–41.

Brouwer, L[uitzen] E[gbertus] J[an]: Die Struktur des Kontinuums. In: Ibid.: *Collected Works. Vol. 1. Philosophy and Foundations of Mathematics.* Ed. by A Heyting. Amsterdam et al. 1975, pp. 429–440.

Buchbinder, [Friedrich]: Verhandlung der Sektionen für mathematischen und naturwissenschaftlichen Unterricht auf der diesjährigen Versammlung deutscher

Philologen und Schulmänner, Vom 1.–4. Oktober 1884 in Dessau. In: *Zeitschrift für mathematischen und naturwissenschaftlichen Unterricht* 16:1 (1885), pp. 66–76.

Büchel, Gregor: *Geometrie und Philosophie. zum Verhältnis beider Vernunftwissenschaften im Fortgang von der Kritik der reinen Vernunft zum Opus postumum.* Berlin et al. 1987.

Bullynck, Maarten: Erhard Weigel's Contributions to the Formation of Symbolic Logic. In: *History and Philosophy of Logic* 34 (2013), pp. 25–34.

Burge, Tyler: *Truth, Thought, Reason. Essays on Frege.* Oxford 2005.

Burton, Jim/ Stapelton, Gem/ Delaney, Aidan/ Howse, Jon/ Chapman, Peter: Visualizing Concepts with Euler Diagrams. In: Diagrammatic Representation and Inference. Diagrams 2014. Lecture Notes in Computer Science, vol 8578. Ed. by T. Dwyer, H. Purchase, A. Delaney. Berlin, Heidelberg 2014, pp. 54–56.

Busche, Hubertus: Leibniz' Weg ins perspektivische Universum. Eine Harmonie im Zeitalter der Berechnung. Hamburg 1997.

Cajori, Florian: A Review of Three Famous Attacks upon the Study of Mathematics as a Training of the Mind. In: *Popular Science* 80:22 (1912), pp. 360–372.

Cakmak, Cengiz: Schopenhauer & Wittgenstein. The Unsayable. In: *Philosophical Inquiry* 25:1/2 (2003), pp. 115–124.

Camp, Elisabeth: A Language of Baboon Thought?. In: *The Philosophy of Animal Minds.* Ed. by Robert W. Lurz. Cambridge 2009, pp. 108–128.

Canfield, John V.: Wittgenstein versus Quine. The Passage into Language. In: *Wittgenstein and Quine.* Ed. by Hans-Johann Glock, Robert L. Arrington. London 1996, pp. 116–144.

Cantor, Moritz: Ferdinand Schweins und Otto Hesse. In: *Heidelberger Professoren aus dem 19. Jahrhundert* 2 (1903), pp. 221–242.

Caramuel Lobkowitz, Juan: *Logica vocalis, mentalis et obliqua.* [Vigevano: s.n., 1680.]

Carboncini, Sonia/Finster, Reinhard: Das Begriffspaar Kanon-Organon. In: *Archiv für Begriffsgeschichte* 26 (1982), pp. 25–59.

Carnap, Rudolf: *Meaning and Necessity. A Study in Semantics and Modal Logic.* Chicago 1947.

Carnap, Rudolf: The Old and the New Logic. In: A.J. Ayer (ed.): *Logical Positivism.* Repr. Westport 1978, pp. 133–146.

Carnap, Rudolf: *Logische Syntax der Sprache.* Wien 1934.

Carnap, Rudolf: *Philosophical Foundations of Physics. An Introduction to the Philosophy of Science.* Ed. by Martin Gardner. New York, London, 1966.

Carnap, Rudolf: Über die Abhängigkeit der Eigenschaften des Raumes von denen der Zeit. In: *Kant-Studien* 30 (1925), pp. 331–345.

Carroll, Lewis: What the Tortoise Said to Achilles. In: *Mind* 4:14 (1895), pp. 278–280.

Carruthers, Mary: *The Book of Memory. A Study of Memory in Medieval Culture*. 2nd ed. Cambridge 2008.

Cartwright, David E.: Locke as Schopenhauer's (Kantian) Philosophical Ancestor. In: *Schopenhauer-Jahrbuch* 84 (2003), pp. 147–156.

Carus, David G.: *Die Gründung des Willensbegriffs. Die Klärung des Willens als rationales Strebevermögen in einer Kritik an Schopenhauer und die Ergründung des Willens in einer Auseinandersetzung mit Aristoteles*. Wiesbaden 2016.

Centrone, Stefania: Der Reziprozitätskanon in den Beyträgen und in der Wissenschaftslehre. In: *Zeitschrift für philosophische Forschung* 64:3 (2010), pp. 310–330.

Chalmers, Margaret; McGonigle, Brandan O.: Are Monkeys Logical?. In: *Nature* 267 (1977), pp. 694–696.

Chasles, Michel: *Geschichte der Geometrie. Hauptsächlich mit Bezug auf die neueren Methoden*. Transl. by. L. A. Sohncke. Halle 1839.

Chemla, Karine: *The History of Mathematical Proof in Ancient Tradition*. Cambridge 2012.

Chomsky, Noam: *Knowledge of Language. Its Nature, Origin, and Use*. New York et al. 1986.

Churchill, John: Wittgenstein's Adaption of Schopenhauer. In: *The Southern Journal of Philosophy* 21 (1983), pp. 489–502.

Ciracì, Fabio/ Fazio, Domenico/ Koßler, Matthias (Eds.): *Schopenhauer und die Schopenhauer-Schule*. Würzburg 2009.

Clegg, Jerry S.: Logical Mysticism and the Cultural Setting of Wittgenstein's Tractatus. In: *Schopenhauer-Jahrbuch* 59 (1978), pp. 29–47 [Transl.: Der logische Mystizismus und der kulturelle Hintergrund von Wittgensteins "Tractatus". In: *Schopenhauer*. Ed. by J. Salaquarda. Darmstadt 1985, pp. 190–218].

Clegg, Jerry S.: Schopenhauer and Wittgenstein on Lonely Languages and Criterialess Claims. In: *Schopenhauer. New Essays in Honor of His 200th Birthday*. Ed. by Eric v. Luft. Lewiston et al. 1988, pp. 82–100.

Clemmius, Henr[icus] Gvil[ielmus]: *Novae amoenitates literariae. Fascicvlvs Qvartvs*. Stvtgardiae 1764.

Cleve, James Van: *Problems from Kant*. Oxford et al. 1999.

Condoravdi, Cleo/Lauer, Sven: Anankastic conditionals are just conditionals. In: *Semantics & Pragmatics* 9:8 (2016), pp. 1–60.

Corcoran, John: Aristotle's Natural Deduction System. In: *Ancient Logic and Its Modern Interpretations. Proceedings of the Buffalo Symposium on Modernist Interpretations of Ancient Logic. 21. and 22. April, 1972*. Ed. by J. Corcoran. Dordrecht 1974, pp. 85–131.

Coseriu, Eugenio: Der Fall Schopenhauer. Ein dunkles Kapitel in der deutschen Sprachphilosophie. In: *Integrale Linguistik. FS für Helmut Gipper*. Ed. by Edeltraut Bülow, Peter Schmitter. Amsterdam 1979, pp. 13–19.

Bibliography

Costa, Newton da: *Logiques classiques et non classiques. Essai sur les fondements de la logique.* Paris 1997.

Costanzo, Jason M.: Schopenhauer on Intuition and Proof in Mathematics. In: *Language, Logic, and Mathematics in Schopenhauer.* Ed. by Jens Lemanski. Cham 2020, pp. 287–305

Costanzo, Jason M.: The Euclidean Mousetrap. Schopenhauer's Criticism of the Synthetic Method in Geometry. In: *Journal of Idealistic Studies* 38:3 (2008), pp. 209–220.

Coumet, E[rnest]: Sur l'histoire des diagrammes logiques, 'figures géométriques'. In: *Mathematiques et Sciences Humaines* 60 (1977), pp. 31–62.

Cox, R./ Stenning, K./ Oberlander, J.: Graphical Effects in Learning Logic. Reasoning, Representation and Individual Differences. In: *Proceedings of the 16th Annual Conference of the Cognitive Science Society, August 13–16, 1994, Cognitive Science Program, Georgia Institute of Technology.* Ed. by A. Ram, K. Eiselt. Hillsdale/ N.J. 1994, pp. 188–198.

Cromwell, Peter/ Beltrami, Elisabetta/ Rampichini, Marta: The Borromean Rings. In: *Mathematical Intelligencer* 20:1 (1998), pp. 53–62.

Croy, Marvin J.: Problem Solving, Working Backwards, and Graphic Proof Representation. In: *Teaching Philosophy* 23 (2000), pp. 169–187.

Crusius, Christian August: *Entwurf der nothwendigen Vernunft=Wahrheiten, wiefern sie den zufälligen entgegen gesetzt werden.* 3rd revised ed. Leipzig 1766.

Davidson, Donald: Quotation. In: *Theory and Decision* 11:1 (1979), pp. 27–40.

Davidson, Donald: Truth and Meaning. In: *Synthese* 17:1 (1967), pp. 304–323 (Repr. in *Inquiries into Truth and Interpretation.* Oxford 1967, pp. 17–42).

Davidson, Donald: What Metaphors Mean. In: *Critical Inquiry* 5:1 (1978), pp. 31–47.

Davies, Hank: Transitive Inference in Rats (Rattus norvegicus). In: *Journal of Comparative Psychology* 106:4 (1992), pp. 342–349.

de Jong, Willem R.: Kant's Analytic Judgments and the Traditional Theory of Concepts. In: *Journal of the History of Philosophy* 33:4 (1995), pp. 613–641

De Soto, C. B.; London, M.; Handel, S.: Social Reasoning and Spatial Paralogic. In: *Journal of Personality and Social Psychology* 2 (1965), pp. 513–521.

Demey, Lorenz: Between Square and Hexagon in Oresme's *Livre du Ciel et du Monde.* In: *History and Philosophy of Logic* 41:1 (2020), pp. 36–47.

Demey, Lorenz: From Euler Diagrams in Schopenhauer to Aristotelian Diagrams in Logical Geometry. In: *Language, Logic, and Mathematics in Schopenhauer.* Ed. by Jens Lemanski. Basel 2020, pp. 181–205.

Denzinger, Ignatius: *Institutiones logicæ.* Vol. II. Leodii 1824.

Deussen, Paul: *Allgemeine Geschichte der Philosophie mit besonderer Berücksichtigung der Religionen. Die neuere Philosophie von Descartes bis Schopenhauer. Vol. II/3: Die neuere Philosophie von Descartes bis Schopenhauer.* Leipzig 1917.

Dierksmeier, Claus: *Der absolute Grund des Rechts. Karl Christian Friedrich Krause in Auseinandersetzung mit Fichte und Schelling.* Stuttgart-Bad Cannstatt 2003.

Diesterweg, F[riedrich] A[dolph] W[ilhelm]: *Leitfaden für den ersten Unterricht in der Formen-Größen- und räumlichen Verbindungslehre oder Vorübungen zur Geometrie für Schulen.* Elberfeld 1822.

Dilthey, Wilhelm: Kants Aufsatz über Kästner und sein Antheil an einer Recension von Johann Schnitz in der Jenaer Literatur-Zeitung. In: *Archiv für Geschichte der Philosophie* 3:2 (1890), pp. 275–281.

Dobrzański, Michał: *Begriff und Methode bei Arthur Schopenhauer.* Würzburg 2017.

Doyle, Tim/ Kutler, Lauren/ Miller, Robin/ Schueller, Albert: Proofs Without Words and Beyond – A Brief History of Proofs Without Words. In: *Convergence* 11 (August 2014).

Dragalin, Albert G.: Proof Theory (art.). In: *Encyclopaedia of Mathematics.* Ed. by M. Hazewinkel. Intern. Ed. in 6 Vols. Dordrecht 1995, Vol. 4, pp. 596–599.

Drobisch, Mauritius Guilielmus: *De calculo logico.* Lipsae, s.a. [1827].

Drobisch, Moritz Wilhelm: *Neue Darstellung der Logik nach ihren einfachsten Verhältnissen. Nebst einem logisch=mathematischen Anhange.* Leipzig 1836.

Dümig, Sascha: Lebendiges Wort? Schopenhauers und Goethes Anschauungen von Sprache im Vergleich. In: *Schopenhauer und Goethe. Biographische und philosophische Perspektiven.* Ed. by Daniel Schubbe, Søren R. Fauth. Hamburg 2016, pp. 150–183.

Dümig, Sascha: The World as Will and I-Language. Schopenhauer's Philosophy as Precursor of Cognitive Sciences. In: *Language, Logic, and Mathematics in Schopenhauer.* Ed. by Jens Lemanski. Basel 2020, pp. 85–95.

Dummett, Michael: *Frege. Philosophy of Language.* New York et al. 1973.

Dummett, Michael: Frege. *Philosophy of Mathematics.* Cambridge/Mass 1991.

Dummett, Michael: *The Interpretation of Frege's Philosophy.* Cambridge/Mass 1980.

Dummett, Michael: The Justification of Deduction. In: Ibid.: *Truth and Other Enigmas.* Duckworth 1978, pp. 290–318.

Dummett, Michael: *The Logical Basis of Metaphysics.* Cambridge/Mass. 1993.

Dummett, Michael: What is a theory of meaning? (II). In: *Truth and Meaning.* Ed. by G. Evans & J. McDowell. Oxford 1976, pp. 34–93.

Dutilh Novaes, Catarina: Surprises in Logic. In: *Logica Yearbook 2009.* Ed. by Michal Peliš. London 2010, pp. 47–63.

Eberhard, Johann August: Berichtigungen einer Stelle in dem phil. Mag. B. I. St. 2. S. 159. mit Beziehung auf H. Prof. Kants Schrift über eine Entdeck. […]. In: *Philosophisches Magazin* 3:2 (1790), pp. 205–211.

Eberhard, Johann August: Ueber die apodiktische Gewisheit. In: *Philosophisches Magazin* 2:2 (1789), pp. 129–186.

Eberhard, Johann August: Von den Begriffen des Raums und der Zeit in Beziehung auf die Gewißheit der menschlichen Erkenntniß. In: *Philosophisches Magazin* 2:1 (1789), pp. 53–92.

Eberhard, Johann August: Ueber die logische Wahrheit oder die transscendentale Gültigkeit der menschlichen Erkenntniß. In: *Philosophisches Magazin* 1:2 (1788), pp. 150–175.

Edwards, A[nthony] W[illiam] F[airbank]: An Eleventh-Century Venn Diagram. In: *BSHM Bulletin: Journal of the British Society for the History of Mathematics* 21:2 (2006), pp. 119–121.

Eidam, Heinz: *Dasein und Bestimmung. Kants Grund-Problem.* Berlin 2000.

Einarson, Benedict: On Certain Mathematical Terms in Aristotle's Logic: Part II. In: *The American Journal of Philology* 57:2 (1936), pp. 151–172.

Ende, Helga: *Der Konstruktionsbegriff im Umkreis des deutschen Idealismus.* Meisenheim am Glan 1973.

Engel, S. Morris: Schopenhauer's Impact on Wittgenstein. In: *Journal of the History of Philosophy* 7:3 (1969), pp. 285–302 [Repr.: *Schopenhauer. His Philosophical Achievement.* Ed. by Michael Fox. Brighton 1980, pp. 236–254].

Englebretsen, George: *Figuring it Out. Logic Diagrams. In Cooperation with José Martin Castro-Manzano and José Roberto Pacheco-Montes.* Berlin, Boston 2020.

Erasmus Roterodamus: *Moriae encomium.* S.l., s.a. [1511].

Erdmann, Benno: *Die Axiome der Geometrie. Eine philosophische Untersuchung der Riemann-Helmholtz'schen Raumtheorie.* Leipzig 1877

Esteve, Maria Rosa Massa: The symbolic treatment of Euclid's Elements in Hérigone's Cursus mathematicus (1634, 1637, 1642). In: *Philosophical Aspects of Symbolic Reasoning in Early Modern Mathematics.* Ed. by. Albrecht Heeffer, Maarten Van Dyck. London 2010, pp. 165–191.

Euler, Leonard: *Lettres à une princesse d'Allemagne sur divers sujets de physique & de philosophie.* 2 vols. Saint Petersbourg 1768.

Feder, Johann Georg Heinrich: *Uber Raum und Caussalität, zur Prüfung der Kantischen Philosophie.* Göttingen 1787.

Ferber, Johann Carl Christoph: *Vernunftlehre.* Helmstädt, Magdeburg 1770.

Fichte, Johann Gottlieb: *Die Grundzüge des gegenwärtigen Zeitalters in Vorlesungen, gehalten zu Berlin, im Jahre 1804–5.* Berlin 1806.

Fine, Kit: *The Limits of Abstraction.* Oxford 2002.

Fischer, Kuno: *Schopenhauers Leben, Werke und Lehre.* (Geschichte der neuern Philosophie IX) 3rd ed. Heidelberg 1908.

Flannery, Kevin L.: *Ways into the Logic of Alexander of Aphrodisias.* Leiden et al. 1995.

Fodor, Jerry A./ LePore, Ernest: *The Compositionality Papers.* Oxford 2002.

Follesa, Laura: From Necessary Truths to Feelings: The Foundations of Mathematics in Leibniz and Schopenhauer. In: *Language, Logic, and Mathematics in Schopenhauer*. Ed. by Jens Lemanski. Cham 2020, pp. 315–326.

Forbes, Morgan: Peirce's Existential Graphs. A Practical Alternative to Truth Tables for Critical Thinkers. In: *Teaching Philosophy* 20 (1997), pp. 387–400.

Forster, Michael N.: Herder's Doctrine of Meaning as use. In: *Linguistic Content: New Essays on the History of Philosophy of Language*. Ed. by Margaret Cameron, Robert J. Stainton. Oxford 2015, pp. 201–222.

Forster, Michael N.: *Wittgenstein on the Arbitrariness of Grammar*. Princeton/N.J 2004.

Fouqueré, Christophe/ Quatrini, Myriam: Ludics and Natural Language. First Approaches. In: *Logical Aspects of Computational Linguistics. LACL 2012*. Ed. by D. Béchet, A. Dikovsky. Berlin, Heidelberg 2012, pp. 21–44.

Frampton, Michael: *Embodiments of Will. Anatomical and Physiological Theories of Voluntary Animal Motion from Greek Antiquity to the Latin Middle Ages, 400 B.C.–A.D. 1300*. Saarbrücken 2008.

Franck, Sebastian: *Das Theür vnd künstlich Büchlin Morie Encomion*. S.l.: s.n., s.a. [ca. 1543].

Frauenstädt, Julius: Eine beachtenswerthe Erscheinung in der Mathematik. In: *Blätter für literarische Unterhaltung* 1852, Nr. 35 (28. August 1852), p. 836.

Frege, Gottlob: *Nachgelassene Schriften und wissenschaftlicher Briefwechsel*. Ed. by Friedrich Kaulbach. 2 vols, 2. rev. ed. Hamburg 1976ff.

Frege, Gottlob: *Logische Untersuchungen*. Ed. by Patzig, 4th ed. Göttingen 1993.

Frege, Gottlob: *Philosophical and Mathematical Correspondence*. Ed. by G. Gabriel, H. Hermes, F. Kambartel, C. Thiel, A. Veraart; trans. by H. Kaal, Chicago, 1980.

Frege, Gottlob: Über die wissenschaftliche Berechtigung einer Begriffschrift [sic!]. In: *Zeitschrift für Philosophie und philosophische Kritik* 81 (1882), pp. 48–56 (Repr. in G. Frege: *Begriffsschrift und andere Aufsätze*. 2nd ed. Hildesheim 2007, pp. 106–114).

Fries, Jakob Friedrich: Brief an Jacobi 10.12.1807. In: *Hegel in Berichten seiner Zeitgenossen*. Ed. by Günther Nicolin. Hamburg 1970.

Fries, Jakob Friedrich: *Die Geschichte der Philosophie dargestellt nach den Fortschritten ihrer wissenschaftlichen Entwicklung*. Vol. 1. Halle 1837.

Fries, Jakob Friedrich: *System der Logik. Ein Handbuch für Lehrer und zum Selbstgebrauch*. Heidelberg 1811.

Gabriel, Gottfried: Vorwort. In: Hermann Lotze: *Logik III. Vom Erkennen (Methodologie)*. Ed. by Gottfried Gabriel. Hamburg 1989.

Gabriel, Gottfried: Windelband und die Diskussion um die Kantischen Urteilsformen. In: *Kant im Neukantianismus. Fortschritt oder Rückschritt?*. Ed. by Marion Heinz, Christian Krijnen. Würzburg 2007, pp. 91–109.

Gardner, Martin: *Logic Machines and Diagrams*. New York, Toronto et al. 1958.

Gardner, Martin: *Sixth Book of Mathematical Games from Scientific American*. New York 1975.

Garewicz, Jan: Erkennen und Erleben: Ein Beitrag zu Schopenhauers Erlösungslehre, in: 70. *Schopenhauer-Jahrbuch* (1989), pp. 75–83.

Geach, Peter: Mental Acts. In: Ibid.: *Their Content and their Objects*. London 1957.

Gentzen, Gerhard: Investigation into Logical Deduction. In: *The Collected Papers of Gerhard Gentzen*. Ed. by M. E. Szabo. North-Holland, Amsterdam 1969, pp. 68–131.

Gillan, Douglas: Reasoning in the Chimpanzee: II. Transitive Inference. In: *Journal of Experimental Psychology: Animal Behavior Processes* 7:2 (1981), pp. 150–164.

Glock, Hans-Johann: Schopenhauer and Wittgenstein. Representation as Language and Will. In: *The Cambridge Companion to Schopenhauer*. Ed. by Christopher Janaway. Cambridge 1999, pp. 422–458.

Göcke, Benedikt Paul: Karl Christian Friedrich Krause Einfluss auf Arthur Schopenhauers "Die Welt als Wille und Vorstellung". In: *Archiv für Geschichte der Philosophie* 103:1 (2021), pp. 148–168.

Göcke, Benedikt Paul: *The Panentheism of Karl Christian Friedrich Krause (1781–1832). From Transcendental Philosophy to Metaphysics*. Berlin 2018.

Goldin, Owen: *Explaining an Eclipse. Aristotle's Posterior Analytics 2.1–10*. Ann Arbor 1996.

Goodman, Nelson: The New Riddle of Induction. In: Ibid.: *Facts, Fiction, and Forecast*. 4th ed. Cambridge/ Mass. 1983, pp. 59–84.

Goodman, Russell: Schopenhauer and Wittgenstein on Ethics. In: *The Journal of the History of Philosophy* (1979), pp. 437–447.

Gottsched, Johann Chr.: *Erste Gründe der gesammten Weltweisheit: darinn alle philosophische Wissenschaften in ihrer natürlichen Verknüpfung abgehandelt werden, zum Gebrauche academischer Lectionen*. 2nd ed. Leipzig 1735.

Goy, Ina: *Architektonik oder die Kunst der Systeme*. Paderborn 2007.

Grattan-Guinness, Ivor: Numbers, Magnitudes, Ratios, and Proportions in Euclid's Elements: How Did He Handle Them?. In: *Historia Matematica* 23 (1996), pp. 355–375.

Greaves, Mark: *The Philosophical Status of Diagrams*. Stanford 2002.

Grice, H[erbert] P[aul]/ Strawson P[eter] F[rederick]: In Defense of a Dogma. In: *The Philosophical Review* 65:2 (1956), pp. 141–158.

Griffiths, A. Phillips: Wittgenstein, Schopenhauer, and Ethics. In: *Royal Institute of Philosophy Lectures* 7 (1973), pp. 96–116 (Repr.: *Understanding Wittgenstein*. Ed. by G. Vesey. Ithaca 1974, pp. 96–116).

Griffiths, A. Phillips: Wittgenstein and the Four-Fold Root of the Principle of Suffucient Reason. In: *Aristotelian Society Supplementary Volume* 50:1-2 (1976), pp. 1–20.

Großer, Samuel: *Gründliche Anweisung zur Logica [...]*. Budißin, Görlitz 1697.

Grosserus, Samuel: *Pharus Intellectus, sive Logica Electiva.* Lipsiae 1697.

Gruppe, Otto Friedrich: *Wendepunkt der Philosophie im neunzehnten Jahrhundert.* Berlin 1834.

Gullberg, Ebba/ Lindström, Sten: Semantics and the Justification of Deductive Inference. In: *Hommage à Wlodek. Philosophical Papers Dedicated to Wlodek Rabinowicz.* Ed. by T. Rønnow-Rasmussen, B. Petersson, J. Josefsson, D. Egonsson. S.l. 2007. (www.fil.lu.se/hommageawlodek [letzter Abruf 11.01.2017]).

Gusserow, Carl: *Leitfaden für den Unterricht in der Stereometrie mit den Elementen der Projektionslehre.* Berlin 1885.

Haack, Susan: The Justification of Deduction. In: *Mind* 85:337 (1976), pp. 112–119.

Hacker, P[eter] M[ichael] S[tephan]: *Insight and Illusion. Themes in the Philosophy of Wittgenstein.* Revised ed. Oxford 1986.

Hacker, Peter M[ichael] S[tephan]: The Rise of Twentieth Century Analytic Philosophy. In: *Ratio* 9:3 (1996), pp. 243–268.

Hahn, Hans: The Crisis in Intuition. In: *Empiricism, Logic and Mathematics. Vienna Circle Collection, Vol 13.* Ed. by B. McGuinness. Dordrecht, 1980, pp. 73–102.

Hallett, Gareth: *Companion to Wittgenstein's Philosophical Investigations.* Ithaca 1977.

Hamblin, C[harles] L[eonhard]: An Improved Pons Asinorum?. In: *Journal of the History of Philosophy* 14:2 (1976), pp. 131–136.

Hamilton, William: Discussions on Philosophy and Literature, Education and University Reform. Chiefly from the Edinburgh Review; Corrected, Vindicated, Enlarges in Notes and Appendices. 2nd revised ed. London 1853.

Hamilton, William: *Lectures on Metaphysics and Logic.* Ed. by H. L. Mansel, J. Veitch. 4 vols. London 1860.

Hammer, Eric M.: *Logic and Visual Information.* Stanford 1995.

Hammer, Eric M./Shin, Sun-Joo: Euler's Visual Logic. In: *History and Philosophy of Logic* 19:1 (1998), pp. 1–29.

Han, Linhe: Wittgenstein and Schopenhauer. In: *Wittgenstein and the Future of Philosophy. A Reassessment after 50 Years / Wittgenstein und die Zukunft der Philosophie. Eine Neubewertung nach 50 Jahren.* Ed. by R. Halle, K. Puhl. Wien 2002, pp. 112–121.

Hanna, Robert: *Kant and the Foundations of Analytic Philosophy.* Oxford et al. 2001.

Harari, Orna: John Philoponus and the Conformity of Mathematical Proofs to Aristotelian Demonstrations. In: *The History of Mathematical Proof in Ancient Tradition.* Ed. by Karine Chemla. Cambridge 2012, pp. 206–228.

Hartmann, Eduard von: *Phänomenologie des sittlichen Bewusstseins. Prolegomena zu jeder künftigen Ethik.* Berlin 1879.

Hasse, Heinrich: [Review of] Schopenhauer, Arthur. Handschriftlicher Nachlass: "Philosophische Vorlesungen." Arthur Schopenhauers sämtliche Werke, hrsg. v.

Paul Deussen. München 1913. Bd. IX und X. In: *Kant-Studien* 19 (1914), pp. 270–272.

Hasse, Heinrich: *Schopenhauers Erkenntnislehre als System einer Gemeinschaft des Rationalen und Irrationalen. Ein historisch-kritischer Versuch*. Leipzig 1913.

Hauswald, Rico: Umfangslogik und analytisches Urteil bei Kant. In: *Kant-Studien* 101:3 (2010), pp. 283–308.

Heinemann, Anna-Sophie: 'Horrent with Mysterious Spiculæ'. Augustus De Morgan's Logic Notation of 1850 as a 'Calculus of Opposite Relations'. In: *History and Philosophy of Logic* 39:1 (2018), pp. 29–52.

Heinemann, Anna-Sophie: Schopenhauer and the Equational Form of Predication. In: *Language, Logic, and Mathematics in Schopenhauer*. Ed. by Jens Lemanski. Basel 2020, pp. 165–181.

Hennigfeld, Jochem: Metaphysik und Anthropologie des Willens. Methodische Anmerkungen zur Freiheitsschrift und zur *Welt als Wille und Vorstellung*. In: *Die Ethik Arthur Schopenhauers im Ausgang vom Deutschen Idealismus (Fichte/Schelling)*. Ed. by Lore Hühn. Würzburg 2006, pp. 459–472.

Heßler, Martina/ Mersch, Dieter: Bildlogik oder Was heißt visuelles Denken?. In: *Logik des Bildlichen. Zur Kritik der ikonischen Vernunft*. Ed. by Martina Heßler, Dieter Mersch. Bielefeld 2009, pp. 8–62.

Hilbert, David: Neubegründung der Mathematik. Erste Mitteilung. In: *Abhandlungen aus dem Mathematischen Seminar der Universität Hamburg* 1 (1922), pp. 157–177.

Hilbert, David/ Ackermann, Wilhelm: The Principles of Mathematical Logic. Transl. by L.M. Hammond, G.G. Leckie, F. Steinhardt. New York 1950.

Hintikka, Jaakko: On the Logic of Perception. In: *Models for Modalities. Selected Essays IV*. Ed. by Jaakko Hintikka. Dordrecht et al. 1969, pp. 151–183.

Hodges, Wilfrid: A Correctness Proof for al-Barakāt's Logical Diagrams. In: *The Review of Symbolic Logic* (forthc.).

Hodges, Wilfrid: Formalizing the Relationship between Meaning and Syntax. In: *The Oxford Handbook of Compositionality*. Ed. by Markus Werning, Wolfram Hinzen, Edouard Machery. Oxford 2012, pp. 245–261.

Hodges, Wilfrid: Medieval Arabic Notions of Algorithm. Some Further Raw Evidence. In: *Fields of Logic and Computation III*. Ed. by A. Blass, P. Cégielski, N. Dershowitz, M. Droste, B. Finkbeiner (LNCS, vol. 12180). Cham 2020, pp. 133–146.

Hodges, Wilfrid: Remarks on Compositionality. In: *Dependence Logic: Theory and Applications*. Ed. by Samson Abramsky, Juha Kontinen, Jouko Väänänen, Heribert Vollmer. Cham 2016, pp. 99–107.

Hodges, Wilfrid: Two Early Arabic Applications of Model-Theoretic Consequence. In: *Logica Universalis* 12 (2018), pp. 37–54.

Höffding, Harald: *Geschichte der neueren Philosophie. Eine Darstellung der Geschichte der Philosophie von dem Ende der Renaissance bis zum Schlusse des 19. Jahrhunderts. Vol. II.* Leipzig 1896.

Höfler, Alois/ Meinong, Alexius: *Logik*. Prague, Vienna, Leipzig 1890.

Höfler, Alois: *Logik*. 2nd rev. ed. Vienna, Leipzig 1922.

Hoffmann, Volkmar: Hebung eines Missverständnisses. In: *Zeitschrift für mathematischen und naturwissenschaftlichen Unterricht* 16:4 (1885), pp. 263–264.

Hoffmann, Volkmar: Schopenhauer, der Philosoph, über den Wert des Calcüls. In: *Zeitschrift für mathematischen und naturwissenschaftlichen Unterricht* 16:4 (1885), p. 186.

Hoffmann, Volkmar: Schopenhauer, der Philosoph, über die Euklidische Methode und die 'Mausefallenbeweise'. In: *Zeitschrift für mathematischen und naturwissenschaftlichen Unterricht* 16:3 (1885), pp. 105–107.

Holland, Georg Jonathan: *Abhandlung über die Mathematik, die allgemeine Zeichenkunst und die Verschiedenheit der Rechnungsarten. Nebst einem Anhang, worinnen die von Hrn. Prof. Ploucquet erfundene logikalische Rechnung gegen die Leipziger neue Zeitungen erläutert und mit Hrn. Prof. Lamberts Methode verglichen wird*. Tübingen 1764.

Hörnig, Robin: *Eigennamen referieren – Referieren mit Eigennamen. Zur Kontextinvarianz der namentlichen Bezugnahme*. Wiesbaden 2003.

Hösle, Vittorio: Zum Verhältnis von Metaphysik des Lebendigen und allgemeiner Metaphysik. Betrachtungen in kritischem Anschluss an Schopenhauer. In: *Metaphysik. Herausforderungen und Möglichkeiten*. Ed. by Vittorio Hösle. Stuttgart-Bad Cannstatt 2002, pp. 59–97.

Hübscher, Arthur: *Denker gegen den Strom. Schopenhauer: Gestern – Heute – Morgen*. Bonn 1973.

Hübscher, Arthur: *Schopenhauer. Biographie eines Weltbildes*. Stuttgart 1952.

Ierodiakonou, Katerina: Psellos' Paraphrasis on De interpretation. In: *Byzantine Philosophy and its Ancient Sources*. Ed. by Katerina Ierodiakonou. Oxford 2004, pp. 157–183.

Ingenkamp, Heinz G.: Plutarch und das Leben der Heiligen. In: *Valori letterari delle opere di Plutarco*. Ed. by Aurelio Pérez Jiménez, Frances Bonner Titchener. Málaga 2005, pp. 225–242.

Jacobi, Friedrich H.: *Über die Lehre des Spinoza in Briefen an den Herrn Moses Mendelssohn*. Hamburg 2000.

Jacquette, Dale: Schopenhauer's Philosophy of Logic and Mathematics. In: *A Companion to Schopenhauer*. Ed. by Bart Vandenabeele. Hoboken 2012, pp. 41–59.

Jahnke, Hans Niels: *Mathematik und Bildung in der Humboldtschen Reform*. Göttingen 1990.

Jamnik, Mateja/ Bundy, Alan/ Green, Ian: On Automating Diagrammatic Proofs of Arithmetic Arguments. In: *Journal of Logic, Language, and Information* 8:3 (1999), pp. 297–321.

Janaway, Christopher: Introduction. In: *The Cambridge Companion to Schopenhauer*. Ed. by Christopher Janaway. Cambridge 1999, pp. 1–17.

Janaway, Christopher: *Self and World in Schopenhauer's Philosophy*. Clarendon 1989.

Janik, Allan S.: Schopenhauer and the Early Wittgenstein. In: Ibid.: *Essays on Wittgenstein and Weininger*. Amsterdam 1985, pp. 26–48 (Orig.: *Philosophical Studies* 15 (1966), pp. 76–95).

Janik, Allan S.: On Schopenhauer's Relationship to Wittgenstein. In: *Zeit der Ernte. Studien zum Stand der Schopenhauer-Forschung*. Ed. by Wolfgang Schirrmacher. Stuttgart-Bad Cannstatt 1982, pp. 271–279.

Janik, Allan S.: Wie hat Schopenhauer Wittgenstein beeinflußt?. In: *Schopenhauer-Jahrbuch* 73 (1992), pp. 69–78.

Janssen, Theo M. V.: Compositionality. Its Historic Context. In: *The Oxford Handbook of Compositionality*. Ed. by Markus Werning, Wolfram Hinzen, Edouard Machery. Oxford 2012, pp. 19–46.

Janssen, Theo M. V.: Frege, Contextuality and Compositionality. In: *Journal of Logic, Language, and Information* 10 (2001), pp. 115–136.

Janzen, Oscar: Schopenhauers. Auffassung des Verhältnisses der mathematischen Begründung zur logischen. In: *Archiv für Geschichte der Philosophie* 22 (1909), pp. 342–364.

Jenson, Otto: *Die Ursache der Widersprüche im Schopenhauerschen System*. Rostock 1906.

Jong, Willem R. de: Kant's Analytic Judgments and the Traditional Theory of Concepts. In: *Journal of the History of Philosophy* 33:4 (1995), pp. 613–641.

Johnson-Laird, P. N.: Reasoning without Logic. In: *Reasoning and Discourse Processes*. Ed. by T. Myers, K. Brown, B. McGonigle. London 1986, pp. 13–50.

[Ps.-]Joslenus Suessionensis: De generibus et speciebus. In: *Ouvrages inédits d'Abélard*. Ed. by Victor Cousin. Paris 1836.

Juhl, Cory/ Loomis, Eric: *Analyticity*. New York et al. 2010.

Kaestner, Abraham Gotthelf: *Geschichte der Mathematik seit der Wiederherstellung der Wissenschaften bis an das Ende des achtzehnten Jahrhunderts. Vol. 1. Arithmetik, Algebra, Elementargeometrie, Trigonometrie, Praktische Geometrie bis zum Ende des sechzehnten Jahrhunderts*. Göttingen 1796.

Kästner, Abraham Gotthelf: Ueber den mathematischen Begriff des Raums. In: *Philosophisches Magazin* 2:4 (1790), pp. 403–429.

Kästner, Abraham Gotthelf: Was heißt, in Euklids Geometrie möglich?. In: *Philosophisches Magazin* 2:4 (1790), pp. 391–402.

Kakridis, Ioannis: *Codex 88 des Klosters Dečani und seine griechischen Vorlagen. Ein Kapitel der serbisch-byzantinischen Literaturbeziehungen im 14. Jahrhundert*. München et al. 1988.

Kant, Immanuel: *Correspondence*. Transl. and ed. by Arnulf Zweig. Cambridge/Mass. 1999.

Kant, Immanuel: *Kant's Critical Philosophy, Vol. II. The Prolegomena*. Ed. and Transl. by J. P. Mahaffy, J. H. Bernard. London 1989.

Kant, Immanuel: *Logik. Ein Handbuch zu Vorlesungen*. Ed. by G. B. Jäsche. Königsberg 1800.

Kant, Immanuel: *Lectures on Logic*. Ed. and transl. by J. Michael Young. Cambridge 1992.

Katz, Jerrold J.: Analyticity and Contradiction in Natural Language. In: *The Structure of Language*. Ed. by Jerry A. Fodor, Jerrold J. Katz. Prentice-Hall 1964, pp. 519–543.

Katz, Jerrold J.: *Cogitations. A Study of the Cogito in Relation to the Philosophy of Logic and Language and a Study of Them in Relation to the Cogito*. Oxford et al. 1988.

Katz, Jerrold J.: Some Remarks on Quine on Analyticity. In: *The Journal of Philosophy* 64:2 (1967), pp. 36–52.

Kautsky, Karl: Arthur Schopenhauer (Schluß). In: *Die neue Zeit. Revue des geistigen und öffentlichen Lebens* 6:3 (1888), pp. 97–109.

Kawamura, Katsutoshi: Eine Wurzel der Vierfachen Wurzel des Satzes vom zureichenden Grund Schopenhauers. Schopenhauer und Crusius. In: *Schopenhauers Wissenschaftstheorie. Der "Satz vom Grund"*. Ed. by Dieter Birnbacher. Würzburg 2015, pp. 59–74.

Keckermannus, Bartholomæus: *Systema Logicæ. Compendiosa methodo [...]*. Hanoviae 1601.

Keckermannus, Bartholomæus: *Systema Logicæ. Tribus Libris Adornatvm [...]*. Hanoviae 1611.

Kellert, Stephen H.: *Borrowed Knowledge. Chaos Theory and the Challenge of Learning Across Disciplines*. Chicago 2008.

Keutner, Thomas/ Gehring, Petra (Eds.): *Diagrammatik und Philosophie. 1. Interdisziplinäres Kolloquiums der Forschungsgruppe Philosophische Diagrammatik, 15./16.12.1988 an der FernUniversität/Gesamthochschule Hagen*. Amsterdam 1992.

Kewe, Adolf: *Schopenhauer als Logiker*. Bonn 1907.

Kienzler, Wolfgang: *Begriff und Gegenstand. Eine historische und systematische Studie zur Entwicklung von Gottlob Freges Denken*. Frankfurt/ Main 2009.

Kiesewetter, Johann Gottfried: *Grundriß einer reinen allgemeinen Logik nach kantischen Grundsätzen [...]*. Frankfurt 1793.

Klamp, Gerhard: Das Streitgespräch zwischen Becker und Schopenhauer. In: *Schopenhauer-Jahrbuch* 39 (1958), pp. 38–75.

Klamp, Gerhard: Die Architektonik im Gesamtwerk Schopenhauers. In: *Schopenhauer-Jahrbuch* 41 (1960), pp. 82–98.

Klamp, Gerhard: Vom Symbolgebrauch geometrischer Figuren in der Logik. In: *Schopenhauer-Jahrbuch* 33 (1949/-50), pp. 39–65.

Klamp, Gerhard: Zur Zeit- und Wirkungsgeschichte Schopenhauers. In: *Schopenhauer-Jahrbuch* 40 (1959), pp. 1–23.

Kleemeier, Ulrike: *Gottlob Frege. Kontext-Prinzip und Ontologie*. Freiburg i. Br. 1997.

Kleemeier, Ulrike/ Weidemann, Christian: Brandom and Frege. In: *Robert Brandom. Analytic Pragmatist*. Ed. by Bernd Prien, David P. Schweikard. Heusenstamm 2008, pp. 115–125.

Klein, Felix: *Elementarmathematik vom höheren Standpunkte aus. Vol. II: Geometrie*. Ed. by E. Hellinger. Berlin 1909.

Kneale, William/Kneale, Martha: *The Development of Logic*. Revised ed. Oxford et al. 1971 (Repr.).

Knobloch, Eberhard: Leonhard Euler als Theoretiker. In: *Mathesis & Graphe. Leonhard Euler und die Entfaltung der Wissensysteme*. Ed. by Wladimir Velminski, Horst Bredekamp. Berlin 2010, pp. 19–36.

Knorr, Wilbur Richard: On the Early History of Axiomatics. The Interaction of Mathematics and Philosophy in Greek Antiquity. In: *Theory Change, Ancient Axiomatics, and Galileo's Methodology. Proceedings of the 1978 Pisa Conference on the History and Philosophy of Science. Vol. 1*. Ed. by Jaakko Hintikka, D. Gruender, Evandro Agazzi. London et al. 1982, pp. 145–187.

Кобзарь, Владимир Иванович: Элементарная логика Л. Эйлера. In: *Логико-философские штудии* [*Logiko-filosofskie studii*] 3 (2005), pp. 130–152.

Кобзарь, Владимир Иванович: Гносеология и логика Л. Эйлера в "Письмах к немецкой принцессе о разных физических и философских материях". In: *Логико-философские штудии* [*Logiko-filosofskie studii*] 8 (2010), pp. 98–120.

Koetsier, Teun: Arthur Schopenhauer and L.E.J. Brouwer. A Comparison. In: *Mathematics and the Divine. A Historical Study*. Ed. by L. Bergmans, T. Koetsier. Amsterdam et al. 2005, pp. 571–595.

Köhnke, Klaus Christian: *Entstehung und Aufstieg des Neukantianismus. Die deutsche Universitätsphilosophie zwischen Idealismus und Positivismus*. Frankfurt/ Main 1986.

Körber, C[hristian] A[lbrecht]: *Archimedes defensus. Das ist Gründlicher Beweiß Daß das Theorema Archimedis Von der Verhältniß der Kugel zum Cylinder, So beyde einerley Höhe und Grund-Fläche haben, nicht solo oculorum usu, wie einige meynen, könne erfunden werden*. […]. Halle 1731.

Koriako, Darius: *Kants Philosophie der Mathematik. Grundlagen – Voraussetzungen – Probleme*. Hamburg 1999.

Körner, Stephen: *Kant*. Baltimore/Maryland 1955.

Kosack, C[arl] R[udolf]: Beiträge zu einer systematischen Entwickelung der Geometrie aus der Anschauung. In: *Zu der öffentlichen Prüfung sämmtlicher Klassen des Gymnasiums zu Nordhausen* […]. Nordhausen 1852, pp. 1–31.

Koßler, Matthias: Die eine Anschauung – der eine Gedanke. Zur Systemfrage bei Fichte und Schopenhauer. In: *Die Ethik Arthur Schopenhauers im Ausgang vom Deutschen Idealismus (Fichte/Schelling)*. Ed. by Lore Hühn. Würzburg 2006, pp. 349–364.

Koßler, Matthias: *Empirische Ethik und christliche Moral. Zur Differenz einer areligiösen und einer religiösen Grundlegung der Ethik am Beispiel der Gegenüberstellung Schopenhauers mit Augustinus, der Scholastik und Luther*. Würzburg 1999.

Koßler, Matthias: Schopenhauer als Philosoph des Übergangs. In: *Nietzsche und Schopenhauer. Rezeptionsphänomene der Wendezeiten*. Ed. by Marta Kopij, Wojciech Kunicki. Leipzig 2006, pp. 365–379.

Krämer, Sybille: Tatsachenwahrheiten und Vernunftwahrheiten. In: *Gottfried Wilhelm Leibniz: Monadologie*. Ed. by Hubertus Busche. Berlin 2009, pp. 95–111.

Kratochwil, Stefan: Johann Christoph Sturm und Gottfried Wilhelm Leibniz. In: *Johann Christoph Sturm (1635 – 1703)*. Ed. by Hans Gaab, Pierre Leich, Günter Löffladt. Frankfurt/ Main 2004, pp. 104–119.

Krause, Karl Christian F[riedrich]: *Die Lehre vom Erkennen und von der Erkenntniss, als erste Einleitung in die Wissenschaft*. Ed. by Hermann Karl von Leonhardi. Göttingen 1835.

Krause, Karl Christian Friedrich: *Grundriss der historischen Logik für Vorlesungen*. Jena et al. 1803.

Kreiser, Lothar: *Gottlob Frege. Leben, Werk, Zeit*. Hamburg 2004.

Krewet, Michael: *Zum Wissenstransfer in Ammonios' Kommentierung des neunten Kapitels von Aristoteles'* De Interpretatione. (Working Paper des SFB 980 Episteme in Bewegung). Berlin 2019.

Krug, Wilhelm Traugott: *Briefe über den neuesten Idealismus. Eine Fortsetzung der Briefe über die Wissenschaftslehre*. Leipzig 1801.

Krug, Wilhelm Traugott: *Logik oder Denklehre (System der theoretischen Philosophie I)*. Königsberg 1806.

Krüger, Lorenz: Wollte Kant die Vollständigkeit seiner Urteilstafel beweisen?. In: *Kant-Studien* 59:4 (1968), pp. 333–356.

Künne, Wolfgang: *Die philosophische Logik Gottlob Freges. Ein Kommentar*. Frankfurt/ Main 2010.

La Grange, M. de: *Méchanique analytique*. Paris 1788.

Bibliography

Lakatos, Imre: *Proofs and Refutations. The Logic of Mathmatical Discovery*. Ed. by John Worrall, Elie Zahar. Repr. Cambridge/Mass. 2015.

Lakoff, George/ Johnson, Mark: *Metaphors We Lived By*. Chicago 1980.

Lakoff, George: *Women, Fire, and Dangerous Things. What Categories Reveal About the Mind*. Chicago 1987.

Lambert, Johann Heinrich: *Anlage zur Architectonic, oder Theorie des Einfachen und des Ersten in der philosophischen und mathematischen Erkenntniß*. 2 vols. Riga 1771.

Lambert, Johann Heinrich: *Joh[ann] Heinrich Lamberts deutscher gelehrter Briefwechsel*. Ed. by Joh[ann II.] Bernoulli. Berlin s.a. [1782].

Lambert, J[ohann] H[einrich]: *Neues Organon oder Gedanken über die Erforschung und Bezeichnung des Wahren und dessen Unterscheidung vom Irrthum und Schein*. 2 vols. Leipzig 1764.

Lando, Giorgio: Assertion and Affirmation in the Early Wittgenstein. In: *Wittgenstein-Studien* 2 (2011), pp. 21–49.

Lange, Ernst Michael: *Wittgenstein und Schopenhauer. Logisch-philosophische Abhandlung und Kritik des Solipsismus*. Cuxhaven 1989.

Lange, Friedrich Albert: *Logische Studien. Ein Beitrag zur Neubegründung der formalen Logik und der Erkenntnistheorie*. Iserlohn 1877.

[Lange, Johann Christian:] Ausführliche Vorstellung von einer neuen und gemeinersprießlichen zu beßtem Behuf und Auffnahm Aller wahren und rechtschaffenen Gelehrtheit gereichenden Anstalt [...]. Idstein: Lyce, 1720.

Langius, Iohannes Christianus: *Inuentum Nouum Quadrati Logici Vniversalis*. Gissae-Hassorum 1714.

Langius, Iohannes Christianus: *Nvclevs Logicae Weisianae*. [...] *illustrates* [...] *per varias schematicas* [...] *ad ocularem evidentiam deducta* [...]. Editus antehac Avctore Christiano Weisio, Gissae-Hassorum 1712.

Larkin, Jill H./ Simon, Herbert A.: Why a Diagram is (Sometimes) Worth Ten Thousand Words. In: *Cognitive Science* 11:1 (1987), pp. 65–100.

Legg, Catherine: What is a Logical Diagram?. In: *Visual Reasoning with Diagrams*. Ed. by Sun-Joo Shin, Amirouche Moktefi. Basel 2013, pp. 1–18.

Lehmann, Rudolf: *Schopenhauer. Ein Beitrag zur Psychologie der Metaphysik*. Berlin 1894.

Leibniz, Gottfried Wilhelm: *Essais de théodicée sur la bonté de Dieu, la liberté de l'homme et l'origine du mal*. Amsterdam 1714.

Leibniz, Gottfried Wilhelm: De formæ logicæ per linearum ductus. In: Ibid.: *Opuscules et fragments inédits de Leibniz. Extraits des manuscrits de la Bibliothegue royale de Hanovre*. Ed. by Louis Couturat. Paris 1903, pp. 292–321.

Leibniz, Gottfried Wilhelm: *Sämtliche Schriften und Briefe*. Ed. by Preußische/ Deutsche/ Göttinger/ Berlin-Brandenburgische Akademie der Wissenschaften. Darmstadt et al. 1923ff.

Lemanski, Jens/ Demey, Lorenz: Schopenhauer's Partition Diagrams and Logical Geometry. In: *Diagrammatic Representation and Inference. Diagrams 2021.* LNCS, vol. 12909. Ed. by A. Basu, G. Stapleton, S. Linker, C. Legg, E. Manalo, P. Viana. Cham, 2021, pp. 149–165.

Lemanski, Jens/ Jansen, Ludger: Calculus *CL* as a Formal System. In: *Diagrammatic Representation and Inference. Diagrams 2020.* LNCS, vol. 12169. Ed. by A.-V. Pietarinen, P. Chapman, L. Bosveld-de Smet, V. Giardino, J. Corter, S. Linker. Cham 2020, pp. 445–460.

Lemanski, Jens: Calculus *CL* – From Baroque Logic to Artificial Intelligence. In: *Logique et Analyse* 249 (2020), pp. 111–129.

Lemanski, Jens: Calculus *CL* as Ontology Editor and Inference Engine. In: *Diagrammatic Representation and Inference 10th International Conference, Diagrams 2018, Edinburgh, UK, June 18–22,* 2018, Proceedings. Ed. by P. Chapman, G. Stapleton, A. Moktefi, S. Perez-Kriz & F. Bellucci. Cham 2018, pp. 752–756.

Lemanski, Jens: Concept Diagrams and the Context Principle. In: *Language, Logic, and Language in Schopenhauer.* Hrsg. v. Jens Lemanski. Cham 2020, pp. 47–73.

Lemanski, Jens: 'Cur potius aliquid quam nihil' von der Frühgeschichte bis zur Hochscholastik. In: *Warum ist überhaupt etwas und nicht nichts? Wandel und Variationen einer Frage.* Ed. by Daniel Schubbe, Jens Lemanski, Rico Hauswald. Hamburg 2013, pp. 23–65.

Lemanski, Jens: *Christentum im Atheismus. Spuren der mystischen Imitatio Christi-Lehre in der Ethik Schopenhauers.* 2 Vols. London 2009/ 2011.

Lemanski, Jens: Christentum und Mystik. In: *Schopenhauer-Handbuch. Leben – Werk – Wirkung.* Ed. by Daniel Schubbe, Matthias Koßler. Weimar 2014, pp. 201–207.

Lemanski, Jens: Die 'Evolutionstheorien' Goethes und Schopenhauers. Eine kritische Aufarbeitung des wissenschaftsgeschichtlichen Forschungsstandes. In: *Schopenhauer und Goethe. Biographische und philosophische Perspektiven.* Ed. by Daniel Schubbe, Søren R. Fauth. Hamburg 2016, pp. 247–295.

Lemanski, Jens: Die Königin der Revolution. Zur Rettung und Erhaltung der Kopernikanischen Wende. In: *Kant-Studien* 103:4 (2012), pp. 448–471.

Lemanski, Jens: Die neuaristotelischen Ursprünge des Kontextprinzips und die Fortführung in der fregeschen Begriffsschrift. In: *Zeitschrift für philosophische Forschung* 67:4 (2013), pp. 566–587.

Lemanski, Jens: Die Rationalität des Mystischen. In: *Schopenhauer-Jahrbuch* 91 (2010), pp. 93–120.

Lemanski, Jens: Euler-type Diagrams and the Quantification of the Predicate. In: *Journal of Philosophical Logic* 49 (2020), pp. 401–416.

Lemanski, Jens: Extended Syllogistics in Calculus *CL*. In: *Journal of Applied Logics* 8:2 (2021), pp. 557–577.

Lemanski, Jens: Galilei, Torricelli, Stahl – Zur Wissenschaftsgeschichte der Physik in der B-Vorrede zu Kants *Kritik der reinen Vernunft*. In: *Kant-Studien* 107:3 (2016), pp. 451–484.

Lemanski, Jens: Geometrie. In: *Schopenhauer-Handbuch. Leben – Werk – Wirkung*. Ed. by Daniel Schubbe, Matthias Koßler. 2nd ed. Weimar, Stuttgart 2018, pp. 331–335.

Lemanski, Jens: Logic Diagrams, Sacred Geometry and Neural Networks. In: *Logica Universalis* 13 (2019), pp. 495–513.

Lemanski, Jens: Logic Diagrams in the Weise and Weigel Circles. In: *History and Philosophy of Logic* 39:1 (2018), pp. 3–28.

Lemanski, Jens: Logik und Eristische Dialektik. In: *Schopenhauer-Handbuch. Leben – Werk – Wirkung*. Ed. by Daniel Schubbe, Matthias Koßler. 2nd ed. Weimar, Stuttgart 2018, pp. 160–169.

Lemanski, Jens: Logikdiagramme und Logikmaschinen aus der Zittauer Schule um Christian Weise. In: *Neues Lausitzische Magazin* 141:1 (2019), pp. 39–57.

Lemanski, Jens: Means or End? On the Valuation of Logic Diagrams. In: *Логико-философские штудии* [*Logiko-filosofskie studii*] 14:2 (2017), pp. 98–122.

Lemanski, Jens: Periods in the Use of Euler-Type Diagrams. In: *Acta Baltica Historiae et Philosophiae Scientiarum* 5:1 (2017), pp. 50–69.

Lemanski, Jens: Philosophia in bivio – Über die Bedeutung des Fragmentenstreits für die Ausdifferenzierung von Rationalismus und Irrationalismus. In: *Georg Lukács. Kritiker der unreinen Vernunft*. Ed. by Britta Caspers, Christoph J. Bauer. Duisburg 2009, pp. 85–107.

Lemanski, Jens: Schopenhauer's World: The System of The World as Will and Presentation I. In: *Schopenhaueriana. Revista española de estudios sobre Schopenhauer* 2 (2017), pp. 297–315.

Lemanski, Jens: Schopenhauers Gebrauchstheorie der Bedeutung und das Kontextprinzip. Eine Parallele zu Wittgensteins Philosophischen Untersuchungen. In: *Schopenhauer-Jahrbuch* 97 (2016), pp. 171–196.

Lemanski, Jens: Schopenhauers hagioethischer Konsequentialismus im System der Welt als Wille und Vorstellung. In: 93. *Schopenhauer-Jahrbuch* (2012), pp. 485–503.

Lemanski, Jens: *Summa und System. Historie und Systematik vollendeter bottom-up- und top-down-Theorien*. Münster 2013.

Lemanski, Jens: The Denial of the Will-to-Live in Schopenhauer's World and his Association of Buddhist and Christian Saints. In: *Understanding Schopenhauer through the Prism of Indian Culture. Philosophy, Religion and Sanskrit Literature*. Ed. by Arati Barua, Michael Gerhard, Matthias Koßler. Berlin 2013, pp. 149–187.

Lemanski, Jens: Vom Alles zum Nichts oder die Überwindung des dogmatischen Spinozismus in der Ethik Schopenhauers. In: *Schopenhauer-Jahrbuch* 90 (2009), pp. 19–44.

Lemanski, Jens: Wissen, Wissenschaft, Wissenschaftslehre. In: *Philosophie als Wissenschaft*. Ed. by Nora Schleich et al. Hildesheim 2021, pp. 113–133.

Lemanski, Jens/Alogas, Konstantin: The Function of Decadence and Ascendance in Analytic Philosophy. In: *Decadence in Literature and Intellectual Debate since 1945*. Ed. by Diemo Landgraf. New York 2014, pp. 49–65.

Lemanski, Jens/ Dobrzański, Michał: Reism, Concretism and Schopenhauer Diagrams. In: Studia Humana 9:3/4 (2020), pp. 104–119 (also in: Judgments and Truth: Essays in Honour of Jan Woleński. (Tributes, Vol. 43). Ed. by Andrew Schumann. London 2020, pp. 105–131).

Lemanski, Jens/ Moktefi, Amirouche: Making Sense of Schopenhauer's Diagram of Good and Evil. In: Francesco Bellucci, Peter Chapman, Gem Stapleton, Amirouche Moktefi, Sarah Perez-Kriz: *Diagrammatic Representation and Inference. 10th International Conference, Diagrams 2018, Edinburgh, UK, June 18–22, 2018, Proceedings*. Cham 2018, pp. 721–724.

Lenzen, Wolfang: Caramuel's Pentagon of Opposition and his Vindication of the Principle Ex contradictorio quodlibet. In: *History of Logic and its Modern Interpretation*. Ed. by Ingolf Max, Jens Lemanski. London (forthc.)

Leonhard, Heinrich: *Beitrag zur Kritik der Schopenhauer'schen Erkenntnistheorie, insbesondere in ihrer Anwendung auf das Euklidsche Beweisverfahren*. Bonn 1891.

Levi, Salomon: *Das Verhältnis der 'Vorlesungen' Schopenhauers hrsg. von P. Deussen Bd IX u. X zu der 'Welt als Wille und Vorstellung'*. Gießen 1922.

Lidner, Gustav Adolph: *Lehrbuch der formalen Logik*. 2nd revised ed. Wien 1867.

Locke, John: *An Essay Concerning Human Understanding*. London 1690.

Longuenesse, Béatrice: *Kant and the Capacity to Judge. Sensibility and Discursivity in the Transcendental Analytic of the Critique of Pure Reason*. Princeton 2001.

Lotze, Hermann: *Logic in Three Books of Thought, of Investigation and of Knowledge*. Transl. and ed. by Bernard Bosanquet. 2nd ed. in 2 Vols. Oxford 1888.

Lovejoy, Arthur O.: Schopenhauer as an Evolutionist. In: *The Monist* 21:2 (1911), pp. 195–222.

Lu-Adler, Huaping: *Kant's Conception of Logical Extension and Its Implications*. California 2012.

Łukasiewicz, Jan/ Tarski, Alfred: Investigations into the Sentential Calculus. In: Jan Łukasiewicz: *Selected Works*. Ed. by L. Borkowski. Amsterdam et al. 1970, pp. 131–152.

Łukasiewicz, Jan: *Aristotle's Syllogistic from the Standpoint of Modern Formal Logic*. 2nd ed. Oxford 1957.

Lyons, John: *Semantics*. 2 Vols. Cambridge 1977.

Maaß, Johann Gebhard Ehrenreich: *Grundriß der Logik, zum Gebrauche bei Vorlesungen*. Halle 1793.

Maaß, Johann Gebhard: Neue Bestätigung des Satzes: daß die Geometrie aus Begriffen beweise. In: *Philosophisches Archiv* 1:3 (1792), pp. 96–99.

Maaß, J[ohann] G[ebhard] E[hrenreich]: Ueber den höchsten Grundsatz der synthetischen Urtheile; in Beziehung auf die Theorie von der mathematischen Gewisheit. In: *Philosophisches Magazin* 2 (1789), pp. 186–231.

Maaß, Johann Gebhard: Ueber den Unterschied der Philosophie und der Mathematik, in Rücksicht auf ihre Gewisheit. In: *Philosophisches Magazin* 2:2 (1789), pp. 316–341.

Macbeth, Danielle: *Frege's Logic*. Cambridge 2009.

Macbeth, Danielle: *Realizing Reason. A Narrative of Truth and Knowing*. Oxford 2014.

MacFarlane, John: Frege, Kant, and the Logic in Logicism. In: *The Philosophical Review* 111:1 (2002), pp. 25–65.

Magee, Bryan: *The Philosophy of Schopenhauer*. Oxford 1983.

Mager [, Karl Wilhelm E.]: *Die Encyklopädie, oder das System des Wissens. Teil II*. Zürich 1847.

Malink, Marko: Aristotle on Principles as Elements. In: *Oxford Studies in Ancient Philosophy* 53 (2017), pp. 163–214.

Malter, Rudolf: *Arthur Schopenhauer. Transzendentalphilosophie und Metaphysik des Willens*. Stuttgart-Bad Cannstatt 1991.

Malter, Rudolf: *Der eine Gedanke. Hinführung zur Philosophie Arthur Schopenhauers*. Darmstadt 2010.

Malter, Rudolf: Schopenhauer und die Biologie. Metaphysik der Lebenskraft auf empirischer Grundlage. In: *Berichte Zur Wissenschaftsgeschichte* 6 (1983), pp. 41–58.

Mancosu, Paolo: Aristotelian Logic and Euclidean Mathematics. Seventeenth-Century Developments of the Quaestio de Certitudine Mathematicarum. In: *Studies in History and Philosophy of Science Part A* 23:2(1992), pp. 241–265.

Mancosu, Paolo: *Abstraction and Infinity*. Oxford 2016.

Mansfeld, Jaap: *Heresiography in Context. Hippolytus' Elenchos as a Source for Greek Philosophy*. Leiden et al. 1992.

Marc-Wogau, Konrad: Kants Lehre vom analytischen Urteil. In: *Theoria* 17 (1951), pp. 140–157.

Margolius, Hans: System und Aphorismus. In: *Schopenhauer-Jahrbuch* 41 (1960), pp. 117–124.

Märtens, [Hermann]: Schopenhauer über den 'Mausefallenbeweis'. In: *Zeitschrift für mathematischen und naturwissenschaftlichen Unterricht* 16:4 (1885), pp. 181–186.

Martini, Jacobus: *Institutionum Logicarum Libri VII*. Wittebergae 1610.

Masthoff, Judith/ Flower, Jean/ Fish, Andrew/ Southern, Jane: Automated Theorem Proving in Euler Diagram Systems. In: *Journal of Automated Reasoning* 39:4 (2007), pp. 431–470.

Matsuda, Katsunori: Spinoza's Redundancy and Schopenhauer's Concision. An Attempt to Compare Their Metaphysical Systems Using Diagrams. In: *Schopenhauer-Jahrbuch* 97 (2016), pp. 117–131.

Max, Ingolf: Wittgensteins Philosophieren zwischen Kodex und Strategie. Logik, Schach und Farbausdrücke. In: *Realism – Relativism – Constructivism. Proceedings of the 38th International Wittgenstein Symposium in Kirchberg.* Ed. by Christian Kanzian, Sebastian Kletzl, Josef Mitterer & Katharina Neges. Berlin, New York 2017, pp. 409–424.

McCulloch, Warren S.: *Embodiments of Mind.* Cambridge/ Mass. 1965.

McCulloch, Warren S.: Machines that Think and Want. In: *Brain and Behavior. A Symposium.* Ed. by W.C. Halstead. Berkeley/ CA 1950, pp. 39–50.

McDowell, John H.: *Mind and World: With a New Introduction.* 5th ed. Cambridge/Mass. et al. 2000.

McKirahan Jr., Richard D.: *Principles and Proofs. Aristotle's Theory of Demonstrative Science.* Princeton 1992.

McLaughlin, Peter/ Schlaudt, Oliver: Kant's Antinomies of Pure Reason and the 'Hexagon of Predicate Negation'. In: *Logica Universalis* 14 (2020), pp. 51–67.

Meier, Georg F.: *Auszug aus der Vernunftlehre.* Halle 1752.

Meier-Oeser, Stephan: *Die Präsenz des Vergessens. Zur Rezeption der Philosophie des Nicolaus Cusanus vom 15. bis zum 18. Jahrhundert.* Münster 1989.

Meulen, Ross Vander: Using Venn Diagrams to Represent Meaning. In: *Die Unterrichtspraxis / Teaching German* 23:1 (1990), pp. 61–63.

Mellin, Georg Samuel Albert: *Encyclopädisches Wörterbuch der kritischen Philosophie,* Vol. 2:2. Jena, Leipzig 1799.

Menne, Alfred: Arthur Schopenhauer. In: *Klassiker des philosophischen Denkens.* Vol. 2. Ed. by N. Hoerster. 7th ed. München 2003, pp. 194–230.

Mill, John Stuart: *A System of Logic, Ratiocinative and Inductive.* New York 1848.

Millet, Julián Marrades: Subject, World and Value (Some Hypotheses on the Influence of Schopenhauer in the Early Wittgenstein). In: *Doubt, Ethics and Religion. Wittgenstein and the Counter-Enlightenment.* Ed. by L. Perissinotto, V. Sanfélix. Heusenstamm 2011, pp. 63–83.

Misch, Georg: *Geschichte der Autobiographie. Vol. 4, 2. Hälfte: Von der Renaissance bis zu den autobiographischen Hauptwerken des 18. und 19. Jahrhunderts.* Frankfurt/ Main 1969.

Moktefi, Amirouche/ Shin, Sun-Joo: A History of Logic Diagrams. In: *Logic. A History of its Central Concepts.* Ed. by Dov M. Gabbay, John Woods. Oxford et al. 2012, pp. 611–682.

Moktefi, Amirouche/ Bellucci, Francesco/ Pietarinen, Ahti-Veikko: Continuity, Connectivity and Regularity in Spatial Diagrams for N Terms. In: *Diagrams, Logic and Cognition*. Ed. by J. Burton, L. Choudhury. CEUR Workshop Proceedings 1132 (2013), pp. 23–30.

Moktefi, Amirouche: Diagrams as Scientific Instruments. In: Visual, Virtual, Veridical. Ed. by Andras Benedek, Agnes Veszelszki. Frankfurt/ Main 2017, pp. 81–89.

Moktefi, Amirouche: Schopenhauer's Eulerian Diagrams. In: *Language, Logic, and Mathematics in Schopenhauer*. Ed. by Jens Lemanski. Basel 2020, pp. 111–129.

Monge, Gaspard: *Géométrie descriptive. Lecons données aux écoles normales, l'an 3 de la République*. Paris 1798.

Moretti, Alessio: Arrow-Hexagons. In: *The Road to Universal Logic. FS for the 50th Birthday of Jean-Yves Béziau. Vol. 2*. Ed. by A. Koslow und A. Buchsbaum. Cham 2015, pp. 417–489.

Moretti, Alessio: *The Geometry of Logical Opposition*. Neuchâtel 2009.

Moss, Larry: Completeness Theorems for Syllogistic Fragments. In: Logics for Linguistic Structures. Ed. by F. Hamm, S. Kepser. Berlin 2008, pp. 143–175.

Mostowski, Andrzej: On a Generalization of Quantifiers. In: *Fundamenta Mathematicae* 44:2 (1957), pp. 12–36.

Mugnai, Massimo: Logic and Mathematics in the Seventeenth Century. In: *History and Philosophy of Logic* 31 (2010), pp. 297–314.

Mühlethaler, Jacob: *Die Mystik bei Schopenhauer*. Berlin 1910.

Natterer, Paul: *Systematischer Kommentar zur* Kritik der reinen Vernunft. *Interdisziplinäre Bilanz der Kantforschung seit 1945*. Berlin 2003.

Neidert, Rudolf: *Die Rechtsphilosophie Schopenhauers und ihr Schweigen zum Widerstandsrecht*. Tübingen 1966.

Newen, Albert/Horvath, Joachim: Apriorität und Analytizität: Zwei Grundbegriffe der Philosophie und ihre Entwicklung – Eine Einleitung. In: *Apriorität und Analytizität*. Ed. by Albert Newen, Joachim Horvath. Paderborn 2007, pp. 9–33.

Nietzsche, Friedrich: The Case Wagner. In: The Complete Works, Vol. VIII. Ed. by Oscar Levy. Edinburgh, London 1911, pp. 1–53.

Nolan, Catherine: Music Theory and Mathematics. In: *The Cambridge History of Western Music Theory*. Ed. by T. Christensen. Cambridge 2002, pp. 272–304.

Nowak M[artin] A., Plotkin J. B., Jansen V. A.: The Evolution of Syntactic Communication. In: *Nature* 404:6777 (20. März 2000), pp. 495–498.

Nowak, Martin A.; Krakauer, David C.: The Evolution of Language. In: *Proceedings of the National Academy of Sciences* 96:14 (1999, Juli 6), pp. 8028–8033.

Nuchelmans, Gabriel: *Geulincx Containment Theory of Logic*. Amsterdam 1988.

Nyman, Alf: *Rumsanalogierna inom Logiken. En Undersökning av den Logiska Evidensens Natur och Hjälpkällor*. Lund, Leipzig 1926.

O'Meadhra, Uaininn: Medieval Logic Diagrams in Bro Church, Gotland, Sweden. In: *Acta Archaeologica* 83 (2012), pp. 287–316.

Panizza, Letizia: Learning the Syllogisms. Byzantine Visual Aids in Renaissance Italy – Ermolao Barbaro (1454–93) and others. In: *Philosophy in the Sixteenth and Seventeenth Centuries. Conversations with Aristotle*. Ed. by Constance Blackwell, Sachiko Kusukawa. London, New York 1999, pp. 22–48.

Pap, Arthur: *Semantics and Necessary Truth. An Inquiry into the Foundations of Analytic Philosophy*. New Haven 1958, pp. 59–62.

Patschovsky, Alexander: *Die Bildwelt der Diagramme Joachims von Fiore. Zur Medialität religiös-politischer Programme im Mittelalter*. Ostfildern 2003.

Pears, David: *The False Prison. A Study of the Development of Wittgenstein's Philosophy*. Vol. 1. Oxford 1987.

Peckhaus, Volker: *Logik, mathesis universalis und allgemeine Wissenschaft. Leibniz und die Wiederentdeckung der formalen Logik im 19. Jahrhundert*. Berlin 1997.

Peirce, Charles Sanders: Book II. Existential Graphs: In: *Collected Papers of Charles Sanders Peirce. Vol. 4. The Simplest Mathematics*. Ed. by Charles Hartshorne, Paul Weiss. 5th ed. Cambridge/MA 1980 (Repr. 1933), pp. 293–470.

Peirce, Charles Sanders: Logic of the Future. Peirce's Writings on Existential Graphs. Ed. by Ahti Pietarinen. Berlin, Boston *[forthc.]*.

Peregrin, Jaroslav: *Inferentialism. Why Rules Matter*. New York 2014.

Perrett, Roy W. (ed.): *Indian Philosophy: Logic and philosophy of language*. New York 2001.

Pietzker, [Friedrich]: Ein Jünger Schopenhauers in der Geometrie. In: *Zeitschrift für mathematischen und naturwissenschaftlichen Unterricht* 16:4 (1885), pp. 187–190.

Pitts, Walter H./ McCulloch, Warren S.: A Logical Calculus of the Ideas Immanent in Nervous Activity. In: *The Bulletin of Mathematical Biophysics* 5:4 (1943), pp. 115–133.

Ploucquet, Gottfredus: *Fvndamenta Philosophiæ Speculativæ*. Tübingae 1759.

Pluder, Valentin: "Skitze einer Geschichte der Lehre vom Idealen und Realen". In: *Schopenhauer-Handbuch. Leben – Werk – Wirkung*. Ed. by Daniel Schubbe, Matthias Koßler. Weimar 2014, pp. 124–129.

Pluder, Valentin: Schopenhauer's Logic in its Historical Context. In: *Language, Logic, and Mathematics in Schopenhauer*. Ed. by Jens Lemanski. Basel 2020, pp. 129–145.

Pluder, Valentin: The Limits of the Square. Hegel's Opposition to Diagrams in its Historical Context. In: *The Exoteric Square of Opposition*. Ed. by Jean-Yves Beziau and Ioannis Vandoulakis. Basel [forthc.]

Pollok, Konstantin: *Kant's Theory of Normativity. Exploring the Space of Reason*. Cambridge 2017.

Prade, Henry/ Marquis, Pierre/ Papini, Odile: Elements for a History of Artificial Intelligence. In: *Guided Tour of Artificial Intelligence Research. Vol. 1: Knowledge Representation, Reasoning and Learning*. Heidelberg et al. 2020, pp. 1–45.

Prantl, Carl: *Geschichte der Logik im Abendlande*. Vol. 1. Leipzig 1855.

Prantl, Carl: Ueber die mathematisierende Logik. In: *Sitzungsberichte der Bayerischen Akademie der Wissenschaften, Philosophisch-Philologische und Historische Classe* 4 (1886), pp. 497–515.

Prawitz, Dag: The Philosophical Position of Proof Theory. In: *Contemporary Philosophy Scandinavia*. Ed. by R. E. Olson, A. M. Paul. Baltimore, London 1972, pp. 123–134.

Prien, Bernd: *Kants Logik der Begriffe. Die Begriffslehre der formalen und transzendentalen Logik Kants*. Berlin et al. 2006.

Pringsheim, Alfred: Über Wert und angeblichen Unwert der Mathematik. In: *Jahresbericht der Deutschen Mathematiker-Vereinigung* 13 (1904), pp. 357–382.

Prior, Arthur: The Runabout Inference-Ticket. In: *Analysis* 21:2 (1960), pp. 38–39.

Proops, Ian: Kant's Conception of Analytic Judgment. In: *Philosophy and Phenomenological Research* 70:3 (2005), pp. 588–612.

Puntel, Lorenz B.: *Grundlagen einer Theorie der Wahrheit*. Berlin et al. 1990.

Putnam, Hilary: The Analytic and the Synthetic. In: *Minnesota Studies in the Philosophy of Science* 3 (1962), pp. 358–397.

Quatrini, Myriam: Une relecture ludique des stratagèmes de Schopenhauer. In: *Influxus* (2013), No. 65.

Quine, Willard Van Orman: A Postscript on Metaphor. In: *Critical Inquiry* 5:1 (1978), pp. 161–162.

Quine, Willard Van Orman: Epistemology Naturalized. In: *Ontological Relativity and Other Essays*. New York 1969, pp. 69–90.

Quine, Willard Van Orman: *Methods of Logic*. 4th revised ed. Cambridge/Mass. 1982.

Quine, Willard Van Orman: Ontological Relativity. In: *The Journal of Philosophy* 65:7 (1968), pp. 185–212.

Quine, Willard Van Orman: Two Dogmas of Empiricism. In: *From a Logical Point of View*. 2nd revised ed. New York et al. 1963, pp. 20–47.

Quine, Willard Van Orman: Use and its Place in Meaning. In: Ibid.: *Theories and Things*. Cambridge/Mass., London 1981, pp. 43–55.

Rabus, Leonhard: *Logik und Metaphysik. Band 1: Erkenntnisslehre, Geschichte der Logik, System der Logik, nebst einer chronologisch gehaltenen Uebersicht über die logische Literatur und einem alphabetischen Sachregister*. Erlangen 1868.

Radbruch, Knut: Anschauung und Beweis in der Mathematik. Skeptische Anmerkungen zum Optimisten Schopenhauer. In: *Schopenhauer-Jahrbuch* 69 (1988), pp. 199–226.

Randolph, John F.: Cross-Examining Propositional Calculus and Set Operations. In: *The American Mathematical Monthly* 72 (1965), pp. 117–127.

Raymarus Vrsvs Dithmarsivs, Nicolaus: *Metamorphosis Logicae* […]. Argentorati 1589.

Read, Stephen: John Buridan's Theory of Consequence and His Octagons of Opposition. In: *Around and Beyond the Square of Opposition*. Ed. by Jean-Yves Béziau, Dale Jacquette. Basel 2012, pp. 93–110.

Reed, Delbert: *The Origins of Analytic Philosophy. Kant and Frege*. London 2007.

Rehberg, A[ugust] W[ilhelm]: Beantwortung von Herrn Eberhards Duplik, meine Rezension des philosophischen Magazins in der A.L.Z. 1789. No. 10 und 90 betreffend, im 2ten Bande 4tes Stück No. X seines philosophischen Magazins. In: *Neues Deutsches Museum* 4 (1791), pp. 299–305.

Rehberg, August Wilhelm: Ueber die Natur der geometrischen Beweise. In: *Philosophisches Magazin* 4:4 (1792), pp. 447–461.

Reich, Klaus: *Die Vollständigkeit der Kantischen Urteilstafel*. Berlin 1948.

Reidemeister, Kurt: Anschauung als Erkenntnisquelle. In: *Zeitschrift für philosophische Forschung* 1 (1946), pp. 197–210.

Reimarus, H[ermann] S[amuel]: *Vernunftlehre, als eine Anweisung zum richtigen Gebrauche der Vernunft* […]. Hamburg 1756.

Risi, Vincenzo de: *Leibniz on the Parallel Postulate and the Foundations of Geometry. The Unpublished Manuscripts*. Cham et al. 2016.

Risse, Wilhelm: *Die Logik der Neuzeit*. 2 vols. Stuttgart-Bad Cannstatt 1964/1970.

Ritschl, Otto: *System und systematische Methode in der Geschichte des wissenschaftlichen Sprachgebrauchs und der philosophischen Methodologie*. Bonn 1906.

Robinson, Richard: Necessary Propositions. In: *Mind* 67 (1958), pp. 289–304.

Rooij, Robert van: Extending Syllogistic Reasoning. In: *Logic, Language and Meaning*. Ed. by M. Aloni, H. Bastiaanse, T. de Jager, K. Schulz. Berlin, Heidelberg 2010, pp. 124–132.

Rorty, Richard: *Philosophy and the Mirror of Nature*. Princeton 1980.

Rorty, Richard: *Truth and Progress*. Philosophical Papers Vol. 3. Cambridge 1998.

Rose, Lynn E.: *Aristotle's Syllogistic*. Springfield 1968.

Rosenkoetter, Timothy: Are Kantian Analytic Judgments About Objects?. In: *Recht und Frieden in der Philosophie Kants. Vol. 5*. Ed. by Valerio Rohden, Ricardo R. Terra, Guido A. Almeida, Margit Ruffing. Berlin et al. 2008, pp. 191–202

Rosenkranz, Johann Carl Friedrich: Zur Charakteristik Schopenhauer's. In: *Deutsche Wochenschrift* [by Karl Goedeke] 22 (1854), pp. 673–684.

Rosenzweig, Franz: *Stern der Erlösung*. Frankfurt/ Main 1921.

Rostand, François: Schopenhauer et les démonstrations mathématiques. In: *Revue d'histoire des sciences et de leurs applications* 6:3 (1953), pp. 202–230.

Rott, Hans: Vom Fließen theoretischer Begriffe. Begriffliches Wissen und theoretischer Wandel. In: *Kant-Studien* 95:1 (2004), pp. 29–51.

Ruffing, Margit: Die 1, 2, 3/4-Konstellation bei Schopenhauer. In: *Die Macht des Vierten. Über eine Ordnung der europäischen Kultur*. Ed. by Reinhard Brandt. Hamburg 2014, pp. 329–347.

Russell, Bertrand: An Inquiry into Meaning and Truth. 5th ed. London 1956.

Russell, Bertrand/ Whitehead, Alfred Nort: *Principia Mathematica I*. 2nd ed. Cambridge 1927.

Russell, B[ertrand]: On Denoting. In: *Mind, New Series* 14:56 (Oct. 1905), pp. 479–493.

Ryle, Gilbert: Categories. In: Ibid.: *Collected Papers Vol II. Collected Essays 1929–1968*. 2nd ed. London, New York 2009, pp. 178–194.

Ryle, Gilbert: Philosophical Arguments. In: Ibid: *Collected Papers Vol II. Collected Essays 1929–1968*. 2nd ed. London, New York 2009, pp. 203–222.

S.a.: De Audito Kabbalistico seu Kabbala. In: *Raymundi Lulli Opera ea quae ad adinventam ab ipso artem universalem* [...]. Argentinae 1598.

S.a. [maybe Pietro Bruno]: *Opvscvlvm Raymvndinvum de avditv Kabbalistico Sive ad omnes scientias introdvctorivm*. S.l., s.a. [1518].

S.a. [Reinhold]: Philosophisches Magazin, Ed. by J.A. Eberhard. Drittes und Viertes Stück. Fortsetzung (Rev.). In: *Allgemeinen Literatur-Zeitung* 175, (12ten Junius 1789:2), Col. 585–592.

S.a. [Johannes Schultz]: Philosophisches Magazin herausgegeben von Johann August Eberhard [Rez.]. In: *Literatur-Zeitung*, Nr. 283 (26. Sept. 1790), pp. 801–802.

Sæbø, Kjell Johan: *Notwendige Bedingungen im Deutschen. Zur Semantik modalisierter Sätze. Arbeitspapiere des Sonderforschungsbereiches 99, Nr. 108*. Konstanz 1985.

Santozki, Ulrike: *Die Bedeutung antiker Theorien für die Genese und Systematik von Kants Philosophie. Eine Analyse der drei Kritiken*. Berlin 2006.

Sauter-Ackermann, Gisela: *Erlösung durch Erkenntnis? Studien zu einem Grundproblem der Philosophie Schopenhauers*. Cuxhaven 1994.

Savigny, Eike von: *Die Philosophie der normalen Sprache. Eine kritische Einführung in die "ordinary language philosophy"*. 2nd revised ed. Frankfurt/ Main 1974.

Schang, Fabien/ Lemanski, Jens: A Bitstring Semantics for Calculus CL. In: *The Exoteric Square of Opposition*. Hrsg. v. Jean-Yves Beziau, Ioannis Vandoulakis. Basel (forthc.)

Schellenbauerus, Jo[annes] Henricus: *Compendium logices*. Stuttgardiae 1715.

Schelling, Friedrich Wilhelm Joseph von: Zur Geschichte der neueren Philosophie. (Aus dem handschriftlichen Nachlaß). In: Ibid.: *SämmtlicheWerke, Vol. 1/10*. Ed. by K.W.A. Schelling. Stuttgart et al. 1861, pp. 1–201.

Schepers, Heinrich: Eselsbrücke (Art.). In: *HWPh*, Vol. 2, pp. 743–745.

Schepers, Heinrich: Logisches Quadrat (Art.). In: *HWPh*, Vol. 7, pp. 1733–1736.

Scheybel, Johann: *Das sibend/ acht vnd neunt buch/ des hochberümbten Mathematici Euclidis Megarensis* [sic!] [...]. S.l. [Augsburg] 1555.

Schleiermacher, Friedrich: *Kurze Darstellung des theologischen Studiums zum Behuf einleitender Vorlesungen*. Berlin 1811.

Schleiermacher, Friedrich: Schriften aus der Berliner Zeit, 1800–1802. In: *Kritische Gesamtausgabe*. Ed. by Hans-Joachim Birkner et al. Berlin et al. 1988.

Schleiermacher, Friedrich: Rezension. In: *Jenaische Allgemeine Literatur-Zeitung* 1 (1804), Vol. 2, Nr. 96–97, Col. 137–151.

Schlüter, Robert: *Schopenhauers Philosophie in seinen Briefen*. Leipzig 1900.

Schmeißer, Friedrich: *Kritische Betrachtung einiger Grundlehren der Geometrie, wie sie meistens in Lehrbüchern vorkommen*. Frankfurt/ Oder 1851.

Schmicking, Daniel: Zu Schopenhauers Theorie der Kognition bei Mensch und Tier – Betrachtungen im Lichte aktueller kognitionswissenschaftlicher Entwicklungen. In: *Schopenhauer-Jahrbuch* 86 (2005), pp. 149–176.

Schmidt, Alfred: *Die Wahrheit im Gewande der Lüge. Schopenhauers Religionsphilosophie*. München 1986.

Scholz, Heinrich: *Abriß der Geschichte der Logik*. 3rd ed. Freiburg et al. 1967.

Scholz, Heinrich: *Geschichte der Logik*. Berlin 1931.

Schreiber, Alfred: Vorsicht, Mausefalle!. In: *Mitteilungen der DMV* 11:1 (2003), pp. 58–59.

Schröder, Ernst: *Vorlesungen über die Algebra der Logik (Exakte Logik)*. Vol. 1, Leipzig: Teubner, 1890.

Schroeder, Severin: Schopenhauer's Influence on Wittgenstein. In: *A Companion to Schopenhauer*. Ed. by Bart Vandenabeele. Chichester et al. 2012, pp. 367–385.

Schubbe, Daniel: Formen der (Er-)kenntnis. Ein morphologischer Blick auf Schopenhauer. In: *Der Besen, mit dem die Hexe fliegt. Wissenschaft und Therapeutik des Unbewussten. Vol. 1: Psychologie als Wissenschaft der Komplementarität*. Ed. by Günter Gödde, Michael B. Buchholz, Gießen 2012, pp. 359–385.

Schubbe, Daniel/ Lemanski, Jens: Konzeptionelle Probleme und Interpretationsansätze der Welt als Wille und Vorstellung. In: *Schopenhauer-Handbuch*. Ed. by Daniel Schubbe, Matthias Koßler. Stuttgart 2014, pp. 36–44.

Schubbe, Daniel: *Philosophie des Zwischen. Hermeneutik und Aporetik bei Schopenhauer*. Würzburg 2010.

Schubbe, Daniel: "…welches unser ganzes Wesen in Anspruch nimmt" – Zur Neubesinnung philosophischen Denkens bei Jaspers und Schopenhauer. In: *Schopenhauer-Jahrbuch* 89 (2008), pp. 19–40.

Schüler, Hubert Martin/ Lemanski, Jens: Arthur Schopenhauer on Naturalness in Logic. In: Language, Logic, and Mathematics in Schopenhauer. Ed. by Jens Lemanski. Cham 2020, pp. 145–165.

Schulthess, Peter: *Relation und Funktion. Eine systematische und entwicklungsgeschichtliche Untersuchung zur theoretischen Philosophie Kants*. Berlin, New York 1981.

Schultz, Johann: *Entdeckte Theorie der Parallelen nebst einer Untersuchung über den Ursprung ihrer bisherigen Schwierigkeit*. Königsberg 1784.

Schultz, Johann: *Prüfung der Kantischen Critik der reinen Vernunft*. 2 vols. Königsberg 1789/ 1792.

Schumann, Gunnar: A Comment on Lemanski's 'Concept Diagrams and the Context Principle'. In: *Language, Logic, and Mathematics in Schopenhauer.* Ed. by Jens Lemanski. Basel 2020, pp. 73–85.

Schwab, Johann Christoph: Einige Bemerkungen über den zweyten Theil der Schulzischen Prüfung der Kantischen Vernunftkritik. – (Königsberg, 1792. bey Nicolovius.). In: *Philosophisches Archiv* 1:3 (1792), pp. 1–21.

Schwab, Johann Christoph: Einige Bemerkungen über vorstehenden Aufsatz. In: *Philosophisches Magazin* 4:4 (1792), pp. 461–469.

Schweins, Ferdinand: *Mathematik für den ersten wissenschaftlichen Unterricht systematisch entworfen.* 2 vols. Darmstadt und Gießen 1810.

Seebach, Heinrich Ernst: *Introductio in iuris et politices utrium per viam logices.* Wittebergae 1697.

Segala, Marco: Schopenhauer and the Mathematical Intuition as the Foundation of Geometry. In: *Language, Logic, and Mathematics in Schopenhauer.* Ed. by Jens Lemanski. Cham 2020, pp. 261–285.

Seifert, Arno: *Logik zwischen Scholastik und Humanismus. Das Kommentarwerk Johann Ecks.* München 1978.

Sellars, Wilfrid: Empiricism and the Philosophy of Mind. In: *The Foundations of Science and the Concepts of Psychoanalysis (Minnesota Studies in the Philosophy of Science, Vol. I).* Ed. by H. Feigl, M. Scriven. Minneapolis 1956), pp. 253–329.

Sellars, Wilfrid: Is there a Synthetic a Priori?. In: *Philosophy of Science* 20 (1953), pp. 121–138 (Repr. in: *Science, Perception and Reality.* Atascadero 1991, pp. 298–321).

Sellars, Wilfrid: Sensibility and Understanding. In: Ibid.: *Science and Metaphysics: Variation on Kantian Themes.* London 1968, pp. 1–31.

Sellars, Wilfrid: Truth and 'Correspondence'. In: *Journal of Philosophy* 59:2 (1962), pp. 29–56 (Repr. in *Science, Perception and Reality.* Atascadero 1991, pp. 197–224).

Seydel, Rudolf: *Schopenhauers philosophisches System.* Leipzig 1857.

Sfondrati, Celestino: *Cursus Philosophicus I. Logica Major.* S. Galli 1696.

Shapshay, Sandra: Schopenhauer's Aesthetics (Art.). In: *The Stanford Encyclopedia of Philosophy (Summer 2018 Edition).* Ed. by Edward N. Zalta, URL = https://plato.stanford.edu/archives/sum2018/entries/schopenhauer-aesthetics/.

Shaver, P.; Pierson, L.; Lang, S.: Converging Evidence for the Functional Significance of Imagery in Problem Solving. In: *Cognition* 3 (1975), pp. 359–375.

Shera, Jesse H.; Rawski. Conrad H.: The Diagram is the Message. In: *Journal of Typographic Research* 2:2, (1968), pp. 171–188.

Shimojima, Atsushi: *On the Efficacy of Representation* (PhD thesis). Indiana 1996.

Shin, Sun-Joo: *The Logical Status of Diagrams.* Cambridge/ Mass. 1994.

Siegel, Steffen: *Tabula. Figuren der Ordnung um 1600.* Berlin 2009.

Siever, Holger: *Übersetzen und Interpretation. Die Herausbildung der Übersetzungswissenschaft als eigenständige wissenschaftliche Disziplin im deutschen Sprachraum von 1960 bis 2000*. Frankfurt/ Main 2008.

Sloman, Steven A./ Fernbach, Philip M./ Ewing, Scott: A Causal Model of Intentionality Judgment. In: *Mind & Language* 27:2 (2012), pp. 154–180.

Sluga, Hans Dietrich: Frege and the Rise of the Analytic Philosophy. In: *Inquiry* 18 (1975), pp. 471–498.

Sluga, Hans Dietrich: *Gottlob Frege. The Arguments of the Philosopher*. London 1980.

Smiley, T[imothy] J.: What is a Syllogism?. In: *Journal of Philosophical Logic* 2 (1973), pp. 136–154.

Barry Smith: Boundaries. An Essay in Mereotopology. In: *The Philosophy of Roderick Chisholm*. Ed. by Lewis H. Hahn. Chicago, Ill. 1997, pp. 534–561.

Smith, Kenny; Kirby, Simon: Compositionality and Linguistic Evolution. In: Handbook of

Compositionality. In: *Oxford Handbook of Compositionality*. Ed. by W. Hinzen, E. Machery, M. Werning. Oxford 2012, pp. 493–509.

Soler, Albert: Els manuscrits lul·lians de primera generació als inicis de la primera generacio. In: *Estudis Romànics* 32 (2010), pp. 179–214.

Sommers, Fred: *The Logic of Natural Language*. Oxford 1982.

Sowa, John F.: *Knowledge Representation. Logical, Philosophical, and Computational Foundations*. Pacific Grove/ Calif. 1999.

Spierling, Volker: *Arthur Schopenhauer. Philosophie als Kunst und Erkenntnis*. Frankfurt/ Main 1994.

Spierling, Volker: *Schopenhauers transzendentalidealistisches Selbstmißverständnis. Prolegomena zu einer vergessenen Dialektik*. Diss. München 1977.

Spierling, Volker: Zur Neuausgabe. In: Arthur Schopenhauer: *Theorie des gesammten Vorstellens, Denkens und Erkennens. Philosophische Vorlesungen Teil I. Aus dem handschriftlichen Nachlaß*. Ed. by Volker Spierling. München et al. 1986, pp. 11–14.

Spinoza, Baruch de: *Tractatus Theologico-Politicus. Continens Dissertationes aliquot, Quibus ostenditur Libertatem Philosophandi non tantum salva Pietate, & Reipublicæ Pace posse concedi: sed eandem nisi cum Pace Reipublicæ, ipsaque Pietate tolli non posse*. Hamburgi [i.e. Amsterdam] 1670.

Stapulensis, Jacobus Faber: *Libri logicorum, ad archetypos recogniti* [...]. Parisius 1503.

Stattler, Benedikt: *Anti-Kant*. Vol. 2. München 1788.

Steinbart, Gotthelf Samuel: Gemeinnützige Anleitung des Verstandes zum regelmäßigen Selbstdenken. 2nd ed. Züllichau 1787.

Stekeler-Weithofer, Pirmin: *Formen der Anschauung. Eine Philosophie der Mathematik*. Berlin et al. 2008

Stekeler-Weithofer, Pirmin: *Grundprobleme der Logik. Elemente einer Kritik der formalen Vernunft*. Berlin et al. 1986.

Steppi, Christian R.: *Der Mensch im Denken Arthur Schopenhauers. Eine Anatomie der fundamentalen Aspekte philosophischer Anthropologie in des Denkers Konzeption als kritische und systematische Würdigung*. Frankfurt/ Main et al. 1987.

Stevenson, J. T.: Roundabout the Runabout Inference-Ticket. In: *Analysis* 21:6 (1961), pp. 124–128.

Strawson, Peter F.: On Referring. In: *Mind* 59 (1950), pp. 320–344.

Strub, Christian: *Weltzusammenhänge. Kettenkonzepte in der europäischen Philosophie*. Würzburg 2011.

Stuhlmann-Laeisz, Rainer: *Eine Interpretation auf der Grundlage von Vorlesungen, veröffentlichten Werken und Nachlaß*. Berlin et al. 1976.

Sturmius, Joh[ann] Christopherus: *Universalia Euclidea [...]. Accedunt ejusdem XII. Novi Syllogizandi Modi in propositionibus absolutis, cum XX. aliis in exclusivis, eâdem methodo Geometricâ demonstrates*. Hagæ-Comitis 1661.

Suessionensis, Joslenus: De generibus et speciebus. In: *Ouvrages inédits d'Abélard*. Ed. by Victor Cousin. Paris 1836.

Swinbourne, Alfred: *Picture Logic. Or, The Grave Made Gay; An Attempt to Popularise the Science of Reasoning by the Combination of Humorous Pictures with Examples of Reasoning Taken from Daily Life*. 2nd ed. London 1875.

Szabó, Árpád: *The Beginnings of Greek Mathematics*. Transl. by A. M. Ungar. Dodrecht 1978.

Szabó, Árpád: Die Philosophie der Eleaten und der Aufbau von Euklids Elementen. In: *Philosophia* 1 (1971), pp. 194–228.

Takemura, Ryo: Proof Theory for Reasoning with Euler Diagrams. A Logic Translation and Normalization. In: *Studia Logica* 101:1 (2013), pp. 157–191.

Tejedor, Chon: The Ethical Dimension of the Tractatus. In: *Doubt, Ethics and Religion. Wittgenstein and the Counter-Enlightenment*. Ed. by L. Perissinotto, V. Sanfélix. Heusenstamm 2011, pp. 85–103.

Tennant, Neil: Aristotle's Syllogistic and Core Logic. In: *History and Philosophy of Logic* 35:2 (2014), pp. 120–147.

Thibaut, Bernhard Friedrich: *Grundriß der reinen Mathematik zum Gebrauch bey academischen Vorlesungen*. Göttingen 1809.

Thiel, Christian: Das Verhältnis von Syntax und Semantik bei Frege. In: *Philosophie und Logik. Frege-Kolloquien, Jena, 1989/1991*. Ed. by Werner Stelzner. Berlin 1993, pp. 3–16.

Thiel, Christian: Die Quantität des Inhalts. Zu Leibnizens Erfassung des Intensionsbegriffs durch Kalküle und Diagramme. In: *Die intensionale Logik bei Leibniz und in der Gegenwart*. Ed. by Albert Heinekamp, Franz Schupp. Wiesbaden 1979.

Thiel, Christian: *Sinn und Bedeutung in der Logik Gottlob Freges*. Meisenheim a.G. 1965.

Thomas, Ivo: The Later History of the Pons Asinorum. In: *Contributions to Logic and Methodology. In Honor of J. M. Bochenski*. Ed. by Anna-Teresa Tymieniecka. Amsterdam 1965, pp. 142–151.

Tiedemann, Dietrich: Ueber die Natur der Metaphysik. Zur Prüfung von Hrn Professor Kants Grundsätzen. In: *Hessische Beiträge zur Gelehrsamkeit und Kunst* 1 (1785), pp. 113–130, pp. 233–248, pp. 464–474.

Tolley, Clinton: *Kant's Conception of Logic*. Chicago (Diss.) 2007.

Tonelli, Giorgio: Die Voraussetzungen zur Kantischen Urteilstafel der Logik des 18. Jahrhunderts. In: *Kritik und Metaphysik. Studien. Heinz Heimsoeth zum achtzigsten Geburtstag*. Ed. by Friedrich Kaulbach und Joachim Ritter. Berlin 1966, pp. 134–158.

Tortoriello, Francesco Saverio: Schopenhauer e la didattica della matematica. In: *Archimede: Rivista per gli insegnanti e i cultori di matematiche pure e applicate* 2 (2014), pp. 86–91.

Trapezvntius, Gregorius: *De re dialectica* […]. Colonia 1538.

Tremblay, Frédérick: *La rationalité d'un point de vue logique. Entre dialogique et inférentialisme, étude comparative de Lorenzen et Brandom*. Nancy 2008.

Trendelenburg, Friedrich Adolf: *Elementa Logices Aristotelicae. In usum scholarium. Ex Aristotele excerpsit convertit illustravit*. Berlin 1836.

Trendelenburg, Friedrich Adolf: *Erläuterungen zu den Elementen der aristotelischen Logik. Zunächst für den Unterricht in Gymnasien*. Berlin 1842.

Trendelenburg, Friedrich Adolf: *Geschichte der Kategorienlehre. Zwei Abhandlungen*. Berlin 1846.

Trendelenburg, Friedrich Adolf: *Logische Untersuchungen*. 2 vols, 2nd revised ed. Leipzig 1862.

Ueberweg, Friedrich: *System der Logik und Geschichte der logischen Lehren*. [1st ed.] Bonn 1857.

Ueberweg, Friedrich: System of Logic. Transl. by Thomas M. Lindsay. London 1871.

Vlrich, Io[annes] Avg[vstvs] Henr[icus]: *Institvtiones logicae et metaphysicae. Scholae svae scripsit perpetva Kantianae disciplinae ratione habita*. Ienae 1792.

Unguru, Sabetai: On the Need to Rewrite the History of Greek Mathematics. In: *Archive for History of Exact Sciences* 15 (1976), pp. 67–114.

Urbas, Matej/ Jamnik, Mateja: Heterogeneous Proofs. Spider Diagrams Meet Higher-Order Provers. In: *Interactive Theorem Proving 6898: Second International Conference, ITP 2011, Proceedings*. Berlin et al. 2011, pp. 376–382.

Vaihinger, H[ans]: *Kommentar zur Kritik der reinen Vernunft*. Ed. by Raymund Schmidt. 2nd ed. Stuttgart 1922.

van Inwagen, Peter/Sullivan, Meghan: Metaphysics (Art.). In: *The Stanford Encyclopedia of Philosophy* (Spring 2016 Edition). Ed. by Edward N. Zalta. URL =

http://plato.stanford.edu/archives/spr2016/entries/metaphysics/ (letzter Zugriff am: 29.03.2017)

Vanheeswijck, Guido: Otto Friedrich Gruppe. The Linguistic Turn and the End of Metaphysics. In: *1830–1848. The End of Metaphysics as a Transformation of Culture*. Ed. by Herbert de Vriese. Louvain 2003.

Venn, John: On the Employment of Geometrical Diagrams for the Sensible Representation of Logical Propositions. In: *Proceedings of the Cambridge Philosophical Society* IV (Oct. 25, 1880 – May 23, 1883), pp. 47–59.

Venn, John: *Symbolic Logic*. 1st ed. London 1881.

Venn, John: *Symbolic Logic*. 2nd ed. London et al. 1894.

Verboon, Annemieke Rosalinde: *Lines of Thought. Diagrammatic Representation and the Scientific Texts of the Arts Faculty, 1200–1500*. S.l. 2010. http://hdl.handle.net/1887/16029

Verburg, P. A.: Hobbes' Calculus of Words. In: *Statistical Methods in Linguistics* 6 (1970), pp. 60–65.

Vives, Ioannes Ludovicus: De censura veri et falsi. In: Ibid.: *De disciplinis Libri XX, Tertio tomo de artibus libri octo*. Antverpia 1531.

Vmg.: Philosophisches Magazin, [...] Dritten Bandes zweytes und drittes Stück [Rez.]. In: *Oberdeutsche, allgemeine Litteraturzeitung* IX (21sten Jäner 1791), Col. 129–136.

Volkelt, Klaus Thomas: *Die Krise der Anschauung. Eine Studie zu formalen und heuristischen Verfahren in der Mathematik seit 1850*. Göttingen 1986.

von Plato, Jan: The Development of Proof Theory. In: *The Stanford Encyclopedia of Philosophy (Winter 2018 Edition)*. Ed. by Edward N. Zalta, URL = <https://plato.stanford.edu/archives/win2018/entries/proof-theory-development/>.

Wallace, John: Only in the Context of a Sentence do Words have any Meaning. In: *Midwest Studies in Philosophy* 2 (1977), pp. 144–164.

Walsh, William H.: *Reason and Experience*. London 1947.

Webb, Judson: Immanuel Kant and the Greater Glory of Geometry. In: *Naturalistic Epistemology. A Symposium of Two Decades*. Ed. by D. Nails, A. Shimony. Dordrecht et al. 1987, pp. 17–70.

Weber, Jürgen: *Begriff und Konstruktion. Rezeptionsanalytische Untersuchungen zu Kant und Schelling*. Diss. Göttingen 1995.

Weigelt, Georg: *Zur Geschichte der neueren Philosophie. Populäre Vorträge*. Hamburg 1855.

Weigelus, Erhardus: *Analysis Aristotelica ex Euclide restituta*. Jena 1658.

VVeigelus, Erhardus: *Idea Matheseos universæ cum speciminibus Inventionum Mathematicarum*. Jenae 1669.

VVeigelus, Erhardus: *Philosophia Mathematica*. Jenæ 1693.

Weimer, Wolfgang: Ist eine Deutung der Welt als Wille und Vorstellung heute noch möglich? Schopenhauer nach der Sprachanalytischen Philosophie. In: *Schopenhauer-Jahrbuch* 76 (1995), pp. 11–53.

Weiner, David Avraham: *Genius and Talent. Schopenhauer's Influence on Wittgenstein's Early Philosophy*. Rutherford 1992.

Weiner, Joan: *Frege in Perspective*. Ithaca 2008.

Weißhaupt, Adam: *Ueber die Kantischen Anschauungen und Erscheinungen*. Nürnberg 1788.

Welsen, Peter: *Schopenhauers Theorie des Subjekts. Ihre transzendentalphilosophischen, anthropologischen und naturmetaphysischen Grundlagen*. Würzburg 1995.

Wesoły, Marian: Αναλυσις περι τα σχηματα. Restoring Aristotle's Lost Diagrams of the Syllogistic Figures. In: *Peitho. Examina Antiqua* 1:3 (2012), pp. 83–114.

White, Morton: The Analytic and the Synthetic. An Untenable Dualism. In: *Semantics and the Philosophy of Language*. Ed. by Leonard Linsky. Urbana 1952, pp. 272–286.

Winterstein, Daniel/ Bundy, Alan/ Gurr, Corin: Dr. Doodle. A Diagrammatic Theorem Prover. In: *International Joint Conference on Automated Reasoning* (2004), pp. 331–335.

Wirgman, Thomas: Logic (art.). In: *Enyclopædia Londinensis*, Vol. XIII. London 1815, pp. 1–51.

Wittgenstein, Ludwig: *Werkausgabe in 8 Bänden*. Ed. by R. Rhees. Frankfurt/ Main 1984.

Wittgenstein, Ludwig: *Culture and Value. A Selection from the Posthumous Remains*. Ed. by Georg Henrik von Wright et al. Rev. 2nd Ed. London et al. 1998.

Wolff, Michael: *Abhandlung über die Prinzipien der Logik*. Frankfurt/ Main 2004.

Wolff, Michael: *Die Vollständigkeit der kantischen Urteilstafel. Mit einem Essay über Freges Begriffsschrift*. Frankfurt/ Main 1995.

Wolfram, Stephen: *The Mathematica Book*. 5th ed. Champaign, Ill., London 2003.

Worthington, B. A.: Ethics and the Limits of Language in Wittgenstein's 'Tractatus'. In: *Journal of the History of Philosophy* 19 (1981), pp. 481–496.

Wright, Georg H. v.: *Norm and Action. A Logical Enquiry*. London 1963.

Xhignesse, Michel-Antoine: Schopenhauer's Perceptive Invective. In: *Language, Logic, and Mathematics in Schopenhauer*. Ed. by Jens Lemanski. Basel 2020, pp. 95–107.

Young, Julian: *Willing and Unwilling. A Study in the Philosophy of Arthur Schopenhauer*. Dordrecht 1987.

Zekl, Hans Günter: Einleitung. In: Aristoteles: *Erste Analytik. Zweite Analytik. (Organon Vol. 3/4)*. Hamburg 1998, pp. IX–CXXI.

Ziehen, Theodor: *Lehrbuch der Logik auf positivistischer Grundlage mit Berücksichtigung der Geschichte der Logik*. Bonn 1920.

Zint, Hans: Das Religiöse bei Schopenhauer, in: 17. *Jahrbuch der Schopenhauer-Gesellschaft* (1930), pp. 3–76.

Žunjić, Slobodan: Logički dijagrami u srpskim srednjovekovnim rukopisima. In: *Theoria* 54:4 (2011), pp. 127–160.

Abbreviated Sources

Abbreviations of Greek and Latin authors and works are based on: Der Neue Pauly. Enzyklopädie der Antike. 12 vols. Ed. by Hubert Cancik et al. Stuttgart et al. 1996, vol. 1, pp. XXXIX–XLVII and Henry George Liddell/ Robert Scott: A Greek-English Lexicon. Edited and expanded by Henry Stuart Jones et al. 9th ed. Oxford 1996, pp. XVI–XXXVIII. *Greek sources* are cited according to the bibliography of the Thesaurus Linguae Graecae Canon of Greek Authors and Works. Ed. by Luci Berkowitz. New York et al. 1986. *Latin texts* are cited according to the bibliography of the Thesaurus linguae Latinae. Index. 5th ed. Leipzig 1990.

AA	Immanuel Kant: *Gesammelte Schriften* (Akademie-Ausgabe). Ed. by the Preußischen/Deutschen/Göttinger/Berlin-Brandenburgischen Akademie der Wissenschaften. Berlin 1900ff.
CN	Gottlob Frege: Conceptual Notation. In: Conceptual Notation and Related Articles. Transl. by Tereell Ward Bynum. Oxford 1972, pp. 101–204
CpR	Immanuel Kant: Critique of Pure Reason. Transl. by Paul Guyer & Allen W. Wood. Repr. Cambridge et al. 2000.
ElA	Friderich[us] Adolph[us] Trendelenburg: Elementa Logices Aristotelicae. In usum scholarum. Ex Aristotele excerpsit convertit illustravit. Berolini 1836.
FR	Arthur Schopenhauer: On the Fourfold Root of the Principle of Sufficient Reason. In: On the Fourfold Root of the Principle of Sufficient Reason and Other Writings. Transl. by David Cartwright, Edward Erdmann, and Christopher Janaway, Cambridge 2015, pp. 1–198.
FW	Arthur Schopenhauer: Prize Essay On the Freedom of the Will. Transl. by Christopher Janaway. In: The Two Fundamental Problems of Ethics (1841/1860). Cambridge 2009, pp. 31–112.
FoA	Gottlob Frege: The Foundations of Arithmetic. A Logico-Mathematical Enquiry Into the Concept of Number. Transl. by J.L. Austin. 2nd rev. ed. New York 1960.
HWPh	Historisches Wörterbuch der Philosophie. Ed. by Joachim Ritter, Karlfried Gründer et al. Basel 1971ff.

Abbreviated Sources

MR	Arthur Schopenhauer: Manuscript Remains in Four Volumes. Edited by Arthur Hübscher, Transl. by E.F.J. Payne, Oxford, New York, Hamburg 1988.
PP I	Arthur Schopenhauer: Parerga and Paralipomena. Short Philosophical Essays, Volume I. (The Cambridge Edition of the Works of Schopenhauer.) Transl. by Sabine Roehr, Christopher Janaway. Cambridge 2014.
PP II	Arthur Schopenhauer: Parerga and Paralipomena. Short Philosophical Essays, Volume 2. (The Cambridge Edition of the Works of Schopenhauer.) Transl. by Adrian del Caro, Christopher Janaway. Cambridge 2015.
PI	Ludwig Wittgenstein: Philosophical Investigations. English & German. 2nd ed. rept. Transl. by G.E.M. Anscombe. Oxford, Malden/Mass. 1999.
SW	Arthur Schopenhauer: Sämtliche Werke. 16 Vols. Ed. by Paul Deussen et al. München 1911–1941.
Tlp	Ludwig Wittgenstein: Tractatus logico-philosophicus. German text with English translation. London, New York 2015.
WWR I	Arthur Schopenhauer: The World as Will and Representation, Volume 1. (The Cambridge Edition of the Works of Schopenhauer.) Transl. by Judith Norman, Alistair Welchman, Christopher Janaway. Cambridge et al. 2010.
WWR II	Arthur Schopenhauer: The World as Will and Representation, Volume 2. (The Cambridge Edition of the Works of Schopenhauer.) Transl. by Judith Norman, Alistair Welchman, Christopher Janaway. Cambridge et al. 2018.
WWR2	Arthur Schopenhauer: *Philosophische Vorlesungen.* Ed. by F. Mockrauer. In: Ibid.: Sämtliche Werke. Ed. by Paul Deussen. Vol. 9–10. München 1913.

List of Abbreviations

A(1-4)	Weak argument (1-4)	Chap 2.3.5, 2.3.6
(aI)	Affirmative interpretation	Chap. 1.1
(AJ)	Analytic judgement	Chap. 2.2
(AJ_K)	Kant's (AJ)	Chap. 2.2
AS([+cypher])	Abstraction step	Chap. 3.2
B(1-2)	Strong argument (1-2)	Chap. 2.3.5, 2.3.6
B (I-IV)	Book (I-IV) of WWR	Chap. 1.1, 1.2
(BA)	Backing	Chap. 2.3.6
(CTP)	Context principle	Chap. 2.1
(CTP_F)	Frege's (CTP)	Chap. 2.1
(CTP_S)	Schopenhauer's (CTP)	Chap. 2.1
(CTP_W)	Wittgenstein's (CTP)	Chap. 2.1
(CPP)	Compositionality principle	Chap. 2.1
(CPP_F)	Frege's (CPP)	Chap. 2.1
(CPP_W)	Wittgenstein's (CPP)	Chap. 2.1
($\mathcal{D}$[+cypher])	Diagram	Chap. 2.3, 3.2
(J[+cypher])	Judgment	Chap. 2.3.5
(K1-3)	Arguments of the Kantians	Chap. 2.3
L	Level of abstraction	Chap. 3.2.2
(L1-3)	Arguments of the Leibnizians	Chap. 2.3
M	Matrix	Chap. 3.2.2
(NSA)	Negative strong argument	Chap. 2.3.5
(nI)	Negative interpretation	Chap. 1.1
O	Object	Chap. 3.2.2
P1-4	Part 1-4 of WWR2	Chap. 1.3, 2.2.5
(PC)	Priority of concept	Chap. 2.1
(PJ)	Priority of judgement	Chap. 2.1
(PSA)	Positive strong argument	Chap. 2.3.5
R	Region	Chap. 3.2.2
(RTM)	Representational/ picture theory of language	Chap. 2.1
(RTM_S)	Schopenhauer's (RTM)	Chap. 2.1
(RTM_W)	Wittgenstein's (RTM)	Chap. 2.1
S	Space	Chap. 3.2.2
(SJ)	Synthetic judgements	Chap. 2.2
(SJ_K)	Kant's (SJ)	Chap. 2.2
Syn	Syntax	Chap. 3.2.2
T	Time	Chap. 2.3.6
(UTM)	Use theory/ thesis of meaning	Chap. 2.1
(UTM_S)	Schopenhauer's (UTM)	Chap. 2.1
(UTM_W)	Wittgenstein's (UTM)	Chap. 2.1

www.ingramcontent.com/pod-product-compliance
Lightning Source LLC
Chambersburg PA
CBHW071327190426
43193CB00041B/899